# Volcanic Activity
# and Human Ecology

# Volcanic Activity and Human Ecology

Edited by

### PAYSON D. SHEETS

*Department of Anthropology*
*University of Colorado*
*Boulder, Colorado*

### DONALD K. GRAYSON

*Department of Anthropology*
*University of Washington*
*Seattle, Washington*

ACADEMIC PRESS

*A Subsidiary of Harcourt Brace Jovanovich, Publishers*

New York   London   Toronto   Sydney   San Francisco

ACADEMIC PRESS, INC.
111 Fifth Avenue, New York, New York 10003

*United Kingdom Edition published by*
ACADEMIC PRESS, INC. (LONDON) LTD.
24/28 Oval Road, London NW1 7DX

**Library of Congress Cataloging in Publication Data**
Main entry under title:

Volcanic activity and human ecology.

    Includes bibliographies and index.
    1.  Volcanoes.  2.  Man--Influence of environment.
I.  Sheets, Payson D.  II.  Grayson, Donald K.
GF58.V64       301.31         79--51701
ISBN  0--12--639120--3

PRINTED IN THE UNITED STATES OF AMERICA

79 80 81 82    9 8 7 6 5 4 3 2 1

To Charles and Charline Sheets,
and Barbara and Tug Grayson

# Contents

# List of Contributors

Numbers in parentheses indicate the pages on which the authors' contributions begin.

GLENN A. ANDERSON (487), Department of Biology, Northern Arizona University, Flagstaff, Arizona 86001

ERIC BLINMAN (393), Department of Anthropology, Washington State University, Pullman, Washington 99164

FRED M. BULLARD (9), Geological Sciences, University of Texas, Austin, Texas 78712

DWIGHT R. CRANDELL (195), U.S. Geological Survey, Denver, Colorado 80225

DON E. DUMOND (373), Department of Anthropology, University of Oregon, Eugene, Oregon 97403

DONALD K. GRAYSON (1, 427, 623), Department of Anthropology, University of Washington, Seattle, Washington 98195

RICHARD H. HEVLY (487), Department of Biological Sciences, Northern Arizona University, Flagstaff, Arizona 86001

DAVID HODGE (221), Department of Geography, University of Washington, Seattle, Washington 98195

WILHELMINA F. JASHEMSKI (587), Department of History, University of Maryland, College Park, Maryland 20742

ROGER E. KELLY (487), National Park Service, Western Region, Department of the Interior, San Francisco, California 94102

LAURENCE R. KITTLEMAN (49), 170 East 39th Avenue, Eugene, Oregon 97405

MARION MARTS (221), Department of Geography, University of Washington, Seattle, Washington 98195

PETER J. MEHRINGER, JR. (393), Departments of Anthropology and Geology, Washington State University, Pullman, Washington 99164

C. DAN MILLER (195), U.S. Geological Survey, Denver, Colorado 80225

DONAL R. MULLINEAUX (195), U.S. Geological Survey, Denver, Colorado 80225

MARY LEE NOLAN (293), Department of Geography, Oregon State University, Corvallis, Oregon 97331

STANLEY J. OLSEN (487), Department of Anthropology, University of Arizona, Tucson, Arizona 85721

PETER J. PILLES, JR. (459), Coconino National Forest, Box 1268, Flagstaff, Arizona 86001

JOHN D. REES (249), Department of Geography, California State University, Los Angeles, California 90032

COLIN RENFREW (565), Department of Archaeology, University of Southampton, Southampton, England

VIRGINIA SHARP (221), Department of Geography, University of Washington, Seattle, Washington 98195

PAYSON D. SHEETS (1, 525, 623), Department of Anthropology, University of Colorado, Boulder, Colorado 80309

JOHN C. SHEPPARD (393), Department of Chemical Engineering, Washington State University, Pullman, Washington 99164

SIGURDUR THORARINSSON (125), Science Institute, University of Iceland, Dunhaga 3, 107 Reykjavik, Iceland

F. C. UGOLINI (83), College of Forest Resources, University of Washington, Seattle, Washington 98195

RICHARD A. WARRICK (161), Graduate School of Geography, Clark University, Worcester, Massachusetts 01610

WILLIAM B. WORKMAN (339), Department of Anthropology, University of Alaska, Anchorage, Alaska 99504

R. J. ZASOSKI (83), College of Forest Resources, University of Washington, Seattle, Washington 98195

# Foreword

The chapters of this volume deal with dating, chronology, stratigraphy, volcanic activity, and with the impacts of volcanism on animals, plants, human populations, and the environment. Some of these chapters explain how such findings must be weighed against other causes that influence human behavior and survival, such as factors of social customs, climatic change, shifting biogeographic patterns, disease, and the ability to adapt.

Each of the chapters that assesses the possible human response to volcanism does so by searching for multiple explanations of the archaeological record, avoiding the simple argument that people were dramatically and inevitably overcome by catastrophic geologic events. Indeed, in many of the cases studied, overwhelming catastrophe has been hard to demonstrate, and the human populations often have been resilient, although perhaps surviving in altered form and numbers.

The multifaceted approach used in this volume emphasizes the numerous interrelations that must be evaluated in striving for a balanced judgment about man and nature. Thus, it is satisfying to find that the search for the human response to geological events brings out the need to integrate many scientific disciplines. In this, many scholars must share their theory, their methods, and their results, and we will surely see in the years ahead more research teams that cooperate in such efforts.

I would like to mention briefly only a few topics for anthropologists, geologists, geographers, and biologists.

As shown here, the fragmentary archaeological record is often difficult to interpret in terms of identifying the impacts of volcanic events. The residues of mankind are often incomplete. To gain a better understanding of how the human response correlates with volcanism, anthropologists might consider how the people of Iceland have reacted in the face of this adversity. The volcanic history of Iceland is documented in a detailed literature extending back a thousand years, and this written record also deals with human affairs. Parts of these accounts have been reviewed by geologists, notably by the volcanologist Sigurdur Thorarinsson (see Chapter 5 of this volume), but I think not by anthropologists. Many though not all of the Icelandic eruptions were characterized by eruptions of volcanic ash, and some were associated with devastating amounts of sulfur dioxide and other volcanic gases. The 1783 eruption of Lakagígar, for example, the greatest of historic times, released volcanic gas that lay over the entire country during the summer and fall. By stunting the grass crop, this resulted in famine in which 24% of the population and 80% of the livestock perished. As a further region for research, the archives of Mexican churches might also be a fruitful ground for anthropologists who seek records of the response to volcanism.

Geologists, for their part, have the task of assembling the tangible history of volcanic regions, not merely in stratigraphic terms, but by giving attention to the geographic dispersal of volcanic materials and the effect of these materials on surface geologic processes. Much is known about processes close to volcanoes—for example, the destructive mudflows from Irazu, the floods at Parícutin, and the geologic record of lahars on Mount Rainier—but substantially less is known about the geologic effects at distant places. The role of an ashfall in defining the landscape and environment of its time is a largely unexplored subject. Also, compilation of adequate, detailed volcanic histories has barely begun, although much is known about individual volcanoes in some places—the Pacific Northwest, for instance. For areas in North America other than the Northwest, and I have in mind Central America and Mexico, much more must be learned about individual volcanic events before an historical summary can have much substance, but a beginning should be made. This will involve detailed petrographic studies to identify the products of eruptions, as well as the building of chronologies.

Much has been published on biological responses to volcanism, and this remains a productive topic for research. The biological observations at Parícutin, for example, although nonquantitative, are particularly vivid in pointing out the interdependence of mammals, birds, insects, and reptiles—all of which were ultimately exterminated near the vent. The fate of certain animals that live on land is cited in some of the papers given here, but little is said about the effect of ashfalls on fish and shellfish, even though these must be counted as staples in many volcanic regions. The differential tolerance of plants to variable levels of volcanic exposure, and under differing climates, must also be evaluated.

I trust that these comments are not so general as to be wholly without meaning. The contributors to this volume are to be congratulated on their diverse and provocative remarks, and I hope their findings will be a prelude to many other research projects and publications of similar excellence.

*HAROLD E. MALDE*

# Volcanic Activity
# and Human Ecology

1

# Introduction

PAYSON D. SHEETS
DONALD K. GRAYSON

## THE STUDY OF NATURAL DISASTERS

The famous and graphic letters Pliny the Younger wrote to the historian Tacitus about the eruption of Pompeii in A.D. 79 represent one of the earliest known written records of a natural disaster. Since that famous eruption we have come a long way toward understanding the processes which produce natural disasters. But this increase in understanding has been due primarily to the work of physical scientists who have elucidated the causes of the geophysical events that often have devastating impacts on human groups. It has only been within the past two decades that the responses of human groups to these geophysical events have come under serious study by social scientists. Only recently have sociologists, economists, and, especially, geographers begun to contribute substantially to the understanding of human behavior under hazard or disaster conditions; the contribution of anthropologists has been meager (Vayda and McCay 1975; Hardesty 1977).

The focus of this volume is on the effects of one external agent of change, volcanism, on human populations. Of all natural hazards, only volcanism has been studied in any detail by *both* earth scientists and social scientists and in both the past and in the present. We attempt to provide a series of studies that will clarify the nature of volcanism and the nature of human responses to the hazards—threatened or actual—which volcanism presents to human societies. The chapters in this volume strongly reflect the interdisciplinary approach needed if the complex interrelationship between people and volcanism is to be

1

understood. It is our hope that these papers will provide the geological background necessary to understand what volcanism is, why it occurs, why it can be beneficial, why it has such tremendous potential for disrupting human societies, and why it can be difficult to cope with impending or actual volcanic disasters.

## VOLCANIC PHENOMENA

What are the specific features of volcanic eruptions that can cause environmental damage or cultural disruption? Lava flows cause environmental damage and destroy cultural facilities, but they rarely result in a loss of life because their onset is comparatively gradual. Ash falls, because of their more sudden onset and their tendency to blanket extensive areas, may have more severe although less permanent consequences for people. Depending on their chemistry and thickness, as well as upon local climate (and hence weathering rate and resulting speed of soil recovery) and the society affected, ash falls can be exceedingly disruptive (see Thorarinsson's review in Chapter 5 of this volume).

People have, directly or indirectly, been affected by volcanic activity for millions of years. From the time of the earliest hominids, whose fossils have been found interbedded between volcanic ash layers (Jolly and Plog 1976), to today, people have been enjoying the long-term benefits of volcanism as well as paying a high short-term price for those benefits in the form of volcanic disasters.

The beneficial aspects of volcanism include the fertile soils that can develop from weathered lava or ash, processes of mineralization which concentrate elements into economically important deposits, the production of obsidian and other raw materials for tool manufacture, and the modification of climates, landforms, and, therefore, floral and faunal assemblages. It is not surprising, as a result, that volcanic areas have often been favored for human settlement. And it is also not surprising that people, along with the flora and fauna, of these areas have occasionally fallen victim to volcanic events.

Volcanic eruptions can be ranked along a spectrum from gentle to violent; from non-explosive slow-moving lava flows to explosive eruptions that blast many cubic kilometers of material high into the atmosphere. In all cases, the driving force behind volcanic eruptions is gas. Magmas that are fluid with low gas content and have access to the surface will rise slowly toward the surface and produce lava flows that flow gently downhill. But magmas that are viscous with a high gas content can explode violently and blast lava fragments many kilometers into the air to settle over large expanses of countryside.

More fluid magmas tend to erupt more quietly, resulting in gentle lava flows, while more viscous magmas tend to erupt more violently. At first glance, this might seem contradictory, as one might expect more fluid magmas to rush out of the vent faster and thus be more violent. The apparent contradiction is explained by differing gas content and viscosity: In more viscous magmas the gasses are trapped, thus forcing an explosive rush out of a vent following a sudden release of pressure.

The more fluid magmas tend to be more basic (mafic) in composition, in that they have a lower silica content and more iron, magnesium, and other metallic gases. The more viscous, explosive-prone magmas generally have a higher silica content, are low in iron and magnesium, and are called acidic or silicic magmas. Silicic volcanic materials weather more slowly and have fewer initial nutrients for plant growth than do mafic deposits. Silicic rocks may therefore, in the decades or centuries following an eruption, be more damaging to agriculture than mafic materials. Floral, faunal, and human recovery often takes longer from silicic than from mafic eruptions.

The volcanic activity of Hawaii stands in marked contrast to that which occurs in the circum-Pacific arc systems as exemplified by the Cascade Range and the Aleutian arc of Alaska (see, for instance, Chapter 2 by Bullard; Chapter 12 by Dumond; Chapter 8 by Hodge, Sharp, and Marts; Chapter 6 by Warrick; and Chapter 11 by Workman in this volume). Mild earthquakes and ground tilt in Hawaii frequently signal the opening of a vent, or rift system, and lava rises to the surface, flows out of the vent, and slowly moves down slope. Rarely does such an eruption last for more than a few months. The broad, gently sloping terrain of the Hawaiian Islands is the characteristic result of this kind of eruption. Airfall deposits form a very minor proportion of the products of Hawaiian eruptions.

In contrast, Alaskan and Cascade eruptions usually produce a higher proportion of pyroclastic ("fire-broken") debris with much of the material erupted high into the air to settle as airfall tephra or flowing laterally to inundate valleys as hot pyroclastic flows. The potential for environmental and social damage is thus far greater with explosive eruptions, but it must be remembered that in the long run even such explosive eruptions can also be beneficial. Near the source, the deep, valley-filling pyroclastic or mudflow deposits can create a flat fertile basin after decades or centuries of weathering and plant recolonization. Such is the case with the valley-filling ash flow deposits in the Maya Highlands of Central America and elsewhere.

Pyroclastic flows, the clouds of hot volcanic gases, blocks, and ash particles that flow downhill from certain kinds of explosive eruptions, are extremely destructive. Their destructiveness derives from their speed (generally between 80 and 320 km/hr), their high temperatures (often 600 to 1000°C), their turbulence, and the fact they may occur with little or no warning. In 1902, for instance, a pyroclastic flow erupted at Mt. Pelée and killed 30,000 people on the Caribbean island of Martinique. Only two people in the town of St. Pierre survived.

Ash falls are the result of explosive eruptions which eject rock particles high into the air above a volcano. Viscous, volatile-rich magmas are exploded through the vent and fracture into fine molten or solid particles. Except for those falling near the vent, these fragments cool in the air and settle to earth cold to blanket the landscape. Where ash layers are thinner than a centimeter or two, they often have little detrimental effect on people, animals, or vegetation, and in small amounts they can provide soil nutrients and can assist in aerating dense, highly weathered tropical soils. But in depth of a few centimeters or more ash layers can cause considerable damage to flora and fauna. Plants are damaged by structural overloading, suffocation, abrasion, gases, or by chemicals absorbed on the ash particles. Animals may have difficulties breathing because of the ash, and often experience difficulties in browsing or grazing because of the ingestion of ash-laden vegetation as well as because of diminished supplies of food. The gills of aquatic animals are susceptible to damage by the angular, glassy particles which constitute volcanic ashes (see Chapter 3 by Kittleman in this volume). And people, relying on vegetal and animal sources of food, may in turn be seriously affected.

Mudflows, also called lahars, often result from explosive eruptions. Deep accumulations of unconsolidated pumice and ash accumulate on the slopes of volcanoes. When they become wet from rainfall or as a result of melting of snowfields, they become saturated and rush downhill as a wet mudflow. It was such a mudflow from Mt. Vesuvius that buried Herculaneum in A.D. 79, while air-fall ash and pumice from the same eruptive episode buried Pompeii.

There are numerous other phenomena directly or indirectly associated with volcanism which can affect people and their environment. Lightning and torrential rainfall usually occur with explosive eruptions, and can result in forest fires and floods. Volcanic gases can directly damage animals and plants down wind from a vent. Lava flows or pyroclastic flows can start forest fires around their perimeters. And the earthquakes associated with volcanic eruptions occasionally can cause structural damage to human settlements or trigger landslides.

## THE STUDY OF MODERN VOLCANIC HAZARDS AND DISASTERS

It is worth repeating that most volcanologic research has been conducted by earth scientists (geologists, geophysicists, geochemists, seismologists, and tephrochronologists) with a clear focus on natural events and processes. The knowledge generated by these studies has been considerable (for example, Bullard 1976; and Chapter 2 in this volume). The same degree of sophistication of research and methods has not yet been matched by social scientists studying volcanic phenomena, in part because social scientists have only recently begun their studies.

The study of natural hazards and disasters by social scientists focuses on the end point within a spectrum of potential and actual stresses affecting human groups. Such studies analyze the adjustments people make to the threat, or occurrence, of natural hazards or disasters. Recent social science approaches to these situations have emphasized:

1. hazard awareness and preparedness before the disaster strikes
2. the varying degrees of detrimental and beneficial impact of the disaster on people within and beyond the devastated area
3. the long term social, political, and economic effects of disasters
4. the implications of these matters for the formulation of public policy

Throughout, hazard and disaster research within the social sciences has been directed toward understanding the mental and behavioral adjustments which people make to extreme environmental perturbations or to the threat of such perturbations.

The impact an extreme geophysical event has on a society is dependent on a number of attributes related to that event, to the environment, and to the nature of the society affected. White (1974) has argued that there are three key attributes common to all disasters, including volcanism, which determine the degree of societal impact:

1. the magnitude of the disaster, or its intensity within a given area
2. the frequency of similar disasters, or, how often similar disasters had occurred in the direct experience or in the written or oral history of the group involved
3. the length of warning available before the disaster strikes, and the speed with which the impacts occur

Similarly, Mileti, Drabek, and Haas (1975) have compiled a set of natural disaster characteristics which allow such disasters to be ranked in terms of severity of impact on human societies. They suggest that the most disruptive events are marked by: (a) a high degree of uncertainty; (b) sudden occurrence with little warning; (c) prolonged duration; (d) broad scope of damage (environmental, cultural, and human physical damage); (e) occurrence at night; and, (f) survivor's exposure to the dead and injured. The variables determining severity of impact specified by White (1974) and by Mileti, Drabek, and Haas (1975), when applied to specific cases of volcanic impacts such as those that follow in this volume, assist in understanding why some instances of volcanic impact on human societies have negligible long-term social effects, while others caused considerable and lasting social and adaptive dislocations.

Social scientists have far to go in understanding the impacts of natural disasters in general, and of volcanic disasters in particular, on human societies. Even such a basic question as "what kinds of adjustments do people make to the threat of volcanic activity" has yet to be answered in any detail. Can patterns in responses to volcanic hazards and disasters by human societies be seen, and are these patterns related in a predictable way to the nature of that volcanism and to the kind of society affected? Is it reasonable to hope for prediction of potentially hazardous volcanic events, and, if so, how can such prediction be used in formulating public policy so as to lessen the impact of the predicted event? This volume does not attempt to answer such questions in a definitive way: Answers to questions relating to major aspects of the interrelationships between volcanism and human societies are many years in the

future. Instead it is our hope that this volume can contribute some understanding to these issues, and can stimulate interest in at least those areas in
which deficiencies are glaringly obvious.


### Prehistoric Volcanic Disasters

A review of the archaeological literature (Sheets 1977) noted that archaeologists have often encountered evidence of what apparently were ancient
natural disasters: floods, earthquakes, mudflows, and volcanism have all been
detected in the record provided by archaeological sites. But most such discoveries were made incidental to each project's research design, and resulted in
little or no detailed consideration of the event's ecological or social effects. All
too often archaeologists have noted the presence of an ash layer or flood
deposit within an archaeological site and left it at that. Granted, these deposits
can be extremely useful as horizon markers, but they rarely have been examined for their ecological or social implications. We suggest that the incorpora-
tion of natural hazards research into archaeological investigations of the past
can provide a long-term view of the effects of disasters that is unavailable from
other sources. Coupled with the investigations of the impacts of such disasters
on modern groups, the analysis of prehistoric responses to disasters by human
groups should provide results invaluable to the understanding of why and how
human societies respond to volcanic events.


## AN OVERVIEW OF THE CONTRIBUTIONS

The next four chapters dicuss volcanism as geologists and pedologists see
it. Bullard (Chapter 2) provides a general overview of volcanoes and volcanism,
an overview that is followed by Kittleman's thorough review (Chapter 3) of the
production, dispersal, and properties of tephra and of the geologic methods
used to study tephra. Ugolini and Zazowski (Chapter 4) then present a detailed
look at the nature of volcanic soils as well as a review of the economic impact of
those soils. Thorarinsson's contribution (Chapter 5) looks closely at the kinds
of damage that volcanic activity may cause, with special emphasis on the
damage caused by tephra and gases.

In looking at the effects of volcanism, Thorarinsson's chapter provides a
transition to the next set of contributions in the volume, a group of papers that
uses the geologic and modern records to examine volcanoes as hazards to
people. Warrick's paper (Chapter 6) is the most general of these, for he
provides a conceptual framework within which volcanic hazards can be
viewed, a framework in which risk assessment plays a key role. Crandell,
Mullineaux, and Miller (Chapter 7) then provide the first of a series of more
regionally oriented studies to be found in this volume. These authors examine

the Holocene record of volcanism in the Cascade Range of the western United States, and demonstrate how that record can be used to assess the kinds of volcanic hazards that threaten human communities in this region today. Hodge, Sharp, and Marts (Chapter 8) present two case studies of peoples from different parts of the world—Washington and Hawaii—who live exposed to danger from volcanic activity, and analyze the responses of these people and of their governments to this danger. Finally, Rees (Chapter 9) and Nolan (Chapter 10) provide separate contributions in which the effects of the volcano Parícutin on landforms, vegetation and human communities are assessed.

The final series of papers deals with the interrelationships between volcanism and human occupations as seen through the archaeological, paleobotanical, and paleozoological records. Workman (Chapter 11) reviews the possible impacts of past volcanism on the native inhabitants of subarctic western North America and suggests that such volcanism may have had variable and occasionally drastic effects on those inhabitants. Dumond (Chapter 12), on the other hand, focusing on the archaeological record of the Alaskan Peninsula, cautions that the long-term impacts of volcanism may be exceedingly difficult, or even impossible, to detect in the archaeological record of hunters and gatherers. Blinman, Mehringer, and Sheppard (Chapter 13) provide exacting analyses of pollen from two bog and lake sites in Montana and Washington, analyses which provide insight into the nature of the timing and impact of ashfalls on the vegetation of the areas studied; their conclusion that there are, at least as of yet, no clear examples of human response to volcanic events in the interior Northwest of the United States seems inescapable. Similar conclusions are reached by Grayson (Chapter 14) in his examination of the impact of the eruption of Mount Mazama on the Fort Rock Basin of southcentral Oregon.

The eruption of Sunset Crater in Arizona in A.D. 1064–1065 has long provided the classic example of the beneficial impacts of volcanism on human communities as seen through the archaeological record. Pilles (Chapter 15) on the one hand, and Hevly (Chapter 16) and his colleagues on the other, provide new data and new interpretations of the effects of this eruption on human settlements in the area which indicate that the traditional interpretation may be in need of modification. Sheets (Chapter 17) provides the final New World archaeological study, a study in which the effects of the eruption of Volcan Ilopango in El Salvador on the prehistoric occupants of the region are examined in detail. Finally, Renfrew (Chapter 18) provides a review of the effects of the eruption of Thera on Minoan Crete, while Jashemski (Chapter 19) examines perhaps the most famous of all examples of the impacts of volcanic activity on human settlements—Pompeii and Mount Vesuvius.

We recognize fully that many volcanically active parts of the world are not represented in this volume, and that others are but incompletely represented. We offer this volume not as a summary of all that is known, but instead as an indication of the kinds of approaches to volcanic hazards that have been taken, and, hopefully, as a stimulus to further work of this sort.

## REFERENCES

Bullard, F. M.
  1976  *Volcanoes of the earth*. Austin: University of Texas Press.
Jolly, C., and F. Plog
  1976  *Physical anthropology and archaeology*. New York: Alfred A. Knopf.
Hardesty, D. L.
  1977  *Ecological anthropology*. New York: Wiley.
Mileti, D., T. Drabek, and G. Haas
  1975  Human systems in extreme environments: a sociological perspective. *Program on Technology, Environments, and Man, Institute of Behavioral Science, Monograph* 21. Boulder: University of Colorado.
Sheets, P. D.
  1977  Natural disasters in prehistory. Paper presented at the 76th Annual Meeting of the American Anthropological Association, Houston.
Vayda, A. P., and B. J. McCay
  1975  New directions in ecology and ecological anthropology. *Annual Review of Anthropology* 4:293–306.
White, G. F. (ed.)
  1974  *Natural hazards: local, national, global*. London: Oxford University Press.

# 2

# Volcanoes and Their Activity[1]

FRED M. BULLARD

## INTRODUCTION

There is a tendency to view volcanoes as agents of destruction and to overlook their beneficial contributions to the development of the earth and to our present environment. Their most obvious effect is the destruction caused by eruptions, in which cities and the countryside are often buried by ash or lava. Equally important, and probably more widespread, is the destruction resulting from mudflows generated by volcanic eruptions. Many volcanoes have crater lakes that may be suddenly ejected in an eruption, sending a flood of water and mud racing down the side of the cone. Some volcanoes support glaciers that may be melted in an eruption and also may generate huge mudflows. The word *volcano* itself, in everyday language, is synonymous with something explosive and violent. However, it seems appropriate to enumerate some of the useful contributions of volcanoes, or, more appropriately, of *volcanism*, to our environment. *Volcanism* is a general term that includes the process by which magma (molten rock below the Earth's crust) and associated gases rise into the crust and are extruded on the Earth's surface or into the atmosphere. Thus, *volcanoes*, as the term is commonly used, are the surface manifestations of volcanism. We shall use the term *volcano* with the understanding that, in some cases, other aspects of volcanism are also involved.

[1] This chapter is based on the author's *Volcanoes of the Earth*, 579 pages, 24 color plates, 184 black-and-white figures (Austin: University of Texas Press, 1976).

*VOLCANIC ACTIVITY AND HUMAN ECOLOGY*

1. Eruptions of ash periodically renew the fertility of the soil in many areas, and this must rank as one of the more important contributions of volcanoes to our environment.
2. Volcanoes are an important factor in changes in the Earth's climate. Whether these are useful changes is open to question.
3. Volcanoes have been a major contributor to the building of the continents as well as the ocean floor.
4. With few exceptions, all of the oceanic islands owe their existence, directly or indirectly, to volcanoes. The Hawaiian island chain is a good example. Volcanoes are also adding new land to existing islands; for example, the 1973 eruption of Helgafell on Heimaey, off the coast of Iceland, added 1.3 km² (.5 mi²) of land area to the island.
5. The water of the oceans and the gases of the atmosphere are believed to have been derived from the cooling of magma, and to have reached the surface through hot springs and volcanoes. Indeed, a good case can be made that volcanoes and related features are responsible for the Earth being a habitable planet!
6. Hydrothermal fluids associated with magmas are responsible for many of the ore deposits on which our civilization depends.
7. Magmas are responsible for tremendous reserves of geothermal energy, which are now becoming increasingly important.
8. Volcanoes provide magnificent scenery and recreational areas, not only in the United States but throughout the world. Many of these areas have been set aside as national parks, such as Crater Lake in Oregon and Tongariro in New Zealand.

Specific examples of some of the destructive effects and also the useful contribution of volcanism are described in the following chapters.

The effect of a volcano on the surrounding region depends on the type of eruption. Volcanic eruptions vary between two extremes. In one, the lava rises more or less quietly to the surface and overflows the crater. The gases bubble through the lava and escape, or in some instances they rush out with sufficient force to form lava fountains hundreds of feet in height. Nevertheless, the lava is not disrupted but flows away as a river of lava, with little damage except to objects in its path of flow. On the other extreme, tremendous explosions occur in the chimney of the volcano, and as the lava rises to the surface, with the consequent decrease in pressure, it "froths" because of the admixture with rapidly expanding gases and is ejected as a cloud of gas, filled with particles of ash. It was this type of eruption that buried Pompeii in the classic eruption of Mount Vesuvius in A.D. 79. The essential difference between the two types is in the gas content and the manner in which the gases are released when the magma reaches the surface. The great majority of volcanoes of the world are intermediate between the two extremes described, yielding both lava and fragmental products. In the pages that follow, some of the basic aspects of volcanology will be summarized with special attention to the types of eruptions. For a more comprehensive coverage the reader is referred to Bullard's *Volcanoes of the Earth* (1976).

## WHAT IS A VOLCANO?

Perhaps no other phenomenon of the physical world has been surrounded by as much mystery and as little truth as have volcanoes. In ancient mythology volcanoes were believed to be the home of the gods and were regarded with suspicion and awe; it would have been considered wicked, or even dangerous, to attempt to investigate them. In Roman mythology Vulcan was the god of fire, especially terrestrial fire, volcanic eruptions, and the glow of the hearth and the forge. Vulcan was the blacksmith of the gods and his forge at Mount Olympus was equipped with anvils and all the implements of the trade. Poets have identified Vulcan's workshop with various active volcanoes in the belief that the smoking mountain was the chimney of Vulcan's forge. Most frequently, in ancient writings, Vulcan's forge was located on the island of Vulcano, one of the Aeolian Islands off the coast of Sicily. The word *volcano*, derived from Vulcanus or Volcanus, the Latin for the Roman god of fire, Vulcan, came to be applied to all mountains that give off "smoke and fire." The popular idea that a volcano is a burning mountain from the top of which issue fire and smoke contains few elements of truth. The "smoke" is not smoke but gases, largely steam, containing ash particles that give them a dark, smokelike appearance; and the fire is the reflection of red-hot material on the vapor cloud above the volcano. To describe what a volcano is not is much easier than to give a concise definition of what it is. As now defined, a volcano is a vent or chimney that connects a reservoir of molten matter, known as *magma*, in the depths of the crust of the Earth, with the surface. The material erupted through the vent usually accumulates around the opening, building up a cone, called the *volcanic edifice*. Some of the loftiest mountains on earth are volcanic edifices. The material ejected consists of liquid lava (essentially magma after it reaches the surface and loses some or most of its gas content), broken fragments of partially or completely solidified rock (pyroclastic debris such as ash and cinders), and great quantities of gases. The gases are the motivating force and the most important factor in volcanic action. Some authors maintain that the only feature common to all volcanoes is the channel through which the material reaches the surface, and therefore a volcano should be defined as the vent through which the material is erupted. This leaves us in the difficult position of trying to explain that Vesuvius is not really a volcano but merely a mountain built around one! As now used, the term *volcano* includes both the vent and the accumulation (cone) around it.

## PRODUCTS OF VOLCANOES

### Gases

When water changes to a gaseous state, its volume increases instantly about 1000 times, and here is the force necessary to produce a volcanic eruption. It was early recognized that the chief gas given off in a volcanic eruption is water vapor, or steam. The source of the water was the subject of a long controversy and it was not until the early part of the twentieth century that convincing evidence was obtained to show that water is a primary constituent of magma. The proof that granitic magma can contain water in solution was furnished by R. Goranson (1931) in a series of experiments on artificial silicate melts. The amount of water that can be contained in a magma depends on the temperature and pressure. Geologic studies indicate that most granitic magmas have crystallized at depths of 1220–3650 m, and at a temperature not exceeding 870°C. Under such conditions Goranson showed that a granitic magma was capable of holding from 6 to 9% of water in solution. Continued crystallization must increase the pressure and here, Goranson writes, "is a mechanism for developing all the pressure a volcanologist may desire [1938, p. 272]." A growing mass of evidence indicates that the Earth's atmosphere and the waters of the ocean may have been derived from magmatic gases, rather than from the dense, primitive atmosphere that was once believed to have enveloped the Earth. William Rubey (1951) in a study of the geologic history of the sea, points out that the more volatile materials, such as water, carbon dioxide, chlorine, nitrogen, and sulfur, are much too abundant in our present atmosphere and hydrosphere (oceans) and ancient sediments, to be accounted for as the product of rock weathering alone. He also points out that the relative abundance of these "excess" volatiles is similar to the relative amounts of these same materials in volcanic gases. He concludes that the crystallization of a shell of igneous rocks 40 km thick would suffice to account for the Earth's hydrosphere.

The collecting of volcanic gases must be done with great caution in order to avoid contamination by atmospheric gases. In the collections of gases from active volcanoes, the amount of water varies from around 50% to more than 80% of the total. In addition to water vapor, other volcanic gases include carbon dioxide, carbon monoxide, hydrogen, sulphur dioxide, sulphur, chlorine, nitrogen, and minor amounts of several other gases.

## Volcanic Solids (Pyroclastics)

All of the fragmental material ejected by a volcano is described as *pyroclastic* (*pyro*, fire, plus *clastic*, broken) material. A simple classification of pyroclastic material is based on size. The larger fragments, consisting of fragments of the crustal layers beneath the volcano or of older lavas broken from the walls of the conduit or from the surface of the crater, are called *blocks*. *Volcanic bombs* are masses of new lava blown from the crater and solidifying in flight, becoming rounded or spindle-shaped as they are hurled through the air. The smaller fragments from the size of walnuts to the size of peas are called *lapilli*, and then, in order of decreasing size, *cinders*, *ash*, and *dust*. Originally the terms *cinders* and *ash* were used because the material resembled the cinders and ashes from grates of the fire in the home, and since volcanoes were then believed to be due to the burning of underground coal beds, the names seemed appropriate. Cinders and ash are pulverized lava, broken up by the force of the rapidly expanding gases contained in it or by the grinding together of the fragments in the crater as they are repeatedly blown out and fall back into the crater after each explosion. *Pumice* is a product of acidic magmas in which the gas content is so great as to cause the magma to froth as it rises toward the surface and into zones of decreasing pressure. When the explosion occurs, the rock froth is expelled as pumice or is shattered into ash-size particles. Pumice will float on water because of the many entrapped air spaces formed by the expanding gases. It should be noted that this material is as much magma as any lava flow, but, because of the high gas content, it is expelled as pumice and ash rather than as a liquid. It was pumice and ash that buried Pompeii in the A.D. 79 eruption of Vesuvius.

In a volcanic eruption the coarser pyroclastics usually fall on or near the base of the cone, but the ash and dust particles may be carried by air currents and spread over thousands of square miles. The eruption of Krakatoa Volcano in 1883 provided an opportunity to study the distribution of ash in an eruption. Krakatoa is located on a small island between Java and Sumatra, in Indonesia. Since this was one of the most violent eruptions of historic time, the distribution of ash was more extensive than is usually the case. It is estimated that in the eruption 4.17 billion m³ (about 1 mi³) of material was blown to a height of 27 km and that the dust was carried around the world several times. Dust fell in quantity on the decks of vessels 2500 km away 3 days after the eruption. Measurements of the sun's rays reaching the Earth's surface for the year following the eruption were only 87% of normal, attesting to the effect of the dust in the atmosphere.

## Lava

DEFINITIONS

The term *lava*, from the Italian *lavare*, meaning "to wash," was used to denote anything that "washes away." It was first applied, in Neapolitan dialect, to lava streams from Vesuvius and was adopted into Italian literature, from which it has developed its present meaning. The root of the verb *lavare*, common to Latin, French, Spanish, and Italian, is found in the English language in such words as *lavatory*, a place for washing. Lava, as now used, refers to the liquid product, or molten rock, that issues from a volcano. Magma differs essentially from lava in its gas content. While magma is confined under sufficient pressure, the gaseous content remains in solution, but, as the magma rises toward the surface, with a consequent reduction in pressure, the gases escape, sometimes with explosive violence. The expansive force of a gas-charged liquid is familiar to anyone who has opened a bottle of beer or uncorked a bottle of champagne. The frothing of the liquid and the overflowing of the bottle are analogous, on a small scale, to the flashing of the magma into pumice (rock froth) when the pressure is suddenly relieved by the blowing out (uncorking) of the plug filling the crater of the volcano.

## COMPOSITION

Lava is a mixture of several oxides, of which $SiO_2$ (silicon dioxide) is the most abundant. The constituents are virtually the same in all types of lavas but vary considerably in their relative porportions. In addition to $SiO_2$, the principal constituents of lava are the oxides of aluminum, calcium, sodium, potassium, magnesium, and iron. The range in $SiO_2$ content is the basis for grouping lavas (and in fact all igneous rocks) into three categories. Those which contain 66% or more of $SiO_2$ are the *acidic* rocks, because $SiO_2$ acts as an acid, combining with the remaining oxides to form silicate compounds. The lavas with $SiO_2$ content between 52% and 66% are *intermediate*, and those with $SiO_2$ content less than 52% are the *basic* lavas. The composition affects the viscosity, thereby influencing the ease with which the gases escape, and as a consequence, determines the type of eruption, the rate of flow, and other characteristics. In general, the acidic lavas, which are more stiff and viscous even at high temperatures, thus causing the gases to escape with difficulty, result in explosive-type eruptions. Such lavas, known as *rhyolites*, are generally light in color, frequently gray or pink. The basic lavas are dark-colored and, being quite fluid, flow readily, thus allowing the gases to escape with ease. Their eruptions are usually of the so-called quiet type, and they yield a lava known as *basalt*. Between the two extremes are the intermediate lavas, known as *andesites*, from the Andes Mountains of South America, but they occur abundantly throughout the western United States and elsewhere.

## TEMPERATURE AND RATE OF FLOW

The temperature at which a lava issues at the surface varies from around 800°C to about 1200°C. The highest temperatures are found in the basic lavas, which are commonly around 1100°C, but temperatures in excess of 1200°C have been recorded.

After pouring out on the surface, the lava spreads as tongues or sheets that flow over the countryside, often finding their way into stream valleys, along which they may extend for miles. Some sheets of lava, such as those forming the great lava plateaus, cover thousands of square miles and extend tens or even hundreds of miles from their source. The thickness of an individual flow may vary from a few inches to hundreds of feet.

The mobility of molten lava depends on its composition and temperature. The stiff, viscous, acidic lavas usually congeal before they have traveled far, whereas the basic lavas, being more fluid, tend to flow for long distances before they come to rest. The speed of a flow depends not only on the slope of the land surface but also on the lava's viscosity. In unusual cases speeds of 16–40 km (10–25 mi) per hour have been attained by lava flows on Mauna Loa Volcano, Hawaii, but more commonly the speed is measured in meters or feet per hour. At Parícutin Volcano, Mexico, velocities up to 15 m (50 ft) per minute were observed near the source, but at a distance of 1.6 km (1 mile), where the lava spread out on a more gentle slope, the velocity decreased to 15 m (50 ft) per hour. At Parícutin a tongue of lava often would continue to move for several months, finally decreasing its forward advance to only about a meter per day in its final stage.

## LAVA SURFACES

The upper part of a lava flow is usually porous because of the escape of the contained gases or the expansion of the gases to form bubbles just prior to the freezing of the flow. The surface of lava flows commonly develops into one of two contrasting types, for which the Hawaiian names of *pahoehoe* and *aa* are used (Figure 2.1). In the pahoehoe type the surface is smooth and billowy and frequently molded into forms that resemble coils of rope (ropy lava). Such lava surfaces commonly develop on basic lavas, in which a skinlike surface covers the still liquid lava below; as the flow continues to move, the smooth skin is wrinkled into ropy and billowy surfaces that are preserved when the mass congeals. In the aa type the surface of the flow consists of a confused mass of angular, crustal blocks of lava. No continuous surface forms in this type of flow, and the blocks, being carried on the surface, remain when the flow freezes (Figure 2.2).

## RATE OF COOLING

Lava, like all rock, is a poor conductor of heat, and it cools very slowly. The scoriaceous nature of the surface layer, with its many cavities and entrapped air spaces, provides a splendid insulation to prevent the heat of the lava from escaping. In the 1959 eruption of Kilauea Iki, a crater adjacent to Kilauea Volcano, Hawaii, a pool of lava 91.5 m thick accumulated in the crater. After 7 months of cooling, a diamond drill hole revealed that the crust was only 5.6 m thick, by 1961 it had increased to 19.8 m (Macdonald 1972:73). The eruption in 1973 of Alae Crater, one of the pit craters on the east rifts zone of Kilauea Volcano, Hawaii, left a pool of lava 15.2 m thick in the crater. Drilling and temperature measurements 6 days after the eruption recorded the growth of the crust as it increased from 1 m to 5.8 m at the end of 6 months. The temperature at the base of the crust was 1067°C. From these and other studies it was determined that the thickness of the crust (cooling rate) was proportional to the square root of the time elapsed since the crust began to form (Peck *et al*. 1964; Peck *et al*. 1966). Thus, it is apparent that it requires many years for a thick lava flow to cool to surface temperatures.

## Conclusion

Although we have discussed the products of volcanoes under separate headings of gases, liquids, and solids, it must be remembered that this separation does not exist until the material approaches the surface. In the magma chamber the gases are dissolved in the magma. When the magma reaches the surface, the gases may escape more or less quietly, in which case the material will flow out as lava, or they may escape with explosive violence, disrupting the magma into fragmental material, such as cinders and ash. Thus, a study of the

FIGURE 2.1.  Church at San Juan Parangaricutiro, buried by a lava flow from Parícutin Volcano in June, 1944. The lava flow is about 9 m thick and exhibits a typical *aa* surface. The cone of Parícutin Volcano is in the background. (Photo by Fred M. Bullard.)

FIGURE 2.2.  Lava flows on the floor of Mokuaweoweo, summit caldera of Mauna Loa Volcano, Hawaii. The dark area, center right, is an *aa* flow; the area to the left and far right is a *pahoehoe* flow. (Photo by Fred M. Bullard.)

products ejected by a volcano will reveal the type of eruption, even though it may have occurred thousands of years ago.

## CONES, CRATERS, AND CALDERA

### Cones

A volcanic cone is the result of the accumulation of debris around the vent, and its shape is determined by the proportion of lava and pyroclastic material. Typically, a cup-shaped depression, the crater, occupies the apex of the cone. This is the surface connection of the volcanic conduit through which the erupted material reaches the surface. The slope of the cone is determined by the angle of repose of the material. Ash and cinders come to rest on slopes of 30–35°, and hundreds of such cones, with a truncated summit occupied by the crater, are found in the western United States.

Most of the large volcanoes of the world are composite cones, consisting of layers of ash and cinders, alternating irregularly with tonguelike lava flows. The lava flows issue through breaches in the crater wall or through fissures on the flanks or at the base of the cone. A volcano of the composite type is also known as a *strato-volcano*.

When the ejected material consists predominantly of lava, as in the Hawaiian volcanoes, a lava cone is formed. The successive flows of lava spread out in thin sheets to form a wide-spreading dome with gentle slopes, rarely more than 5–10°. Such cones are known as *shield volcanoes*, because the flat dome has the profile of an ancient Roman shield.

### Craters and Calderas

DEFINITIONS

Some volcanoes have a depression in the truncated summit, much larger than the usual crater, giving the impression that in a violent eruption the top of the cone had been blown away. Such a depression is known as a *caldera*. If a caldera was, in fact, formed by the explosive disruption of the cone, the fragments of the missing part of the cone should be prominent in the debris produced by the eruption. When, as is often the case, such fragments are rare, the only alternative is that the vanished material must have collapsed into the magma chamber. Two of the greatest volcanic eruptions of historic time, Krakatoa Volcano, Indonesia, in 1883, and Coseg uina Volcano, Nicaragua, in 1830, produced calderas. The term *caldera*, as now used, was defined by Howel Williams (1941a) as a large depression, more or less circular in form, the diameter of which is many times greater than that of the included vent or vents. The crater is the vent through which ash, cinders, lava, and other ejecta are erupted, and though craters may be enlarged by the force of the explosions, they rarely, if ever, exceed 1.2–1.6 km in diameter. On the other hand, many calderas with diameters of 8–16 km, and even more, are known. Thus, Williams (1941b) concludes that "almost all volcanic depressions more than one mile [1.6 km] in diameter are produced, for the most part, by collapse [p. 375]."

CRATER LAKE, OREGON

One of the best examples of a caldera and one that has been studied in detail is Crater Lake, one of the volcanoes in the Cascade Range in southern Oregon. Crater Lake caldera is nearly circular with a diameter of about 9.7 km and a maximum depth of 1.22 km. It is surrounded by cliffs that rise from 762 m feet to 1219 m above the caldera floor. Crater Lake itself, which fills the lower portion of the caldera, is about 610 m deep. The cinder cone of Wizard Island, rising about 238 m above the lake level, represents a brief renewal of volcanic activity following the caldera collapse (Figure 2.3).

The ancestral cone, the summit of which collapsed to form Crater Lake caldera, is known as Mount Mazama (Figure 2.4). Like other large volcanoes in the Cascade Range, Mount Mazama was active during and following the Pleistocene (Ice Age), and much glacial debris is found interbedded with volcanic ejecta. Some of the last glaciers to occupy Mount Mazama were more than 16 km long and .33 km thick, and when the top of the cone collapsed, the lower ends of the glaciers were left stranded on the slopes. Today, the wide, U-shaped cross section of these glacial valleys is preserved as notches in the caldera rim.

The first explosions of the caldera-making eruption were probably not catastrophic, but they increased gradually to a climax and then diminished rapidly. After the initial activity, the vents must have enlarged, and the pressure on the magma was reduced so the gases boiled off with increasing rapidity. Finally, the frothy magma (pumice and ash) were ejected as great ash-filled clouds (called *nuées ardentes*) that rushed down the side of the cone spreading out on the surrounding plains up to a distance of 56 km. The temperature of the *nuées ardentes* must have been high, for large trees embedded in the pumice, as far as 48 km from the caldera, are carbonized. Judging from observations of caldera-making eruptions of Krakatoa and Coseg̈uina (mentioned earlier), it is likely that the climactic phase lasted for only 2 or 3 days, although intermittent ash eruptions doubtless continued for some time. A study of pollen influx on samples from a site about 750 km to the northeast of Mount Mazama (Mehringer *et al.* 1977) showed that ash from an eruption of Mount Mazama, which, according to radiocarbon dating, occurred about 4750 B.P., first fell in the autumn and 4.6 cm accumulated before the following spring; about 1 cm of ash fell the following year and 1.7 cm the year after. It should be noted that eruptions of the magnitude of Mount Mazama eject large quantities of volcanic ash (dust) into the upper atmosphere, where it may remain for many months and even years before finally settling to the earth (for more details see Lamb 1970, and Mitchell 1970).

Williams and Goles (1968) in a revision of an earlier calculation by

FIGURE 2.3. Crater Lake caldera, Oregon. *Upper:* Western end of the caldera with Wizard Island, a cinder cone that rises about 238 m above the lake level. Two lava flows are visible extending from the left side of the cone. *Lower:* The northeastern section of the caldera. Note that Mount Thielsen, the triangular peak in the background, is also visible in the upper picture. (Photo by Fred M. Bullard.)

Williams (1941a) concluded that the total volume of liquid magma (equal to about half the volume of the inflated pumice and ash) plus the volume of the old rock fragments blown out during the climactic eruption was about 42 km³ (10 mi³). This exceeds by about half the volume of the mountain top that collapsed. If the cone collapsed into this space, there would still remain about 21 km³ to be accounted for and this Williams and Goles (1968) do by postulating "that some of the space necessary to permit the collapse of the top of Mount Mazama was provided by subterranean withdrawal of magma, either through fissures in the walls of the reservoir or by recession at still greater

FIGURE 2.4. Restoration of Mount Mazama, ancestral cone of Crater Lake. (After Atwood 1935.)

depths [p.41]." Normally, in caldera-making eruptions the space made available by pumice eruptions should accommodate the total volume of material that disappeared in forming the caldera. The discrepancy that exists at Crater Lake may be due, in part, to difficulties in estimating the volume of the cone that disappeared as well as the volume of the material ejected. The caldera-making eruptions of Krakatoa Volcano in 1883, Coseqüina Volcano in 1835, and others are described by Bullard (1976:85–112).

## TYPES OF VOLCANIC ERUPTIONS

### Classification of Volcanoes

Classification is an effort to simplify a complex subject by grouping together elements with similar characteristics. Although many attempts have been made to classify volcanoes, no entirely satisfactory scheme has yet been devised. In the late nineteenth century, volcanoes were commonly classified as (a) explosive, (b) intermediate, and (c) quiet (Dana 1891). During this same period Professor A. Stoppani (1871–1873) made an effort to devise a more exact classification, yet one still based on the type of eruption. He used as types the names of Italian volcanoes that display different types of eruptions, such as Ischia, Stromboli, and Solfatara. A few years later, Professor Mercalli (1907) modified and amplified Stoppani's classification but retained the general plan. Meanwhile, the French geologist A. Lacroix, who had made extensive studies at Mount Pelée following the tragic eruption of 1902, found that none of the then existing classifications had a category that would fit the eruptions he had observed at Mount Pelée. Accordingly, Lacroix (1908) proposed a classification that, using terms introduced for the most part by others, recognized four types of eruptions: (a) Peléan, (b) Vulcanian, (c) Strombolian, and (d) Hawaiian. In order to make Lacroix's classification more complete, one new type is added, the Icelandic, and a "stage," the Solfataric, is recognized. Both of these terms have been in use for many years and the inclusion of them in the classification makes possible a more comprehensive view of volcanic activity.

Although Lacroix's classification has apparent defects, as was early recognized, it is widely used today and is firmly entrenched in geologic literature. The real difficulty inherent in such a classification is that any one particular type of volcano may at times exhibit other types of eruptions. Stromboli, for example, though it normally conforms to the Strombolian-type eruptions, does at times exhibit Vulcanian activity. This led Rittmann (1944) to propose that a new nomenclature, based on whether the vent was open or closed, be adopted to replace Lacroix's terms based on specific volcanoes. It does not seem desirable to abandon Lacroix's classification at this time since it provides a suitable framework for a description of the various types of volcanic eruptions. The identification of a volcanic eruption with one of Lacroix's types, even though it is not entirely consistent, will enable the reader to anticipate the kind of material to be ejected as well as the general character of the eruption.

### Peléan-type Eruption

The Peléan is the only new term introduced by Lacroix in his classification, since all of the other terms had been used previously by other writers. The name is from Mount Pelée, on the island of Martinique in the West Indies, which erupted in 1902, destroying the city of St. Pierre with the loss of more than 30,000 lives. The Peléan type is characterized by extreme explosiveness, resulting from a highly viscous magma. The distinguishing feature of the Peléan eruption is the *nuée ardente*, or "glowing cloud" (Figure 2.5). It is a highly heated gas, so charged with incandescent ash particles that it resembles a mobile emulsion, yet dense enough to maintain contact with the ground as it rushes down the side of the cone with hurricane force. It was such a cloud that overwhelmed St. Pierre and resulted in the destruction of everything and everyone in its path. In Peléan eruptions the crater is frequently blocked with a lava plug and the explosions break out from beneath the plug. The magma is expelled as a highly heated, ash-charged gas, and no lava issues except that pushed up as a viscous plug (or dome) in the crater. A brief description of the 1902 eruption of Mount Pelée will serve as an example of the Peléan-type eruption.

#### THE ERUPTION OF MOUNT PELÉE IN 1902

Mount Pelée is a roughly circular cone at the north end of the island of Martinique in the Lesser Antilles of the West Indies. Before the eruption, the summit of Mount Pelée was occupied by a bowl-shaped basin, the floor of a caldera that had existed since prehistoric time. On the floor of the caldera was a lake, L'Etang de Palmistes, a popular picnic spot. The crater that gave rise to

FIGURE 2.5. Mayon Volcano, Philippines, showing *nuées ardentes* in the eruption on April 24, 1968. In the lower right is the tower of the church of Cagsaua, which was destroyed by mudflows in the 1814 eruption, killing several hundred people who had sought refuge in the church. (Photo by SIX-SIS Studio, Legaspi City, Albay, Philippines. Courtesy Philippine Commission of Volcanology.)

the 1902 eruption, known as L'Etang Sec, was somewhat below the summit on the south side, overlooking the city of St. Pierre. It was an oval-shaped depression, about 0.8 km in diameter at the top, and surrounded on all sides, except the southwest, by precipitous cliffs. On the southwest, through a great gash, fully 300 m below the rim, the crater drained into the canyon of Rivière Blanche. This great cleft was an important element in the 1902 eruption, because it was responsible for directing the explosions toward St. Pierre. Prior to 1902 only two minor eruptions of Mount Pelée are recorded, one in 1792 and the other in 1851, neither of which was very serious; there was no loss of life.

The first signs of activity were observed on 2 April 1902, when several steaming fumaroles appeared in the upper valley of Rivière Blanche. On 23 April a slight fall of ash and a strong odor of sulfur were noted in the streets of St. Pierre, and a few minor earthquakes were felt. On 25 April explosions occurred in the basin L'Etang Sec, sending ash clouds filled with rocks into the air. Visitors who climbed to the rim of the crater on 27 April reported that the normally dry bed of L'Etang Sec was occupied by a lake, at least 180 m in diameter, and that a small cinder cone, hardly more than 9 m high and with a column of steam "boiling" from its summit, had formed at one side. Light falls of ash covered the streets of St. Pierre, giving it a wintry look. Shortly after noon on 5 May, the sugar mill at the mouth of Rivière Blanche, 3.2 km north of St. Pierre, was destroyed by a torrent of boiling mud that swept down the valley with express-train speed. Thirty or more workmen were entombed in the boiling mud, which left only the chimney visible to mark the location of the mill. The mud swept into the sea, extending the coastline and creating a huge wave that overturned boats tied at anchor and washed the adjacent coast, flooding the lower portion of St. Pierre. The tragedy at the sugar mill was the first loss of life due directly to the volcano. Apparently, accumulations of ash in the basin of L'Etang Sec formed a dam, which blocked the gorge, permitting a large quantity of water to accumulate in the old lake basin. The water was heated by volcanic action, and shortly after noon on 5 May the dam broke, and the water rushed from its height of nearly 914 m as a deluge of mud and boulders down the Rivière Blanche. On 6 May the eruptions were particularly violent, the explosions were heard in the neighboring islands, and many of the inhabitants attempted to leave the city. Although the volcano was in almost constant eruption, most of the people were expecting an earthquake or a tidal wave, and the danger from the volcano was not anticipated. In fact, the authorities stationed soldiers to guard the roads to prevent the people from leaving the city. There was an important election scheduled for 10 May, and it is suspected that the authorities were eager to keep the people from leaving in order to assure a good turnout for the election. In order to reassure the population that there was no danger, the governor from Fort de France visited St. Pierre with his wife for a personal inspection, and both were victims of the disaster that overwhelmed St. Pierre. The eruptions continued on 7 May, with flashes of lightning piercing the dense cloud of ash that rose with each outburst. It should be noted here that for more than a month there had been warning signals of an impending eruption. The fact that Peléan-type eruptions had not been previously recognized and the character of such eruptions was not known prevented the inhabitants (and the authorities) from appreciating the impending danger.

The morning of 8 May dawned bright and sunny, with nothing in the appearance of Mount Pelée to excite suspicion, other than the cloud of vapor rising from the crater. About 6:30 a.m. the ship Roraima, its decks covered with gray ashes, came into port at St. Pierre and tied up alongside the 17 other vessels in port. At 7:50 a.m. the volcano exploded with four tremendous cannonlike blasts, discharging upward from the crater a black cloud pierced with lightning flashes. Another blast shot out laterally and with hurricane speed rolled down the mountain slope, keeping contact with the ground. In a short 2 minutes it had overwhelmed St. Pierre and spread fanlike out to sea—and almost at once the entire population of about 30,000 was destroyed. The blast consisted largely of superheated steam, filled with even hotter ash particles, traveling at a velocity of around 160 km/hr. The force of the blast is shown by the fact that walls of stone and cement over 1 m thick were torn to pieces as though they were made of cardboard. Since the windows in most of the houses in that tropical city were covered only with shutters, the highly heated gases quickly penetrated all parts of the buildings, and almost instantly the city was in flames. All but two of the ships in port were capsized. The British steamer *Roddam*, set free by the parting of its anchor chain, escaped to St. Lucia with 12 of its crew dead and 10 so badly burned that they had to be hospitalized. The other ship to survive was the *Roraima*, which had arrived only an hour before the blast. Its mast, bridge, funnel, and boats were swept away and it took fire fore and aft, and more than half of its crew of 47 perished. Accounts of the eruption by survivors on the two ships provide the only eyewitness reports of the catastrophe (see Bullard 1976:125–127).

Of the entire population of St. Pierre, only two survived the catastrophe. Auguste Ciparis, a 25-year-old black stevedore, was a prisoner in an underground dungeon. His cell was windowless and the only opening was a small grated space above the door. He was badly burned on his back and legs, but he survived and gave a coherent account of his experience. The other survivor was Leon Compere-Leandre, a black shoemaker, about 28 years old and strongly built. He lived in the southwest part of St. Pierre, and though six people in the house with him were killed and he was burned, he saved himself by running to Fonds-Saint-Dennis, 6 km from St. Pierre. The temperature of the blast can only be estimated from its effect on objects where no conflagration took place. Glass objects were softened (650–700°C) but the melting point of copper (1058°C) was not reached. Death was apparently the result of inhaling the highly heated gases, and in many cases it appears to have been almost instantaneous.

The *nuée ardente* eruption that destroyed St. Pierre on 8 May, was followed by intermittent *nuée ardente* eruptions for several months. Many followed much the same path as the 8 May blast, but one on 30 August extended somewhat to the east of the previous ones and destroyed five villages, adding 2000 victims to the death toll of Mount Pelée (Heilprin 1903).

About the middle of October 1902, a domelike mass of lava, too stiff to flow, formed in the crater of L'Etang Sec, and from its surface a spine protruded (Heilprin 1904). This remarkable "Tower of Pelée," with a diameter of 107–152 m, in a few months reached a maximum height of 311 m above the crater floor. Its rise was irregular, but it averaged about 9 m per day. Huge blocks were continually breaking from the top, exposing a red-hot surface, and in time it was reduced to only a stump in the middle of a heap of rubble. The lava dome continued to grow on a diminishing scale until October 1905 (Robson and Tomblin 1966).

The next major eruption of Mount Pelée, following the 1902 outbreak, began on 16 September 1929, and lasted until nearly the end of 1932, or a little more than 3 years (Perret 1935). The eruption was similar to that of 1902, although somewhat less intense. St. Pierre, which by that time had grown to have a population of about 1000, and other villages on the western slope of the volcano were evacuated, so there was no loss of life.

Other Peléan eruptions that are well documented are Mount Lamington, Papua, New Guinea in 1951 (Taylor 1958); Bezymianny Volcano, Kamchatka, USSR in 1955–1956 (Gorshkov 1959); and Mayon Volcano, Philippines in 1968 (Moore and Melson 1969).

## SIGNIFICANCE OF LAVA DOMES

As was true of *nuées ardentes*, the nature of lava domes was not appreciated prior to the 1902 eruption of Mount Pelée. Since then such features have been recognized in many volcanoes and their significance is better understood. In volcanoes such as Mount Pelée that produce acidic material, the lava is very stiff and viscous and the chimney is commonly sealed by viscous lava as the eruption ends. The situation is quite different in volcanoes that produce basic-type lavas, such as Stromboli, where the lava in contact with the air remains liquid and the gases escape more or less freely. In the Peléan type, with the chimney sealed, the gases accumulate until the upper part of the magma is gas-saturated. In such cases, the initial stage of a new eruption consists of the explosive expulsion of the highly gas-charged magma, which is shattered into ash particles by the expanding gas, forming *nuées ardentes*. Following the expulsion of the highly gas-charged material, which in the 1902 eruption of Mount Pelée required about 4 months (May–August), there is a decrease in activity. The magma, now relatively gas-free and highly viscous, begins to rise in the conduit and slowly accumulates in the crater as a lava dome (also known as a *tholoid*). Occasionally spines or, as in the case of Mount Pelée, a giant monolith, is forced through cracks in the surface of the dome. Dome formation, like caldera collapse, indicates that the volcano is in the final stage of its eruptive life. The dome may be disrupted or destroyed and rebuilt by succeeding eruptions, often numerous times, before the activity ceases.

## ASH-FLOW ERUPTIONS

Covering large areas in the western United States, western Mexico, Central America, southern Peru, North Island of New Zealand, and elsewhere are thick sheets of volcanic rocks that have characteristics of lava flows and also of air-fall ash deposits. The obviously fragmental rock, with pieces of pumice and bits of other rock types, often grades downward into denser rock resembling lava. Near the base of individual sheets the pumice fragments are commonly flattened, the vesicles more or less closed or drawn out into rods, and mixed with lenses of obsidian. It appears that the material was hot enough to soften the pumice and in extreme cases to melt it to lenses of glass (obsidian); that is, it has been "welded." The origin of such ash-flow deposits was in doubt until the 1912 eruption of Mount Katmai in Alaska produced such a deposit and scientists were able to study it firsthand. The material in an ash flow is the product of a *nuée ardente*-type eruption that issues from a fissure or a series of fissures, and the flows often extend for tens of kilometers and more from their source, leaving a deposit covering many thousands of square kilometers. Ash-flow deposits range in thickness from about 1 m up to hundreds of meters and extend, in some cases, more than 80 km from the source. In southwestern Utah and adjacent Nevada, individual ash-flow beds have been traced for more than 160 km (Mackin 1960:95). The fact that ash flows can cover such large areas and still retain temperatures high enough to produce welding of the fragments after the flow comes to rest implies very rapid extrusion of the material and rapid spreading of the deposit with little time for loss of heat. The Valley of Ten Thousand Smokes ash flow (to be described later) was emplaced in 20 hours, and the volume of some of the ash flow deposits of the western United States is staggering. The Bandelier tuff (ash-flow deposit) in the Jemez Mountains of New Mexico has a volume of about 200 km$^3$ (Smith and Bailey 1966). Ash-flow deposits in the San Juan Mountains of Colorado had an original volume, before erosion, of nearly 21,000 km$^3$ (Lipman and Stevens 1969). Ash flows often cover the fissure from which they are erupted, and frequently it is difficult to locate the actual source of the material.

There is some confusion in the terminology applied to ash-flow deposits. In describing the ash-flow deposits that cover large areas in the central part of North Island, New Zealand, P. Marshall (1935) named them *ignimbrites* (literally, glowing cloud-rocks) to indicate their similarity to the deposits of *nuées ardentes*. Marshall pointed out that these rocks must have been formed by the welding or sticking together of still plastic bits of pumice as a result of the high temperature of the material when it came to rest. The rocks called ignimbrites by Marshall are also known as "welded tuffs." The use of the term *ignimbrite* has been criticized because it is not clear whether Marshall's original definition was based on the mode of origin of the material or on its composition and physical nature. It seems preferable to use the term *ash flow* for the eruptive mechanism and limit the term *ignimbrite* to the deposit of an ash flow, whether or not it is welded. An exhaustive study of ash flows and ash-flow deposits was made by Smith (1960) and by Ross and Smith (1961).

Ash-flow eruptions have been included with the Peléan-type eruptions as a matter of convenience. The *nuée ardente*, basic to the Peléan-type eruption, is the eruptive mechanism for ash flows, and the type of material is the same in each case. It is advantageous, especially where space is limited, to discuss the two as a unit. However, the extensive deposits from ash flows, covering thousands of square miles, are in sharp contrast to the limited extent of ash flows from the central vent Peléan-type eruption. Macdonald (1972:211) includes such eruptions under a separate category that he terms "Rhyolitic flood eruptions."

## THE 1912 ERUPTION OF MOUNT KATMAI, ALASKA

The eruption of Mount Katmai on 6 June 1912 was one of the most violent eruptions of historic times. It is also one of the landmarks in the development of volcanology, for like the 1902 eruption of Mount Pelée, it had features that were previously unknown or unrecognized. It was the first eruption in which it could be shown that pyroclastic flows of pumice and ash were discharged from fissures, and it led to the recognition that vast sheets of volcanic rocks, formerly regarded as lava flows, were actually products of such eruptions. Space limitations permit only a brief account of this eruption, but Bullard (1976:159–176) gives a detailed account of the event.

Mount Katmai, located near the eastern end of the Alaskan Peninsula, opposite Kodiak Island, had not been active in historic times and was among the least known of the many Alaskan volcanoes. The first big explosion, at 1:00 p.m. on 6 June, was heard at Juneau, Alaska, 1200 km to the east, and at Dawson, 1046 km to the north across the Alaskan Range. A second tremendous explosion occurred at 11:00 p.m. on 6 June, and a third at 10:40 p.m. on 7 June. Much of the ash and fragmental material discharged in the eruption is believed to have accompanied these three blasts. The strong explosive activity ceased on 8 June, although minor explosions continued through the summer. Following the initial explosion on 6 June, a dense ash cloud enveloped the area, producing total darkness, which at Kodiak, 160 km away, lasted for 60 hours. The ash was carried by the wind, and ash falls were reported at Juneau, 1200 km away, and in the Yukon valley, 1600 km to the north. A reddish haze from fine dust in the upper atmosphere was noted throughout the world, being reported from Algiers (North Africa) on 19 June. The National Geographic Society promptly organized an expedition to investigate the eruption, and under the direction of Dr. George C. Martin (1913) the party arrived at Mount Katmai 4 weeks after the eruption. They found that the top of Mount Katmai had disappeared and a huge basin now occupied the summit (Figure 2.6). This enormous pit, or caldera, was nearly 4.8 km in diameter and from 610 to 1128 m in depth. The walls of the caldera were nearly vertical, and about one-third of the western rim was an ice wall, formed by remnants of glaciers that had been beheaded by the caldera collapse. The National Geographic Society sent several expeditions to Mount Katmai in the succeeding years under the direction of Dr. Robert F. Griggs (see Griggs 1917, 1918, and 1922). It was on the 1916 expedition that Griggs and his associates discovered and named the Valley of Ten Thousand Smokes, which he explored the following year.

THE VALLEY OF TEN THOUSAND SMOKES, ALASKA

Prior to the 1912 eruption, there was a broad, wooded valley that extended for some 16 km to the northwest of Mount Katmai. In the early morning hours of 6 June, before the explosive eruption at 1:00 p.m., an ash flow issued from a series of concentric and radial fissures at the head of the valley. The emplacement of this mammoth ash flow was complete before the second great explosion at 11:00 p.m., since the pumice characteristic of this eruption does not occur beneath the flow. Every detail of the former topography was covered, and the surface now forms a gently sloping plain, covering an area of about 128 km². Thousands of fumaroles and steam jets were issuing from this deposit when it was discovered by Griggs in 1916, and these prompted him to name it

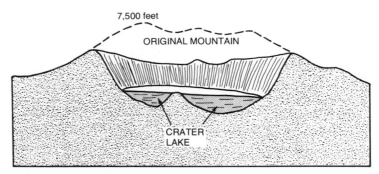

FIGURE 2.6. Mount Katmai before and after the 1912 eruption. (After Griggs 1917.)

FIGURE 2.7. Astronauts examining the ash flows in the Valley of Ten Thousand Smokes (part of their training program), July 1967. (Photo courtesy Dr. U. Clanton, NASA.)

The Valley of Ten Thousand Smokes. The material filling the valley (first called volcanic sand) consists of fragments of pumice, andesite, obsidian, and similar materials, embedded in a matrix of fine ash. It fills the main valley, and tongues project into the side valleys like a flood deposit. The ash flow, a typical *nuée ardente*, moved with hurricane velocity, as is attested by fallen trees at the lower end of the valley (Figure 2.7). The velocity must be attributed to gases liberated by the ash flow and to the expansion of entrapped air during the turbulent advance, for with a gradient of only slightly more than 1° in a distance of 20 km, the velocity cannot be attributed to gravity. Studies by Curtis (1968:187) indicate a thickness of 213–274 m and a volume of 10.96 km³. The pumiceous material is highly inflated, and when reduced to its solid volume, or the space occupied in the magma chamber, it amounts to 6.25 km³. The evidence indicates that the ash flow erupted at an ever-increasing rate during the daylight hours and up to 11:00 p.m. on 6 June, with an average rate of expulsion during this period of 1 billion m³ per hour! Curtis (1968:183) was unable to detect any welding in the ash exposed in the walls of gorges cut into the flow, but he believes that at greater depth in the upper part of the valley the flow is welded.

### Vulcanian-type Eruption

The term *Vulcanian*, first used by Mercalli *et al.* (1891), comes from the island of Vulcano, in the Aeolian Island group north of Sicily. In mythology Vulcano was the home of Vulcan, the god of fire. The lava from Vulcano forms a thick, solid crust over the crater between infrequent eruptions. Gases accumulate beneath the congealed crust, and in time the upper part of the magma column becomes gas-saturated. Finally, with strong explosions, sometimes sufficient to disrupt the cone, the obstructions are blown out, and the broken fragments of the crater crust, together with some new lava, are ejected. Such eruptions are accompanied by huge cauliflower-shaped eruption clouds containing an abundance of ash. After the vent is cleared, lava flows may issue from the crater or from fissures on the side of the cone. Indeed, the initial explosion may split the cone from top to bottom, and lava flows may issue from various points along the opening.

The Vulcanian type of eruption was named for the volcano Vulcano, which from ancient times until late in the nineteenth century was one of the most active volcanoes in the Mediterranean area. However, the last eruption of Vulcano was in 1888–1890, and it would perhaps be more meaningful to describe Mount Vesuvius as our example of the Vulcanian-type eruption.

Mount Vesuvius, on the shore of the Bay of Naples in central Italy, is but one of several volcanoes in the Naples area, some of which were in vigorous eruption in ancient times, long before Vesuvius was even known to be a volcano. Because the ancients made no reference to Vesuvius as a volcano, it must be assumed that a long period of repose intervened between its early activity and the eruption of A.D. 79, which initiated the present cone. At that time (Figure 2.8) it appeared from a distance as a broad, flat-topped mountain, and the presence of a crater would not have been revealed except from the summit. The first sign of renewal of activity at Vesuvius was a severe earthquake on 5 February A.D. 63, which did considerable damage at Pompeii and Herculaneum. Earthquakes continued intermittently for the next 16 years, finally culminating on the night of 24 August A.D. 79, in extremely violent shocks followed immediately by the historic eruption.

**FIGURE 2.8. Vesuvius before the eruption of A.D.79.**

**FIGURE 2.9. Vesuvius after the eruption of A.D. 79, partially encircled by Mount Somma, a remnant of the pre-A.D. 79 cone.**

In the eruption of A.D. 79 about half of the cone was destroyed and a new cone, the present Vesuvius, was started (Figure 2.9). The remnants of the old cone, which now partly encircles Vesuvius, are known as Mount Somma. The term *somma* is now applied to the remnant of any cone that bears to a later cone the relationship of Mount Somma to Vesuvius. In this eruption Pompeii was buried by pumice and ash, Herculaneum was covered by a mudflow, and several other cities were badly damaged; perhaps the most notable of these was Stabiae, where Pliny the Elder lost his life. The only eyewitness account of the great eruption of 24 August A.D. 79 that has survived today is given by Pliny the Younger in two letters, written 6 years later, to the historian Tacitus. English translations of these letters are included in Bullard's (1976:192–198) account of the history of Vesuvius. The debris that buried Pompeii was primary pumice, with insignificant amounts of material from the old cone of Mount Somma. This indicates that the disappearance of a portion of the cone of Mount Somma during the eruption was largely a result of collapse rather than the force of the explosions. Williams (1941a:311) uses the term *sector graben* for this type of crater, in which a segment of the cone collapses into the magma chamber as a result of the rapid expulsion of pumice, as occurred in the eruption of Vesuvius. Thus, a sector graben is similar in origin to a caldera.

Some have speculated that the deaths in Pompeii were due to *nuées ardentes* similar to those which destroyed St. Pierre in the 1902 eruption of Mount Pelée. Although there are many similarities between the two disasters, the evidence does not support the conclusion that Pompeii was overwhelmed by *nuées ardentes*. The 4.5–7.6 m of pumice and ash that buried Pompeii is uniformly stratified, indicating that it was an air-fall deposit. Although the material was primary magma, it would lose much of its heat in falling through the air, and doubtless some of the residual gases would continue to escape for some time after it had fallen. Many of the skeletons at Pompeii were found about 60 cm above the base of the initial pumice layer, indicating that they endured the eruption for a time before succumbing. In one case, several bodies were found 3 m above the base of the pumice layer, separated from the upper surface by only a thin layer of ash. Many bodies were found near the walls that surrounded the city, indicating that the victims were fleeing, often carrying some of their most precious possessions. It is possible that suffocation due to the great quantity of fine dust in the air was a cause of death, or it may have been a contributing factor along with asphyxiation by gases given off by the pumice and ash (see Alfano and Friedlander 1920, for further information on the causes of death).

The A.D. 79 eruption produced no lava; in fact, the first lava flow in historic time from Vesuvius was in the year 1036. The eruption of A.D. 79 is designated a *Plinian* eruption, named for Pliny the Younger, who provided the only surviving eyewitness account. The term denotes an exceptionally violent phase, usually the initial stage, of a Vulcanian-type eruption.

Prior to the great eruption of 1631 the activity of Vesuvius was intermittent with long periods of repose. Beginning with the eruption of 1631 the activity of Vesuvius has followed, at least up to 1944, a fairly predictable cyclic pattern. The last two cycles, 1872–1906 and 1906–1944, encompassed 34 and 38 years, respectively. Space does not permit a thorough review of this activity, but it can be briefly stated that the end of a cycle is marked by a grand eruption, such as that of 1872 or 1906, in which there were large outflows of lava and the upper part of the cone was destroyed and the crater enlarged. Vesuvius then lapses into a period of repose that, on an average, has lasted for about 7 years. The renewal of activity begins with explosive activity forming a cinder cone in the bottom of the crater, frequently accompanied by small lava flows spreading across the crater floor. In some cases explosive activity may throw out scoria and incandescent lava. When the summit crater has been filled by cone building and lava flows, the stage is set for the culminating eruption. The column of lava, now at a high level in the throat of the volcano, is under great pressure and is gas-saturated. Accompanied by earthquakes and strong explosions, the cone splits, frequently from the crater rim to the base. The initial explosions, which produce huge cauliflower-shaped eruption clouds filled with ash, clear the vent, and shortly lava begins to issue from the fracture on the side of the cone. This fracture taps the lava in the conduit at a lower level, and saturated with gas and under the pressure of the column of lava, great floods of lava issue from the fracture and flow in streams down the side of the cone. This is the grand eruption that signals the end of a cycle. The last grand eruption of Vesuvius was in 1944, when lava flows destroyed the towns of Massa and San Sebastiano. Since the 1944 eruption Vesuvius has been in a state of repose, which is now several times longer than the average length of this period, and it is apparent that a renewal of activity is long overdue (Figure 2.10).

Vulcanian is perhaps the most common type of activity among the volcanoes of the world. An example of a recent Vulcanian eruption is Parícutin Volcano, Mexico, which was active from 1943 to 1952.

### Strombolian-type Eruption

The name of this type is from Stromboli, an active volcano in the Aeolian Islands off the coast of Sicily. Stromboli is an almost perfect cone with twin peaks rising from the floor of the Mediterranean Sea, in water 2011 m deep, to a height of 909 m above sea level (Figure 2.11). Stromboli is in a more or less constant state of activity and since ancient times has been known as the "Lighthouse of the Mediterranean" because of the red glow that flashes from its summit after each explosion. The red glow, reflected on the steam cloud above the crater, quickly fades, only to reappear in about 30 minutes when the next explosion occurs. A typical eruption of Stromboli consists of more or less regular explosions of moderate intensity, which throw out a pasty, incandescent lava (scoria), accompanied by a white vapor cloud. The lava in the crater crusts over lightly, and at intervals of about 30 minutes, the pent-up gases produce mild explosions, hurling out clots of liquid lava and fragments of the crater crust. Though the activity described is the typical "Strombolian eruption," at infrequent intervals lava flows of considerable volume issue from the crater and flow down the side of the cone and into the sea. Furthermore, the so-called "normal" Strombolian activity is interrupted at intervals by more vigorous explosions that are comparable to those of the Vulcanian type of eruption. These explosions, which shake the entire island, are accompanied by a dark eruption cloud, which rises over the crater and spreads red-hot scoria

**FIGURE 2.10. Mount Vesuvius and the harbor of Naples. (Photo by Fred M. Bullard.)**

**FIGURE 2.11. Stromboli Volcano. The active crater is at the head of the Sciarra del Fuoco, some 182 m below the peak labeled observation point. San Vincenzo, one of the settlements on the island and the route of ascent to the observation point, is shown. (Photo by Fred M. Bullard.)**

and cinders over the island, forcing the inhabitants to seek shelter indoors to escape the falling debris. As long as the lava column is kept open by frequent explosions, normal activity prevails. If the vent becomes clogged, then more vigorous explosions are required to remove the obstruction. Such explosions pulverize the material, producing the dark, ash-filled eruption cloud, and eruptions of this type, even though they are at Stromboli, are properly classified as Vulcanian in type. Similarly, Strombolian activity may occur, particularly at parasitic cones, in Vulcanian eruptions.

Some examples of volcanoes displaying Strombolian-type eruptions are Yasour, New Hebrides; Medake, central cone of Akita-Komagatake Volcano, Japan; and Izalco, El Salvador.

## Hawaiian-type Eruption

In the Hawaiian eruption basaltic lava issues more or less quietly, either from the crater or from a fissure on the flank of the cone. Usually the activity begins in the crater (caldera) and then shifts to some point on the fissure (rift zone) lower on the flank of the cone. A series of earthquakes usually precedes the eruption, as the fissure opens to allow the magma to reach the surface. The lava is quite fluid, and the gases escape without disrupting the lava into ash or cinders. In the initial stage of an eruption, spectacular lava fountains play at many points along the fissure, hurling jets of lava hundreds of feet into the air. The lava fountains, known as "curtains of fire," are caused by the frothing at the top of the lava column when the pressure is suddenly reduced by the opening of the fissure and the contained gases explosively expand. Great floods of lava issue from the fissure or are produced by the lava fountains and flow in rivers down the mountainside. Eruptions usually last only a few weeks, at most. The curtain of fire, the spectacular part of an eruption, usually lasts for only a few days, for when the bulk of the gases are released, the fountains cease. Commonly, from one-half to two-thirds of the total volume of lava produced in an eruption is expelled in the first few days. The location of the vents from which the lava issues shifts from one eruption to another along the fissure, so that eventually an elongate dome-shaped mass of lava is built over the fissure by the accumulation of successive lava flows. These large, flat lava domes, known as *shield volcanoes*, are characteristic of the Hawaiian-type eruption.

Volcanic activity in the Hawaiian Islands appears to have started at the northwest end of the archipelago and to have shifted progressively to the southeast, and today the only active volcanoes in the entire chain are on the island of Hawaii, the southeasternmost island. The island of Hawaii, known as the "Big Island," consists of five volcanoes that have combined to form the island (Figure 2.12). Of the five, Mauna Loa and Kilauea are still active; Hualalia, the only other to erupt in historic time, poured out a large lava flow in 1800–1801 but has shown no sign of activity since then.

### MAUNA LOA VOLCANO

Mauna Loa is an oval-shaped lava dome about 97 km long and 48 km wide, rising from a base 4572 m below sea level to 4197 m above sea level, making it one of the world's largest active volcanoes (if not the largest). At the summit of Mauna Loa is an oval-shaped caldera known as Mokuaweoweo, 4.8 km long and 2.4 km wide. Mauna Loa is located on a well-defined rift zone that trends in a northeast–southwest direction.

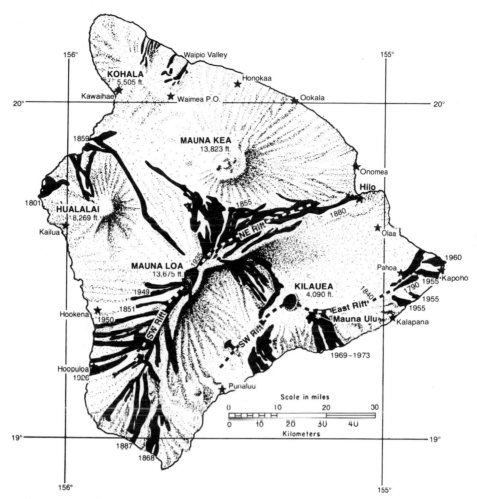

FIGURE 2.12. The island of Hawaii, showing the principal volcanic mountains and the general pattern of lava flows.

A typical eruption of Mauna Loa is signaled by a series of earthquakes that accompany the opening of fissures along the rift zone. The eruptions usually begin on the floor of the caldera and then shift to a section of the rift zone on the flank of the cone. As the eruption begins, lava fountains from many points along the fissure, playing to heights of 160 m and more, produce a veritable curtain of fire. Great volumes of lava are poured out during the first few days. The rivers of lava, a meter or less in thickness at the source, flow rapidly, slowing down and becoming thicker on the lower slopes. During prolonged eruptions the rivers of pahoehoe lava crust over to form lava tubes, and thus insulated beneath the crust the flow may continue with only slight loss of heat for many miles down the mountainside. The longest flow in historic time was that of 1859, which continued for 10 months, reached the coast 53 km away, and continued its flow an unknown distance into the sea. The 1881 flow stopped on the outskirts of Hilo, after flowing about 47 km.

## KILAUEA VOLCANO

Kilauea Volcano is located on the southeast slope of Mauna Loa and about 3125 m below the summit (Figure 2.13). It creates the impression of being a crater on the side of the higher mountain, although in reality it is a separate shield volcano, approximately 80 km long and 22 km wide, built against the side of Mauna Loa. Kilauea is located on two rift zones that extend southeastward and southwestward from the summit caldera. The summit caldera of Kilauea is an oval-shaped depression 4 km long and 3.2 km wide. On the floor of the caldera, near the southwestern edge, is the "fire pit," known as Halemaumau, a circular, pitlike depression that has varied from 600 to 900 m in diameter and up to 365 m in depth. Its dimensions vary from time to time, since lava fills it up from the bottom and the diameter is enlarged by avalanching from the walls. Since 1924 a lake of liquid lava has been present in Halemaumau for only brief periods.

A typical eruption of Kilauea consists of lava flows forming lava lakes in Halemaumau, or of short-lived lava flows within the caldera, and flank flows from fissures along the rift zones on which Kilauea is situated. Flank flows yield

**FIGURE 2.13. Mauna Loa (shield volcano) in profile with Kilauea caldera in the foreground. (Photo by Fred M. Bullard.)**

the greatest volume of lava and are often far removed from the central caldera; sometimes they even occur as submarine eruptions.

At both Kilauea and Mauna Loa, magma rising beneath the mountain preceding an eruption causes the mountain to "swell up" or inflate (a condition known as tumescence) by an amount that can be measured by special instruments known as tilt meters. Measurements of this type appear to be one of the most promising methods of predicting eruptions. As long as the magma is rising and the ground surface in the vicinity of the crater is being inflated, the tilt is away from the crater. Following the eruption and the outpouring of huge volumes of lava, the area collapses, which is indicated by a reversal in the tilt (i.e., toward the crater). Interestingly, the amount of surface collapse following an eruption is often roughly comparable to the volume of the lava erupted.

Some examples of volcanoes exhibiting Hawaiian-type eruptions are Nyamlagira and Nyiragongo in equatorial Africa and the several volcanoes in the Galapagos Islands.

## Icelandic-type Eruption

The great lava plateaus of the world, such as the Columbia River Plateau of the northwestern United States, are believed to have been formed by outpourings of lava from fissures. The Columbia River Plateau and the adjoining Snake River Plain were built up by innumerable lava flows that spread over parts of the states of Washington, Oregon, and Idaho, covering an area of about 512,000 km². The combined thickness of the flow reaches a maximum of over 900 m; yet individual flows rarely exceed 9–30 m in thickness. Similar areas, perhaps even larger, are the Deccan Plateau of India and the Paraná region of South America. The only fissure eruption in historic time, the type believed to have produced the great lava plateaus of the world, occurred in Iceland; such eruptions are thus appropriately termed Icelandic. In the Icelandic type, lava rises from a fissure, or a zone of fissures, several kilometers in length and flows as great sheets for long distances to either side of the fissure. Sometimes the outpourings are in the form of numerous tongues of lava that issue at many points along the fissure. In some of the prehistoric fissure eruptions the upwelling lava sealed the fissure, leaving no indication of the source of the flows. The individual flows are relatively flat-lying, and the vents from which the lava issued must have shifted location from time to time to permit the lava to cover such an extensive area.

The Icelandic-type eruption has many features in common with the Hawaiian type. The lava in both cases is primary basalt, and the eruptions are quite similar. However, in the Hawaiian type the lava piles up in great dome-shaped masses (shield volcanoes), whereas in the Icelandic type the lava forms plateaus with nearly flat-lying beds. It follows, then, that in building a shield volcano the system of fissures that supplies magma must remain open in the same place for long periods of time, whereas in plateau building, the activity shifts frequently from one fissure system to another.

## ICELAND AND THE MID-ATLANTIC RIDGE

The island of Iceland, roughly the size of Ohio, is the exposed part of the northernmost extension of the Mid-Atlantic Ridge. Only a few of the peaks on the Mid-Atlantic Ridge, all of which are volcanoes, emerge to form islands. Iceland consists entirely of volcanic material, largely successive outpourings of basaltic lava that rise from the ocean floor along the Mid-Atlantic Ridge—here between 900 and 1500 m below sea level—to nearly 2133 m above sea level. The *trough*, or *graben*, which is a significant part of the Mid-Atlantic Ridge, crosses Iceland as a prominent surface feature. The sea-floor-spreading hypothesis, which is discussed later, maintains that the trough at the crest of the Mid-Atlantic Ridge is where the crust is pulling apart; in Iceland this zone is on dry land and is available for close study. The active volcanoes of Iceland are restricted to this belt, which covers about one-third of the area of Iceland. Though Iceland is best known for fissure-type eruptions, it should be noted that Hawaiian-type eruptions, with accompanying shield volcanoes, are also common. Since 1100, eruptions have occurred in Iceland, on an average, at 5-year intervals, and a study has shown that nearly one-third of the lava produced on earth since 1500 is to be found in Iceland (Thorarinsson 1967:9).

## LAKI FISSURE ERUPTION OF 1783

The Laki eruption of 1783 is a historic example of a fissure eruption and is also considered the greatest lava eruption of historic time. The Laki fissure in southern Iceland has a length of about 25 km and is divided into two nearly equal segments by Mount Laki. The eruption, preceded by 8 days of severe earthquakes, began on 8 June 1783 (Thorarinsson 1970). During the first days of the eruption, explosive activity generated enormous lava fountains which produced vast quantities of ash that fell over a wide area. From 8 June to 29 July, the section of the fissure southwest of Mount Laki was active. Floods of lava issued from as many as 22 vents along a 16-km-long fissure and flowed in a southeasterly direction. On reaching the glacial river Skaftá, the lava filled the stream bed and overflowed onto the surrounding countryside as it moved rapidly down the valley; in 1 day the lava front advanced 14.5 km. On reaching the coastal area the lava spread out like a giant fan 19–24 km wide. The total length of the flow was about 80 km. On 29 July, the section of the fissure to the northeast of Mount Laki opened, and great volumes of lava poured down the bed of the glacial stream Hverfisfljot, which is parallel to the Skaftá River. Thereafter, to the end of the eruption in November, activity was confined almost exclusively to this area. The lava produced in the section southwest of Mount Laki during the first 50 days of the eruption covered an area of 370 km², with a volume of nearly 10 km³. Thus the average production of solidified lava (specific gravity of 2.4) during these 50 days was about 2200 m³ per second. This would be equal to a discharge of about 5000 m³ per second of fluid lava, or about double the discharge of the Rhine River near its mouth (Thorarinsson 1970). The total area covered by lava in the eruption was 565 km², and the volume of lava produced, with the ash calculated as lava equivalent, was around 12.5 km³. Today the Laki fissure is marked by a line of cinder and spatter cones, many of which are from 40 to 70 m in height.

In the 1783 eruption, rivers flooded by the melting of the glaciers and damming of streams by lava produced floods that destroyed many farms and livestock. Although the lava flows caused great damage, even more serious was the bluish haze (probably containing $SO_2$) that enveloped the country during the summer of 1783. It stunted the growth of the grass, causing a serious famine, still referred to as the Haze Famine. Hunger and disease following the eruption took their toll of human life. As a result of the eruption, Iceland lost nearly one-fifth of its population, about three-fourths of its sheep and horses, and one-half of its cattle (230,000 head). It was a national disaster from which it took years for the country to recover. (See Chapter 5 by S. Thorarinsson in this volume.)

### CRATERS OF THE MOON NATIONAL MONUMENT AND THE GREAT RIFT, IDAHO

This is an area of recent lava flows and cinder and spatter cones associated with a fissure type (Icelandic) eruption from the Great Rift, a southwest-trending fracture zone with a length of about 77 km. The Craters of the Moon National Monument is at the north end of the Great Rift. Following the initial explosive eruptions, rivers of basaltic lava issued from the Great Rift and spread to the northeast and to the southwest. There were several eruptive episodes and the combined lava flows covered an area of about 1500 km² with an estimated volume of about 37 km³. Tongues and sheets of lava spread up to 45 km to the southwest and up to 21 km to the northeast of the Great Rift. The last activity at the Great Rift was about 2100–2300 years ago. This estimate is based on radiocarbon age determinations from charcoal obtained from the roots of trees and shrubs buried by the last lava flow (Bullard 1971). No date for the initial activity has been obtained, but the degree of weathering of the early flows, in contrast to the fresh appearance of the last flows, indicates that the activity from the Great Rift extended over a considerable period of time.

## DISTRIBUTION AND GEOLOGIC RELATIONSHIPS OF VOLCANOES

Although the number of volcanoes known in the world runs into the thousands, only about 500 have been active in historic time and are therefore considered "active volcanoes." Actually, this is somewhat misleading, for some of the volcanoes included in the list of active volcanoes have not erupted for centuries and appear to be extinct. However, in considering the distribution of volcanoes as a clue to their origin, the present listing is quite satisfactory, for any eruption in historic time, regardless of how remote, is certainly in the "present," if considered as a part of geologic time. It is obvious that the list is incomplete, for there must be many submarine eruptions that are not included. Even if a volcanic cone is built up so that it projects above sea level, it is frequently destroyed by wave action and may go unnoticed. A recent list of the active volcanoes of the world (Macdonald 1972; Bullard 1976) contains 516 entries, of which only 69 are submarine eruptions. In order to present a better idea of the day-to-day volcanic activity of the world, the writer (Bullard 1976:448–452) prepared a tabulation of volcanic eruptions covering a 5-year period from 1969 to 1973. The number of eruptions per year varied from a minimum of 17 to a maximum of 27. It can be assumed that some eruptions were not reported, but it is believed that the list is reasonably complete.

## Zones of Active Volcanoes

The great majority of active volcanoes in the world, as well as the recently active, are concentrated in a belt bordering the Pacific Ocean. So marked is the concentration of active volcanoes in this belt that it is known as the "fire girdle of the Pacific." The volcanoes in this belt are associated with geologically young or still growing mountains that are arranged in a series of arcs around the margin of the Pacific Ocean. A second belt of active volcanoes extends from southeastern Eruope, through the Mediterranean area and southern Asia, and into Indonesia. This belt is poorly defined, with significant gaps, such as the area from the Middle East to Indonesia, where there are few, if any, active volcanoes. In addition, more or less isolated volcanoes occur in the three great oceanic regions of the world, the Pacific, the Atlantic, and the Indian oceans.

In the circum-Pacific belt, the volcanoes are either near the margin of the bordering continent or associated with island arcs that lie along the continental margin. In striking contrast, the lands bordering the Atlantic Ocean are relatively free of volcanoes, earthquakes, or growing mountains. The volcanic activity in the Atlantic Ocean is limited largely to islands on the Mid-Atlantic Ridge, such as Iceland and the Azores.

## Continental Drift

Space limitations permit only a few brief observations on the relationship of volcanism to global tectonics. However, the topic has been covered, in some detail, in references available to the reader (Bullard 1976, Chapter 16).

The remarkable jigsaw-puzzle fit of the Atlantic coasts of Africa and South America was apparent from the time that continental outlines first appeared on world maps, leading to speculation that the continents had once been joined. Although it was an old idea and a number of workers had developed hypotheses involving continental drift, it received scant attention until Alfred Wegener, a German meteorologist, presented his proposal in 1912 (see Wegener 1966). It should be noted that the idea of continental drift was in direct opposition to one of the then basic tenets of geology—namely, that the continents and ocean basins were a part of the general framework of the earth and had been essentially stable throughout geologic time. It is not surprising, then, that the idea became the subject of a heated controversy, which has been going on for many years. A major difficulty, from the standpoint of the geologist, was the inability to identify forces capable of moving the continents. The controversy was revived in the 1950s when research on *paleomagnetism*, the magnetism preserved in the rocks at the time of formation, began to reveal new evidence on continental drift. The record of the magnetic field "frozen" in the rocks at the time they were formed is called *remanent magnetism*. When a lava flow cools below the Curie point (temperature above which substances lose their magnetism), it becomes magnetized parallel to the local magnetic field, so that the direction and inclination of the earth's magnetic field at the time the lava solidified is permanently preserved. Methods have also been developed for measuring remanent magnetism in sedimentary rocks. Remanent magnetic data for rocks of different ages have been collected from each of the continents. Analyses of these data show that the magnetic poles, for each geologic period on each continent, consistently fall in about the same area. However, each geologic period has a different location for the magnetic poles, and the conclusion is that the continents have moved in relation to one another.

### Sea-floor Spreading

With the publication of paleomagnetic evidence in the mid-1950s, which seemed to support continental drift, and the discovery that a rift valley or graben (a trough-like structure resulting from tension) follows the crest of the Mid-Atlantic Ridge, there were renewed efforts to find an adequate explanation. Interest centered on a thermal convection hypothesis, elements of which had been advocated by Holmes (1928) and others many years earlier. The hypothesis held that material heated in the subcrustal zone (and hence lighter) rises and is replaced by cold (heavier) crustal material, resulting in a convection cell. In 1960, the late H. H. Hess, a Princeton University professor, presented a concept, which later became known as sea-floor spreading, that was to revolutionize geologic thinking. Hess (1962) postulated that the continents do not "plow" through the oceanic crust, as was held by the early advocates of continental drift, but are carried passively on the mantle that is overturning because of thermal convection. Thus, convection currents bring new oceanic crust from the mantle to the crest of the mid-oceanic ridge, where it spreads laterally, carrying the older oceanic crust as if on a conveyor belt. The sea floor is thus spread apart at the mid-oceanic ridge, and the tension gap is continually filled with new oceanic crust rising from the mantle below. Thus, the continents move **with** the adjacent sea floor and not through it. Where convection cells converge, at the margin of plates (to be described later) a slab of lithosphere is dragged down into the mantle and assimilated. These descending currents, known as *subduction zones*, are the site of compression, characterized by mountain ranges, deep sea trenches, and associated volcanic arcs.

For this overall process R. S. Deitz (1961) coined the term *sea-floor spreading*. The continents are a part of the lithosphere, a layer approximately 100 km thick that is "floating," so to speak, on the asthenosphere, or low-velocity zone in the upper mantle; and as the lithosphere moves, so do the continents. This concept postulates the disruption of the continents by rifting over rising convection currents, followed by the progressive separation of the continental fragments as new oceanic crust is formed. Thus, the ocean basins, instead of being permanent features, are youthful and constantly changing.

## Reversals of the Earth's Magnetic Field

In the early part of the twentieth century Bernard Brunhes (1906), a French physicist, accidentally discovered that the remanent magnetism of a lava flow in central France was exactly opposite to the present magnetic field—that is, the magnetic poles were reversed. Later it was found that many of the Pleistocene volcanic rocks were reversedly magnetized. The cause of field reversals is not understood, but they indicate that the electric currents in the Earth's core, which produces the magnetic field, must, at times, flow "backward" in relation to their present motion. Potassium–argon dating of rocks reveals that there have been long epochs during which the magnetic field was the same as today, alternating with equally long epochs when the field was reversed. Thus, a time scale based on magnetic reversals can be established.

Ships exploring the ocean basins in the 1950s began routinely to measure the strength of the Earth's magnetic field. This became possible with the development of self-recording magnetometers that could be towed behind the ships. Some of the early measurements were made in the eastern Pacific Ocean, and when the results were plotted they showed "ridges" of high magnetic intensity, alternating with "valleys" of low intensity. These came to be known as magnetic "stripes," which are more properly described as a series of *magnetic anomalies*. A magnetic anomaly is the difference between the observed magnetic value and the calculated value; it is said to be positive if the observed value is stronger than the predicted value, and negative if the reverse is true. Similar magnetic stripes were obtained in profiles across the Mid-Atlantic Ridge, and in fact in most of the ocean basins, but the origin of the magnetic stripes remained a puzzle. In 1963 Vine and Mathews developed an explanation by combining Hess's sea-floor-spreading hypothesis and the research then being done on the time scale for magnetic reversals. They proposed that, as new oceanic crust rises along the crest of the oceanic ridges, it cools, passing through the Curie point, at which time the magnetic-field strength and polarity are "frozen" in. If the magnetic field reverses its polarity intermittently, then stripes of alternate polarity will be formed as new crust rises and moves away from the axis of the ridge. When the stripe is magnetized in the direction of the Earth's present magnetic field, the effect is additive, producing a positive magnetic anomaly. When the stripe is magnetized in the opposite direction, it subtracts from the present magnetic field, producing a negative magnetic anomaly. The stripes can be correlated with normal and reverse polarity epochs that have been established from a study of lava flows on the continents. Thus, the distance of an anomaly stripe from the crest of the oceanic ridge can be used to calculate the rate of sea-floor spreading. Studies on the northern part of the Mid-Atlantic Ridge, southwest of Iceland, indicate a spreading rate of 2 cm per year. There is considerable variation in rates: For example, in the South Atlantic a separation rate of around 4 cm per year is indicated, whereas in the Pacific rates up to 9 cm have been calculated. The time scale based on polarity reversals is valid (at this writing) for only about 4 million years, but, if it is assumed that the rate remained relatively constant, the linear spacing of the stripes can be used to extrapolate the time scale back into geologic time. Such calculations indicate that the spreading from the Mid-Atlantic Ridge began between 150 and 200 million years ago (i.e., this is the age of the Atlantic Ocean).

## Plate Tectonics

With the confirmation that mid-oceanic ridges were present in all the major oceans and that sea-floor spreading was a global process, W. J. Morgan (1968) and X. Le Pichon (1968) suggested an overall explanation that has come to be known as *plate tectonics*. Without going into the reasoning or the evidence they used, suffice it to say that they showed that the distribution of fracture zones and observed rates of spreading at oceanic ridges could be explained by the relative motion of a few large plates of lithosphere, which move more or less independently (Figure 2.14); these blocks are jostled about much like gigantic ice floes. Where two blocks move apart, an ocean ridge forms and new oceanic crust is produced; where two blocks move toward each other, compression produces a fold mountain belt, or crust is destroyed as one block is thrust under the other. Using sea-floor-spreading rates for each ocean and seismic evidence, Le Pichon (1968) divided the earth's surface into six large plates, which he postulated were in motion relative to one another. Since that time some smaller plates, such as the Cocos and the Nazca, have been recognized. The relatively rigid plates, which include both continent and oceanic crust, are believed to be about 100 km in thickness and to include the upper part of the mantle, which, together with the crustal layer of the earth, is termed the lithosphere. The lithosphere is resting on (or moving on) the asthenosphere, a relatively mobile, although solid, zone in the mantle capable of movement by slow deformation or creep, in contrast to the lithosphere, which fractures if deformed.

During the relative motion of the lithospheric plates, the area of the earth's crust must be conserved; so the rate at which new crust is being formed at oceanic ridges must be balanced by the rate at which it is being absorbed at plate margins. The margins of the plates, where crust is being consumed, fall into three types: (a) oceanic against continental, as with the Pacific and South American plates; (b) oceanic against oceanic, as with the western two-thirds of the Aleutian arc (Alaska); and (c) continental against continental, as with the Himalayas at the juncture of the African and Eurasian plates. It should be

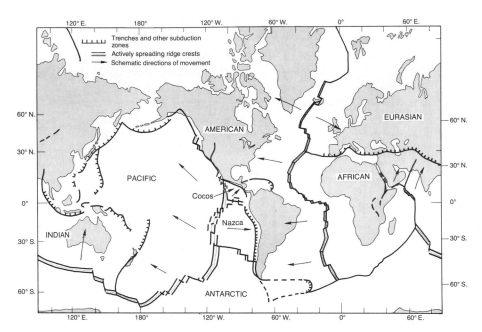

**FIGURE 2.14.** Major lithospheric plates. Mid-oceanic ridges, where plates move apart, are represented by double lines. Trenches and other subduction zones are shown by solid lines with teeth on one side; the teeth point in the direction of the descending slab. The subduction zone around the margin of the Pacific plate includes most of the island arcs and also most of the active volcanoes of the world. It is also the site of most of the intermediate and deep-focus earthquakes. The subduction zone that marks the juncture of the African and Eurasian plates is the site of the Himalaya Mountains. (Modified from Le Pichon 1968.)

noted that two plates may slide past each other, as the Pacific and American plates do on the California coast, the San Andreas rift being the plane of slippage. Since earthquakes would be expected to occur where two plates are in contact, the seismic belts of the earth have been used to identify plate boundaries.

The volcanic island arcs, which are prominent features on the western margin of the Pacific plate, are the site not only of active volcanism but also of most of the intermediate and deep-seated earthquakes. They are also associated with deep sea trenches on the ocean floor. If crustal material does descend into mantle, as postulated by plate tectonics, the island arcs are the most likely site of the sinks, and the ocean trenches are interpreted as the surface expression of the subduction zone along which the plate margin is carried into the mantle.

## Volcanism and Plate Tectonics

Magma is formed at the front and the rear of the moving lithospheric plates, at the oceanic ridges where new crust is being formed, and along the subduction zone where crust is apparently being consumed. The new crust formed at the active oceanic ridges is basaltic lava, and the volcanoes associated with oceanic ridges are usually of the Icelandic or Hawaiian types. Volcanoes associated with the subduction zone, in which rhyolitic and andesitic material are the chief products, are of the Peléan and Vulcanian types. It might appear that the introduction of a relatively cold slab of lithospheric material into the subduction zone would hinder the development of magma, but the active volcanoes associated with this zone and the absence of volcanoes where the crust is not being consumed attest to the fact that magma is being developed. Space does not permit a review of the various suggestions that have been made to account for the heat necessary to produce magma in the subduction zone. There is also the problem of accounting for the rhyolitic and andesitic materials produced by volcanoes associated with the subduction zone. As early as 1962, Coats realized that the Aleutian arc had been formed by underthrusting of the oceanic crust, and he presented evidence to show that the magmas of andesitic volcanoes were formed by basaltic magma from the mantle mixing with sediment being dragged down by the plate. A layer of sediment, which averages about 1 km in thickness, overlies the basaltic oceanic crust. At the subduction zone this sediment is either scraped off the down-going slab and piled up along the edge of the continent or is carried down into the mantle; or some of each may occur. When Gilluly (1971) was unable to find evidence of such volumes of material piled against the continent, he concluded that "a lot of rocks that went down the subduction zone were sandstone, chert, and radiolarian ooze. . . . Such a secondary derivation might account for much of the great volume of andesitic magma of the circum-Pacific [p.2386]." The source of the huge granitic batholiths and the tremendous volume of rhyolitic ash flows of the western United States has long been a problem. If they are the end product of crystal differentiation of primary basaltic magma, then gigantic volumes of intermediate rocks would have to have been formed in the process. Since such huge volumes have not been recognized, some other explanation for the existence of granitic and rhyolitic rocks is required. The recycling of sediments and the crustal layers of the earth, as postulated by the plate tectonic hypothesis, is an acceptable alternative.

## REFERENCES

Alfano, G. B., and I. Friedlander
    1928  La storia del Vesuvio. Naples: K. Holm.
Atwood, W. W., Jr.
    1935  The glacial history of an extinct volcano, Crater Lake National Park. Journal of Geology
          43:142–168.
Brunhes, B.
    1906  Recherches sur la direction d'aimantation des roches volcanique. Journal de Physique
          (Series 4) 5:705–724.
Bullard, F. M.
    1971  Volcanic history of the Great Rift, Craters of the Moon National Monument, south-
          central Idaho. Geological Society of America Abstracts with Programs 3 (3):234.
    1976  Volcanoes of the earth. Austin: University of Texas Press.

Coats, R. R.
    1962   Magma type and crustal structure in the Aleutian arc. In *Crust of the Pacific basin*, edited by Gordon A. Macdonald and H. Kuno. American Geophysical Union Monograph No. 6. Washington, D.C.: American Geophysical Union.

Curtis, G. H.
    1968   The stratigraphy of the ejecta of the 1912 eruption of Mt. Katmai and Novarupta, Alaska. *Geological Society of America Memoir* 116:153–211.

Dana, J. D.
    1891   *Characteristics of volcanoes*. New York: Dodd, Mead.

Deitz, R. S.
    1961   Continent and ocean basin evolution by spreading of the ocean floor. *Nature* 190:845–857.

Gilluly, J.
    1971   Plate tectonics and magmatic evolution. *Geological Society of America Bulletin* 82:2383–2396.

Goranson, R. W.
    1931   The solubility of water in granitic magmas. *American Journal of Science* (Series 5) 22:483–502.
    1938   High temperature and phase equilibria in albite-water and orthoclase-water systems. *Transactions American Geophysical Union* 29:271–273.

Gorshkov, G. S.
    1959   Gigantic eruption of volcano Bezymianny. Bulletin Volcanologique (Series 2) 20:77–109.

Griggs, R. F.
    1917   The Valley of Ten Thousand Smokes. *National Geographic Magazine* 31, 13–18.
    1918   *National Geographic Magazine* 33:115–169.
    1922   *The Valley of Ten Thousand Smokes (Alaska)*. Washington, D.C.: National Geographic Society.

Heilprin, A.
    1903   *Mount Pelée and the tragedy of Martinique*. Philadelphia: J. B. Lippincott.
    1904   *The tower of Pelée*. Philadelphia: J. B. Lippincott.

Hess, H. H.
    1962   History of ocean basins. In *Petrologic studies*, A volume in honor of A. F. Buddington, edited by A. E. J. Engel, H. L. James, and B. F. Leonard. Boulder, Colo.: Geological Society of America.

Holmes, A.
    1928   Radioactivity and earth movements. *Transactions of the Geological Society of Glasgow* 18:559–606.

Lacroix, A.
    1908   *La Montagne Pelée après ses éruptions, avec observations sur les éruptions du Vésuve in 79 et en 1906*. Paris: Masson.

Lamb, H. H.
    1970   Volcanic dust in the atmosphere; with a chronology and assessment of its meteorological significance. *Philosophical Transactions of the Royal Society of London* (Series A) 266:425–533.

Le Pichon, X.
    1968   Sea floor spreading and continental drift. *Journal of Geophysical Research* 73:3661–3697.

Lipman, P. E., and T. A. Stevens
    1969   Petrologic evolution of the San Juan volcanic field, southwestern Colorado, U.S.A. (abst.) International Association of Volcanology and Chemistry of Earth's Interior. In *Symposium on volcanoes and their roots*. Oxford University, Eng. Abstracts volume, p. 254.

Macdonald, G.
    1972   *Volcanoes*. Englewood Cliffs, N.J.: Prentice-Hall.

Mackin, J. H.
 1960   Structural significance of Tertiary volcanic rocks in southwestern Utah. *American Journal of Science* 258:81–131.
Marshall, P.
 1935   Acid rocks of the Taupo-Rotorua district. *Royal Society of New Zealand*, Transaction Vol. 64:323–366.
Martin, G. C.
 1913   The Katmai eruption. *National Geographic Magazine* 24:131–181.
Mehringer, P., E. Blinman, and K. L. Peterson
 1977   Pollen influx and volcanic ash. *Science* 198:257–261.
Mercalli, G.
 1907   *I vulcani attivi della terra.* Milan: Hoepli.
Mercalli, G., O. Silvestri, G. Grablovitz, and V. Clerici
 1891   La eruzione dell'Isola di Vulcano. *Annali dell'Ufficio centrale del metèorologicae geodinamica* (Pt. 4) 10. A review and summary of the conclusions is given in English by G. M. Butler, Eruption of Vulcano, August 3, 1888 to March 22, 1890. *Nature* 46:117–119.
Mitchell, J. Murray
 1970   A preliminary evaluation of atmospheric pollution as a cause of global temperature fluctuations of the past century. In *Global effects of environmental pollution*, edited by S. Fred Singer. New York: Springer-Verlag; Holland. D. Reidel.
Moore, J. G., and W. C. Melson
 1969   Nuées ardentes of the 1968 eruption of Mayon Volcano, Philippines. *Bulletin Volcanologique* (Ser. 2) 33 (Pt. 2):600–620.
Morgan, W. J.
 1968   Rises, trenches, great faults, and crustal blocks. *Journal of Geophysical Research* 73:1959–1982.
Peck, D. L., J. G. Moore, and G. Kojima
 1964   Temperature in the crust and melt of the Alac lava lake, after the August 1963 eruption of Kilauea Volcano—a preliminary report. U.S. Geological Survey Prof. Paper 501 D,1–7.
Peck, D. L., T. L. Wright, and J. G. Moore
 1966   Crystallization of tholeitic basalt in Alac lava lake, Hawaii. *Bulletin Volcanologique* (Ser. 2) 29, 629–655.
Perret, F. A.
 1935   Eruption of Mt. Pelée 1929–1932. Carnegie Institution of Washington Publication 458, Washington, D.C.
Rittmann, A.
 1944   *Vulcani attività e genesi.* Naples: Editrice Politecnica.
Robson, G. R., and J. F. Tomblin
 1966   *Catalogue of active volcanoes of the World, including Solfatara Fields. Pt. 20, West Indies.* Rome: International Association of Volcanology.
Ross, C. S., and R. L. Smith
 1961   Ash-flow tuffs: Their origin, geologic relations, and identification. U.S. Geological Survey Prof. Paper 366.
Rubey, W.
 1951   Geologic history of the sea. *Bulletin of the Geological Society of America* 62:1111–1148.
Smith, R. L.
 1960   Ash flows. *Bulletin of the Geological Society of America* 71:795–842.
Smith, R. L., and R. A. Bailey
 1966   The Bandelier Tuff. A study of ash flow eruption cycles from zoned magma chambers. *Bulletin Volcanologique* (Ser. 2) 29:83–104.
Stoppani, A.
 1871–1873   *Corso di geologia* (3 vols.). Milan: Bernardoni e G. Brigola.

Taylor, G. A.
   1958   The 1951 eruption of Mount Lamington, Papua. *Bulletin of the Bureau of Mineral Resources; Geology and Geophysics (Australia)*, No. 38.
Thorarinsson, S.
   1967   *Surtsey*. New York: Viking.
   1970   The Lakagigar eruption of 1783. *Bulletin Volcanologique* (Series 2) 33:910–929.
Vine, F. J., and D. H. Mathews
   1963   Magnetic anomalies over oceanic ridges. *Nature* 199:947–949.
Wegener, A.
   1966   The origin of continents and oceans. Translated by John Biram from the German (4th ed., 1929). New York: Dover.
Williams, H.
   1941a  Calderas and their origin. University of California Publications. *Bulletin of the Department of Geological Sciences* 25:239–346.
   1941b  Volcanology. In *Geology, 1888–1938*. New York: Geological Society of America 50th Anniversary Volume, pp. 367–390.
Williams, H., and G. Goles
   1968   Volume of the Mazama ash-fall and the origin of Crater Lake caldera, Oregon. Andesite Conference Guidebook, Oregon Department of Geology and Mineral Industries, Bulletin 62, pp. 37–41.

# 3

# Geologic Methods in Studies of Quaternary Tephra

LAURENCE R. KITTLEMAN

## INTRODUCTION

On 26 August 1883, the Indonesian volcano Krakatoa erupted after several months of preliminary activity. Macdonald (1972) summarizes contemporary accounts:

> The climax came suddenly. At approximately 1 P.M. . . . . there occurred a violent explosion, the noise of which was heard 100 miles away. . . . At 2 P.M. came an even greater explosion that threw ash to a height of 17 miles above the volcano. . . . For 100 miles around the volcano the country was shrouded in falling ash. Day brought no end to the darkness. Even with lamps and torches the visibility was not more than a few feet. More than 100 miles away this stygian blackness persisted for 22 hours, and on the near end of Sumatra it lasted for 2½ days. . . . Ash rose more than 50 miles into the air and rained down heavily, mixed with chunks of pumice, over an area of 300,000 square miles. The noise of the explosion has been called "the loudest noise on earth." It was heard 3000 miles away . . . [p.238].[1]

The words above describe an explosive volcanic eruption during which at least 16 km³ of volcanic debris fell throughout a vast area around the volcano (Macdonald 1972:239). There is no need to speculate about the effects of the eruption upon people living nearby. The accompanying great oceanic waves

[1] Reprinted by permission of Prentice-Hall, Inc., Englewood Cliffs, New Jersey.

*VOLCANIC ACTIVITY AND HUMAN ECOLOGY*

(*tsunami*) caused by undersea earthquakes killed more than 36,000 people. Some places were buried by more than 70 m of erupted fragments (Macdonald 1972:239), and lesser consequences were noticed nearly everywhere, as volcanic dust spread around the world.

The awesome example of Krakatoa is exceptional—in its magnitude and in that there is a historical record—but its consideration evokes questions. How shall we recognize the deposits or the consequences of similar great eruptions in the prehistoric record? How different are signs of the more numerous lesser eruptions? What are the products of explosive volcanism, and how are they produced, dispersed, and deposited? How can they be distinguished and used to help establish chronology of human culture and to elucidate the relationships among humans, their environment, and the effects of volcanism? How can we predict and plan to deal with the effects of a great eruption upon a modern city? Such questions can be answered or approached through attention to the geology of tephra.

The name *tephra* is given to fragmental volcanic matter transported from the crater through the air during a volcanic eruption (Thorarinsson, in Westgate and Gold 1974:xvii). Tephra is produced mainly by eruptions of the Plinian, Peléean, and Vulcanian types (Bullard, Chapter 2 of this volume) that collectively may be called *tephragenic* eruptions. Most tephra comes from new *magma* (Bullard, Chapter 2) rising in a volcanic conduit, where gas dissolved in the magma escapes explosively, sundering the liquid and creating a cloud of fragments and hot gas. The fragments become solid as they are ejected into the air, whence they are carried leeward by winds. Usually tephra is mostly glassy fragments created by sudden quenching of liquid magma, together with crystalline minerals that already were present in the magma. The particles range in size from a few micrometers to more than a meter, a gradation of more than a millionfold. Tephra usually falls to earth throughout a teardrop-shaped area that has the volcano near its narrow end. Size, shape, and volume of this mantle, and thickness, grain-size, and composition of the layer at a particular place depend on the magnitude and history of the eruption, chemical composition of the magma, speed and direction of winds at various altitudes, and sizes of particles produced at the orifice.

The tephra mantle so formed contains varied clues from which can be read volcanologic and meteorologic conditions at the time of the eruption and perhaps even the year or season of the event. These mantles have three important features: They cover a large area, they intrude upon whatever environment or landscape happened to lie beneath the eruption cloud, and they form quickly. It is these features especially that have attracted attention, not only among geologists, but also among anthropologists, botanists, climatologists, geographers, historians, oceanographers, sociologists, soil scientists, and zoologists. In the pages that follow, I will describe the origin and main properties of tephra and discuss some of the ways of studying this fascinating product of cataclysm.

# PRODUCTION AND DISPERSAL OF TEPHRA

## Tephragenic Eruptions

Nearly every kind of volcanic eruption produces some tephra. Eruptions of the Hawaiian type, for example, are relatively quiet outpourings of very mobile lava, accompanied occasionally by jets or fountains that may spout hundreds of meters into the air, forming a spray of molten lava that solidifies into a kind of tephra. The amount of tephra produced in this way usually is small, and deposition is restricted to the neighborhood of the vent. Several kinds of eruptions, however, produce great quantities of tephra, but little or no liquid lava flows out at the surface. Amount and character of tephra ejected during an eruption are determined by properties of the magma from which the tephra formed, the most important of which are abundance of dissolved gas and viscosity, a measure of the ability of a fluid to flow. Gas-rich, viscous magmas usually are associated with eruptions that produce mainly tephra, whereas gas-poor, fluid magmas yield quiet flows accompanied by little tephra. Viscosity of magma is determined largely by the chemical behavior of the element silicon (Si), after oxygen the most abundant of the 10 or so predominant elements in lava. For example, *mafic* basaltic magma contains about 45% silica (silicon dioxide, $SiO_2$), and its viscosity is around $10^4$ P (10,000 poises) at 900°C; *silicic* rhyolitic magma has about 72% silica and viscosity near $10^{12}$ P at about the same temperature (Macdonald 1972:64), eight orders of magnitude greater. In comparison, the viscosity of water is about 0.01 P, and that of motor oil around 0.1 P.

There is, however, a special kind of eruption, called *hydromagmatic*, in which water from an external source gains access to magma in a conduit. Such eruptions owe their great explosivity to steam, not to properties of the magma.

Observations in an active volcanic vent are not now possible, but studies of the products of eruptions have yielded ideas of how tephra is formed. As silicic magma rises in its conduit, it gets cooler and pressure on it decreases. These changes cause viscosity to increase and cause dissolved gases, mainly steam, carbon dioxide, and sulfur dioxide, to form bubbles. The two effects progressively act together so that bubbles are growing most rapidly and forcefully at the very time that the enclosing liquid is becoming ever more stiff. Eventually gas pressure ruptures the films of liquid between adjacent bubbles, and the froth disintegrates into a cloud of fragments and gas at a temperature near 1000°C that is expelled from the vent, often with unimaginable violence, by continuously expanding gas. Innumerable variations in this process—intricate interactions among composition and amount of lava, rate of ascent, and rates of change of pressure and viscosity—account for the great diversity of tephra, and it is that diversity which allows us to distinguish the products of one eruption from those of another and to guess the nature of long-past eruptions.

Variations in eruptive process account also for several kinds of tephragenic activity: (*a*) *pyroclastic flows*, (*b*) *base-surges*, and (*c*) *tephra-falls*. These phenomena are not sharply delimited and in fact may occur together, but each tends to produce consistent chains of events and distinctive deposits.

## PYROCLASTIC FLOWS

Pyroclastic flows are streams of tephra and gas that form at volcanic vents and flow outward, sometimes at great speed and to considerable distance. They can form in various ways and have various proportions of tephra to gas, but their common features are mobility, travel close to the ground, and inclusion of hot fragments of tephra. Some are caused by the disintegration and slumping of newly formed solid or very stiff protrusions of lava that may form at volcanic orifices. Slumped fragments, still very hot, gather as an avalanche and fall down the slopes of the volcano, accelerated by gravity and increasingly mobilized by gas given off by the fragments. Other kinds of pyroclastic flows originate in the upper part of the conduit, where boiling of magma creates froth that either may overflow and spread outward, or, if the orifice is restricted somehow, shoot out as a jet.

*Ash-flows* (Ross and Smith 1960; R. Smith 1960a), a variety of pyroclastic flow, are eruptions of disrupted froth—mixtures of hot fragments and gas—that flow close to the ground, often conforming to stream valleys and other depressions. They may reach outward tens of kilometers, traveling at hundreds of kilometers per hour. The active flow is a mobile pyroclastic mixture, but, when motion stops, the flow may form variously either a mass of unconsolidated tephra, a hard rock composed of fragments welded together by retained heat, or deposits with characteristics between those extremes, depending upon the thickness and extent of the flow and upon temperature at the time of eruption. Hot, thick flows are likely to form rock called *welded ash-flow tuff* (Figure 3.1), in which tephra particles are greatly compressed and thoroughly fused together. Near the base of some such flows, where pressure and temperature were greatest, the tephra is largely remelted and transformed into a rock that resembles obsidian. Newly formed ash-flow deposits may contain much volatile material, mainly steam, which fumes away as the flow cools, cooking the mass in its own juices. It is such hydrothermal alteration, along with welding, that accounts for lithification and other features of ash-flows (R. L. Smith 1960b). Prehistoric welded ash-flow tuffs are abundant in the western United States, where flows or series of flows cover hundreds or thousands of square kilometers.

Pyroclastic flows, of whatever kind, are a menace, because of their suddenness and extent. A number have occurred during historic time. For example, at the volcano Bezymianny, Kamchatka, an eruption in 1956 hurled a jet of tephra obliquely to an altitude of nearly 40 km, and accompanying ash-flows reached more than 20 km beyond the volcano (Gorshkov 1959).

In 1912, vents near Mount Katmai, Alaska, erupted an ash-flow, together with windborne tephra (see Figure 3.4C on page 58). The flow filled a nearby valley to a distance of 22 km and a depth of 30 m (Curtis 1968). For years

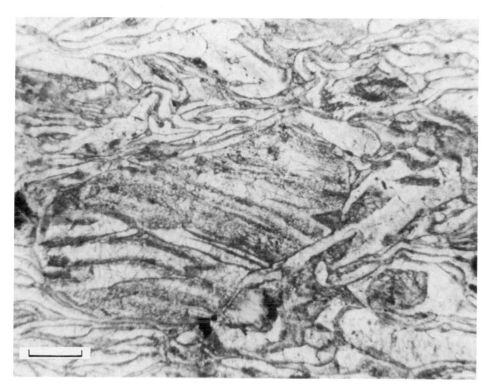

FIGURE 3.1. Welded ash-flow tuff. Irregular gray bodies are glassy shards of tephra welded together, contorted, and compressed by pressure directed vertically, coincident with the short direction of the picture. [Dinner Creek Tuff, Miocene Epoch, Oregon; photomicrograph by the author.] (Scale equals 100 μm.)

afterward, the deposit gave off plumes of steam, for which the place was named Valley of Ten Thousand Smokes.

Mount Lamington, New Guinea, in 1951 erupted pyroclastic flows that swept down the slopes of the volcano on all sides to distances of 12 km. At the same time, a column of tephra and gas rose to an altitude of 12 km. The eruption took nearly 3000 lives (Macdonald 1972:227).

An eruption of Mount Pelée, Martinique, in 1902 (Bullard 1962:100 and Chapter 2, this volume) featured a horizontally ejected cloud of hot gas and tephra mixed with scalding mud that, moving at hundreds of kilometers per hour, swept into the nearby city of St. Pierre, destroying it and its 30,000 inhabitants. The cloud had a temperature of about 700°C by the time it reached the city. Such phenomena have been named *glowing avalanches*, or, in the language of the first witnesses, *nuées ardentes* (glowing clouds). Recent interest in tales of disaster has inspired a novel based on the eruption of Pelée (Thomas and Witts 1969). The book relates, apparently with historical accuracy, that residents were dissuaded from fleeing the early mild tephra showers by politicians who did not want forthcoming elections disrupted (Macdonald 1972:143; Bullard, Chapter 2 of this volume).

## BASE-SURGES

A base-surge is a turbulent, expanding, doughnut-shaped cloud of tephra and gas that rolls out laterally from vents of the most violently explosive eruptions in which water from the ocean, a lake, or groundwater has come into contact with magma in the conduit (Macdonald 1972:250). Such *hydromagmatic eruptions* are characterized by entrainment of copious steam with solid ejecta, and. in some, production of new lava may be slight, the explosion being attributable to the external source of water. The accompanying base-surge moves outward at speeds near 150 km/hr (Moore 1967; Thorarinsson, Chapter 5 of this volume), and tephra deposited from it may form concentrically arranged ridges that resemble aeolian dunes and have bedded internal structures (Crowe and Fisher 1973; Sheridan and Updike 1975). This phenomenon is not restricted to volcanic events, but also is associated with large explosions of all kinds, including nuclear detonations, in which the ring around the base of the familiar mushroom-shaped cloud is a base-surge. Base-surges are thought to develop when great pressure within an explosion crater blows away the rim, allowing material previously confined inside to escape sideways (Moore 1967).

During the eruption of the Philippine volcano Taal in 1965 (Moore 1967; Moore *et al*. 1966), a series of base-surges rolled outward. All trees were obliterated within a radius of 1 km. Within 6 km, sandblasting removed as much as 15 cm of wood from tree trunks on the side that faced the blast. Beyond 6 km, trees were first sandblasted, then coated by a layer of mud as much as 40 cm thick.

## TEPHRA-FALLS

Tephra-falls are produced mostly by explosive volcanic eruptions that are accompanied by the rise of new magma. Much fragmental debris is propelled high above the volcano, whence it is carried by winds to distances of perhaps thousands of kilometers. The phenomenon also is called ash-fall or air-fall, but here I use the term *tephra-fall* because it reveals the nature of the event—falling of ejecta—without specifying the size of fragments or exactly the means of transportation.

Properties of the magma and movements of magma in the conduit before eruption determine whether there will be an ash-flow or a tephra-fall (Macdonald 1972:165). If temperature and pressure are such that bubbles form abruptly in magma that is near the top of the conduit, froth boils from the vent as an ash-flow. If, however, boiling occurs when magma is at some depth below the orifice, the conduit confines the cloud of gas and tephra formed from the disrupted froth, directing it like the barrel of a gun, so that it attains considerable velocity within the conduit and leaves the orifice more or less vertically, propelled by expanding gas. Part or all of some pyroclastic flows may come from material falling back from a collapsing eruption column (R. L. Smith 1960a:804; Wright and Walker 1977) that is very rich in solid fragments; the fallen tephra then rushes down all sides of the volcano, propelled by gravity and lubricated by entrapped gas. The eruption of Mount Lamington in 1951 (Macdonald 1972:227) perhaps is an example.

## Dispersal of Tephra

It is convenient to recognize two kinds of tephra particles, which are distinguished by their fluid-dynamic behavior: projectiles and windborne particles. *Projectiles* are those larger particles, measured in centimeters or meters, whose flight is determined more by initial velocity than by the particles' resistance to motion through the air. *Windborne particles* are those smaller particles, measured in micrometers or millimeters, whose flight is determined more by aerodynamic resistance than by initial velocity. There is, however, no real distinction between them, but rather a continuous gradation of behavior dependent upon size and density. Particles massive enough that they do not acquire significant velocity owing to wind during their flight can be regarded as projectiles.

Projectiles that are thrown out vertically fall back near the vent, but those that are expelled at some angle to vertical fall at a distance from the vent determined by the angle and velocity of ejection. Conversely, that distance can be measured and used to calculate velocity of ejection, from which can be estimated in turn some conditions in the conduit during eruption. Wood and Dakin (1975) examined the distribution and sizes of projectiles around Ara Shatan maar in Ethiopia. They found a systematic relation between sizes of projectiles and distance from the crater (Figure 3.2). Their analysis is based on the relation,

$$V_e = (g/\sin 2\,\alpha)^{1/2}\, D^{1/2}, \tag{1}$$

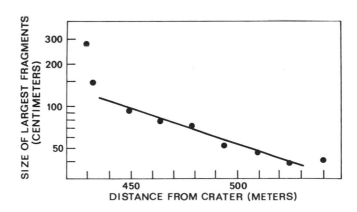

FIGURE 3.2. Distances thrown and sizes of projectiles at Ara Shatan, Ethiopia. [From Wood and Dakin 1975:103; used by permission.]

where $V_e$ is velocity of ejection, $g$ is the gravitational constant, $\alpha$ is the angle of ejection measured from horizontal, and $D$ is distance. For the sake of obtaining conservative estimates, $\alpha$ is assumed to be 45°. This simple fundamental relation neglects air resistance, but, using tables elaborated from Eq. (1), taking air resistance into account (Wilson 1972), Wood and Dakin calculated ejection velocities for their examples that ranged from 70 m/sec, for particles about 100 cm in diameter, to 90 m/sec, for particles about 40 cm in diameter (Wood and Dakin 1975:103). The authors reason that particles are thrown out by a stream of gas of velocity sufficient to suspend large particles and carry them upward at high velocity against gravity. The velocity of the gas can be found from

$$V_f = V_e + V_t, \tag{2}$$

where $V_f$ is the velocity of the gaseous fluid moving upward in the conduit, and $V_t$ is *terminal velocity* (Lorenz 1970:1825), the rate at which a solid particle will sink gravitationally in a still fluid [see Eq. (3)]. Physical conditions in a volcanic conduit are poorly known, but, using reasonable assumptions, Wood and Dakin estimated (1975:103) the gas that expelled projectiles at the relatively small volcano Ara Shatan may have been moving at about 125 m/sec. Combined observation and calculation, as in this example, provide some insight into how tephra particles can be hurled to substantial distances beyond the volcano. Fragments as large as 13 cm fell 40 km from the crater during the eruption of the Indonesian volcano Tambora in 1815 (Bolt *et al.* 1975:118). Nevertheless, distances traveled by projectiles are small compared with those traveled by windborne tephra, and scattered large particles among finer tephra usually indicate that a vent is nearby.

### Tephra Mantles, Thickness, and Grain Size

Windborne tephra particles are those that are small enough to be carried by winds and deposited throughout a large area. Such particles, and deposits of them, are the most abundant products of explosive volcanism and are what is understood commonly by the term, "volcanic ash"—those widespread *tephra mantles* or *tephra layers* that serve as the chronologic and stratigraphic markers so interesting and useful in a variety of endeavors.

The cloud of solid particles and gas propelled aloft from a volcanic vent rises rapidly, both gas and individual particles having large ejection velocities. Observations of eruption clouds yield estimates of rates of ascent. For example, the first cloud above Mount Lamington, New Guinea, during the eruption of 1951, rose 12 km at an average rate of about 100 m/sec (Macdonald 1972:228). That above Mount Hekla, Iceland, in 1947, rose 30 km at an average rate of about 20 m/sec, but it rose the first 20 km at perhaps 70 m/sec (Thorarinsson 1954:PlXV). Minakami (1942:100) observed that the eruption cloud from Mount Asama, Japan, in 1936, rose at a rate of more than 100 m/sec. He calculated also that particles 5 cm in diameter were carried nearly to 15 km altitude during the eruption of Asama in 1783 (Minakami 1942:103).

Wind blowing across the eruption cloud moves particles horizontally, eventually at a velocity equal to that of the wind. For example, tephra from the eruption of Mount Hekla in 1947 was carried first southward and then northeastward about 3800 km to Helsinki, Finland, at an average speed of 75 km/hr, or 21 m/sec (Thorarinsson 1954:58). Figure 3.3 shows the distribution of wind velocity with altitude near Mount Hekla during the eruption. The example brings out that tephra clouds may be carried thousands of kilometers horizontally, that they may follow curved paths, and that the direction or rate of drift may vary with position or altitude. These factors combine to influence the form of the tephra mantle (Figure 3.4). It is important to recognize that variations in shape, thickness, or grain size in the tephra mantle may have little or nothing to do with happenings at the volcano, but merely reflect variable meteorologic conditions. Steady winds will produce a narrow tephra mantle that will be longer for strong winds than for weak ones. Winds that vary in direction will produce a broad, perhaps lobate, mantle, and the more prolonged the eruption, the greater the possible variation.

Historic tephragenic eruptions generally have been brief (Table 3.1), with greatest production of tephra being confined to a short intense phase. Ledbetter (1977) examined evidence in deep-sea sediments of the duration of a prehistoric eruption, using a technique based on terminal velocities of tephra particles in the ocean, and concluded that the 8-cm-thick tephra layer he studied may have been deposited in about 21 days. Mehringer et al. (1977; Blinman et al., Chapter 13 of this volume) studied a layer of Mazama (Crater Lake), Oregon, tephra in a bog in western Montana, wherein they measured the absolute rate of influx of pollen. They estimated from that rate, along with indicators of season obtained from pollen spectra, that the Mazama eruption began in the fall and lasted 3 years, although nearly two-thirds of the layer fell in the first 5 months. This estimated duration is long for a paroxysmal eruption and needs to be confirmed by further work.

Volumes of tephra mantles may be enormous. Williams and Goles

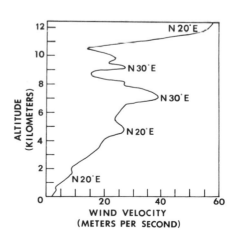

FIGURE 3.3. Variation of wind velocity with altitude near Mount Hekla, Iceland, during the eruption of 1947. Notations along the curve are directions of wind at those altitudes. (Observations at Keflavik, Iceland, 29 March 1947) [From Thorarinsson 1954:47; used by permission.]

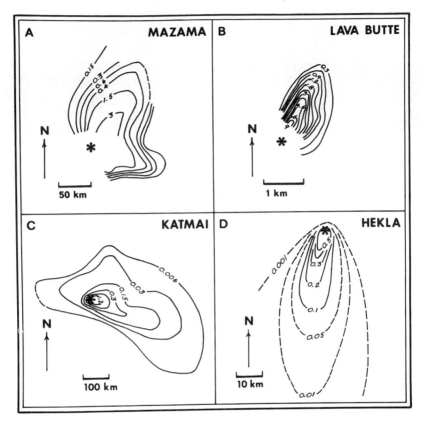

FIGURE 3.4. Thickness–contour maps of some tephra mantles. Thickness in meters. A. Mount Mazama (Crater Lake), Oregon, 7000 B.P. (Williams 1942:16). B. Lava Butte, Oregon, 6100 B.P. (Chitwood *et al.* 1977:59). C. Mount Katmai, Alaska, 1912 (Wilcox 1959a:416). D. Mount Hekla, Iceland, 1947 (Thorarinsson 1970:24). [Used by permission.]

(1968:40) estimated that the volume of the tephra from the eruption of Mount Mazama, Oregon, is about 30 km³; the mantle is still 12 cm thick 200 km from the volcano. Krakatoa in 1883 threw out nearly 16 km³ (Macdonald 1972:239).

Since the orientations of extensive tephra mantles depend mainly upon winds at high altitude that are part of global patterns, study of tephra mantles can reveal directions of ancient windflow that may give clues to climatic change, motions of continents, or wandering of the Earth's axis of rotation (Eaton 1964; Ninkovich and Shackleton 1975). Slaughter and Earley (1965) studied tephra of Cretaceous Age in Wyoming. These rocks, now geochemically transformed into the clay-rich rock bentonite, show the same depositional patterns as do younger tephra mantles, and the patterns generally are oriented in accord with present atmospheric circulation.

The paths of windborn tephra particles depend upon the altitude at which the particles left the eruption column, the horizontal wind velocity, and the rate at which the particles fall. Rate of fall, called *terminal velocity*, is that

**TABLE 3.1**
Durations of Some Tephragenic Eruptions

| Volcano | Date (year A.D.) | Duration (days) | Remarks | Reference[a] |
|---|---|---|---|---|
| Vesuvius, Italy | 79 | 5 | | 1:140 |
| Hekla, Iceland | 1693 | 210 | Most tephra in first hour | 2:104 |
| Asama, Japan | 1783 | 60 | Two phases | 3 |
| Coseguina, Nicaragua | 1835 | 8 | Most tephra in three days | 1:90 |
| Hekla, Iceland | 1845 | 200 | 40% of tephra in first hour | 2:148 |
| Krakatoa, Indonesia | 1883 | 90 | Most tephra in first two days | 4:236 |
| Katmai, Alaska | 1912 | 3 | | 5:427 |
| Kelut, Indonesia | 1919 | 2 | | 5:434 |
| Quizapu, Chile | 1932 | 4 | | 5:432 |
| Hekla, Iceland | 1947 | 180 | 86% of tephra in first day | 6:37 |
| Lamington, New Guinea | 1951 | 45 | Most tephra in one day | 4:230 |
| Bezymianny, Kamchatka | 1955 | 150 | Most tephra in one day | 4:226 |

[a] References
1. Bullard 1962.
2. Thorarinsson 1967.
3. Minakami 1942.
4. Macdonald 1972.
5. Wilcox 1959a.
6. Thorarinsson 1970.

constant velocity attained by a particle falling freely in a fluid, in this case air. It can be found from

$$V_t = \left(\frac{4g}{3C_D} \cdot \frac{\rho_p - \rho_f}{\rho_f}\right)^{1/2} d^{1/2} \tag{3}$$

(Kittleman 1973:2971), where $g$ is the gravitational constant, $C_D$ is a measure of fluidal resistance called *coefficient of drag*, $\rho_p$ is the density of the solid particles, $\rho_f$ is the density of the fluid, and $d$ is the diameter of particles. Terminal velocity is directly proportional to the square root of particle diameter, $d$. The coefficient of drag, $C_D$, has been difficult to evaluate in practice, because it is itself dependent upon both grain-size and velocity, as well as other variables, but approximate solutions can be found (Walker *et al.* 1971; Kittleman 1973:2972). The distance from the volcano at which a tephra particle reaches the ground is whatever distance the particle can move horizontally with the wind in the time required for it to fall from its starting altitude at a rate $V_t$. Table 3.2 gives some hypothetical values based upon reasonable assumptions about physical conditions. Examples 2 through 8 in the table show that particles of various sizes and densities all can arrive together at one spot 75 km distant, and Examples 4 and 7 demonstrate that grains of different sizes or densities nevertheless may have equal terminal velocities, may begin at the same altitude, and may arrive at the same place. The numbers must not be taken literally, for they are based on some simplifying assumptions that are not satisfied in nature; nevertheless, accuracy is sufficient to allow insight into characteristics of real tephra deposits.

**TABLE 3.2**
**Dispersal of Tephra Particles**[a]

| No. | Grain-size (mm) | Terminal velocity (m/sec) | Particle density (gm/cm³) | Distance (km) | Starting altitude (km) |
|---|---|---|---|---|---|
| 1 | 0.05 | 0.20 | 2.65 | 1000 | 7 |
| 2 | 0.05 | 0.20 | 2.65 | 75 | 1 |
| 3 | 0.25 | 0.58 | 0.50 | 75 | 3 |
| 4 | 0.25 | 2.00 | 2.65 | 75 | 11 |
| 5 | 0.50 | 1.35 | 0.50 | 75 | 7 |
| 6 | 0.50 | 4.00 | 2.65 | 75 | 22 |
| 7 | 0.75 | 2.00 | 0.50 | 75 | 11 |
| 8 | 3.00 | 6.00 | 0.50 | 75 | 33 |
| 9 | 10.00 | 11.00 | 0.50 | 10 | 8 |
| 10 | 100.00 | 35.00 | 0.50 | 3 | 8 |

[a] Assumptions: $\rho_a = 1.25 \ 10^{-3}$ gm/cm³; viscosity of air $= 1.76 \times 10^{-4}$ P; atmospheric pressure $= 760$ mm Hg; wind velocity $= 50$ km/hr.

Tephra at a particular locality in a mantle is an assemblage of grains whose sizes and compositions are diverse, for reasons illustrated by Table 3.2. Grain-size characteristics of individual assemblages can be examined through the technique of size–frequency analysis, whereby abundances of grains of various sizes are determined quantitatively, usually by sieving through a graded series of calibrated screens (Kittleman 1973:2961). Analyses may be portrayed as cumulative size–frequency distributions (Figure 3.5). The graph shows that median grain-size is 0.53 mm, and that 5% of all grains are larger than about 3 mm. The range of grain sizes in the samples, called *sorting*, is measured by the standard deviation of the distribution. Tephras commonly are poorly sorted, that is, they have a large standard deviation. The size–frequency distribution shown in Figure 3.5 has two approximately linear segments, a characteristic of heterogeneous (polymodal) distributions, which contain more than one population of grains. The sample illustrated does, in fact, have two different populations, one of pumice, the other of minerals.

Grain-size–frequency characteristics of tephras have been studied by Murai (1961), Sheridan (1971), Walker (1971), Kittleman (1973), and Lirer *et al.* (1973) among others. The log-normal frequency distribution commonly is used for tephra, as it is for analyses of fragmental sediments and rocks generally, but there is evidence that another frequency distribution, Rosin's distribution, better fits data from tephra (Kittleman 1964; 1973).

Studies of tephra mantles show that both thickness and grain-size decrease systematically outward from the source and vary more or less regularly with azimuth. Williams and Goles (1968:39) studied variations in thickness of the tephra mantle from Mount Mazama, Oregon, and found that thickness decreases exponentially with increasing distance out to at least 1000 km, according to:

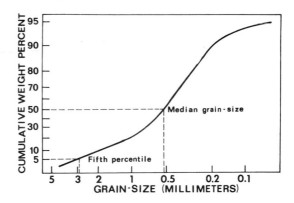

FIGURE 3.5. Size–frequency distribution of tephra. Numbers on the vertical axis give percentage of grains in the sample larger than size shown on the horizontal axis, according to a probability scale for Rosin's distribution. Standard deviation, or sorting, is proportional to the slope of the curve. (Mazama tephra, near Crescent, Oregon, 75 km from the source.)

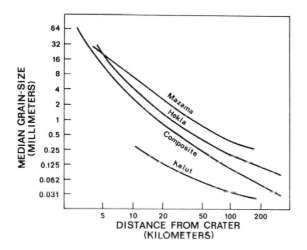

FIGURE 3.6. Variation in median grain-size of tephra with distance from the volcano. Lines represent trends of data for tephra mantles of Mount Mazama, Oregon (Kittleman 1973:2970); Mount Hekla, Iceland (Thorarinsson 1954:Table VII); and Kelut, Indonesia (Baak 1949:Table VIII). The line labeled "composite" is a visual-best-fit trend of data from sources compiled by Fisher (1964: Figure 9).

$$Z = kb^{D}, \qquad (4)$$

where $Z$ is thickness, $D$ is distance, and $k$ and $b$ are parameters that are different for each tephra mantle. Thorarinsson (1954:26) found the same relation for the mantle from the eruption of Hekla in 1947.

Variation of grain-size with distance in several tephra mantles is shown in Figure 3.6, where trends of individual examples are compared with a general composite trend of a large number of analyses compiled by Fisher (1964). The curves have similar slopes and various positions that probably are influenced by characteristics of the eruption and accompanying weather, as well as the relation between grain-size and terminal velocity (Kittleman 1973:2973).

Fisher (1964) studied areal variation of grain-size in the Mazama tephra (Figure 3.7). Contours of equal grain-size are roughly concordant with the northeasterly lobe of the mantle, as defined by contours of equal thickness, called *isopach lines*, but not with the southeasterly lobe, which instead coin-

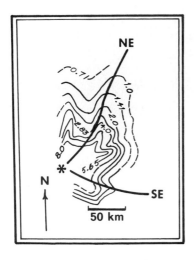

FIGURE 3.7. Contours of equal median grain-size (in millimeters) in the tephra mantle of Mount Mazama, Oregon. Heavy lines mark the northeastern and southeastern lobes of the mantle, as defined by contours of equal thickness (Figure 3.4A). [From Fisher, *Journal of Geophysical Research*, 69:347, 1964. Copyrighted by the American Geophysical Union.]

cides with a direction in which median grain-size decreases abruptly. Fisher (1964) discusses some possible causes of the pattern seen.

Tephra mantles need to be interpreted with due regard for their complexity and with judgment of the statistical adequacy of the data gathered to describe them. Generally, the spacing of samples must be no greater than the spacing of the smallest details that are expected to be resolved. Even when that condition is satisfied, Whitten (1961) found, in a computer-assisted simulation of areal sampling of geologic data, that the machine could be induced to draw greatly different contour patterns from the same data simply by varying the sampling scheme it was told to follow.

Several workers have developed mathematical models of tephra mantles (Slaughter and Earley 1965; Scheidegger and Potter 1968; Shaw *et al*. 1974; Handy 1976), using various approaches based on fundamental principles like that illustrated by Eq. (3), and at least one (Knox and Short 1964) draws on empirical data from studies of nuclear fallout. Comparisons of predictions from these models with actual mantles are scant.

Fine-grained tephra that reaches into the stratosphere, above 9–10 km in high latitudes or 15–18 km in low latitudes, is caught up in the strong global stratospheric circulation. Particles that even in still air would need 5 weeks to fall 12 km remain suspended for months or years in the turbulent stratosphere, while they are carried around the world as *volcanic dust veils*. This stratospheric dust causes diverse effects that range from brilliant sunsets (Meinel and Meinel 1963) and halos around the sun (Pernter 1889) to attenuation of solar radiation reaching the Earth's surface (Lamb 1970). Waxing and waning of the dust veils is thought to be implicated in climatic change and perhaps even the initiation of glaciation (Cronin 1971; Bryson 1974; Schneider and Mass 1975; Bray 1977; Hansen *et al*. 1978; Hein *et al*. 1978), but some workers doubt that dust veils have significant influence on climate (Mason 1976). Dust veils have been investigated by Lamb (1970), who defined the *dust-veil index* as a comprehensive measure of the veil's opacity produced by individual eruptions.

## PROPERTIES OF TEPHRA

Tephra is a sediment whose fragments have been carried through the air from a volcanic vent to a site of deposition. To such sediments is applied the adjective *pyroclastic*, literally, fire-broken. Tephra belongs among the *volcaniclastic* sediments and rocks, a class of materials whose main components are fragmental and come from a volcanic source (Fisher 1960, 1961).

### Components

Tephra can be described as mixtures of glassy, mineralic, or rocky fragments, and varieties are called respectively *vitric tephra*, *crystal tephra*, or *lithic tephra*, according to whichever is the dominant component. Volcanic glass itself is a rock and therefore is a lithic fragment, but it is considered separately in this scheme because of its importance and special properties. Fragments in tephra are called *essential* if they came from newly erupted lava, *accessory* if they came from older rock of the volcano, or *accidental* if they came from some other rock in the walls of the conduit.

Voluminous mantles from great tephragenic eruptions commonly are essential vitric or crystal tephra with scant lithic material, but hydroeruptions that are accompanied by little new lava produce tephra composed mostly of accessory or accidental lithic fragments. Proportions of vitric, crystal, and lithic fragments can be used to distinguish tephras from one another. Lithic fragments sometimes may be particularly useful, because their abundances depend more on physical characteristics of individual eruptions than do abundances of vitric or crystal components, which depend on compositions of magmas that may be erupted successively from a single reservoir of fairly uniform composition (Lirer *et al.* 1973:764).

The name tephra is applied generally to pyroclastic sediment of any grain-size (Table 3.3). It is important here to recognize that classification by

TABLE 3.3
Classification of Tephra According to Grain-Size[a]

| Grain-size (mm) | Tephra fragments | |
|---|---|---|
| | Coarse | Blocks and Bombs |
| — 256 — | | |
| | Fine | |
| — 64 — | Lapilli | |
| — 2 — | | |
| | Coarse | |
| — 0.062 — | | Ash |
| — 0.004 — | Fine | |

[a] From Fisher (1961:1411), slightly modified; used by permission of the Geological Society of America.

size is independent of composition, so, for example, there may be either vitric ash, crystal ash, or lithic ash. A distinction is made between *blocks*, which were solid at the moment of eruption, and *bombs* (Macdonald 1972:125), which were molten.

The glassy component of tephra readily decomposes chemically, forming mainly clays and zeolite minerals (for example, Marshall 1961; Hay 1966), in a process that contributes to the transformation of unconsolidated volcaniclastic sediment into consolidated volcaniclastic rock. Among the products formed is the clay mineral montmorillonite (or smectite), whose special properties dictate the character of soil and rock throughout volcanic terranes. The widespread rock bentonite is tephra in which the glassy component is wholly changed to montmorillonite (Slaughter and Earley 1965). Soil-forming processes in regions underlain by tephra (Ugolini and Zososki, Chapter 4 this volume) are influenced greatly by the chemical instability and physical form of glassy components in the parent material (Baak 1949; Mohr and van Baren 1954; Dyrness 1960; Hay 1960; Tidball 1965; Chichester 1967).

Procedures used to prepare samples of tephra for examination and analysis are designed to remove contaminants, remove obscurant coatings, and separate components for individual study. Preliminary ultrasonic cleaning and removal of material finer than about 45 $\mu$m are essential. A common procedure is the separation of glassy materials from crystals by treatment in a solution of bromoform and ethyl alcohol of density 2.4 gm/cm$^3$, in which glass floats and minerals sink. Details of this and other procedures are given by Kittleman (1973:2959) and Steen-McIntyre (1977).

## Forms of Vitric Fragments

Figure 3.8 shows obsidian, a glassy volcanic rock formed from magma in which both small and large crystalline grains grew before the liquid was quenched. If bubbles had formed in the liquid, then the product might have been a vitric tephra with some crystals in it, instead of obsidian.

All vitric tephra is fragments of natural glass, but it is convenient to recognize several forms, realizing that they merge with one another imperceptibly. *Pumice* is froth in which the walls of bubbles are films of volcanic glass. The bubbles may be roughly spherical or drawn out into long, very thin tubes, rather like taffy. Pumiceous fragments are the size of lapilli or blocks (Table 3.3). *Micropumice* is the same, except that fragments are the size of ash. *Shards* are the walls of burst bubbles. They are the size of ash, rarely lapilli, and they come in various shapes, of which the main ones are (*a*) chunky fragments, sometimes with very small bubbles in them; (*b*) thin plates and curved forms that are the walls of bubbles; and (*c*) pointed forms that are the juncture among several bubbles.

Heiken (1972, 1974) studied morphology of shards. He found relationships among style of eruption, composition of lava, and morphology. Vitric shards from basaltic eruptions are mostly chunky fragments of droplets. Shards from

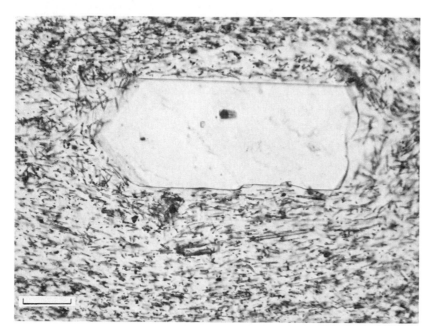

FIGURE 3.8. Photomicrograph of obsidian. In this frozen precursor of tephra, a grain of feldspar swims in a sea of glass that contains tiny mineral grains. Glass shards form when expanding bubbles sunder molten glass. (Llao Rock Dacitic Obsidian, Pleistocene, Crater Lake, Oregon; photomicrograph by the author.) (Scale equals 100 μm.)

basaltic hydroeruptions are chunky fragments with curved faces, but they are not fragments of droplets. Rhyolitic eruptions produce tubular pumiceous shards and tetrahedral bubble-juncture shards. Rhyolitic hydroeruptions, too, yield tubular and tetrahedral shards, along with abundant chunky shards.

Huang and Watkins (1976), using a scanning electron microscope, studied the morphology of fine rhyolitic shards from deep-sea tephra layers. They found on the surfaces of shards tiny pits mostly 0.2–1.0 μm in diameter, in densities as great as 126 pits per 100 μm². They believe that the frequency of pits is related, in a yet unknown way, to the explosivity of the eruption, for the frequencies appear to be related to rate of deposition.

Color of vitric fragments ranges from colorless through pale gray, pink, green, orange, and brown, to nearly black, in a progression determined largely by chemical composition (Table 3.4), but it also depends upon grain-size, and the color of small fragments may differ from that of large ones of the same glass. The color of aggregates differs from that of single grains; pumiceous lapilli may be nearly white but commonly are shades of brown or orange from coatings of iron oxides or humic pigments. Vitric shards and pumiceous lapilli commonly bear coatings of clay minerals that impart to the aggregate a chalky white color or the distinctive silver-gray that distinguishes vitric tephra in many deposits.

**TABLE 3.4**
**Properties of Vitric Fragments**

| Tephra | Color | Silica content (SiO₂, %) | Index of refraction | Reference[b] |
|---|---|---|---|---|
| Glacier Peak, Washington | Colorless | —[a] | 1.490–1.500 | 1 |
| Bishop, California | —[a] | 73 | 1.492–1.499 | 2 |
| Bridge River, Canada | — | — | 1.494–1.499 | 3 |
| Medicine Lake, California | Colorless | — | 1.494–1.499 | 4:2965 |
| St. Helens Y, Washington | — | — | 1.494–1.510 | 3 |
| Newberry Caldera, Oregon | Colorless | 73 | 1.495–1.502 | 4:2965 |
| Valles Caldera, New Mexico | — | 72 | 1.497–1.499 | 5 |
| Mazama, Oregon | Colorless | 68 | 1.502–1.517 | 4:2962 |
| Rainier, Washington | — | — | 1.502–1.585 | 6:72 |
| Thera, Greece | — | — | 1.503–1.513 | 7 |
| Hekla, Iceland | Yellow–white | 65 | 1.510 | 8:173 |
| Hekla, Iceland | Yellow–brown | 60 | 1.530 | 8:173 |
| Hekla, Iceland | Black–brown | 55 | 1.570 | 8:173 |
| Hekla, Iceland | Black | 50 | 1.610 | 8:173 |

[a] Not reported.
[b] References
1. Wilcox 1965.
2. Izett et al. 1970.
3. Nasmith et al. 1967.
4. Kittleman 1973.
5. Izett et al. 1972.
6. Mullineaux 1974.
7. Rapp and Henrickson 1973.
8. Thorarinsson 1967.

## Refractive Index of Vitric Fragments

Volcanic glass, as a transparent solid, has the power to affect light passing through it, and an expression of that power is *refraction*, the bending of light rays. Refraction is a property of glass, defined as *index of refraction*,

$$n = c/c_s \qquad (5)$$

(Bloss 1961:6), where $c$ is the velocity of light in a vacuum and $c_s$ is the velocity of light in a transparent substance. The refractive index of water is 1.333, that of window glass about 1.500, and that of diamond 2.417. The refractive index of volcanic glass depends upon chemical composition, the whole of which tends to vary regularly with silica content. Table 3.4 illustrates that the refractive index of silicic glasses in the neighborhood of 73% silica is relatively low, whereas that of mafic glasses in the neighborhood of 50% silica is relatively high. The whole variation of index is, however, small, practically 1.490 to 1.570, and individual tephras may contain glass within a range of indices that overlaps those of other specimens (Tables 3.4 and 3.6). Thus refractive index of vitric fragments, though useful in combination with other properties, generally is not definitive alone.

The refractive index of vitric grains is measured under a petrographic microscope by observing refractive effects of grains immersed in a liquid of known index. The Becke-line method (Bloss 1961:47) is standard, but the focal-masking technique (Wilcox 1962; Steen-McIntyre 1977) probably is more convenient and precise, though it requires special optical equipment.

## Mineralic Grains

Tephra can contain any of the many minerals that might be found in volcanic rocks generally, as well as those that may be introduced as accidental or accessory fragments, but only a few minerals are common (Table 3.5), and fewer still will occur in one tephra. It is usual for five or fewer minerals to account for more than 90% of the crystal component. Assemblages of minerals tend to be of regional occurrence, so that volcanoes of a chain or province produce mostly certain assemblages. Some minerals rarely occur together, regardless of locale.

It is convenient here to recognize two kinds of mineral grains: One I will call *enclosed crystals*, which are mineral grains that remain enclosed in fragments like pumiceous lapilli; the other, *free crystals*, which are separate individual grains. This distinction is useful, because abundances of free crystals can vary progressively with distance from the volcano through settling of larger or denser grains, and because free crystals can be confused easily with mineral grains that have entered a tephra deposit through contamination, say from accompanying alluvium.

Free crystals, having been released from vitric material during eruption, bear jackets of adherent glass, which appear in the microscope (Figure 3.9) as jagged fringes of edges of grains and as netlike patterns, called bubble-wall texture (Fisher 1963), on faces. These glassy jackets help to distinguish pyro-

TABLE 3.5
Important Minerals of Tephra[a]

| Feldspar minerals | Silica minerals | Feldspathoid minerals |
|---|---|---|
| Anorthoclase | Cristobalite | Leucite |
| Orthoclase | *Quartz* | *Nepheline* |
| *Plagioclase* | | Sodalite |
| Sanidine | | |
| **Ferromagnesian minerals** | **Oxide minerals** | **Other minerals** |
| *Biotite* | Hematite | Allenite |
| *Clinopyroxene* | Ilmenite | *Apatite* |
| Cummingtonite | *Magnetite* | Chevkinite |
| *Hornblende* | Titanomagnetite | Titanite |
| Olivine | | Zircon |
| *Orthopyroxene* | | |

[a] The most common minerals are in *italic*.

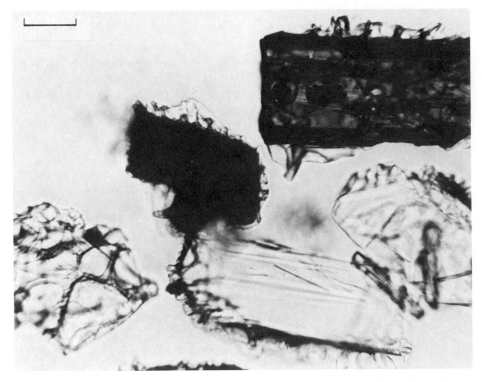

FIGURE 3.9. Mineral grains in tephra. The oblong grain at the upper right is hypersthene, a variety of orthopyroxene; the black grain is magnetite; and clear grains are feldspar. All have jackets of adherent glass, seen as jagged fringes around the grains and as bubble-wall texture, the netlike pattern on the grain at the lower left. (Mazama tephra, Holocene, Oregon; photomicrograph by the author.) (Scale equals 100 $\mu$m.)

clastic grains from contaminants, and they mark pyroclastic grains that have found their way into sediments of other environments. Enclosed crystals, on the other hand, are protected from loss and contamination and so are more reliably characteristic of an individual tephra. They can be extracted for examination by gently crushing glassy lapilli and separating crystal grains from the debris.

The crystal component, particularly dark-colored iron–magnesium silicates, contributes to the color of tephra in aggregate, and coatings formed through chemical decomposition of the grains impart their own hues of yellow, orange, brown, red, or green.

Abundances of minerals are measured by separating crystals from other components, mounting a sample of them on a glass slide, and counting grains under a microscope. As the slide is moved systematically beneath the lens, each grain is identified and recorded, until a previously fixed number of grains, usually between 500 and 1000, has been counted.

Figure 3.10 shows abundances of three minerals in several tephras, in a

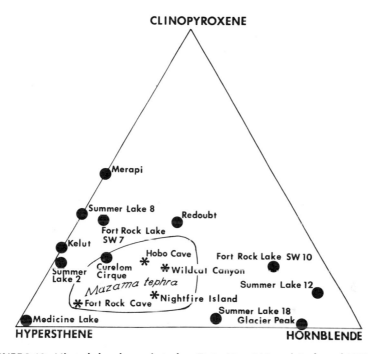

FIGURE 3.10. Mineral abundances in tephra. Dots: Mount Merapi, tephra of 1930, and Mount Kelut, tephra of 1919, Indonesia (Baak 1949); Mount Redoubt, Alaska, tephra of 1966; Glacier Peak, Washington; Medicine Lake Caldera, California; tephras in lacustrine deposits of pluvial Fort Rock Lake and Summer Lake, Oregon (Kittleman 1973). The outline represents the range of 20 analyses of Mazama tephra. Tephra at Curelom Cirque, Utah, is Mazama tephra. Stars: Mazama tephra in archaeologic sites—Hobo Cave, Wildcat Canyon, and Fort Rock Cave, Oregon, and Nightfire Island, California (Kittleman 1973).

diagram so constructed that each tephra is represented by a symbol whose nearness to an apex of the triangle is proportional to the relative abundance of the mineral named at that apex. The tephras contain more than just the three minerals named, but abundances of those three are recalculated to sum to 100%. The irregular outline encloses an area occupied by about 20 analyses of tephra from Mount Mazama, Oregon, which are not shown individually. Such replicate analyses are more informative than single analyses because they show the range of variation in the tephra mantle. Stars represent Mazama tephra recognized in several archaeologic sites more than 100 km from the volcano (Kittleman 1973). Mazama tephra at Curelom Cirque, Utah, is recognizable by its mineral assemblage 725 km distant (Kittleman 1973:2965).

Most volcanic minerals are not single compounds with definite chemical compositions; rather, each named mineral represents a gradational series of variable composition but single atomic structure. The feldspar mineral, plagioclase, for example, is a sodium–calcium aluminosilicate, whose chemical composition can range freely from sodium aluminosilicate ($NaAlSi_3O_8$) at one extremity, to calcium aluminosilicate ($CaAl_2Si_2O_8$) at the other; within that

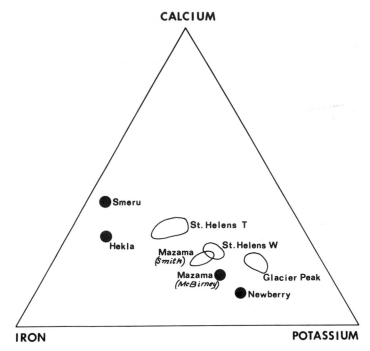

FIGURE 3.11. Abundances of calcium, iron, and potassium in some tephras. Dots: single analyses; outlines: regions of several analyses. Mount Smeru, Indonesia, tephra of 1941 (Baak 1949:14); Mount Hekla, Iceland, tephra of 1947 (Thorarinsson 1954:34); Mount St. Helens, Washington, tephra T, glass (H. W. Smith et al. 1975:213); Mount St. Helens, Washington, tephra W, glass (H. W. Smith et al. 1977b:211); Mount Mazama (Crater Lake), Oregon, glass (McBirney 1968:55); Mount Mazama, glass (H. W. Smith et al. 1977b:214); Glacier Peak, Washington, glass (H. W. Smith et al. 1977a:200); Newberry Caldera, Oregon, obsidian (Higgins 1973:475).

range, any proportion of sodium to calcium is possible. Particular compositions of such minerals are dictated by chemical and physical conditions in the magma during crystallization.

## Optical Properties of Minerals

Color, index of refraction [Eq. (4)], and other optical properties are governed by atomic structure and chemical composition, as in the plagioclase feldspars, wherein index of refraction varies systematically according to proportions of sodium and calcium. Chemical composition thus can be estimated from optical properties. Most minerals, however, are optically complex substances. Unlike glass, for which a single index of refraction serves to describe that property, most minerals require for their determination two or three indices of refraction, because their refractivity depends upon the particular direction taken by light in passing through them. In one large group of minerals, three such directions are defined by convention, and indices of refraction, symbolized $n_x$, $n_y$, and $n_z$, are measured. Color too can act similarly, so that light passing through some minerals emerges with a color determined by its direction of passage. Such minerals are *pleochroic*. In one group of minerals, two special optical directions, called *optic axes*, are defined. The angle between these axes, conventionally symbolized as 2V, is a measurable optical property (Bloss 1961:154).

Optical properties measured from components of some tephras are shown in Table 3.6. Such analyses are a rich source of definitive information, for the table could be expanded to report perhaps 40 properties for each tephra. The method, though, is difficult and time-consuming. Some properties—the pleochroic colors of hornblende, for example—are not definitive for the examples given, and the Bridge River and Mazama tephras could not be distinguished from the properties listed.

Optical properties are measured by means of a petrographic microscope, which is fitted with filters for polarization of light. A special apparatus, the spindle stage (Wilcox 1959b; Steen-McIntyre 1977:105), is necessary for greatest ease of measurement.

**TABLE 3.6**
**Optical Properties of Tephras**

| Property | Bridge River[a] | Mazama[b] | St. Helens Y[c] | Guaje[d] |
|---|---|---|---|---|
| Class | | | | |
| $n$ | 1.494–1.505 | 1.499–1.512 | 1.491–1.496 | 1.497–1.499 |
| Color | White or buff | Buff or white | White | — |
| Feldspar | Plagioclase | Plagioclase | Plagioclase | Sanidine |
| $n_x$ | 1.540–1.547 | 1.545–1.553 | 1.541–1.548 | 1.520–1.521 |
| Clinopyroxene | Present | Present | None | Present |
| Color | Pale green | Pale green | — | — |
| $n_x$ | 1.682–1.686 | 1.684–1.692 | — | 1.726–1.729 |
| 2V | +56–57° | +56–58° | — | +61–64° |
| Orthopyroxene | Present | Present | Present | None |
| Color | | | | |
| X | Pale cinnamon | Pale cinnamon | Pale cinnamon | — |
| Y | Pale yellowish brown | Pale yellowish brown | Pale yellowish brown | — |
| Z | Pale gray green | Pale green | Pale green | — |
| $n_x$ | 1.683–1.690 | 1.686–1.695 | 1.703–1.709 | — |
| 2V | — | — | — | — |
| Hornblende | Present | Present | Present | None |
| Color | | | | |
| X | Pale yellowish brown | Pale yellowish brown | Pale yellowish brown | — |
| Y | Dark olive brown | Dark olive brown | Dark olive brown | — |
| Z | Dark olive green | Dark olive green | Dark olive green | — |
| $n_x$ | 1.643–1.658 | 1.659–1.668 | 1.657–1.666 | — |
| 2V | — | — | — | — |

[a] Bridge River tephra, Canada (R. E. Wilcox in Nasmith *et al*. 1967:165).
[b] Mazama tephra, Pacific Northwest (R. E. Wilcox in Nasmith *et al*. 1967:165).
[c] St. Helens Y tephra, Washington (R. E. Wilcox in Nasmith *et al*. 1967:165).
[d] Guaje tephra, Valles Caldera, New Mexico (Izett *et al*. 1972:560).

## Chemical Properties

About 75 chemical elements are found in volcanic rocks in significant quantities, though the notion of significance changes continually with our ability to detect, measure, and understand the variation of elements. Besides rock, volcanoes erupt important amounts of elements as gases, principally chlorine, fluorine, carbon as carbon dioxide ($CO_2$) and carbon monoxide (CO), and sulfur as hydrogen sulfide ($H_2S$) and sulfur dioxide ($SO_2$). These substances are poisonous in certain concentrations, and some, especially fluorine, can be adsorbed onto surfaces of tephra particles and carried in harmful concentration far from the volcano (Rose 1977; Thorarinsson, Chapter 5 of this volume).

Volcanic rocks generally contain about 10 *major elements*, in abundances of 0.1% or more, and many *trace elements*, in abundances less than 0.1%. Abundances of major elements customarily are expressed as percentages of oxides, silicon dioxide ($SiO_2$), for example. The elements listed in Table 3.7 account for more than 99% of the material analyzed, and oxides of four elements—silicon (Si), aluminum (Al), calcium (Ca), and sodium (Na)—account for more than 80%. Katmai tephra is the most silicic, Smeru tephra, the least, and abundances of the other elements vary concordantly with silica, some increasing, others decreasing, with increasing $SiO_2$.

Not only does chemical composition in general vary systematically, but also volcanoes in a single chain or province all may produce chemically similar magmas, or one volcano may yield similar magmas in successive eruptions (for example, Lirer *et al*. 1973; Keller *et al*. 1978). Conversely, chemical compositions of magmas from some volcanoes may vary regularly through successive

**TABLE 3.7**
**Major Elements in Tephra (in Percent)**[a]

| Oxides of elements | Katmai[b] | Mazama[c] | Hekla[d] | Smeru[e] |
|---|---|---|---|---|
| $SiO_2$ | 76.53 | 73.65 | 61.88 | 52.82 |
| $TiO_2$ | 0.17 | 0.37 | 1.03 | 1.19 |
| $Al_2O_3$ | 12.31 | 14.52 | 16.11 | 17.33 |
| $Fe_2O_3$ | 0.46 | 0.49 | 2.11 | 5.70 |
| FeO | 0.96 | 1.52 | 6.47 | 5.38 |
| MgO | nil | 0.22 | 1.76 | 4.56 |
| CaO | 1.01 | 1.16 | 4.93 | 9.10 |
| $Na_2O$ | 4.15 | 5.08 | 4.21 | 1.89 |
| $K_2O$ | 3.05 | 2.92 | 1.16 | 0.96 |
| $P_2O_5$ | 0.05 | 0.04 | 0.44 | 0.18 |
| Total | 98.69 | 99.97 | 100.10 | 99.11 |

[a] Analyses are of single specimens, not averages for the whole tephra mantle.
[b] Mount Katmai, Alaska, eruption of 1912, vitric tephra (Wilcox 1959a:458).
[c] Mount Mazama (Crater Lake), Oregon, eruption of 7000 B.P., glass (McBirney 1968:55).
[d] Mount Hekla, Iceland, eruption of 1947, crystal–vitric tephra (Thorarinsson 1954:34).
[e] Smeru, Indonesia, eruption of 1941, crystal–lithic tephra (Baak 1949:14).

eruptions. So it is that Mount Hekla, Iceland, is inclined to have violently explosive eruptions of relatively silicic tephra (Table 3.7) after each long interval of quiet, and the longer the repose, the more silicic is the tephra of the succeeding eruption (Thorarinsson 1967:167). Some volcanoes may eject tephras of different compositions during a single eruption (for example, Carey and Sigurdsson 1978).

The analyses in Table 3.7 are of whole tephras—vitric, lithic, and crystal components together—except for Mazama tephra, which is of glass alone. Analyses of whole tephra may more accurately represent the composition of tephragenic lava, but there is the disadvantage that crystal or lithic components can be depleted or contaminated, so analyses of glass alone are preferred for identification of individual tephras.

Variation among some major elements in tephras is illustrated by Figure 3.11. Analyses are arrayed along a line that connects the potassium apex of the triangle with the opposite side, a relationship which indicates that calcium : iron abundance ratios are relatively invariant in these specimens, and the distinctions seen depend mainly on variations in the abundances of potassium. The two samples labeled Mazama, which are of glass analyzed by different methods, do not correspond, but it is unknown whether the discrepancy is caused by sampling error, analytical bias, or misidentification. Such problems need to be resolved before identification of tephras can become routine. Analyses for one of the Mazama samples overlap those of St. Helens W tephra, so those two tephras could not be distinguished reliably by this approach.

Trace elements listed in Table 3.8 together account for 1000 parts per million (ppm) (0.1%) or less of the material analyzed; however, they have

**TABLE 3.8**
**Trace Elements in Tephra (ppm)[a]**

| Element | Mazama[b] | Newberry Caldera[c] | Glacier Peak[d] |
|---------|-----------|---------------------|------------------|
| Rubidium | 43 | 109 | 23 |
| Cesium | 2.09 | 5.1 | 1.4 |
| Barium | 760 | 880 | 220 |
| Lanthanum | 19.2 | 27.4 | 11 |
| Cerium | 46.4 | 62.7 | 18 |
| Samarium | 4.5 | 6.0 | 1.6 |
| Europium | 0.930 | 0.80 | 0.34 |
| Ytterbium | 1.86 | 4.2 | 0.4 |
| Lutecium | 0.51 | 0.77 | 0.12 |
| Thorium | 4.38 | 13.4 | 4.5 |
| Hafnium | 6.00 | 7.25 | 1.7 |
| Cobalt | 5.01 | 1.28 | 1.4 |
| Scandium | 5.03 | 4.58 | 1.8 |

[a] Analyses are of single specimens, not averages for the whole tephra mantle.
[b] Mount Mazama (Crater Lake), Oregon; whole tephra (Randle et al. 1971:265).
[c] Newberry Caldera, Oregon; whole tephra (Randle et al. 1971:265).
[d] Glacier Peak, Washington; glass in tephra (Borchardt 1970:22).

importance far beyond mere abundance, for they furnish many clues to the origin and evolution of magmas and the dating of geologic events. Some trace-elements substitute for particular major elements in the atomic structure of minerals, so they are concentrated in certain minerals; others have less such affinity and instead become concentrated in the glass. Still others occur in more than one atomic form, or isotope, some of which are created or destroyed by radioactive decay. These provide the means for dating rocks and minerals and the events that created them.

Variation among some trace elements is shown in Figure 3.12. The ones illustrated were selected because of their potential for discrimination and their resistance to changes in abundance by weathering (Randle *et al.* 1971:267). The diagram shows large variations, clearly distinguishing among the tephras illustrated. The tephra layer in the Wildcat Canyon archaeological site is confirmed to be Mazama tephra, even though it is contaminated by minerals from surrounding alluvium. Tephra in the Fort Rock Cave archaeological site also is considered to be Mazama tephra from mineralogical analyses (Figure 3.10) and from other trace-element data (Randle *et al.* 1971); however, it falls a little beyond the range of known Mazama samples, perhaps because of deple-tion in thorium by weathering. The Wineglass Tuff is an ash-flow deposit that erupted from Mount Mazama thousands of years before the well-known climactic eruption of 7000 B.P., yet chemically the two are scarcely distinguish-able, either from trace-element abundances on Figure 3.12 or otherwise (Ran-dle *et al.* 1971:265).

On Figure 3.12 are plotted analyses of Mazama tephra and Newberry tephra made in two different laboratories. They do not agree, apparently because of bias in reported abundances of lanthanum and ytterbium and lesser

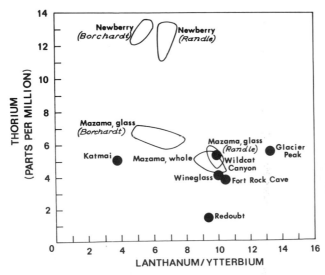

FIGURE 3.12. Variation in the trace-elements thorium, lanthanum, and ytterbium. Dots: Mount Katmai, Alaska, tephra of 1912; Wineglass Tuff, Crater Lake, Oregon; Mazama tephra, Wildcat Ca-nyon, Oregon, archaeological site; Mount Redoubt, Alaska, tephra of 1966; Glacier Peak, Washington, tephra (all Randle *et al.* 1971:265). Outlines: sev-eral analyses each of Mazama tephra, glass; Mazama tephra, whole; Newberry Caldera, Oregon, tephra (Randle *et al.* 1971:265); Mazama tephra, glass; Newberry Caldera, tephra (Borchardt 1970:22).

bias in thorium. Such biases might be eliminated by publication of standard analyses of carefully selected reference samples.

Major elements usually are analyzed through traditional chemical methods of weighing compounds of the elements extracted from solutions. Some elements can be determined with a spectrophotometer, an instrument that measures the intensity of spectral emissions produced when elements are heated in a flame. X-ray-fluorescence spectrometers also are used. This instrument bombards a specimen with a beam of x-rays, which causes elements to fluoresce x-rays of their own. Each element fluoresces x-rays of a particular wavelength with an intensity proportional to its abundance.

Particularly appropriate for analyses of very small samples, such as individual glass shards, is the electron microprobe, which combines an optical microscope with an x-ray spectrometer. Regions to be analyzed as small as one micrometer in diameter are located with the microscope, then bombarded with a beam of electrons, which causes elements to produce x-rays whose wavelengths are characteristic of the element that produced them, and whose intensities are proportional to its abundance.

Most comprehensive trace-element analyses are done by means of neutron-activation analysis. Samples are irradiated with neutrons in an atomic reactor. Nuclear reactions in the sample transmute elements into radioactive nuclides, which emit gamma rays whose properties depend on the identity and abundance of the original element. The gamma rays are detected by a gamma-ray spectrometer, data from which are processed by computer to complete the analysis.

## DISCRIMINATION AMONG TEPHRAS

A windborne tephra falls quickly throughout a large area at the same time, entering whatever environments of deposition lie in the path of the tephra cloud, whether these be hilltops, alluvial fans, stream terraces, human habitations, lakes, bogs, glaciers, or oceans. The tephra layer itself contains minerals and rock-fragments that might be dated by techniques like fission-track dating or potassium–argon dating, and it may contain entrapped organic material that can be dated through radiocarbon analysis. The tephra becomes associated with diverse other sediments that themselves may contain materials that can be dated. Once the age of a tephra layer has been found, by whatever means, that age is valid wherever the layer can be recognized. A determination of age, though, is not a substitute for analysis of properties in seeking to identify one tephra layer with another whose age already is known. Such a substitution not only would be circular reasoning, but also it may go awry, for it rests upon the assumption that a region will not contain two or more tephra mantles of about the same age from different volcanoes.

The versatility of tephra layers as time markers has aroused much interest (Westgate and Gold 1974), and many studies of tephra have been directed toward means of recognition, called *characterization* (Steen-McIntyre 1977), or, with poor analogy, "fingerprinting." The practice of dating events by means of tephra layers is called *tephrochronology* (Thorarinsson 1954:3; Steen-McIntyre 1977).

To be suitable for characterization, properties ideally should have certain attributes:

1. They should have much less variability within a single tephra mantle than among mantles.
2. They should not vary with respect to distance from the source.
3. They should not be affected by contamination.
4. They should not be affected by weathering.

The first of these I will call *discrimination*, and the others together, *persistence*. Steen-McIntyre (1977:6) calls those properties that have the second attribute *essential properties*. Other aspects of tephra mantles, such as stratigraphic associations and relationships to soil profiles, are useful in tephrochronology to supplement characterizations or provide dates, but they do not take the place of characterization, which properly is restricted to the description of tephras by means of intrinsic properties.

No property or class of properties is ideal, so it is necessary to recognize the advantages and limitations of various approaches and to select that approach which is most likely to fulfill requirements in particular circumstances, considering not only properties of the tephra, but also those of the analytical techniques that must be used. Some methods, although they are sophisticated and discriminate well, require costly, rare instruments, special skills, or both. These ideas are summarized in Table 3.9. Some properties generally have poor discrimination and persistence; however, color, thickness, grain-size, and mineralogical properties that can be seen in the field, for example, might adequately characterize a simple assemblage or an assemblage in a small area, although such an approach would be inadequate for definitive characterization of extensive tephra mantles. Properties that generally are discriminatory and persistent nevertheless are not infallible, as examples of large variability within mantles (Figure 3.10), biases, and spurious resemblances (Figures 3.11 and 3.12) illustrate. Resemblances may as well be caused by genetic relationships among magmas as by stratigraphic identity. Characterizations are best founded upon as many properties as possible and upon more than one class of properties.

## SUMMARY

Tephra is fragmental material expelled aerially from a volcanic vent. Commonly it falls to earth throughout a large area around the volcano, forming an extensive tephra mantle that is deposited quickly and pervasively, qualities that have aroused interest in the use of tephra layers as stratigraphic or temporal markers.

**TABLE 3.9**
**Characterization of Tephra**

| Property | Discrimination[a] | Persistence[b] | Skill required for analysis | Cost of instruments for analysis | Availability of instruments for analysis |
|---|---|---|---|---|---|
| Color in aggregate | Fair[c] | Poor | Slight | Slight | Common[d] |
| Thickness | Poor[c] | Poor[c] | Slight | Slight | Universal |
| Grain-Size | Fair to poor[c] | Poor | Moderate | Moderate | Common |
| Proportions of vitric, crystal, and lithic fragments | Fair | Variable[c] | Slight | Slight[e] | Common[e] |
| Mineral abundances | | | | | |
|   Enclosed crystals | Good | Very good | Much | Moderate | Common |
|   Free crystals | Fair | Fair | Much | Moderate | Common |
| Morphology of shards | ?Poor | ?Fair[f] | Some | Moderate[g] | Common[g] |
| Optical properties of minerals | Good | Very good | Much | Moderate | Common[h] |
| Refractive index of glass | Poor | Good | Much | Moderate | Common[h] |
| Chemical composition | | | | | |
|   Bulk composition | Good | Variable | Much | Great[i] | Uncommon[i] |
|   Glass | Very good | Very good | Much | Great[i] | Uncommon[i] |

[a] Refers to the property by itself; may be very useful in combination with others.
[b] With respect to space, time, and contaminations.
[c] May be useful in some cases or in small areas.
[d] Munsell color chart should be used.
[e] Quantitative measurement requires common laboratory instruments.
[f] May persist over moderate distances.
[g] Costly, uncommon scanning electron microscope sometimes used.
[h] Uncommon if focal-masking technique or spindle stage is used.
[i] Some kinds of analyses can be purchased at moderate unit cost.

Most tephra is produced in certain kinds of volcanic eruptions that are associated with the rise of viscous, gas-rich, silicic magma. Physical and chemical conditions combine variously to produce diverse, sometimes mingled, styles of eruption, of which pyroclastic flows, base-surges, and tephra-falls are among those most readily distinguished. Pyroclastic flows and base-surges travel close to the ground, and the aggregate of solid fragments and gases in them behaves somewhat as a coherent fluid. In tephra-falls, fragments that have been propelled to considerable height above the volcano during an explosive eruption are carried by horizontally blowing winds and fall to earth beneath their paths in a manner and in a pattern governed by characteristics of the eruption, meteorological conditions, and the size and density of fragments. The finest tephra commonly is hurled into the stratosphere, where it may be dispersed worldwide as a volcanic dust veil. Dust veils reduce the intensity of sunlight that reaches the earth's surface, and some students believe them to influence climate significantly.

Tephras are mixtures of vitric, crystal, and lithic fragments whose proportions may vary according to the composition of the parental magma, as well as with progressive fall and depletion of larger, denser fragments as the plume of tephra moves leeward from the volcano. Measurable properties of the tephra include grain-size, optical properties of vitric shards and of crystalline grains, abundances of individual components, and chemical composition of the whole tephra or any of its components.

Tephras can be distinguished from one another—characterized—by means of their properties, and an individual tephra may be recognizable at a great distance from its source. A tephra mantle that has been dated through any of various stratigraphic, historical, or chemical techniques carries that date with it wherever the mantle can be recognized, thereby acting as a chronologic marker. Such usage, though, is to be approached with adequate understanding of the geology of tephra and the capabilities of analytical approaches that might be used.

## ACKNOWLEDGMENTS

I thank the following people for their conscientious and helpful reviews of the manuscript: D. R. Crandell, U.S. Geological Survey; D. R. Mullineaux, U.S. Geological Survey; W. B. Purdom, Southern Oregon State College; V. C. Steen-McIntyre, Colorado State University; Sigurdur Thorarinsson, University of Iceland; and R. E. Wilcox, U.S. Geological Survey. The contents of this article nevertheless remain solely the responsibility of the author. Several individuals and firms have kindly permitted me to use excerpts of their published materials as examples. Their contributions are acknowledged at the appropriate places in the text.

## REFERENCES

Baak, J. A.
    1949   A comparative study of recent ashes of the Java volcanoes, Smeru, Kelut, and Merapi. *Medd van het Algemeen Profestation voore de Landbouw* 83:1–60.
Bloss, F. D.
    1961   *An introduction to the methods of optical crystallography*. New York: Holt, Rinehart and Winston.
Bolt, B. A., W. L. Horn, G. A. Macdonald, and R. F. Scott
    1975   *Geologic hazards*. New York: Springer-Verlag.
Borchardt, G. A.
    1970   Neutron activation analysis for correlating volcanic ash soils. Doctoral dissertation, Oregon State University.
Bray, J. R.
    1977   Pleistocene volcanism and glacial initiation. *Science* 197:251–254.
Bryson, R. A.
    1974   A perspective on climatic change. *Science* 184:753–760.
Bullard, F. M.
    1962   *Volcanoes*. Austin: University of Texas Press.

Carey, S. N., and H. Sigurdsson
  1978  Deep-sea evidence for distribution of tephra from the mixed-magma eruption of the
        Soufrière on St. Vincent, 1902: Ash turbidities and air fall. *Geology* 6:271–274.
Chichester, F. W.
  1967  Clay mineralogy and related chemical properties of soils formed on Mazama pumice.
        Doctoral dissertation, Oregon State University.
Chitwood, L. A., R. A. Jensen, and E. A. Groh
  1977  The age of Lava Butte. *Ore Bin* 39:157–164.
Cronin, J. F.
  1971  Recent volcanism and the stratosphere. *Science* 172:847–849.
Crowe, B. M., and R. V. Fisher
  1973  Sedimentary structures in base-surge deposits with special reference to cross-bedding,
        Ubehebe Craters, Death Valley, California. *Geological Society of America Bulletin*
        84:663–682.
Curtis, G. H.
  1968  The stratigraphy of the ejecta of the 1912 eruption of Mount Katmai and Novarupta,
        Alaska. *Geological Society of America Memoir* 116:153–210.
Dyrness, C. T.
  1960  Soil–vegetation relationships within the ponderosa pine type in the central Oregon
        pumice region. Doctoral dissertation, Oregon State University.
Eaton, G. P.
  1964  Windborne volcanic ash: A possible index to polar wanderings. *Journal of Geology* 72:1–
        35.
Fisher, R. V.
  1960  Classification of volcanic breccias. *Geological Society of America Bulletin* 71:973–982.
  1961  Proposed classification of volcaniclastic sediments and rocks. *Geological Society of
        America Bulletin* 72:1409–1414.
  1963  Bubble-wall texture and its significance. *Journal of Sedimentary Petrology* 33:224–235.
  1964  Maximum size, median diameter, and sorting of tephra. *Journal of Geophysical Research*
        69:341–355.
Gorshkov, G. S.
  1959  Gigantic eruption of the volcano Bezymianny. *Bulletin Volcanologique* 20:77–109.
Handy, R. I.
  1976  Loess distribution by variable winds. *Geological Society of America Bulletin* 87:915–927.
Hansen, J. W., Wei-Chyung Wang, and A. A. Lacis
  1978  Mount Agung eruption provides test of a global climatic perturbation. *Science* 199:1065–
        1068.
Hay, R. L.
  1960  Rate of clay formation and mineral alteration in a 4000 year old ash soil on St. Vincent,
        B.W.I. *American Journal of Science* 258:354–368.
  1966  Zeolites and zeolitic reactions in sedimentary rocks. *Geological Society of America Special
        Paper* 81:1–130.
Heiken, G.
  1972  Morphology and petrography of volcanic ashes. *Geological Society of America Bulletin*
        83:1961–1988.
  1974  An atlas of volcanic ash. *Smithsonian Contributions to the Earth Sciences* 12:1–101.
Hein, R. J., D. W. Scholl, and Jacqueline Miller
  1978  Episodes of Aleutian Ridge explosive volcanism. *Science* 199:137–141.
Higgins, M. W.
  1973  Petrology of Newberry Volcano, central Oregon. *Geological Society of America Bulletin*
        84:455–488.
Huang, T. C., and N. D. Watkins
  1976  Volcanic dust in deep-sea sediments: Relationships of microfeatures to explosivity esti-
        mates. *Science* 193:576–579.

Izett, G. A., R. E. Wilcox, and G. A. Borchardt
    1972   Correlation of a volcanic ash bed in Pleistocene deposits near Mount Blanco, Texas, with
           the Guaje pumice bed of the Jemez Mountains, New Mexico. *Quaternary Research*
           2:554–578.
Izett, G. A., R. E. Wilcox, H. A. Powers, and G. A. Desborough
    1970   The Bishop ash bed, a Pleistocene marker bed in the western United States. *Quaternary
           Research* 1:121–132.
Keller, J., W. B. F. Ryan, D. Ninkovich, and R. Altherr
    1978   Explosive volcanic activity in the Mediterranean over the past 200,000 years as recorded
           in deep-sea sediments. *Geological Society of America Bulletin* 89:591–604.
Kittleman, L. R.
    1964   Application of Rosin's distribution in size–frequency analysis of clastic rocks. *Journal of
           Sedimentary Petrology* 34:483–502.
    1973   Mineralogy, correlation, and grain-size distributions of Mazama tephra and other post-
           glacial pyroclastic layers, Pacific Northwest. *Geological Society of America Bulletin*
           84:2957–2980.
Knox, J. B., and N. M. Short
    1964   A diagnostic model using ashfall data to determine eruption characteristics and atmos-
           pheric conditions during a major volcanic event. *Bulletin Volcanologique* 27:5–24.
Lamb, H. H.
    1970   Volcanic dust in the atmosphere, with a chronology and assessment of its meteorologic
           significance. *Royal Society of London, Philosophical Transactions* A266:525–533.
Ledbetter, M. T.
    1977   Duration of volcanic eruptions: Evidence from deep-sea layers. Geological Society of
           America Abstracts with Program, 1977 Annual Meetings.
Lirer, Lucio, Tullio Pascatore, Basil Booth, and G. P. L. Walker
    1973   Two Plinian pumice-fall deposits from Somma-Vesuvius, Italy. *Geological Society of
           America Bulletin* 84:759–772.
Lorenz, Volker
    1970   Some aspects of the eruption mechanism of the Big Hole Maar, central Oregon. *Geologi-
           cal Society of America Bulletin* 81:1823–1830.
Macdonald, G. A.
    1972   *Volcanoes*. Englewood Cliffs, N.J.: Prentice-Hall.
Marshall, R. R.
    1961   Devitrification of natural glass. *Geological Society of America Bulletin* 72:1493–1520.
Mason, B. J.
    1976   The nature and prediction of climatic change. *Endeavour* 35:51–57.
McBirney, A. R.
    1968   Compositional variations of the climactic eruption of Mount Mazama. In *Andesite
           Conference Guidebook*, edited by H. M. Dole. *Oregon Department of Geology and
           Mineral Industries Bulletin* 62:53–56.
Mehringer, P. J., Jr., E. Blinman, and K. L. Petersen
    1977   Pollen influx and volcanic ash. *Science* 198:257–261.
Meinel, M. P., and A. B. Meinel
    1963   Late twilight glow of the ash stratum from the eruption of Agung Volcano. *Science*
           142:582–583.
Minakami, T.
    1942   On the distribution of volcanic ejecta; the distribution of Mount Asama pumice in 1783,
           Part II. *Bulletin of the Earthquake Research Institute, Tokyo* 20:93–105.
Mohr, E. C. J., and F. A. van Baren
    1954   *Tropical soils*. New York: John Wiley/Interscience.
Moore, J. G.
    1967   Base surge in recent volcanic eruptions. *Bulletin Volcanologique* 30:337–363.

Moore, J. G., K. Nakamura, and A. Alcaraz
  1966  The 1965 eruption of Taal Volcano. *Science* 151:955–960.
Mullineaux, D. R.
  1974  Pumice and other pyroclastic deposits in Mount Rainier National Park, Washington. *U.S. Geological Survey Bulletin* 1326:1–83.
Murai, I.
  1961  A study of the textural characteristics of pyroclastic flow deposits in Japan. *Bulletin of the Earthquake Research Institute, Tokyo* 39:133–248.
Nasmith, H., W. H. Matthews, and G. E. Rouse
  1967  Bridge River ash and some other recent ash beds in British Columbia. *Canadian Journal of Earth Science* 4:163–170.
Ninkovich, D., and N. J. Shackleton
  1975  Distribution, stratigraphic position and age of ash layer "L" in the Panama basin region. *Earth and Planetary Science Letters* 27:3034.
Pernter, J. M.
  1889  Zur Theorie des Bishopringes. *Meteorologische Zeitschrift* 6:401–409.
Randle, K., G. G. Goles, and L. R. Kittleman
  1971  Geochemical and petrological characterization of ash samples from Cascade Range volcanoes. *Quaternary Research* 1:261–282.
Rapp, G., Jr., and Eiler Henrickson
  1973  Pumice from Thera (Santorini) identified from a Greek mainland archaeologic excavation. *Science* 179:471–473.
Rose, W. I.
  1977  Scavenging of volcanic aerosol by ash: Atmospheric and volcanological implications. *Geology* 5:621–624.
Ross, C. S., and R. L. Smith
  1960  Ash-flow tuffs: Their origin, geologic relations, and identification. *U.S. Geological Survey Professional Paper* 366:1–81.
Scheidegger, A. E., and P. E. Potter
  1968  Textural studies of grading: Volcanic ash falls. *Sedimentology* 11:163 170.
Schneider, S. H., and C. Mass
  1975  Volcanic dust, sunspots, and temperature trends. *Science* 190:741–746.
Shaw, D. M., N. D. Watkins, and T. C. Huang
  1974  Atmospherically transported volcanic glass in deep-sea sediments: Theoretical considerations. *Journal of Geophysical Research* 79:3087–3094.
Sheridan, M. F.
  1971  Size characteristics of pyroclastic tuffs. *Journal of Geophysical Research* 76:5627–5634.
Sheridan, M. F., and R. G. Updike
  1975  Sugarloaf Mountain tephra—A Pleistocene deposit of base-surge origin in northern Arizona. *Geological Society of America Bulletin* 86:571–581.
Slaughter, M., and J. W. Earley
  1965  Mineralogy and geological significance of the Mowry bentonites, Wyoming. *Geological Society of America Special Paper* 83.
Smith, H. W., R. Okazaki, and C. R. Knowles
  1975  Electron microprobe analysis as a test of the correlation of West Blacktail ash with Mount St. Helens pyroclastic layer T. *Northwest Science* 49:209–215.
  1977a  Electron microprobe data for tephra attributed to Glacier Peak, Washington. *Quaternary Research* 7:197–206.
  1977b  Electron microprobe analysis of glass shards from tephra assigned to Set W, Mount St. Helens, Washington. *Quaternary Research* 7:207–217.
Smith, R. L.
  1960a  Ash flows. *Geological Society of America Bulletin* 71:795–842.
  1960b  Zones and zonal variations in welded ash flows. *U.S. Geological Survey Professional Paper* 354-F:149–159.

Steen-McIntyre, V. C.
  1977  *Manual for tephrochronology*. Idaho Springs, Colo.: Author.
Thomas, Gordon, and M. M. Witts
  1969  *The day the world ended*. New York: Stein and Day.
Thorarinsson, Sigurdur
  1954  The tephra fall from Hekla on March 29, 1947. In *The eruption of Hekla, 1947–1948*,
        Volume 2, No. 3, edited by Trausti Einarsson, Gudmundur Kjartansson, and Sigurdur
        Thorarinsson. Rykjavik: Visindafelag Islendinga.
  1967  The eruptions of Hekla in historical times. In *The eruption of Hekla, 1947–1948*, Volume
        1, edited by Trausti Einarsson, Gudmundur Kjartansson, and Sigurdur Thorarinsson.
        Reykjavik: Visindafelag Islendinga.
  1970  *Hekla*. Reykjavik: Almenna Bokafelagid.
Tidball, R. R.
  1965  A study of soil development on dated pumice deposits from Mount Mazama, Oregon.
        Doctoral dissertation, University of California, Berkeley.
Walker, G. P. L.
  1971  Grain-size characteristics of pyroclastic deposits. *Journal of Geology* 79:696–714.
Walker, G. P. L., L. Wilson, and E. L. G. Bowell
  1971  Explosive volcanic eruptions. I. The rate of fall of pyroclasts. *Royal Astronomical Society
        Geophysical Journal* 22:377–383.
Westgate, J. A., and C. M. Gold (Editors)
  1974  *World bibliography and index of Quaternary tephrochronology*. Edmonton, Alberta, Can.:
        International Union of Quaternary Research.
Whitten, E. H. T.
  1961  Quantitative areal modal analysis of granitic complexes. *Geological Society of America
        Bulletin* 72:1331–1360.
Wilcox, R. E.
  1959a Some effects of recent volcanic ash falls, with especial reference to Alaska. *U.S. Geologi-
        cal Survey Bulletin* 1028-N:409–476.
  1959b Use of the spindle stage for determination of principal indices of refraction of crystal
        fragments. *American Mineralogist* 44:1272–1293.
  1962  *Cherkasov's "focal screening" for determination of refractive index by the immersion
        method*. International Microscopy Symposium, Chicago, 1960. Chicago: McCrone As-
        sociates.
  1965  Volcanic-ash chronology. In *The Quarternary of the United States*, edited by H. E. Wright
        and D. G. Frey. Princeton: Princeton University Press.
Williams, H.
  1942  *The geology of Crater Lake National Park, Oregon*. Washington, D.C.: Carnegie Institu-
        tion Publication 540.
Williams, H., and G. G. Goles
  1968  Volume of the Mazama ash-fall and the origin of Crater Lake caldera. In *Andesite
        Conference Guidebook*, edited by H. M. Dole. Oregon Department of Geology and
        Mineral Industries Bulletin 62:37–41.
Wilson, L.
  1972  Explosive volcanic eruptions. II. The atmospheric trajectories of pyroclasts. *Royal As-
        tronomical Society Geophysical Journal* 30:381–392.
Wood, C. A., and F. M. Dakin
  1975  Ara Shatan maar: Dynamics of an explosion. *Ethiopian Geophysical Observatory Bulletin*
        15:99–106.
Wright, J. V., and G. P. L. Walker
  1977  The ignimbrite source problem: Significance of a co-ignimbrite lag-fall deposit. *Geology*
        5:729–732.

**4**

# Soils Derived from Tephra

F. C. UGOLINI
R. J. ZASOSKI

## INTRODUCTION

Volcanic soils in the context of this chapter are those derived from relatively fresh tephra deposits. Soils that have developed from lava, welded tuff, ignimbrites, and other rocks and sediments containing volcanic glass may display properties similar to the soils derived from volcanic ash, but are not discussed here. Most of this discussion is centered around soils that have developed on Holocene and late Quaternary tephra[1] formed under humid temperate climate. These soils have been recognized under different names (Dudal 1964); however, the term *Ando*, or its derivatives, *Andosols* and *Andepts*, are widely accepted terms. Because of this and because most of the pertinent literature uses these names, our presentation has placed considerable emphasis on the Andosols. Distinguishing these soils exclusively on the basis of their parent material is justified because many features displayed by these soils are directly attributable to the tephra parent material. Soils developed on tephra have a low-bulk density, a friable consistency, a large amount of organic matter on a weight basis, a clay fraction dominated by amorphous material, a high pH-dependent cation exchange capacity, an ability to fix phosphorous, and a high water content at 15-bar suction. These unique characteristics have

---

[1]The terms *pyroclastic deposits*, *volcanic ash*, and *tephra* are used interchangeably. Although it is recognized that *soil derived from volcanic ash* is the correct expression, for the purpose of avoiding redundance, *volcanic soils*, *volcanic-ash soil*, and *ashy soils* are often used.

VOLCANIC ACTIVITY AND HUMAN ECOLOGY

been determinative for classifying ash-derived soils under separate categories by the FAO/UNESCO (1974) and in the United States, Russia, France, Japan, New Zealand, and other countries. Soils derived from volcanic ash are widely distributed and occur under different environmental conditions. A great variety of names has been used for designating these soils. As Dudal (1964) points out, the difference in nomenclature often reflects differences in the criteria used for distinguishing these soils. Volcanic ash and pumice-derived soils, although occurring in several climatic zones, have been most intensively studied in Japan, New Zealand, and Hawaii under humid temperate and tropical climates. The Japanese literature, not fully available to the authors, has been included in recent reviews by Wada and Harward (1974), Wada (1977), and Maeda et al. (1977). Other informative reviews have been provided by Fields and Claridge (1975), by Mitchell et al. (1964), and by the FAO/UNESCO (1964).

## GEOGRAPHICAL DISTRIBUTION

The distribution of soils derived from volcanic ash closely parallels the global distribution of active and dormant volcanoes. Because of this association, the approximate distribution of volcanic soils is geographically predictable. Most of the soils have developed on Pleistocene and Holocene tephra; however, some have developed on lahars and tuff. The majority of these soils occur in the regions of the Pacific ring of volcanism including Kamchatka, Japan, Korea, the Philippines, Indonesia, New Guinea, Malaysia, New Zealand, Australia, and Hawaii. In South America, volcanic soils are found in Chile, Argentina, Bolivia, Peru, Ecuador, and Colombia. In Central America and in the West Indies they have been reported in Panama, Costa Rica, Nicaragua, San Salvador, British Honduras, Guatemala, Mexico, and the Lesser Antilles. In North America, volcanic soils are present in the arid areas of the western United States, but are more prevalent in the states of Washington and Oregon, in northern California, in the Aleutian chain, and in southern Alaska. In Canada, ashy soils are mostly restricted to western Canada, where volcanoes have been active since early Tertiary times. In Europe, volcanic soils are found in Iceland and in Italy, with lesser amounts seen in Greece. In Africa, few are present in Nigeria, wheras they are abundant in Cameroons. Volcanic soils exist in Ethiopia, Somaliland, Kenya, Uganda, Tanzania, Congo, and in Madagascar. Soils derived from volcanic ash are also present in Madeira, one of the Canary Islands. The global distribution of tephra-derived soils is presented in Figure 4.1.

## MORPHOLOGY

The term *Ando*, or modifications of it, has been used in a number of classification systems to identify soils derived from volcanic ash. *Ando*, from

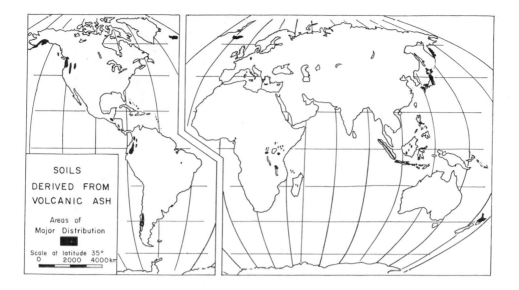

FIGURE 4.1. The distribution of humid-region soils formed in tephra. Certain arid-region soils that do not show characteristics commonly ascribed to tephra-derived soils are excluded (modified from Dudal 1964).

the Japanese word *An* (dark) and *do* (soil or earth) was first introduced by the American soil surveyors who mapped Japan after World War II. The usage of this term originated because many Japanese volcanic-ash soils acquire a characteristic thick, dark-colored A horizon; however, this morphological feature is not universal. The black, humus-rich A horizon is best displayed by soils derived from recent volcanic ash formed under a humid temperate climate. In Japan, Kuroboku soils, equivalent to Ando soils, have developed under grass and show black humus-rich Al horizons generally 30–50 cm thick, but ranging up to 1 m in low-lying areas (Figure 4.2). Buried Al horizons may be present as a result of repeated ash depositions. The humus content of the Al horizon is between 15 and 30% (Ohmasa 1964), and the soils show moderate to strong structure and a friable consistency. The B horizon is not markedly expressed, varying in color according to the degree of development and showing very little evidence of clay illuviation. A feature of the Ando soils, but also displayed by other soils affected by volcanic ash, is a "smeariness." This characteristic is manifested when the soil is crushed between the fingers, and, as it is manipulated, becomes progressively wetter and more slippery. Texturally, Ando soils vary considerably from sandy loam to silt clay loam. As pointed out by many authors and summarized by Maeda *et al.* (1977), particle-size distribution analysis may be vitiated by the nondispersability of the clay and the presence of amorphous colloids with high isoelectric point. On a regional basis, changes in particle-size distribution in volcanic ash-derived soils are related to the distance from the source. Soils 80 km from Crater Lake, Oregon, the source of Mazama

**FIGURE 4.2. Profile of an Andosol, Umbric Vitrandept. (Reproduced from Marbut Memorial Slide Set by permission of the Soil Science Society of America.)**

ash, still display gravelly textures; however, the percentage of fine silt increases with distance up to 320 km, after which it remains constant (Haward and Youngberg 1969).

Wright (1964), in a review of the Ando soils or Humic Allophane soils of South America, discussed changes in soil morphology along a moisture gradient following a latitudinal transect. In spite of common features displayed by the soils along the transect, horizon differentiation was evident; AC profiles progressively changed into A(B)C[2] and ABC profiles with increasing rainfall. Other changes were also apparent: The type of humus changed from mull to mor, the top soils became blacker as the length of the dry season shortened, and organic matter and granular structure tended to increase with increasing precipitation. The subsoil color changed from light reddish brown to yellowish brown and then to brown and dark brown with increasing moisture.

In the tropical areas of Hawaii, according to Swindale (1964), the top soil of ash-derived soils is reddish yellow and does not acquire the black color characteristic of the temperate regions; furthermore, the subsoil ranges from yellowish brown to dark red. In San Salvador, where volcanic ash has been deposited from early Quaternary up to the present, AC soils prevail on young landscape, but on older landscape red clay Latosols are the most common of all soils derived from volcanic ash (Rico 1964).

One of the most outstanding features of Ando soils or Ando-like soils is the low-bulk density, a criterion commonly used in several classification schemes. Values ranging from 0.1 to 0.9 $gm/cm^3$ have been reported worldwide in different ash and pumic soils (Flach 1964; Sherman et al. 1964; Youngberg and Dyrness 1964; Maeda et al. 1977; Shoji and Ono 1978).

## CHEMICAL AND PHYSICAL PROPERTIES

An important characteristic of pumice- and ash-derived soils is the fact that both chemical and physical properties are strongly dependent upon the history of a particular soil sample. Many researchers have noted that significant changes can take place when soils derived from volcanic ash and pumice are dried. Volume decreases, acidity increases, average pore size increases, water transmission increases, and water retention decreases when volcanic soils are dried. These changes are more or less irreversible, depending on the severity of drying (Maeda et al. 1977).

Textural analysis of ash or pumice soils is subject to several errors. Pumice soils may be degraded into finer particles by mechanical dispersion, and sand-size pumice particles may float or settle more slowly than would be expected because of entrapped air (Packard 1957; Youngberg and Dyrness 1964). Consequently, the commonly used techniques do not always work well in ash-derived soils (Maeda et al. 1977; Wada 1977).

A lack of long-range order in colloids of ash-derived soils has necessitated the use of several techniques to distinguish these so-called amorphous colloids. No one technique unambiguously defines these minerals, and data from sev-

---

[2]The symbol (B) is described by Kubiena (1953) and is used to designate B horizons that are not enriched in illuviated substances, but have been altered by soil-forming processes sufficiently to change the original parent material.

eral analytical methods must be pooled to characterize the colloids that are present.

Studies of volcanic and pumice soils have shown that techniques developed for soils dominated by crystalline clay minerals must be applied with caution. On the other hand, methodologies used to study soils dominated by amorphous materials may prove to be useful in examining nonvolcanic soils where amorphous coatings may influence soil properties (Jenne 1968; Loganathan 1971; Jones and Uehara 1973).

Chemical properties of several Andosols and Andepts are presented in Table 4.1. Table 4.2 shows comparative chemical and physical properties of Andepts and other soils.

## Chemical Properties

### COLLOIDS PRESENT

The nature of the colloids and their surface characteristics largely determine the chemical and physical properties of soils. This section will examine the colloids that appear to dominate the clay fraction of ash-derived soils.

Soils are not static systems. The suite of colloids vary from soil to soil depending on both the age of the deposits and the weathering regime. Wada (1977), and Wada and Harward (1974) have summarized the sequence of clay minerals that may be expected in various aged volcanic soils. They suggest that a progression from amorphous aluminosilicates (allophane and imogolite) and opaline silica to crystalline layer silicates (halloysite), and hydrous metal oxides (gibbsite) occurs with increasing soil age. Burial by younger ash falls may interrupt or change weathering sequences (Saigusa et al. 1978).

The unique chemical and physical properties of volcanic ash soils seem to be associated with the early weathering stage in which allophane, imogolite, iron and aluminum gels, and organic materials dominate the chemistry of these soils. A more detailed account of the weathering sequence can be found in a later section.

*Allophane.* The term *allophane* has been variously defined and applied by several authors. Mitchell (1975) stressed the fact that inorganic gel materials in soil are very heterogeneous and suggested that most soil mineralogists accept a broad definition of allophane as a mutual solution of silica, alumina, and water with other bases. In a review of soil allophanes, Fields and Claridge (1975) recognized that allophanes acquire several forms, and suggested that the weathering sequence allophane B → allophane A-B → allophane A represents a series of x-ray amorphous minerals with increasing degrees of organization and varying chemical composition. More recently, however, Campbell et al. (1977) have proposed to eliminate the use of allophane B, suggesting it is most likely finely divided volcanic glass. Wada (1977) and Wada and Harward (1974) defined allophane as a naturally occurring hydrous aluminosilicate of varying

# TABLE 4.1
## Chemical Properties of Clays and Selected Ash and Pumice soils

| Material | Location | Minerals present | Horizon | Total C (percentage) | pH | C.E.C.[a] meq/100gm | BS[b] (percentage) | Al/Al+Si | Fe₂O₃[c] (percentage) | Al₂O₃[c] (percentage) | Acidity[d] | Surface area m²/g | Reference |
|---|---|---|---|---|---|---|---|---|---|---|---|---|---|
| Vitric Andept clay fraction rhyolitic ash | New Zealand | Allophane, some imogolite | — | — | — | — | — | 0.54 | — | — | — | 168 | Rajan 1975 |
| Synthetic Al-Si gels | | | | — | — | — | — | 0.29 | — | — | — | 190 | Rajan and Perrott 1975 |
| | | | | — | — | — | — | 0.56 | — | — | — | 278 | |
| | | | | — | — | — | — | 0.64 | — | — | — | 357 | |
| | | | | — | — | — | — | 0.87 | — | — | — | 264 | |
| Gel film pumice beds | Japan | Imogolite, allophane | — | — | — | 30 | — | | | | | | Henmi and Wada 1974 |
| | | | — | — | — | 135 | — | | | | | | |
| Andosol | Japan | Fe and Al oxides amorphous | A1 | 10.7 | 4.4 | 25.4 | 17.7 | | 3.4 | 3.2 | 23.5 | — | Shoji and Ono, 1978 |
| | | | A2 | 9.0 | 4.3 | 23.8 | 7.7 | | 3.8 | 3.5 | 21.0 | — | 1978 |
| | | | IIBb | 0.5 | 4.8 | 14.6 | 6.8 | | | | 34.1 | — | |
| | | | IICb | tr | 5.1 | 15.3 | 11.1 | | | | 15.8 | — | |
| Dystric Andept (0–25 cm) | Chile | | | | | | | | | | | | |
| Santa Barbara | | Allophane | | 9.4[e] | 6.0 | 18[f]48[f] | 15 | | 6.0 | 4.0 | — | 225 | Espinoza et al. 1975 |
| Arrayan | | Allophane | | 8.6[e] | 5.8 | 22[f]45[f] | 17 | | 7.6 | 5.0 | — | 225 | |
| Umbric Vitrandept Cervantes | Costa Rica | Allophane | A1 | 8.0[e] | 4.5 | 57[f]21[g] | 17 | | 5.1 | 13.2 | 0.75[h] | — | Igue and Fuentes 1972 |
| | | | B2 | 1.9[e] | 5.8 | 38[f]15[g] | 26 | | 6.4 | 15.3 | 0.83[h] | | |
| Vitrandept? Coronado | Costa Rica | Allophane | A | 12.6[e] | 5.3 | 64[f]10[g] | 6.1 | | 5.4 | — | — | — | " |
| | | | B | 10.2[e] | | 63[f]20[g] | 3.9 | | 8.0 | — | — | — | " |
| Andept (Humic allophane) (0–20 cm) | Japan | Allophane | A1 | 16.5 | 4.8 | 51.3 | | | — | — | 4.6 | — | Arai 1975 |

[a]Cation exchange capacity.
[b]Percentage base saturation.
[c]Extracted by citrate-dithionite solutions.
[d]Ml of 0.1 N NaOH used to titrate 125 ml of a N KCl soil extract (soil: N KCl 100:250).
[e]Converted from percentage of organic matter by assuming O.M. to be 58% C.
[f]C.E.C. at pH 3.5 and 7.0, respectively.
[g]C.E.C. determined by pH 7.0 N NH₄OAC and unbuffered N CaCl₂, respectively.
[h]Total acidity displaced from a 5-g sample by 250 ml of N KCl.

**TABLE 4.2**
**Physical and Chemical Properties of Different Soils**[a]

| Soil | Location | Horizon | Depth (in cm) | pH | C.E.C. (meq/100 gm) | BS[b] (percentage) | N (percentage) | C (percentage) | Db (g/cm³) | Water content | |
|---|---|---|---|---|---|---|---|---|---|---|---|
| | | | | | | | | | | 1/3 Bar (percentage) | 15 Bar (percentage) |
| Typic Cryandept[f] | Alaskan Peninsula | A11 | 0–12 | 4.6 | 21.6 | 20 | 0.29 | 3.6 | 0.88(OD)[c] | 43 | 12.6 |
| | | A12 | 12–23 | 4.9 | 34.0 | 11 | 0.33 | 3.7 | 0.79 | 55 | 15.1 |
| | | A11b | 23–32 | 5.3 | 26.8 | 7 | 0.25 | 3.3 | 0.86 | 47 | 12.4 |
| Oxic Dystrandept[f] | Hawaii | Ap | C–20 | 6.2 | 53.7 | 48 | 0.69 | 8.9 | 0.70(FM)[d] | 65 | 41 |
| | | B21 | 20–41 | 6.6 | 40.6 | 23 | 0.30 | 3.81 | 0.70 | 67 | 49 |
| | | B22 | 41–74 | 7.1 | 36.6 | 38 | 0.21 | 2.57 | 0.84 | 51 | 40 |
| Typic Argiustoll | Kansas | Ap | 0–15 | 6.1 | 22.6 | 81 | 0.12 | 1.17 | 1.11(FM) | 32 | 21 |
| | | B21t | 15–30 | 6.9 | 30.7 | 107 | 0.08 | 0.83 | 1.38 | 29.6 | 20 |
| | | B22t | 30–46 | 7.8 | 29.6 | — | 0.07 | 0.64 | 1.49 | | |
| Typic Acrorthox | Puerto Rico | A1 | 0–28 | 5.1 | 25.4 | 8 | 0.39 | 6.04 | 1.30(AD)[e] | 35.0 | 26 |
| | | B1 | 28–46 | 5.0 | 12.1 | tr | 0.13 | 2.04 | 1.32 | 27.0 | 23 |
| | | B21 | 46–71 | 5.0 | 8.2 | — | — | 1.33 | 1.23 | 34.0 | 25 |
| Typic Tropohumult | Puerto Rico | Ap | 0–15 | 4.6 | 23.6 | 25 | 0.41 | 5.2 | 0.90 | 42 | 30 |
| | | B21t | 15–28 | 4.5 | 19.9 | 10 | 0.17 | 2.0 | 1.02 | 43 | 34 |
| | | B22t | 28–48 | 4.6 | 20.2 | 8 | 0.14 | 1.39 | 1.00 | 44 | 35 |

[a]Source: Soil Survey Staff 1975.
[b]Base saturation ($NH_4OAc$).
[c]OD = oven dry wt basis.
[d]FM = field moisture conditions.
[e]AD = air dry conditions.
[f]Soils derived from volcanic ash.

chemical composition, but maintaining within the variable composition a predominance of Si–O–Al bonds. This definition excludes alumina and silica as end members because of a lack of Si–O–Al bonds, and follows the definition adopted by the 1969 United States Japanese Seminar on Amorphous Clays (Wada and Harward 1974).

A chemical formula proposed for allophane is $Al_2O_3 \cdot 2SiO_2 \cdot nH_2O$ (Brown 1955). The Si–Al ratio in this formulation is 1:2; however, Fields and Claridge (1975) indicate the ratio varies from 1:1 to 1:2, and Wada and Yoshinaga (1969) report for several ash and pumice soils a ratio of $SiO_2/Al_2O_3$, in the <2 $\mu$m fraction between 1.3 and 2.0. Based on the weathering sequence proposed (Wada and Harward 1974; Wada 1977) in which allophane is converted over time to halloysite by progressive desilication, minerals with varying Si–Al ratios would be expected. In early work, imogolite was considered a type of soil allophane, and because of its 1:1 Si–Al ratio it would extend the variation in reported allophane composition.

A lack of long-range order and a variable composition dictate that a variety of techniques must be used to characterize allophane. Wada (1977) and Yuan (1974) have summarized the characteristics of allophane determined by x-ray diffraction, electron diffraction, electron microscopy, thermal analysis, infrared spectroscopy, and x-ray fluorescence. No definitive structure for soil allophane emerges from these techniques; however, there are certain characteristics that distinguish allophane from other amorphous or poorly crystalline soil components. X-ray fluorescence measurements have shown that portions of the aluminum exist in both fourfold and sixfold coordination (Cloos et al. 1969; Leonard et al. 1969; Wada 1977). Allophane retains a great deal of water. Structural OH groups and OH groups of absorbed water are completely solution accessible, as demonstrated by the relative ease with which water can be evacuated at room temperature and the ability of $DO^-$ to replace all of the $OH^-$ (Wada 1966; Wada 1977). Infrared spectroscopy has been extensively used to characterize allophane, although the spectra are less distinct than those associated with crystalline material. Adsorption bands of allophane and imogolite are found at 2800–3800, 1400, 1800, and 650–1200 cm$^{-1}$. The first bands correspond to stretching vibrations of structural OH or adsorbed water, the second region is due to adsorbed water, the third region is attributed to Al–O–Si vibrations and SiOH or AlOH deformations (Yuan 1974; Wada 1977). Electron microscopy has shown some allophane to have a spherical form; Wada (1977) suggests that these spherules are hollow with diameters of 35–50 Å. These findings contradict the notion that allophane is completely amorphous and maintains no characteristic morphology.

Based on chemical data from soil allophane and synthetic amorphous aluminosilicates, various structural models of allophane have been proposed (Wada 1967; Cloos et al. 1969; Leonard et al. 1969). At the present time, these models serve as a basis for further refinements, since a definitive structure for allophane cannot be assigned (Wada and Harward 1974; Fields and Claridge 1975; Wada 1977). Campbell et al. (1977) have shown that the fine fraction of a

New Zealand soil contained both comminuted volcanic glass particles and a weathering product with properties similar to synthetic aluminosilicate gels. They suggest that unless these two phases can be separated, chemical data ($SiO_2/Al_2O_3$) and aluminum coordination numbers cannot be utilized to test the applicability of the proposed allophane models.

*Imogolite*. Imogolite is a hydrous aluminosilicate that is found in association with allophane (Wada 1977). Only recently has the term *imogolite* been approved by the Nomenclature Committee of the AIPEA (Bailey *et al*. 1971). Imogolite is defined as a "hydrous aluminosilicate having a thread-like morphology and the diffraction characteristics described by Wada and Yoshinaga (1969) and others [Bailey *et al*. 1971]." Imogolite yields broad x-ray diffraction peaks, and the structure is more definitive than that of allophane (Cradwick *et al*. 1972). The infrared spectrum of imogolite is similar to that of allophane; however, it differs from that of allophane and layer silicate clays in the 900–1100 cm$^{-1}$ region (Wada 1977). Cradwick *et al* (1972) have proposed a structural model in which a portion of the OH groups in a gibbsite structure is replaced by an oxygen of a Si-tetrahedron. This fitting of the Si-tetrahedron in the gibbsite structure constricts the hydroxyl sheet and causes a curvature of the crystal. It is postulated that there are 10, 11, or 12 gibbsite units making up the tubular structure. This model appears to account for most of the observed properties of imogolite (Cradwick *et al*. 1972); however, further refinements of the model are needed (Wada 1977).

One characteristic of imogolite is that aluminum appears to be only in sixfold or octahedral coordination (Henmi and Wada 1976; Wada 1977), an important feature because the coordination number of aluminum has a bearing on the charge characteristics and acidity of imogolite.

*Oxides of Aluminum and Iron*. The presence of allophane and, more recently, imogolite in tephra-derived soils has received considerable attention; however, the unique properties of ash-derived soils have also been observed where allophane and imogolite are absent (Shoji and Ono 1978). Shoji and Ono (1978) examined several well-drained Japanese Andosols and found allophane and imogolite to be present in pumice-derived soils but not in those derived from ash. A characteristic low-bulk density, high moisture and phosphate retention, and other Andosol properties displayed by the ash soils lacking allophane were closely related to the dithionite-citrate soluble constituents, especially extractable aluminum. High extractable aluminum levels are often found in volcanic soils (Arai 1975). Crystalline gibbsite and hematite have been reported as weathering products in several older soils (Wada and Aomine 1966; Wada and Harward 1974; Wada 1977). The mineralogy and chemistry of crystalline iron and aluminum oxides have recently been reviewed by Schwertmann and Taylor (1977) and Hsu (1977); however, during the earlier weathering stages, iron and aluminum oxides appear to be present as amorphous gels (Wada and Harward 1974). Though these amorphous gels may mimic the behavior of crystalline materials (Hsu 1977), they should offer a larger surface area and be more reactive (Jenne 1968; Greenland 1971). Amorphous oxides and gels can coat the surface of solid particles and thus provide an increased surface area (Jenne 1968; Anderson and Jenne 1970; Loganathan 1971; Jones and Uehara 1973). Wada and Harward (1974) concluded that little is known about the amorphous iron and aluminum compounds in soils. Nonetheless, many properties of ash-soils can be correlated with the amount of extractable aluminum and iron oxides (Wada 1977; Shoji and Ono 1978).

*Phyllosilicate Minerals.* Although amorphous minerals are often associated with ash-derived soils, layer silicate clays have also been found in many Andosols and Andepts, provoking discussions on their mode of formation.

Irrespective of their provenance, halloysite, metahalloysite, chlorite, illite, kaolinite, montmorillonite, and vermiculite have been reported in ash-derived soils (Masui and Shoji 1969; Wada and Harward 1974; Wada 1977). These minerals provide cation exchange capacity and a permanent charge that is independent of pH (Babcock 1963; Gast 1977).

*Colloidal Silicon.* Opaline silica has been reported as an early weathering product in volcanic soils (Fields and Swindale 1954; Wada and Harward 1974; Wada 1977). Laminar opaline silica has been reported in the 0.2–20 $\mu$m fraction, but it is most abundant in the 0.4–2.0 $\mu$m fraction of Japanese volcanic-ash soils (Shoji and Masui 1971). Mitchell (1975) suggests that most opaline silica in the surface of Andosols is of biological origin. Complexing of aluminum by organic materials in the surface and a lack of aluminosilicate formation have also been proposed to account for the surface abundance of opaline silica (Wada 1977).

Silica gel has been used as a sorbant in thin-layer chromatography and would be expected to function as a sorbant in soils. Opaline silica and amorphous silica, if finely divided, have a high surface area and can provide another charged surface in ash-derived soils. Parks (1965) summarized data on the isoelectric points (IEP) of metal oxides and concluded that the IEP of the $MO_2$ group should be about 2.0 ± 0.5. Bolt (1957) found that silica soils had a zero point of charge (ZPC) of about 3.5. Dugger *et al.* (1964) examined the exchange behavior of silica gel (BET surface area 498 $m^2$/gm) and found that the weakly acidic silanol group functioned as an effective exchanger for 20 different cations. Aluminum, among other ions, was adsorbed by this surface.

*Organic Colloids and Clay Organic-matter Complexes.* Andepts and Andosols are noted for their accumulation of organic matter (Tables 4.1, 4.2), and it has been suggested that the high organic content of these soils is as important as the inorganic minerals in determining soil properties (Birrell 1964; Tan 1964). A portion of the organic materials in ash-derived soils is complexed by or adsorbed on the inorganic colloids. These so-called clay–organo complexes can alter the surface properties of the inorganic fraction.

Greenland (1971) has reviewed clay–organic matter interactions. He pointed out that at pHs below 8, aluminum and iron oxides can have positive charges, and simple coulombic attraction (anion exchange adsorption) of organic materials is possible although other or additional adsorption mechanisms may also be operative. An abundance of iron and aluminum oxides in Andepts and Andosols combined with a generally low soil pH results in surfaces with positive charges and an ability to adsorb organic materials. The persistence of humic substances in ash-derived soils has been a major area of study. Kononova (1975), reviewing the humus of virgin and cultivated soils, commented on the high humus content of volcanic soils and the relatively more complex (higher percentage of gray humic acids) humic materials contained therein. The reaction and association of humic matter with allophane has been advocated as the reason for the intensely dark surficial horizons.

Allophane has been shown to adsorb larger quantities of humified materials than either montmorillonite or halloysite (Wada and Inoue 1967; Inoue and Wada 1968). Allophane appears both to catalyze polyphenol oxidation, promoting the formation of humic-like substances, and to limit decomposition by adsorbing and deactivating large quantities of enzymes (Kyuma and Kawaguchi 1964; Kobayashi and Aomine 1967). Broadbent et al. (1964) demonstrated a reduced mineralization in highly allophanic soils suggesting that the formation of stable aluminum–organic complexes may limit decomposition.

Although the importance of allophane or allophane-like substances is recognized in the stabilization of organic matter in soils, increasing evidence has suggested that extractable iron and alumina are responsible for organic matter accumulation in ash-derived soils (Kosaka et al. 1962; Wada and Harward 1974; Wada and Higashi 1976; Shoji and Ono 1978). Allophane and imogolite may not function directly to adsorb organic materials; however, they may furnish both aluminum and iron to stabilize organics. Griffith and Schnitzer (1975) suggest that the unusually high resistance to degradation of organic matter in tropical volcanic soils may be due to the formation of stable silica-fulvic acid complexes. Organic carbon and clay–organo complexes, prominent features of Andosols and Andepts, are other amphoteric colloids that influence soil chemical and physical properties.

COLLOIDAL PROPERTIES

*Amorphous Nature.* Often a majority of colloids found in recent volcanic-ash and pumice soils fail to give x-ray diffraction patterns and therefore are considered amorphous, although Wada (1977) points out that they may have short-range order and are not entirely amorphous. These amorphous materials have a large specific surface (Aomine and Otsuka 1968; Maeda *et al* 1977) and, in comparison with more crystalline material, would be more reactive (Stumm and Morgan 1970; Greenland 1971; Hsu 1977). The noncrystalline nature and variable composition also make their study and characterization more difficult.

*Surface Area.* Soil chemistry is essentially a combination of solution and surface chemistry, so that surfaces and their charge characteristics have a very large import on soil chemical properties.

In general, ash-derived soils exhibit a large surface area (Table 4.1). Values ranging from 100 m²/gm to over 300 m²/gm have been reported for Andepts and Andosols (El-Swaify and Sayegh 1975; Espinoza *et al.* 1975; Schalscha *et al.* 1975), whereas the clay fraction of these soils has surface areas as high as 1000 m²/gm (Egashira and Aomine 1974). Calculated surface area of imogolite based on a particle density and morphology produced values as high as 1500 m²/gm (Egashira and Aomine; Wada 1977). Though the surface area of soil colloids depends to a certain extent on the methods of measurement, calculated and measured surface areas of ash-derived soils are high.

*Charge Characteristics.* Clay minerals can exhibit either positive or negative charge. Stumm and Morgan (1970) suggest that surface charge can originate from

1. Ionization or protonation of surface functional groups such as —OH, —COOH, or —Si—OH
2. Lattice imperfections at the surface and isomorphous substitution within the lattice
3. Adsorption of charged ions on surfaces by London-van der Waals forces

Charge resulting from H ionization or protonation of surface functional groups results in either a net positive or negative charge on the surface in relation to pH of the surrounding media. For many soil materials, H and OH concentration determines the surface charge (van Raij and Peech 1972; Schwertmann and Taylor 1977). Colloids in ash-derived soils are dominated by clays that exhibit a predominance of pH-dependent charge (Wada and Harward 1974, Wada 1977, Balasubramanian and Kanehiro 1978); however, Espinoza *et al.* (1975) detected small amounts of permanent charge associated with x-ray amorphous colloids in Andepts from Chile. Crystalline, layer silicates with permanent charge may also occur in ash and pumice soils (Masui and Shoji 1969; Balasubramanian and Kanehiro 1978; Shoji and Ono 1978); nevertheless, pH-dependent surfaces appear to predominate.

The point where a colloid or soil has no net negative or positive charge is called the zero point of charge (ZPC) or isoelectric point (IEP).[3] If solution pH is higher than the ZPC, a soil or particle will acquire a negative charge.

---

[3]Parks (1967) considers the IEP to be the point where only $H^+$, $OH^-$, and $H_2O$ are interacting with the surface, whereas ZPC is a more general term to denote a lack of net charge when ions, in addition to $H^+$ or $OH^-$, are interacting with the surface. This distinction is not made by all authors.

Conversely, if pH values are less than the ZPC, a net positive charge will be displayed by the colloids.

In soils containing allophane and imogolite the ZPC is around pH 7.0 (Wada and Harada 1969; Espinoza et al. 1975). Horikawa (1975) found the IEP of allophane in NaCl solutions to be about 5.6–5.9, whereas imogolite had an IEP between 9 and 10. Pre-acidified samples of allophane had an IEP of 7.3; this higher IEP in acidified samples was not explained. Iron and aluminum oxides have zero points of charge that vary from about 5 to 9, although exceptions do exist (Parks 1965).

*Colloidal Interactions.* Because of their large surface area and their charge characteristics, colloids in ash- and pumice-derived soils have high ion-exchange capacities. Wada and Harward (1974) reported a cation exchange capacity (CEC) of 135 meq/100 gm for allophane ($SiO_2/Al_2O_3 = 2.0$) and 30 meq/100 gm for imogolite. These values were obtained at pH 7.0 using 0.05N $NaCH_3COO$. Constant potential surfaces such as those in volcanic ash and pumice soils should exhibit increased charge as salt concentration (ionic strength) increases (Babcock 1963; van Raij and Peech 1972; Gast 1977). Wada and Harward (1974) and Wada (1977) have reviewed the influence of solute concentration on cation exchange. The point of zero charge for volcanic soils is near neutrality (Wada and Harada 1969; Espinoza et al. 1975), and the majority of volcanic soils have mildly to strongly acid pHs (Table 4.1). In view of this, these soils should exhibit a significant anion exchange capacity; these expectations have been substantiated in studies that examined $NO^{-3}$ and $Cl^-$ adsorption by Andepts (Singh and Kanehiro 1969; Kino and Pratt, 1971a,b; Gebhardt and Coleman 1974a; Espinoza et al. 1975). Gebhardt and Coleman (1974a) reported a marked increase in $Cl^-$ adsorption as pH decreased. Nitrate adsorption has also been shown to increase with decreasing pH (Espinoza et al. 1975).

In addition to the electrostatic reversible attraction between oppositely charged colloidal surfaces and solution ions, certain anions can be specifically adsorbed. Specifically adsorbed ions are more strongly held than exchangeable ions and may form covalent-like bonds with the sorbate surface (Stumm and Morgan 1970). Sulfate and phosphate are well known for this behavior in volcanic soils (Birrell 1964; Swindale 1964; Gebhardt and Coleman 1974b,c; Yuan 1974; Rajan and Fox 1975; Wada 1977), although strong specific adsorption of silicate, arsenate, fluoride, pyro- and tripolyphosphate, selinite, and molybdate have been reported (Gonzales et al. 1974; Wada and Harward 1974; Wada 1977).

On the basis of several lines of evidence, phosphorus adsorption on iron and aluminum oxides has been proposed to account for the large P retention observed (Wada 1977). The first line of evidence relates to the high correlation between extractable iron and aluminum content and the retention of applied phosphorus. Second, a strong phosphorus adsorption has been demonstrated in soils lacking allophane and imogolite but having high aluminum and iron levels (Wada 1977; Shoji and Ono 1978). Strong specific adsorption of P on iron and aluminum oxides in pure systems is a third reason to suggest that iron and aluminum oxides can be responsible for phosphorus fixation (Hingston *et al*. 1968; Parfitt *et al*. 1974; Hsu 1977; Schwertmann and Taylor 1977). In spite of the strong evidence for adsorption on iron and aluminum oxides, allophane, synthetic aluminosilicates, and imogolite have also been shown to adsorb added phosphorus (Parfitt *et al*. 1975; Rajan and Perrott 1975; Veith and Sposito 1977). Precipitation of a separate amorphous aluminum phosphate phase has also been proposed (Veith and Sposito 1977). Allophane decomposition induced by added phosphate and the formation of taranakites has been proposed as another way in which phosphates may interact with colloidal surfaces (Wada 1959; Wada 1977).

Several reactions alone or in concert could account for high P retention in soils and, irrespective of the mechanisms or the adsorbing surfaces, a large P fixation is a prominent feature of volcanic ash- and pumice-derived soils.

*Soil Acidity*. Though volcanic soils occur in a wide variety of climatic settings, substantial rainfall and a leaching environment tend to be common to these settings. Under high rainfall regimes an acidic soil reaction is expected, and these mildly to strongly acid soils have been discussed extensively in the existing literature (Birrell 1964; Mariano 1964; Swindale 1964; Tan 1964; Shoji and Ono 1978).

Imogolite and allophane have weak acid properties, as do iron and aluminum oxides (Parks 1965; Henmi and Wada 1974; Wada and Harward 1974; Wada 1977). A relatively high ZPC of allophanic soils and their weak acid properties mean that these soils are well buffered and have a large pH-dependent cation exchange capacity. Since the weak acid sites have a greater affinity for protons than do the strong acid sites on layer silicate clays, as pH decreases, cation exchange sites are occupied by hydrogen ions and exchangeable aluminum levels are comparatively lower in ash-derived soils.

Arai (1975) examined extractable Al in several acidic Japanese soils. An Andept, Umbrept, and Ochrept had extractable Al content of 4.6, 17.2, and 27.9 meq/100 gm, respectively, at pH values of 4.75, 4.6, and 4.62. Yoshida (1970, 1971) treated layer silicate clays, allophane, and imogolite with $AlCl_3$, and determined exchangeable Al. Layer silicate clays had more than 60% aluminum saturation, whereas allophane and imogolite were hydrogen-saturated. Low exchangeable aluminum and a concomitant low base saturation (Birrell 1964; Saigusa *et al*. 1978; Shoji and Ono 1978) have important fertility consequences.

## Physical Properties

Soils derived from volcanic ash or pumice have several characteristic physical properties such as low-bulk density, high porosity, and unique water retention and transmission characteristics, which are directly related to their chemical properties (Table 4.2). Maeda *et al.* (1977), Wada and Harward (1974), and Wada (1977) have presented excellent review articles that should be consulted for detailed discussions of the topic covered in this section. Furthermore, these reviews discuss soil engineering properties, which are not included in this section. The general allophanic soil profile[4] has been summarized by Maeda *et al.* (1977) as follows:

> Allophane soils generally have a friable surface soil and massive structure in the subsoil which, however has a relatively high permeability. The friable structure of the surface soil is partly due to effects of drying. Often allophane soils have several layers with very different physical properties which affect water movement and water available for plant use. Many properties such as volume or water retention, which decrease on drying, show an irreversible decrease beyond to 10- to 15-bar suction. Physical properties of allophane soils do not show the dependence upon exchangeable cations which is prominent in soils with crystalline minerals [pp. 231–232.].

## SOIL DENSITY

The low density of Andosols, Andepts, and pumice soils is well documented (Table 4.2) (Packard 1957; Birrell 1964; Wright 1964; Youngberg and Dyrness 1964; Shoji and Ono 1978), and a value $<0.85$ gm/cm$^3$ constitutes one of the criteria for defining the Andepts in the U.S. Soil Taxonomy (Soil Survey Staff 1975). Sherman *et al.* (1964) have reported bulk densities as low as 0.1 gm/cm$^3$ in low-Si allophane soils from Hawaii. This low density, generally 0.3–0.8 gm/cm$^3$, arises from a high porosity, since the mineral particles have densities similar to crystalline clays (Youngberg and Dyrness 1964; Kitagawa 1976; Maeda *et al.* 1977). Porosities of 65–85% in the Lapine series, a pumice soil from south-central Oregon, are attributed to a high internal porosity of the pumice particles (Youngberg and Dyrness 1964). Harward and Youngberg (1969) examined the porosity of Mount Mazama particles by mercury intrusion. Coarse sand-size particles had median pore diameters of 7.2–7.4 $\mu$m, with total pore volumes between 0.8 and 1.0 ml/gm. Physiochemical properties of pumice soils are predominantly determined by their large internal porosity (Harward and Youngberg 1969).

Allophanic soils have a large total porosity as well as a large internal porosity. As a result, the void characteristics, more than the surface attributes, influence the physical properties of these soils (Maeda *et al.* 1977).

Because of the characteristic low density of volcanic soils, it may be preferable to express soil properties on a volume basis rather than on a mass (weight) basis (Packard 1957; Broadbent *et al.* 1964; Youngberg and Dyrness 1964).

## PARTICLE-SIZE ANALYSIS

In nonvolcanic soils particle-size analysis provides a measure of the clay content and, with it, an estimation of soil behavior. Allophanic soils, however,

---

[4]Since allophane is a major mineral constituent of Andepts and Andosols, the term allophanic soils is often used to describe soils derived from volcanic ash, although allophane can be found in other soils.

are difficult to disperse, and various techniques give strikingly different clay contents. Kubota (1972) found values ranging from 56% clay to 1% clay, depending on the method of dispersion chosen. A lack of adequate dispersion and the dependence of the sample on its past drying history suggest that other methods of characterization may be more useful in ash-derived soils. Maeda *et al.* (1977) suggest that grain-size analysis is of limited value in these soils. Pumice soils are also subject to measurement errors: Entrapped air may cause slower sedimentation, and the physical energy used in dispersing the samples may break the particles into smaller units (Packard 1957; Youngberg and Dyrness 1964).

## HEAT CONDUCTION

The ability of soil to transfer heat depends on the ability of the particles to conduct heat in relation to the air and water in the soil. Because of the high porosity and the high water contents of volcanic soils, they would be expected to transfer heat slowly (Maeda *et al.* 1977). As a consequence of the low heat conductivity, temperature changes should also occur more slowly. Cochran *et al.* (1967), and Cochran (1975) have suggested that this property has influenced regeneration patterns of ponderosa pine (*Pinus ponderosa*) and lodgepole pine (*Pinus contorta*) in the pumice region of south-central Oregon.

## WATER TRANSMISSION, INFILTRATION, AND RETENTION

Volcanic soils are highly porous; however, it is the pore-size distribution that determines the soils' ability to store and transmit water (Maeda *et al.* 1977). Maeda *et al.* (1977) have stated that a major problem in describing the physical properties of ash-derived soils containing allophane is the lack of understanding of their structure. The behavior of pumice soils, on the other hand, can be explained on the basis of pore size, shape, continuity, and void strength (Sasaki *et al.* 1969).

The decrease in volume that occurs in drying volcanic soils affects the distribution of pore sizes (Maeda *et al.* 1977). Thus, water retention, transmission, and infiltration may reflect the past history of the sample, and laboratory measurements cannot be reliably extrapolated to field water regimes.

Because of the unique internal and external pore-size distribution, the concepts of available water are difficult to define in ash and pumice soils. There is abundant evidence that suggests that the commonly used 0.3-bar suction does not properly define field capacity (Youngberg and Dyrness 1964; Chichester *et al.* 1969). Maeda *et al.* (1977) indicated that the 15-atmosphere moisture content was too high to define the wilting percentage. In pumice soils, the last portion of available water is extracted only very slowly, and at 15 atmospheres, the equilibration time may be as long as 10 days (Will and Stone 1967). Youngberg and Dyrness (1964) also found long time periods necessary to characterize the 15-atmosphere moisture percentage.

Field capacity and wilting percentage are not well defined in ash and pumice soils, and, consequently, available moisture is also poorly defined. In volcanic soils plant-available moisture, even when corrected for the density differences, is usually high (Packard 1957; Geist and Strickler 1978). In spite of a large amount of available moisture in the pumice soils of Oregon, Youngberg and Dyrness (1964) found that sunflowers wilted at moisture contents wetter than the 15-atmosphere percentage. As would be expected, infiltration is rapid in pumice soils. Finer-textured volcanic ash soils also have high infiltration rates (Maeda *et al.* 1977). Forsythe (1975) reported that allophanic soils of South America had infiltration rates of 20–70 cm/hr, making them unsuitable for flood or furrow irrigation.

Layering of volcanic soils is quite common because of repeated ash falls (Saigusa *et al.* 1978), resulting in boundaries between the various depositions that can restrict internal drainage (Maeda *et al.* 1977). Water flow is hindered at these boundaries by large differences in the resistances of the two layers under unsaturated flow conditions.

## SOIL GENESIS AND WEATHERING

### Profile Development

In the regions where soils derived from volcanic ash are found, volcanism has been active throughout the Quaternary. Consequently, intermittent deposition of tephra has occurred, burying or rejuvenating the existing soils, depending on the thickness of the ash showers. Areas near the vent of an explosive volcano may receive tephra sufficient to kill the existing vegetation and to start a new episode of soil formation on the fresh substratum (Smathers and Mueller-Dombois 1974). At sites distant from the vent or receiving small increments of ash, soil formation may not be interrupted, and the general effect is a rejuvenation of an established soil (Gibbs 1968).

In the Cascade region of Washington, where repeated ash showers have occurred throughout the Holocene, soils evidence these events by displaying a stratigraphic sequence of ashes in their profiles (Bockheim and Ugolini 1972; Singer and Ugolini, 1974; Ugolini *et al.* 1977a; Zachara 1979). A typical Podzol (Cryandept) in the central Cascades shows an A2 horizon developed from Mount St. Helens, W and Y ash, respectively, 400 and 4000 years old. The IIB2hir and IIB2ir had developed in the 6600-year-old Mazama ash, the IIIB3 formed on a mixture of Mazama ash and broken bedrock, and a C horizon consisted of fractured andesitic bedrock. Since thickness of the ash layers depends on aspect, topography, susceptibility to erosion, tree throwing, and other disturbances, the aforementioned sequence may not be revealed at all points of the landscape.

Where the ash mantle is thick and time of deposition known, either from historical records or from radiocarbon dating of buried organics, an excellent opportunity is offered to study soil formation as a function of time. Wada and Aomine (1973) report on the work of S. Yamada regarding the rate of horizon differentiation. Considering the time of deposition as time zero, fully developed A horizons appeared within 100 years. An incipient B horizon had formed in the interval between 100 and 500 years, and the profile showed an A(B)C sequence. Between 500 and 1000 years both A(B)C or ABC profiles were present. The ABC profiles were reported to be more than 1000 years old. This sequence was studied in Hokkaido where cool, moist summers and snowy winters prevail. Another soil chronosequence, also cited by Wada and Aomine (1973), showed the progressive building of organic carbon on soils of different ages. Soils near the vent of Mount Aso, rejuvenated by the continuing activity of the central cone, had less than 5% carbon. Soils northeast of the cone, not continuously supplied by fresh ash, contained between 5% and 10% carbon; soils developed on ash estimated to be 4000–5000 years old contained more than 10% carbon. From this information, Wada and Aomine (1973) suggested that Ando soils can acquire maturity within 5000 years in the Mount Aso district in central Kyushu under a warm, humid climate and grassland vegetation.

Studies conducted by Ugolini et al. (1977a) and Zachara (1979) on soil developed on volcanic ash in the cool-to-cold snowy climate of the Washington Cascades have shown that Podzols (Cryandept) with A2, B2hir, and B2ir can develop within approximately 4000 years. On St. Vincent Island, British West Indies, Hay (1960) studied the rate of soil formation and mineral alteration in volcanic ash 4000 years old. This study showed that the original andesitic ash consisting of glass, anorthite, labradorite, hypersthene, augite, and olivine had altered into a clayey soil at the rate of 45–60 cm/1000 years (1.5–2 ft/1000 years).

According to Gibbs (1968), the differences displayed by the volcanic-ash soils in New Zealand are mostly related to soil age and to the initial composition of the ash. Soils developing on ash that had accumulated during the past 500–750 years had only A and C horizons and sandy textures. These soils are the Recent Soils of Fresh Volcanic Ash, equivalent to the Vitrandepts in the United States. Soils formed from rhyolitic pumice 3000–5000 years old had AC or ABC profiles, sandy loam texture, and pieces of pumice in the upper horizon. These are the Yellow Brown Volcanic Ash Soils and are equivalent to the Normandepts of the United States (Gibbs 1968). Soils developed from either rhyolitic or andesitic ash ranging in age from 5000 to 20,000 years showed ABC profiles, loamy texture, and a high percentage of amorphous clay. Soil formed either from rhyolitic or andesitic ash erupted between approximately 20,000 and 100,000 years ago had ABC profiles, a clayey texture, and more crystalline than amorphous clays. These are the Brown Granular Loams and are equivalent to the Ultisols in the United States. Another example of soil development and evolution in relation to the Quaternary volcanic history in the district of Taranaki, New Zealand, is cited by Neall (1977). A detailed tephrochronology and lahar stratigraphy has allowed this author to reconstruct the major stages in the genesis of these Andosols. Under well-drained conditions and rapid tephra accretion, Entic Dystrandepts developed in 2000–3000 years. Under poorly drained situations and slow tephra accumulation a continuous iron pan was formed, and Aeric Andaquepts developed on lahar surfaces within the time range of 3000–25,000 years. However, these Aeric Andaquepts did not persist indefinitely on the landscape because, once buried by thick tephra, a new stage of soil development was initiated under freely drained conditions. When tephra substrata are exposed to humid temperate climates, their friable and loose consistency, vitreous composition, and abundance of available nutrients favor plant growth and rapid soil development.

## Humus Content

Soil-forming processes operating on fresh Holocene ash in a humid temperate climate result in the development of a thick, dark, humus-rich surface horizon and the formation of large amounts of amorphous inorganic material. In a 4000–5000-year-old well-drained Andosol developed under natural grass, Wada and Aomine (1973) reported a humus accumulation of 450–1530 tons/ha in both the surface and buried horizons. Other Japanese investigators stressed the high carbon content of ash-derived soils and the importance of grassland vegetation in humus accumulation (Yamane *et al.* 1958; Shinagawa 1960). Gibbs (1968) cites values from 7 to 15.3% carbon in the surface Al horizon of New Zealand ash soils. Taylor (1964) reported 20% carbon in the first 7.6 cm (3 in.) of a New Zealand soil formed on andesitic ash, and 10% carbon for a soil on pumice erupted about 1700 years B.P. Carbon content of Hawaiian soils derived from volcanic ash varies as a function of climate: Red Desert soils had virtually no carbon, whereas Reddish Prairie (Typic Normandept) and Brown Forest (Typic Normandept) contained, respectively, 13% and 15% carbon in the A horizon (Swindale 1964).

The Andosols of South America, provisionally grouped under "environmental phases" (Wright 1964), showed the general tendency to contain high organic matter in the continuously humid zone (about 2000 mm precipitation) and to have even higher organic matter in the subhumid zone. Andosols of the cool temperate equatorial highland (3000–5000 m) had deep black A horizons with organic matter content of 12–22%. Andosols of the warmer temperate humid and superhumid equatorial environment occurred between 2000 and 3500 m. The thin Al horizon of these soils contained a very high organic content (22–28%). Andosols of temperate, dry summer environments contained between 8 and 15% organic matter.

In the subalpine region of Mount Baker, Washington, well-drained meadow soils, which are derived from mixed Holocene volcanic ash, showed an accumulation of 6.4–14.5% organic carbon in the surface mineral horizon (Bockheim 1971). Similar soils derived from the 2440-year-old "Bridge River" ash in British Columbia contained up to 17.85% organic matter in the Al horizon (Sneddon 1973). A soil in the Aleutian Islands, Alaska, formed under 1722 mm (67.8 in.) of precipitation with a mean annual temperature of 4°F, averaged 8% carbon in the three Al horizons (Flach 1964). Another soil in the Aleutians, a Typic Cryandept, averaged about 3.5% carbon in the upper 109 cm (43 in.) (Rieger 1964).

According to Wada and Aomine (1973), the humus accumulation in the Japanese Andosols is characterized by wide C–N ratio with values over 20. Whereas Yamada (1968) claims a positive correlation between the C–N ratio and the $^{14}C$ age of the humus, Wada and Aomine (1973) contend that the age of the humus alone does not explain the fluctuation of the C–N ratio. According to these authors, both the age and the nature of the humus composition affect the C–N ratio. Walker et al. (1959), in contrasting virgin soils derived from rhyolitic volcanic ash with soils turned into pasture, warned that percentages by weight of carbon and nitrogen in virgin soils are higher because of their low volume weight.

Numerous Japanese researchers have investigated the humus content of ash-derived soils and found that humification generally increased from north to south. Tokudome and Kanno (1965a,b) observed that the humic acids of ash-derived soils had high C–H and C–N ratios, and that fulvic acids extracted from these soils had relatively low carbon content and low C–N ratios, irrespective of bioclimatic conditions. These authors also reported that the young soils had a lower carbon content and more fulvic acids than older soils, and that the amount of fulvic acid increased with depth. Humic acids of ash-derived soils were found to be more complex than those from nonash soils because they contained higher percentages of the more complex gray humic acids (Kononova 1975). The effect of burial on humus composition was studied by Adachi (1963), who reported wider humic–fulvic acid ratios in buried A horizons than in the surface A1 horizons. Wright (1964) contended that the high organic matter content of Andosols often consists mostly of fulvic acids. The humic–fulvic acid ratios have been used by Tan (1964) to separate the Andosols of Indonesia into three groups. Volcanic soils of Kamchatka developed under parkland birch forest showed more active humification than other nonashy soils of the Taiga region. The more active decomposition of the forest floor is due, according to Kononova (1975), to the influence of volcanic ash, which liberates bases as it weathers, thus neutralizing the acid products of organic decomposition.

## Weathering

Soils derived from volcanic ash display distinctive assemblages of minerals originating both from the parent ash and from the weathering of the ash and pumice. Tephra deposits are commonly glassy and permeable, favoring rapid weathering in humid temperate regions.

In Japan, according to Wada (1977) and other authors, the early stage of volcanic-ash weathering is characterized by the presence of opaline silica, humus, hydroxy-Al, and Fe probably complexed with humus, allophane-like minerals, and montmorillonite; allophane and imogolite are absent. Wada

(1977) contended that in the Andepts, opaline silica and allophane do not occur together. This is explained by the reaction of Al, derived from weathering of the ash, with the humic substances, forming Al-humus complexes. The removal of Al from the soil solution prevents the formation of allophane and imogolite and favors instead the formation of opaline silica. Tokashiki and Wada (1975) found allophane only in the B and the buried Al horizons, but it was virtually absent in the present Al horizons. Opaline silica was found only in the young, buried Al horizons that had remained near the present surface. Opaline silica, however, does not persist indefinitely in the Andosols. According to Wada's weathering scheme, opaline silica is lost after about 7000 years (Wada 1977) (Figure 4.3). Allophane and imogolite were prominent in the B horizons of soils 5000–10,000 years old, whereas after 10,000 years both minerals appeared to be transfromed into gibbsite.

Miyauchi and Aomine (1964) found allophane in the fine clay fraction ($<0.2$ $\mu$m) of six soils derived from ashes less than 1000 years old. Birrell (1962) also reported allophane in the early weathering stages in the subsoil of ash soils. Shoji and Ono (1978) proposed that the occurrence of allophane and imogolite in some Andosols is related to the nature of the parent material. In their studies, these authors found that allophane and imogolite were absent from the clay fraction of soils formed from volcanic ash; however, these minerals dominate the clay fraction of the soils derived from andesitic pumice.

Imogolite may form from allophane via desilication or by precipitation from solution (Wada 1977). Burial of ash beds by small addition of volcanic ash appeared to favor the formation of imogolite by providing a limited source of silica (Aomine and Mizota 1973). Miyauchi and Aomine (1966) identified imogolite gellike substances in pumice beds. Kanno *et al.* (1968) found imogolite in humic allophane soils on the slopes of Volcano Unzen and observed that the proportion of imogolite increased with depth, whereas the presence of allophane decreased. Work of Henmi and Wada (1976) disclosed that allophane forms within weathered glass fragments, whereas imogolite forms outside the

**FIGURE 4.3.** Formation and transformation of clay minerals and their organic complexes in soils developed from volcanic ash in humid temperate climatic zones. Abbreviations: A, allophane; A¹, allophane-like constituents; Al(Fe), sesquioxides; Ch, chlorite; Gb, gibbsite; Ht, halloysite; Im, imogolite; Sm, smectite; Hy-Sm, hydroxy interlayered smectite; O.S., opaline silica; Vt, vermiculite. Horizontal bars indicate approximate duration of respective constituents. (Reproduced from Minerals in Soil Environments, p. 620, 1977, by permission of the Soil Science Society of America.)

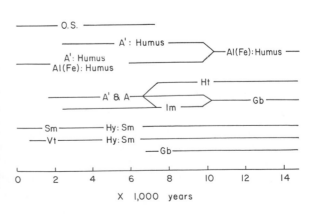

fragments. In a recent study on poorly ordered aluminosilicates in a Vitric Andosol from New Zealand, Campbell *et al.* (1977) suggested that Allophane B may be glass shards.

Although the presence of different forms of allophane is now uncertain, the occurrence of metahalloysite and halloysite in volcanic soils has long been recognized (Sudo 1954; Aomine and Wada 1962). The mechanism of halloysite formation has been discussed by Aomine and Wada (1962), and it has been suggested that the appearance of halloysite is favored by resilication of al-lophane. Resilication is possible in old, buried soils covered by thick ash layers and in poorly drained soils; in both environments, a high silica activity would be expected. The age of ash deposits containing halloysite ranges between 6000 and 10,000 years (Wada and Aomine 1973). Saigusa *et al.* (1978), in discussing the origin of halloysite in Ando soils in northern Honshu, supported the mechanism suggested by Wada and Aomine (1973). They identified halloysite in buried soils developed on 8600- and 10,000-year-old tephra and covered by thick fresh volcanic ash. Halloysite was found to be abundant in humus-rich, buried horizons and in the 10,000-year-old paleosols. Halloysite formation in a buried soil was ostensibly attributed to an influx of silica leached from the fresh ash and accumulated over the buried soil surface. In buried soils, the occur-rence of allophane and imogolite together with halloysite is indicative of a relict condition dating to the time when the buried soils were surface soils undergo-ing intense leaching. Japanese ash deposits containing halloysite range in age between 6000 and 10,000 years (Wada and Aomine 1973); in New Guinea, halloysite is present in ash between 300 and 2000 years old (Bleeker and Parfitt 1974), and in St. Vincent Island, under humid tropical conditions, the age of the ash containing halloysite is 4000 years (Hay 1960).

The occurrence of 2:1 layer silicates in soils derived from volcanic ash has been reported by different investigators under many different climatic condi-tions. Uchiyama *et al.* (1960) detected a montmorillonite-like clay mineral in strongly acidic volcanic-ash soils in north-central Japan. Kanno *et al.* (1959) also reported the presence of 14 KX [sic] minerals in intensively weathered andesitic volcanic ash. Chichester *et al.* (1969) reported numerous 2:1 phyl-losilicates in ashy soils of Oregon. Montmorillonite was also identified by Uchiyama *et al.* (1962) in dacitic volcanic ash. Partially expandable and nonex-pandable chloritized 2:1 minerals were the dominant minerals of Andosols from Kitakami, Japan (Shoji and Ono 1978). Other occurrences of 2:1 layer silicates in volcanic-ash soils were cited by Kawasaki and Aomine (1966), Masui *et al.* (1966), Masui and Shoji (1967), and Zachara (1979). The occurrence of these 2:1 and 2:1:1 phyllosilicates in ash-derived soils was interpreted by some as the result of the synthesis of amorphous inorganic material into crystalline components. Chichester *et al.* (1969), in a study of soils derived from Mazama ash in Oregon, found chloritic intergrades, beidellite, chlorite, vermiculite, montmorillonite, and a micaceous mineral. Other minerals in the $<2$ $\mu$m fraction were gibbsite, plagioclase, feldspars, and quartz. The presence of phyllosilicates among a predominantly amorphous component was explained by Chichester *et al.* (1969) as a result of an in situ process aided by wet and dry episodes and the retention of solutions in the internal pores of the vesicular pumice. An in situ process—conversion of allophane to layer silicates—was also suggested by Loganathan and Swindale (1969) to explain the presence of 2:1 clay minerals in the soils derived from volcanic ash in the Island of Hawaii. Masui et al. (1969) also advanced the in situ synthesis to account for the 2:1 layer silicates detected in the volcanic soils of northeastern Japan. Wada and Aomine (1973), however, reinterpreted Masui's result and concluded that the 2:1 layer silicates were formed by the weathering of the mica and hornblende present in dacitic ash and not by crystallization of the amorphous material. Dudas and Harward (1975a,b) also reconsidered the synthesis hypothesis and suggested that the 2:1 clay minerals had their genesis in the weathering of the lithic fragments incorporated into the ash by superficial mixing. A similar conclusion was reached by Zachara (1979), who examined the clay mineralogy of two soils derived from Holocene tephra and underlain by andesitic rocks. The in situ formation of 2:1 clay suggested by Loganathan and Swindale (1969) was reexamined by Jackson et al. (1971) and Syers et al. (1971). Both authors discovered that eolian dust containing quartz, mica, and vermiculite was the source of the 2:1 clay minerals on the Hawaiian island landscape. According to these findings, it appears that no concrete evidence presently exists on the possibility of the synthesis of 2:1 phyllosilicates from amorphous aluminosilicate weathering products of volcanic ash.

## Soil-forming Processes

As previously mentioned, some of the unique properties of the soils derived from volcanic ash are best manifested when the soils are developed on Holocene tephra and under a humid temperate climate. The striking features of ashy soils developed under these conditions—thick Al horizons associated with the accumulation of organic material and poorly developed B horizons— are not universal. As previously described, time and the environmental conditions are as important for the development of these soils as for other soils derived from nonashy parent material. This situation is very well exemplified in Hawaii, where a number of genetically distinct soils appear on the landscape as the precipitation ranges from 260 to 5200 mm (Flach 1964). An analogous situation exists in New Zealand, where, under a humid and cool climate, the process of podzolization is operating. Bleached A2 horizons, depleted of iron and aluminum, overlie dark brown B2 horizons rich in iron, aluminum, and humus illuviated from above. On the other hand, under a humid but warmer climate the soils are affected by the process of laterization. Here, the organic material does not persist at the surface but is rapidly decomposed. Decomposition of the mineral fraction is also rapid, and clay is produced. Under these conditions, deeply weathered profiles, rich in clay and residual oxides, have developed (Gibbs 1968). Many intermediate stages exist in New Zealand between these two extreme soil-forming processes. According to Gibbs (1968), similar processes occur on the soils derived from sedimentary rocks; however, the rate of soil development is less rapid than on those derived from ash. Other things being equal, soil formation progresses more rapidly on rhyolitic ashes than on andesitic or basaltic ashes. A study of the soil-solution dynamics conducted by Ugolini and his associates in the central Cascades, Washington, provided details on the in situ process of podzolization operating on andesitic Holocene ashes (Ugolini *et al.* 1977a, b; Dawson *et al.* 1978; Singer *et al.* 1978). According to our findings, podzolization proceeds rather rapidly in the cool-to-cold snowy climate of the high elevations of the Cascades covered by Pacific silver fir (*Abies amabilis*) and mountain hemlock (*Tsuga mertensiana*). Fulvic acid originating in the litter layer controls solution pH and the movement of iron and aluminum. These metal–organo complexes are illuviated at the B2hir–B2ir boundary. As a result of this process a bleached A2 and a dark brown B horizon are formed.

## SOIL CLASSIFICATION

Soils derived from volcanic ash have been classified under numerous names: Volcanic Ash soil or Trumao soils (South America); Andosols; Black Dust soils or High Mountain soils (Indonesia); Kuroboku, Black Volcanic Ash soil, Kurotsuchi, Andosols, Humic Allophane soils, or Brown Forest soils (Japan); Alvisol or Yellow Brown Loams (New Zealand); Talpetate soils (Nicaragua); and Andepts (United States) (Dudal 1964). In the United States, ash-derived soils were not recognized as a separate Great Soil Group in the 1938 classification; in the 1949 revision, although Ando Soils were mentioned, they were not incorporated into the system (Flach 1964). In *Soil Taxonomy* (Soil Survey Staff 1975), the soils derived from volcanic ash are placed in the Inceptisols order and in the Andepts suborder. The Inceptisols are soils of the humid regions that may show some translocation of either silicate clay or aluminum and iron in the B horizon but lack sufficient illuviation to acquire diagnostic horizons such as the Spodic or Argillic. Andepts suborder is constituted by soils that "have a low-bulk density and have an appreciable amount of allophane that has a high exchange capacity [Soil Survey Staff 1975]." More specifically, Andepts are Inceptisols that

1. Have, to a depth of 35 cm or more, or to a lithic or paralithic contact if one is shallower than 35 cm, one or all of the following must be present:
    a. A bulk density (at 1/3-bar water retention in the fine earth fraction) of less than 0.85 gm/cm$^3$
    b. Exchange complex that is dominated by amorphous material; or
    c. 60% or more (by weight) vitric volcanic ash, cinder, or other vitric pyroclastic material
2. Do not have an aquic moisture regime or other wetness characteristic displayed by the Aquepts

The suborder of the Andepts is in turn divided into seven Great Groups:

1. Cryandepts: Andepts of the cold regions with mean annual temperature less than 8°C
2. Durandepts: Andepts that have a duripan within 1 m of the soil surface
3. Hydrandepts: Andepts that have clays irreversibly aggregated into sand and gravel size
4. Placandepts: Andepts that display a placic horizon (cemented horizon by iron, manganese, or iron–organic complexes)
5. Vitrandepts: Andepts in which the weighted average 15 atm water retention of the fine-earth fraction is less than 20% for all the horizons between 25 cm and 1 m, or between 25 cm and a lithic or paralithic contact if one is shallower than 1 m
6. Eutrandepts: Andepts that have a base saturation ($NH_4OAc$) of 50% or more
7. Dystrandepts: Andepts that have a low amount of bases

Among the suborder of the Aquepts, the wet Inceptisols, there is a Great Group of Andaquepts. The Andaquepts are the Aquepts that have developed on tephra and have the same characteristics as the Andepts, except they are poorly drained.

In the French Taxonomy (Duchaufour 1970, 1978), the soils are classified on the basis of genetic and dynamic properties dictated by the environmental factors. Soils derived from volcanic ash are named Andosols and are classified under Class II—Sols à profil peu differencié—Immature soils. At the subclass level, the Andosols are divided into

1. Andosols Très Humifères (surtout tempérés): Humic Andosols (mostly temperate)
   Andosols types (Eutrophes-mollic-like; Oligotrophe–acidic)
   Andosol brunifiés (brunified)
   Sols ando podzoliques (Andopodzolic soils)
2. Andosols Peu Humifères (tropicaux): Low Humic Andosols (tropical)
   Andosols eutrophes (mollic)
   Andosols oligotrophes (acidic)

A genetic soil classification developed in New Zealand includes both nontechnical common soil group names for practical purposes and a technical nomenclature for a more strict scientific use (Taylor and Pohlen 1968). When this classification is arranged into a zonal pattern, the soils are distinguished into Zonal, Intrazonal, and Azonal (Taylor and Pohlen 1968; Metson and Lee 1977). The Intrazonal soils are atypicals because they develop on "unusual" sites. Soils formed from volcanic parent material that still retain unusual chemical and physical attributes are classified among the Intrazonal and Azonal soils. Table 4.3 shows the New Zealand genetic soil groups zonally arranged and listed with the common and technical names and the approximate equivalent in the U.S. soil taxonomy.

The modern Japanese soil classification (Matsuzaka 1977; Otowa 1977) distinguishes major groups as the highest category. Major groups are similar to

**TABLE 4.3**
**New Zealand Genetic Soil Groups with Common and Technical Names Zonally Arranged**[a]

| Common names | Technical names | Approximate U.S. taxonomic equivalent |
|---|---|---|
| | Intrazonal soils | |
| Yellow Brown Pumice soils | Subalvic soils | Andepts (vitric) |
| Central Yellow Brown Loams (rhyolitic) | Alvic soils | Andepts (entic) |
| Central Yellow Brown Loams (andesitic) | Alvic soils | Andepts (entic) |
| Central Brown Granular Loams | Spadous | Andepts, Humults Undults |
| Northern Red and Brown Loams (immature) | | Andepts, Humults |
| | Azonal soils | |
| Recent soils from pumice alluvium or from volcanic ash | Volic soils from pumice or volcanic ash | Andepts (vitric) |

[a]Source: Taylor and Pohlen (1968); Metson and Lee (1977).

the Great Soil Group of the 1949 U.S. classification (Thorp and Smith 1949). Andosols are among the major groups and cover about one-sixth of the total area of Japan. The Andosols in turn are divided into seven groups: Regosolic Andosols; Usual Andosols; Brown Andosols; Fluvic Andosols; Gleic Regosolic Andosols, Gleic Usual Andosols; and Fluvic Gleic Andosols. These groups are separated on the basis of parent material, drainage conditions, base status, and morphological characteristics.

For the purpose of providing an inventory of the soil resources of the world, the FAO and the UNESCO jointly prepared a Soil Map of the World. This map consists of soil map units that reflect "the present knowledge of the formation, characteristics, and distribution of the soils covering the earth surface, and their importance as resources for production . . . [FAO–UNESCO 1974]." The soil units are not equivalent to categories of other classification systems, but they are in general comparable to the Great Soil Groups. Soils derived from volcanic ash are in the soil unit Andosols. They are delineated according to the criteria established for the Andepts in the U.S. Soil Taxonomy (Soil Survey Staff 1975). The Andosols are further distinguished into

1. Ochric Andosols: Andosols with a light color A horizon and a cambic B and/or with a texture of silt loam or finer on the weighted average for all the horizons within 100 cm of the surface
2. Mollic Andosols: Andosols having a mollic A horizon (dark and rich in bases)

with a smeary consistency and/or with a texture of silt loam or finer on the weighted average for all the horizons within 100 cm of the surface
3. Humic Andosols: Andosols with an umbric A horizon (dark, but poor in bases) with a smeary consistency and/or with a texture of silt loam or finer on the weighted average for all the horizons within 100 cm of the surface
4. Vitric Andosols: Andosols without a smeary consistency and/or with a texture coarser than silt loam on the weighted average for all the horizons within 100 cm of the surface

The modern Russian soil classification is based on the genesis of the soils in relation to ecological–climatic factors. In the Soviet system, the Soil Type can be correlated at the level of the Order and Suborder in the U.S. Soil Taxonomy. Volcanic ash forest soils (acid) are listed among the soils of the Taiga Forest Boreal Regions and are mostly located in the Kamchatka Peninsula (Soil Survey Staff 1975).

## SOIL FERTILITY

Soil fertility is the ability of soil to grow crops and produce biomass. Fertility, however, is not necessarily synonymous with productivity. Productivity depends upon physical, chemical, and biological soil processes and their interaction with the attendant climate. Fertility can be more strictly constructed to mean the ability of a soil system to supply plant nutrients, whereas productivity may vary among plant species growing in the same soil. For example, productive forest soils may be quite infertile and yield very poorly when cropped to agronomic species. This essentially nutritional construction of soil fertility will be used in the following discussion.

Nutrient supplies in any soil depend upon the nutrients contained in the parent material and its weathering products. Climate, along with soil age, has a large influence on fertility. Highly leached soils are more intensively weathered, and the weathering products (iron and aluminum oxides and allophane) are responsible for the strong adsorption of P and S (Hasan *et al*. 1970; Metson and Lee 1977). Intensive weathering also leaches available bases from the soils. Secondary interactions between nutrients and the soil colloids further regulate available nutrients. Organic-matter accumulation and the rate of decomposition also influence plant nutrient supplies. The sporadic deposition of fresh tephra that renews nutrient supplies must also be considered. In examining the myriad of possible interactions that can influence soil fertility, it is easier to consider elements separately.

### Nitrogen

On a worldwide basis, nitrogen deficiency is a common growth limitation. Ash and pumice soils may also be nitrogen deficient. Responses to nitrogen have been reported in several locations (Oyama 1964; Rico 1964; Youngberg and Dyrness 1965; Geist 1977); however, nitrogen deficiency is surprising in view of the high organic carbon content of the Andosols and Andepts (Tables 4.1, 4.2). A large organic-matter content would indicate a high potential for mineralization and an adequate supply of nitrogen; on the other hand, a large carbon accumulation might suggest that the decomposition process is slow in these soils. High carbon–nitrogen ratios may be responsible for the low decomposition rate. Kanno (1962) reported that humic–allophane soils in Japan have C–N values ranging from 15 to 30 or more in the soil surface. He suggested that the combination of organic matter with either allophane or aluminum is an important factor limiting decomposition. Satho (1976) suggested that aluminum plays an important role in stabilizing humus in Japanese volcanic-ash soils. Sowden et al. (1976) also found high nitrogen accumulation in West Indian soils in spite of their infertility. Broadbent et al. (1964) found little difference between the mineralization rates in pumice and nonvolcanic soils of New Zealand, but they found mineralization rates were slower in ash soils of high allophane content. These authors suggested that on a volume basis the carbon content of the pumice soils was not unusual. Will (1968) reported on organic matter in buried fossil pumiceous soils that were very refractory, ostensibly because of stable allophane–organic matter complexes. Sowden et al. (1976) also suggested that allophane may be responsible for low organic-matter turnover. Munevar and Wollum (1977) reported that low phosphate availability appeared to limit nitrogen mineralization of Columbian Andepts. Dalal (1978) reported that volcanic soils from St. Vincent Island, West Indies, had a lower portion of soil-N present in acid hydrolysate, ammonium, and amino sugars than nonvolcanic soils. These volcanic soils are considered relatively infertile.

## Phosphorus

Phosphorus is by far the most notable elemental deficiency in soils derived from volcanic ash and pumice. Reports from numerous sources have described phosphorus deficiencies or high P fixation in ash and pumice soils (Rodriquez 1962; Birrell 1964; Mariano 1964; Rico 1964; Tan 1964; Gebhardt and Coleman 1974c; Appelt et al. 1975; Rajan and Fox 1975; Metson and Lee 1977; Wada 1977). High phosphate retention and low plant availability are characteristic of volcanic soils, and are used in Japan to distinguish ash soils. Relatively large phosphorus additions are necessary, and there is a rapid conversion to unavailable forms (strong reversion) over time (Appelt and Schalscha 1970; Matsuzaka 1977).

Strong P retention by allophane, imogolite, and iron and aluminum oxides and formation of solid-phase aluminum phosphates have all been proposed to account for this large fixation. Irrespective of the reasons, reversion of applied P and low availability is a major factor in the fertility of ash-derived soils.

## Sulfur

Sulfur deficiency has been reported for several soils derived from volcanic ash (During 1964; Hasan et al. 1970; Geist 1971; Pumphrey 1971). This can be attributed to the same strong adsorption mechanism shown for phosphorus in soils with low pH or to a lack of retention in slightly weathered soils displaying high pH (Metson and Lee 1977). Hasan et al. (1970) found that sulfate adsorption in Hawaiian soils is related to rainfall, which is a major factor in weathering. The more weathered soils have higher stabilized organic-matter content and more sulfate-adsorbing weathering products. Reduced mineralization rates would also limit sulfur availability. Pumice soils of Oregon have shown a response to sulfur fertilization (Youngberg and Dyrness 1965; Youngberg 1975). Native understory species growing on pumice soils of central Oregon have low sulfur status (Nissley and Zasoski, unpublished data; Nissley 1978).

## Potassium, Calcium, and Magnesium

The porous nature of pumice and volcanic-ash soils contributes to an intensive leaching of these soils and a loss of basic cations. Under high-rainfall regimes, base saturation is low (Birrell 1964; Saigusa et al. 1978), although exchangeable aluminum content is low considering the low base status (Igue and Fuentes 1972; Arai 1975). Rapid weathering of glassy fragments tends to resupply bases; however, many ash and pumice soils are low in basic cations.

Amorphous colloids in pumice and ash soils do not have a crystal structure that strongly fixes potassium and ammonium. These ions can be more easily leached than would be expected in soil dominated by crystalline 2:1 phyllosilicates (Kobo 1964). Schalscha et al. (1975) found that Andepts had lower affinity for K than soils dominated by layer silicate clays and, furthermore, affinities of the colloids for K relative to Ca decreased as surface charge density increased (pH increased). As one would expect, Dystrandepts from Chile exhibited a preference for Ca over K (Galindo and Bingham 1977).

Pumice soils of central Oregon developed under moderate rainfall do not appear to be deficient in basic cations (Youngberg and Dyrness 1965; Youngberg 1975; Nissley 1978). On the other hand, Will (1966) reported Mg deficiency in some pumice soils of New Zealand, and Metson and Lee (1977) suggest that rhyolitic tephra, pumice, or pumice alluvium in New Zealand has low Mg reserves and might be expected to develop Mg deficiency when exchangeable Mg is depleted. Calcium seems to be adsorbed by volcanic soils more readily than magnesium (Hunsaker and Pratt 1971; Galindo and Bingham 1977). Liming has been practiced in some instances to raise soil pH and reduce K leaching (Mahilum et al. 1970). The high buffering capacities of volcanic soils require large lime additions, and it may be impractical to raise the pH to 6.5 (Yuan 1974). Kobo (1964) has reported that 5,000 to 20,000 Kg/ha of lime are used in Japan on ash soils. Mahilum et al. (1970) reported on Hawaiian liming studies that applied up to 15 tons/ha. Seven years after application in this high-rainfall area there were minimal effects on pH and exchangeable bases.

## Other Nutrients

Consistent reports of other nutrient deficiencies do not seem to be available; however, isolated deficiencies of several elements have been reported.

Soils from Mexico and Hawaii were shown to retain boron to a greater extent than soils derived from other parent material (Bingham et al. 1971), and boron deficiency has been observed in Japan (Yoshida et al. 1966). Boron has also been reported as deficient in New Zealand ash and pumice soils, but only for brassica root crops (During 1964). Forage growing on volcanic soils of New Zealand was found to be low in copper and cobalt (Birrell 1964; During 1964). Low copper levels are a concern because of the relationship between molybdenum and copper in animal nutrition (Underwood 1971). Copper deficiencies or low levels have been reported in Kenya and New Zealand (Rolt 1962; Pinkerton 1967).

## LAND USE

The previous section has examined the behavior of individual elements important in the fertility of ash soils. The parent material composition, weathering regime, and soils age have a very strong influence on the fertility level. Climate also dictates the suitability of any area for land dedication to crop species, whereas local customs, societal needs, and economics combined with the topographic features determine the actual land use. This section is not exhaustive but only meant to present some land uses that have been practiced on volcanic soils.

Oyama (1964) suggested that crops that can take advantage of low-bulk density, high phosphorus fixation, high water retention, and high to excessive permeability should be grown in ash and pumice soils. In Japan, root crops do well because of a general lack of rocks and gravel in the ash soils. Radishes, sweet potatoes, and carrots are suitable, whereas upland rice can be grown because of the high water-holding capacity. Upland Andosols are used for forestland, grassland, and common upland field crops and orchards. Productivity is generally low because of high P adsorption capacity. Increased productivity can be obtained by heavy phosphate fertilization, supplying bases and minor elements, and by applying organic manures (Matsuzaka 1977). The poorly drained ash soils (Wet Andosols) are used mainly as paddy fields. They can be improved by drainage and heavy P fertilization (Matsuzaka 1977). The better-drained soils are not well suited to lowland rice production unless they are compacted or amended with bentonite to reduce infiltration (Kobo 1964).

Rico (1964) reported that a wide variety of crops are grown on ash soils in El Salvador at different elevations. Coffee is grown at high elevations, whereas cotton and maize are cultured at lower elevations. Young soils vary in their fertility status in relation to the organic-matter content. In these young soils, nitrogen limits growth. As the soils become older (more developed), phosphorus in addition to nitrogen is needed. Tan (1964) noted that phosphorus is also needed in the Andosols of Indonesia, and large quantities of organic matter are added possibly to counteract the high aluminum. Andosols on the

Pengalengan highlands support the best tea plantations in Java, yielding about 1400 kg/ha/yr (Tan 1964).

In Hawaii, ash-derived soils are used extensively for all crops except pineapple. In addition to sugar cane, papaya and coffee are also grown on ash-derived soils of warm climates, whereas the drier soils are used for open cattle range (Sherman and Swindale 1964). Philippine ash-derived soils are acidic (pH 5.4–6.2) and likely to need lime and phosphorus. The soils are moderately fertile and are used for a variety of crops. Cultivated soils are erosive and, because of the torrential rains, must be managed carefully (Mariano 1964). Ash soils in Japan are also erosive, and wind rows or other protective measures are needed at times during the year (Oyama 1964). Yamamoto and Hall (1967) found that ash soils are more erosive than other types in Hawaii. Youngberg *et al*. (1975) reported that stream tubidities are related to the parent materials of forested watershed in Oregon. Pyroclastic material has a higher potential to produce turbidity than pumice, basalt, or morainal materials. Large acreages of pumice soils are used to grow *Pinus radiata* in New Zealand; and ash and pumice soils found in Oregon, Washington, and Idaho are used for commercial forests and range land. The potential economic importance of this may be seen by the fact that the pumice plateau of south-central Oregon encompasses 2,000,000 ha (Franklin and Dyrness 1973).

Volcanic ash and pumice soils can be used for a wide variety of crops. Use depends on a suitable climate but is tempered to some extent by the unique properties of volcanic soils. These properties depend upon the weathering state of the soil (Aomine and Wada 1966; Metson and Lee 1977). Soil age and climate interact to determine these properties.

## REFERENCES

Adachi, T.
  1963  The humus composition of volcanic-ash soils in Kachima, Ibaragi Prefecture, 1. *Journal of the Science of Soil and Manure* (Japan 34:278–290.
  1966  The humus composition of volcanic ash soils in southern Kyushu. Studies on the humus of volcanic ash soils, 2. *Journal of the Science of Soil and Manure* (Japan) 37:505–510.
Anderson, B. J., and E. A. Jenne
  1970  Free iron and manganese oxide content of reference clays. *Soil Science* 109:163–169.
Aomine, S., and C. Mizota
  1973  Distribution and genesis of imogolite in volcanic-ash soils of northern Kanto, Japan. Proceeding International Clay Conference, Madrid, Spain, 1972, 207–213.
Aomine, S., and H. Otsuka
  1968  Surface of soil allophanic clays. Transactions of the 9th International Congress of Soil Science (Adelaide, Australia) 1:731–737.
Aomine, S., and K. Wada
  1962  Differential weathering of volcanic ash and pumice resulting in formation of hydrated halloysite. *American Mineralogist* 47:1024–1048.
  1966  Grade of weathering and fertility of volcanic ash soils of Aso Volcano. *Soil Science and Plant Nutrition* (Tokyo) 12:27–33.

Appelt, H., N. T. Coleman, and P. F. Pratt
  1975   Interactions between organic compounds, minerals, and ions in volcanic-ash-derived soils: II. Effects of organic compounds on the adsorption of phosphate. *Soil Science Society of America Proceedings* 39:628–630.
Appelt, H., and E. B. Schalscha
  1970   Effect of added phosphate on the inorganic phosphorus fractions of soils derived from volcanic ash. *Soil Science Society of America Proceedings* 34:599–602.
Arai, S.
  1975   Extraction of active aluminum from acid soils in Japan with different reagents. *Geoderma* 14:63–74.
Babcock, K. L.
  1963   Theory of the chemical properties of soil colloidal systems at equilibrium. *Hilgardia* 34:417–542.
Bailey, S. W., G. W. Brindley, W. D. Johns, R. T. Martin, and M. Ross
  1971   Summary of national and international recommendations on clay mineral nomenclature 1969–1970. C.M.S. Nomenclature Committee. *Clays and Clay Minerals* 19:129–132.
Balasubramanian, V., and Y. Kanehiro
  1978   Surface chemistry of Hydrandepts and its relation to nitrate adsorption as affected by profile depth and dehydration. *Journal of Soil Science* 29:47–57.
Bingham, F. T., A. L. Page, N. T. Coleman, and K. Flach
  1971   Boron adsorption characteristics of selected amorphous soils from Mexico and Hawaii. *Soil Science Society of American Proceedings* 35:546–550.
Birrell, K. S.
  1962   Surface acidity of subsoils derived from volcanic ash deposits. *New Zealand Journal of Science* 5:453.
  1964   Some properties of volcanic ash soils. In FAO–UNESCO *Meeting on the classification and correlation of soils from volcanic ash* (Tokyo, Japan). *World Soil Resource Report* 14:74–81 FAO–UNESCO.
Bleeker, P., and R. L. Parfitt
  1974   Volcanic ash and its clay mineralogy at Cape Hoskins, New Britain, Papua New Guinea. *Geoderma* 11:123–135.
Bockheim, J. G.
  1971   Effects of alpine and subalpine vegetation on soil development, Mount Baker, Washington. Doctoral dissertation, College of Forest Resources, University of Washington, Seattle.
Bockheim, J. G., and F. C. Ugolini
  1972   Soils and parent materials of Findley Lake, Snoqualmie National Forest, Washington. Internal Report 47. Seattle: USIBP Coniferous Forest Biome.
Bolt, G. H.
  1957   Determination of the charge density of silica sols. *Journal of Physical Chemistry* 61:1166–1169.
Broadbent, F. E., R. H. Jackman, and J. McNicoll
  1964   Mineralization of carbon and nitrogen in some New Zealand Allophane soils. *Soil Science* 98:118–128.
Brown, G.
  1955   Report of the clay minerals group subcommittee on nomenclature of clay minerals. *Clay Mineral Bulletin* 13:294.
Calhoun, F. C., V. W. Carlisle, C. Luna Z.
  1972   Properties and genesis of selected Columbian Andosols. *Soil Science Society of America Proceedings* 36:480–485.
Campbell, A. S., A. W. Young, L. G. Livingstone, M. A. Wilson, and T. W. Walker
  1977   Characterization of poorly ordered aluminosilicates in a Vitric Andosol from New Zealand. *Soil Science* 123:362–368.
Chichester, F. W., C. T. Youngberg, and M. E. Harward
  1969   Clay mineralogy of soils formed on Mazama pumice. *Soil Science Society of America Proceedings* 33:115–120.

Cloos, P., A. J. Leonard, J. P. Moreau, A. Herbillon, and J. J. Fripiat
   1969  Structural organization in amorphous silico-aluminas. *Clays and Clay Minerals* 17:279–
         287.
Cochran, P. H.
   1975  Soil temperatures and natural regeneration in south-central Oregon. In *Forest soils and
         forest land management*, Proceeding of the 4th North American Forest Soils Conference,
         edited by B. Bernier and C. H. Winget. Québec: Les Presses De L'Université Laval.
Cochran, P. H., L. Boersma, and C. T. Youngberg
   1967  Thermal properties of a pumice soil. *Soil Science Society of America Proceedings* 31:455–
         459.
Cradwick, P. D. G., V. C. Farmer, J. D. Russel, C. R. Masson, K. Wada, and N. Yoshinaga
   1972  Imogolite, a hydrated aluminum silicate of tubular structure. *Nature* (Physical Science)
         240:187–189.
Dalal, R. C.
   1978  Distribution of organic nitrogen in organic volcanic and nonvolcanic tropical soils. *Soil
         Science* 125:178–180.
Dawson, H. J., F. C. Ugolini, B. F. Hrutfiord, and J. Zachara
   1978  Role of soluble organics in the soil processes of a podzol, central Cascades, Washington.
         *Soil Science* 126:290–296.
Duchaufour, P.
   1970  *Précis de pedologie*. Paris: Masson et Cie.
   1978  *Ecological atlas of the soils of the world*. New York: Masson.
Dudal, R.
   1964  Correlation of soils derived from volcanic ash. In *FAO–UNESCO Meeting on the
         classification and correlation of soils from volcanic ash* (Tokyo, Japan). World Soils
         Resources Report No. 14, FAO–UNESCO.
Dudas, M. J., and M. E. Harward
   1975a Phyllosilicates in soils developed from Mazama ash. *Soil Science Society of America
         Proceedings* 39:572–577.
   1975b Weathering and augthigenic halloysite in soil developed in Mazama ash. *Soil Science
         Society of America Proceedings* 39:561–571.
Dugger, D. L., J. H. Stanton, R. N. Irby, R. L. McConnell, W. W. Cummings, and R. M.
         Maatman
   1964  The exchange of twenty metal ions with the weakly acidic silanol group of silica gel.
         *Journal of Physical Chemistry* 68:757–760.
During, C.
   1964  The amelioration of volcanic ash soils in New Zealand. In *FAO–UNESCO Meeting on the
         classification and correlation of soils from volcanic ash* (Tokyo, Japan). *World Soil Resource
         Report* 14:129–133, FAO–UNESCO.
Egashira, K., and S. Aomine
   1974  Effects of drying and heating on the surface area of allophane and imogolite. *Clay Science*
         4:231–242.
El-Swaify, S. A., and A. H. Sayegh
   1975  Charge characteristics of an oxisol and an inceptisol from Hawaii. *Soil Science* 120:49–56.
Enright, N. J.
   1978  The interrelationship between plant species distribution and properties of soils undergo-
         ing podzolization in a coastal area of S.W. Australia. *Australian Journal of Ecology*
         3:389–401.
Espinoza, W., R. G. Gast, R. S. Adams, Jr.
   1975  Charge characteristics and nitrate retention by two Andepts from south-central
         Chile. *Soil Science Society of America Proceedings* 39:842–846.
Espinoza, W., R. H. Rust, and R. S. Adams, Jr.
   1975  Characterization of mineral forms in Andepts from Chile. *Soil Science Society of America
         Proceedings* 39:556–561.

FAO-UNESCO
    1964 *Meeting on the classification and correlation of soils from volcanic ash* (Tokyo, Japan). World Soils Resources Report 14. FAO-UNECO.
FAO-UNESCO
    1974 *Soil map of the world* (Vol. 1). Paris: UNESCO.
Fields, M.
    1955 Clay mineralogy of New Zealand soils. Part 2, Allophane and related mineral colloids. *New Zealand Journal of Science and Technology* B37:336–350.
Fields, M., and G. G. C. Claridge
    1975 Allophane. In *Soil components Vol. II. Inorganic components*, edited by J. E. Greseking. New York: Springer-Verlag.
Fields, M., and D. L. Swindale
    1954 Chemical weathering of silicates in soil formation. *New Zealand Journal of Science and Technology B*, 36:140–154.
Flach, K. W.
    1964 Genesis and morphology of ash-derived soils in the United States of America. In *FAO-UNESCO Meeting on the classification and correlation of soils from volcanic ash* (Tokyo, Japan). World Soil Resources Report No. 14:61–70. FAO-UNESCO.
Forsythe, W. M.
    1975 Soil-water relations in soils derived from volcanic ash of Central America. In *Proceedings soil management in tropical America*, edited by E. Bornemisza and A. Alvarado Soil Science Department, North Carolina State University, Raleigh, N.C.
Franklin, J. F., and C. T. Dyrness
    1973 Natural vegetation of Oregon and Washington. USDA Forest Service Gen. Tech. Rep. PNW-8.
Galindo, G. G., and F. T. Bingham
    1977 Homovalent and heterovalent cation exchange equilibria in soil with variable surface charge. *Soil Science Society of America Journal* 41:883–886.
Gast, R. G.
    1977 Surface and colloid chemistry. In *Minerals in soil environments*, edited by J. B. Dixon and S. B. Wood. Madison, Wis.: Soil Sci. Soc. Amer.
Gebhardt, J., and N. T. Coleman
    1974a Anion adsorption by allophanic tropical soils: I. Chloride adsorption. *Soil Science Society of America Proceedings* 38:225–259.
    1974b Anion adsorption by allophanic tropical soils: II. Sulfate adsorption. *Soil Science Society of America Proceedings* 38:259–262.
    1974c Anion adsorption by allophanic tropical soils: III. Phosphate adsorption. *Soil Science Society of America Proceedings* 38:263–266.
Geist, J. M.
    1971 Orchardgrass responses to fertilization of seven surface soils from the central Blue Mountains of Oregon. USDA Forest Service Research Paper PNW 122. Portland, Ore.: Pacific Northwest Forest and Range Experiment Station.
    1977 Nitrogen response relationships of some volcanic-ash soils. *Soil Science Society of America Journal* 41:996–1000.
Geist, J. M., and G. S. Strickler
    1978 Physical and chemical properties of some Blue Mountains soils in northeast Oregon. USDA Forest Service Research Paper PNW 236. Portland, Ore.: Pacific Northwest Forest and Range Experiment Station.
Gibbs, H. S.
    1968 *Volcanic-ash soils in New Zealand*. Information Series No. 65. Wellington, New Zealand: New Zealand Department of Scientific and Industrial Research.
Gonzales, F., B. H. Appelt, E. B. Schalscha, and F. T. Bingham
    1974 Molybdate adsorption characteristics of volcanic-ash-derived soils in Chile. *Soil Science Society of America Proceedings* 38:903–906.

Greenland, D. J.
  1971   Interactions between humic and fulvic acids and clays. *Soil Science* 111:34–41.
Griffith, S. M., and M. Schnitzer
  1975   Analytical characteristics of humic and fulvic acids extracted from tropical volcanic soils. *Soil Science Society of America Proceedings* 39:861–867.
Harward, M. E., and C. T. Youngberg
  1969   Soils from Mazama ash in Oregon: Identification, distribution, and properties. Pedology and Quaternary Research Symposium, University of Alberta, Edmonton, Alberta, Canada, May 13–14. Edited by S. Paluk, University of Alberta, Edmonton, Alberta.
Hasan, S. M., R. L. Fox, and C. C. Boyd
  1970   Solubility and availability of sorbed sulfate in Hawaiian soils. *Soil Science Society of America Proceedings* 34:897–901.
Hay, R. L.
  1960   Rate of clay formation and mineral alteration in a 4000-year-old volcanic-ash soil on St. Vincent, B.W.I. *American Journal of Science* 258:354–368.
Henmi, T., and K. Wada
  1974   Surface acidity of imogolite and allophane. *Clay Minerals* 10:231–245.
  1976   Morphology and composition of allophane. *American Mineralogists* 61:379–390.
Hingston, F. J., A. M. Posner, and J. P. Quirk
  1968   Specific adsorption of anions of geothite. *Transactions of the 9th International Congress of Soil Science* (Adelaide, Australia) 1:669–678.
Horikawa, Y.
  1975   Electrokinetic phenomena of aqueou suspensions of allophane and imogolite. *Clay Science* 4:255–263.
Hsu, P. H.
  1977   Aluminum hydroxides and oxyhydroxides. In *Minerals in soils environments*, edited by J. B. Dixon, and S. B. Weed. Madison, Wis.: Soil Sci. Soc. Amer.
Hunsaker, V. E., and P. F. Pratt
  1971   Calcium magnesium equilibria in soils. *Soil Science Society of America Proceedings* 35:151–152.
Igue, K., and R. Fuentes
  1972   Characterization of aluminum in volcanic-ash soils. *Soil Science Society of America Proceedings* 36:292–296.
Inoue, T., and K. Wada
  1968   Adsorption of humified clover extracts by various clays. *Transactions of the 9th International Congress of Soil Science* (Adelaide, Australia) III:289–298.
Jackson, M. L., T. W. M. Levelt, J. K. Syers, R. W. Rex, R. N. Clayton, G. D. Sherman, and G. Uehara.
  1971   Geomorphological relationships of tropospherically derived quartz in soils of the Hawaiian Islands. *Soil Science Society of America Proceedings* 35:515–525.
Jenne, A. E.
  1968   Controls on Mn, Fe, Co, Ni, Cu, and Zn concentrations in soil and water: The significant role of hydrous Mn and Fe oxides. *Advances in Chemistry Series* Washington, D.C.: Amer. Chem. Soc. 73:337–387.
Jones, R. C., and G. Uehara
  1973   Amorphous coatings on mineral surfaces. *Soil Science Society of America Proceedings*. 37:792–798.
Kanno, I.
  1962   Genesis and classification of humic allophane soil in Japan. In *Transactions Joint Meeting Commissions IV and V International Soil Science Society* (New Zealand). Edited by G. N. Neale. Wellington, New Zealand. Wright and Carman, Limited. Pp. 422–427.
Kanno, I., Y. Honjo, and S. Arimura
  1959   Chemical properties and clay minerals of Andesitic volcanic-ash soils developed on the Kikuchi Table, Kyushu. *Bull. Kyushu Agricultural Experiment Station* 5:277–310.

Kanno, I., Y. Onikura, and T. Higashi
   1968   Weathering and clay mineralogical characteristics of volcanic ashes and pumices in
          Japan. *Transactions of the 9th International Congress of Soil Science* (Adelaide, Aus-
          tralia) III:111–122.
Kawasaki, H., and S. Aomine
   1966   So-called 14 Å minerals in some Ando soils. *Soil Science and Plant Nutrition* (Tokyo)
          12:144–150.
Kino, T., and P. F. Pratt
   1971a  Nitrate adsorption: I. In some acid soils of Mexico and South America. *Soil Science Society of
          America Proceedings* 35:722–725.
   1971b  Nitrate adsorption: II. In competition with chloride, sulfate, and phosphate. *Soil Science
          Society of America Proceedings* 35:725–728.
Kitagawa, Y.
   1976   Specific gravity of allophane and volcanic soils determined with a pycnometer. *Soil
          Science and Plant Nutrition* (Tokyo) 22:199–202.
Kobayashi, Y., and S. Aomine
   1967   Mechanism of inhibitory effect of allophane and montmorillonite on some enzymes. *Soil
          Science and Plant Nutrition* (Tokyo) 13:189–194.
Kobo, K
   1964   Amelioration of volcanic ash soils and their potentiality. In FAO-UNESCO *Meeting on the
          classification and correlation of soils from volcanic ash* (Tokyo, Japan). *World Soil Resource Report*
          14:126–128, FAO-UNESCO.
Kononova, M. M.
   1975   Humus of virgin and cultivated soils. In *Soil components*, Vol. I: *Organic compounds*,
          edited by J. E. Gieseking. New York: Springer-Verlag.
Kosaka, J., C. Honda, and A. Izeki
   1962   Transformation of humus in upland soils, Japan. *Soil Science and Plant Nutrition*
          (Tokyo) 8:23–28.
Kubiena, W. L.
   1953   *The soils of Europe*. London: T. Murby and Co. 1953.
Kubota, T
   1972   Aggregate formation of allophanic soils: Effect of drying on the dispersion of the soils. *Soil
          Science and Plant Nutrition* (Tokyo) 18:79–87.
Kyuma, K., and K. Kawaguchi
   1964   Oxidative changes of polyphenols as influenced by allophane. *Soil Science Society of
          America Proceedings* 28:371–374.
Leonard, A., S. Suzuki, J. J. Fripiat, and C. De Kimpe
   1969   Structure and properties of amorphous silicoaluminium. I. Structure from X-ray fluores-
          cence spectroscopy. *Journal of Physical Chemistry* 73:2608–2617.
Loganathan, P.
   1971   Sorption of heay metals on a hydrous manganese oxide. Doctoral dissertation, Dept. of
          Soils and Plant Nutrition, Univ. of Calif., Davis.
Loganathan, P., and L. D. Swindale
   1969   The properties and genesis of four middle altitude Dystrandept volcanic-ash soils from
          Mauna Kea, Hawaii. *Pacific Science* 23:161–171.
Maeda, T., H. Takenaka, and B. P. Warkentin
   1977   Physical properties of allophane soils. *Advances in Agronomy* 29:229–264.
Mahilum, B. C., R. L. Fox, and J. A. Silva
   1970   Residual effects of liming volcanic-ash soils of the tropics. *Soil Science* 109:102–109.
Mariano, A.
   1964   Volcanic soils of the Philippines. In *FAO-UNESCO Meeting on the classification and
          correlation of soils from volcanic ash* (Tokyo, Japan). *World Soils Resource Report* 14:53 -55,
          FAO-UNESCO.
Masui, J., and S. Shoji
   1967   Some problems on clay minerals of volcanic ash soil. *Pedologist* 11(1):33–45.

1969   Crystalline clay minerals in volcanic ash soils of Japan. International Clay Conference. (Tokyo), edited by L. Heller. Jerusalem: Israel University Press.

Masui, J., S. Shoji, and N. Uchiyama
1969   Clay mineral properties of volcanic-ash soils in the northeastern part of Japan. *Tohoku Journal of Agricultural Research* 17:17–36.

Matsuzaka, Y.
1977   Major soil groups in Japan. In *Proceeding of the international seminar on soil environment and fertility management in intensive agriculture*. The Society of the Science of Soil and Manure, Japan. Kyoto: Showado Press.

Metson, A. J., and R. Lee
1977   Soil chemistry in relation to the New Zealand genetic soil classification. *Soil Science* 123:347–353.

Mitchell, B. D.
1975   Oxides and hydrous oxides of silicon. In *Soil components, Vol. II: Inorganic components*, edited by J. E. Gieseking. New York: Springer-Verlag.

Mitchell, B. D., V. C. Farmer, and W. J. McHardy
1964   Amorphous inorganic materials in soils. *Advances in Agronomy* 16:327–383.

Miyauchi, N., and S. Aomine
1964   Does "Allophane B" exist in Japanese Volcanic-ash soils? *Soil Science and Plant Nutrition* (Tokyo) 10:199–203.
1966   Mineraology of gel-like substance in the pumice bed in Kanuma and Kitakami Districts. *Soil Science and Plant Nutrition* (Tokyo) 12:187–190.

Munevar F., and A. G. Wollum, II
1977   Effects of the addition of phosphorus and inorganic nitrogen on carbon and nitrogen mineralization in Andepts from Columbia. *Soil Science Society of America Journal* 41:540–544.

Neall, V. E.
1977   Genesis and weathering of andosols in Taranaki, New Zealand. *Soil Science* 123(6):400–408.

Nissley, S. D.
1978   Nutrient changes after prescribed burning of Oregon Ponderosa pine stands. Master's thesis, College of Forest Resources University of Washington, Seattle, Wa.

Nissley, S. D. and R. J. Zasoski
1977   Unpublished data on file. College of Forest Resources, University of Washington, Seattle.

Ohmasa, M.
1964   Genesis and morphology of volcanic-ash soils. In *FAO–UNESCO Meeting on the classification and correlation of soils from volcanic ash* (Tokyo, Japan). World Soil Resources Report No. 14:56–60, FAO–UNESCO.

Otowa, M.
1977   A proposal of a soil classification system for soil survey of Japan. (I) Introduction and structure of the system. *Journal of the Science of Soil and Manure* (Japan) 48:201–206. (Abstract in *Soil Science and Plant Nutrition* (Tokyo) 23:396).

Oyama, M.
1964   Land use of volcanic ash soils in Japan. In *FAO-UNESCO Meeting on the classification and correlation of soils from volcanic ash* (Tokyo, Japan). *World Soil Resource Report* 14:125, FAO–UNESCO.

Packard, R. Q.
1957   Some physical properties of Taupo pumice soils of New Zealand. *Soil Science* 83:273–289.

Parfitt, R. L., R. J. Atkinson, and R. St. C. Smart
1975   The mechanism of phosphate fixation by iron oxides. *Soil Science Society of America Proceedings* 39:837–841.

Parfitt, R. L., A. D. Thomas, R. J. Atkinson, R. St. C. Smart
1974   Adsorption of phosphate on imogolite. *Clays and Clay Minerals* 22:455–456.

Parks, G. A.
1965   The isoelectric points of solid oxides, solid hydroxides and aqueous hydroxo complex systems. *Chemical Reviews* 65:177–198.

1967 Aqueous surface chemistry of oxides and complex oxide minerals. In *Advances in chemistry series*. 67:121–160. Washington, D.C.: Amer. Chem. Soc.

Pinkerton, A.
1967 Copper deficiency of wheat in the Rift Valley, Kenya. *Journal of Soil Science* 18:18–26.

Pumphrey, F. V.
1971 Grass species growth on volcanic ash-derived soils cleared of forest. *Journal of Range Management* 24(3):200–203.

Rajan, S. S. S.
1975 Phosphate adsorption and the displacement of structural silicon in an allophanic clay. *Journal of Soil Science* 26(3):250–256.

Rajan, S. S. S., and R. L. Fox
1975 Phosphate adsorption by soils: II. Reactions in tropical acid soils. *Soil Science Society of America Proceedings* 39:846–851.

Rajan, S. S. S., and K. W. Perrott
1975 Phosphate adsorption by synthetic amorphous aluminosilicates. *Journal of Soil Science* 26(3):257 266.

Rico, M.
1964 Report on soils of volcanic-ash origin in El Salvador. In *FAO-UNESCO Meeting on the classification and correlation of soils from volcanic ash* (Tokyo, Japan). World Soil Resources Report No. 14:23–29, FAO-UNESCO.

Rieger, S.
1964 Humods in relation to volcanic ash in southern Alaska. *Soil Science Society of America Proceedings* 38:347–351.

Rodriguez, M.
1962 Soil classification and its application in Chile. In *Transactions, Joint Meeting of Commissions IV and V International Soil Science Society* (New Zealand). edited by G. N. Neale. Wellington New Zealand: Wright and Carman, Limited.

Kolt, W. F.
1962 The copper status of some soils of the North Island, New Zealand. In *Transactions of Joint Meeting of Commissions IV and V International Society of Soil Science* (New Zealand), edited by G. N. Neale. Wellington, New Zealand: Wright and Carman, Limited.

Saigusa, M., S. Shoji, and T. Kato
1978 Origin and nature of halloysite in Ando soils from Towada tephra, Japan. *Geoderma* 20:115–129.

Sasaki, T., T. Maeda, and S. Sasaki
1969 *Transactions of the Japanese Society of Irrigation Drainage and Reclaimation Engineers.* (Nogyo Doboku Gakkai Ronbunshu) 27:57–60. Cited in Maeda et al. 1977.

Satho, T.
1976 Isolation and characterization of naturally occurring organo–mineral complexes in some volcanic-ash soils. *Soil Science and Plant Nutrition* (Tokyo) 22:125–136.

Schalscha, E. B., P. F. Pratt, and L. De Andrade
1975 Potassium–calcium exchange equilibria in volcanic-ash soils. *Soil Science Society of America Proceedings* 39:1069–1072.

Schalscha, E. B., P. F. Pratt, and D. Soto
1974 Effects of phosphate adsorption on the cation-exchange capacity of volcanic-ash soils. *Soil Science Society of America Proceedings* 38:539–540.

Schwertmann, U., and R. M. Taylor
1977 Iron oxides. In *Minerals in soil environments*, edited by J. B. Dixon and S. B. Weed. Madison, Wis.: Soil Science Society of America.

Sherman, G. D., Y. Matsuzaka, H. Ikawa, and G. Uehara
1964 The role of the amorphous fraction in the properties of tropical soils. *Agrochimica* 8:146–163.

Sherman, G. D., and L. D. Swindale
1964 Hawaiian soils from volcanic ash. In *FAO-UNESCO Meeting on the classification and*

*correlation of soils from volcanic ash*. (Tokyo, Japan). *World Soil Resource Report* 14:36–49, FAO-UNESCO.

Shinagawa, A.
  1960   The accumulation of humus in volcanic-ash soil originating from Mt. Sakurajima's ash. *Pedologist* 4:98–111.

Shoji, S., and J. Masui
  1971   Opaline silica of recent volcanic-ash soils in Japan. *Journal of Soil Science* 22:101–108.

Shoji, S. and T. Ono
  1978   Physical and chemical properties and clay mineralogy of Andosols from Kitakami, Japan. *Soil Science* 126:297–312.

Singer, M., and F. C. Ugolini
  1974   Genetic history of two well-drained subalpine soils formed on complex parent materials. *Canadian Journal of Soil Science* 54:475–489.

Singer, M., F. C. Ugolini, and J. Zachara
  1978   In situ study of podzolization on tephra and bedrock. *Soil Science Society of America Journal* 42:105–111.

Singh, B. R., and Y. Kanehiro
  1969   Adsorption of nitrate in amorphous and kaolinitic Hawaiian soils. *Soil Science Society of America Proceedings* 33:681–683.

Smathers, G. A., and D. Mueller-Dombois
  1974   Invasion and recovery of vegetation after a volcanic eruption in Hawaii. National Park Service Science Monograph Series No. 5, Island Ecosystems IRP/IBP. NSF. GB23230; Contribution No. 38. U.S. Government Printing Office.

Sneddon, J. T.
  1973   A study of two soils derived from volcanic ash in southwestern British Columbia and a review and determination of ash distribution in western Canada. Doctoral dissertation, Dept. of Soil Science. University of British Columbia, Vancouver, Canada.

Soil Survey Staff
  1975   *Soil taxonomy. A basic system of soil classification for mapping and interpreting soil surveys*. USDA Soil Con. Service Agr. Handbook No. 436. Washington, D.C. United States Government Printing Office.

Sowden, F. J., S. M. Griffith, and M. Schnitzer
  1976   The distribution of nitrogen in some highly organic tropical soils. *Soil Biology and Biochemistry* 8:55–60.

Stumm, W., and J. J. Morgan
  1970   *Aquatic chemistry*. New York: Wiley.

Sudo, T.
  1954   Clay mineralogical aspects of the alteration of volcanic glass in Japan. *Clay Mineral Bulletin* 2:96–106.

Swindale, L. D.
  1964   The properties of soils derived from volcanic ash. In *FAO–UNESCO Meeting on the classification of soils from volcanic ash* (Tokyo, Japan). World Soil Resources Report No. 14:82–86, FAO–UNESCO.

Syers, J. K., D. L. Mokma, M. L. Jackson, D. L. Dolocater, and R. W. Rex
  1971   Mineralogical composition and cesium-137 retention properties of continental aerosolic dusts. *Soil Science* 113:116–123.

Tan, K. H.
  1964   The Andosols in Indonesia. In *FAO-UNESCO Meeting on the classification and correlation of soils from volcanic ash* (Tokyo, Japan). *World Soil Resources Report* 14:30–35, FAO-UNESCO.

Taylor, N. H.
  1964   The classification of ash-derived soils in New Zealand. In *FAO-UNESCO Meeting on the classification and correlation of soils from volcanic ash* (Tokyo, Japan). *World Soil Resources Report* 14:101–110, FAO-UNESCO.

Taylor, N. H., and I. J. Pahlen
  1968   Classification of New Zealand soils. *New Zealand Soil Bureau Bulletin* 26(1):15–46.

Thorp, J., and G. Smith
   1949   Higher categories of soil classification: Order, suborder, and great soil groups. *Soil Science* 67:117–126.
Tokashiki, Y., and K. Wada
   1975   Weathering implications of the mineralogy of clay fractions of two Ando soils, Kyushu. *Geoderma.* 14:47–62.
Tokudome, S., and I. Kanno
   1965a Nature of the humus of humic allophane soils in Japan. Part 1. Humic acids (Ch)/fulvic acids (Cf) ratios. *Soil Science and Plant Nutrition* (Tokyo) 11:1–8.
   1956b Nature of the humus of humic allophane soils in Japan. Part 2. Some physico–chemical properties of humic and fulvic acids. *Soil Science and Plant Nutrition* (Tokyo) 11:9–15.
Uchiyama, N., J. Masui, and Y. Onikura
   1960   Humus clay pan developed on a volcanic-ash soil. Montmorillonite-like clay mineral in a strongly acidic soil. *Transactions of the 7th International Congress of Soil Science* (Madison, U.S.A.) 4:443–450.
   1962   Montmorillonite in a volcanic-ash soil. *Soil Science and Plant Nutrition* (Tokyo) 8:13–19.
Ugolini, F. C., R. Minden, H. J. Dawson, and J. Zachara
   1977a An example of soil process in the *Abies amabilis* zone of central Cascades, Washington. *Soil Science* 124:291–302.
   1977b Direct Evidence of particle migration in the soil solution of a podzol. *Science* 198:603–605.
Underwood, E. J.
   1971   *Trace elements in human and animal nutrition.* New York and London: Academic Press.
van Raij, B., and M. Peech
   1972   Electrochemical properties of some oxisols and alfisols of the tropics. *Soil Science Society of America Proceedings* 36:587–593.
Veith, J. A., and G. Sposito
   1977   Reactions of aluminosilicates, aluminum hydrous oxides, and aluminum oxide with O-phosphate: The formation of X-ray amorphous analogs of variscite and montebrasite *Soil Science Society of America Journal* 41:870–876.
Wada, K.
   1959   Reaction of phosphate with allophane and halloysite. *Soil Science* 87:325–330.
   1966   Deuterium exchange of hydroxyl groups in allophane. *Soil Science and Plant Nutrition* (Tokyo) 12:176–182.
   1967   A structural scheme of soil allophane. *American Mineralogist* 52:690–708.
   1977   Allophane and imogolite. In *Minerals in soil environments*, edited by J. B. Dixon and S. B. Weed. Madison, Wis.: Soil Science Society of America.
Wada, K., and S. Aomine
   1966   Occurrence of gibbsite in weathering of volcanic materials at Kuroishibaru, Kumamoto. *Soil Science and Plant Nutrition* (Tokyo) 12(4):151–157.
   1973   Soil development on volcanic materials during the Quarternary. *Soil Science* 116:170–177.
Wada, K., and Y. Harada
   1969   Effects of salt concentration and cation exchange species on the measured cation-exchange capacity of soils and clays. In *Proceedings Int. Clay Conf.*, edited by L. Heller, Jerusalem: Israel University Press, 1:561–571.
Wada, K., and M. E. Harward
   1974   Amorphous clay constituents of soils. *Advances in Agronomy* 26:211–260.
Wada, K., and I. Higashi
   1976   The categories of aluminum- and iron–humus complexes in Ando soils determined by selective dissolution. *Journal of Soil Science* 27:357–368.
Wada, K., and T. Inoue
   1967   Retention of humic substances derived from rotted clover leaves in soil containing montmorillonite and allophane. *Soil Science and Plant Nutrition* (Tokyo) 13:9–16.
Wada, K., and N. Yoshinaga
   1969   The structure of imogolite. *American Mineralogist* 54:50–71.

Walker, T. W., B. K. Thapa, and A. F. R. Adams
   1959   Studies on soil organic matter: 3. Accumulation of carbon, nitrogen, sulfur, organic and
          total phosphorus in improved grassland soils. *Soil Science* 87:135–140.
Will, G. M.
   1966   Magnesium deficiency, the cause of spring needle-tip chlorosis in young pines on pumice
          soils. *New Zealand Journal of Forestry* 11:88–94.
   1968   Some aspects of organic matter formation and decomposition in pumice soils growing
          *Pinus radiata* forest. *Transactions of the 9th International Congress of Soil Science*
          (Adelaide, Australia). III:237–246.
Will, G. M., and E. L. Stone
   1967   Pumice soils as a medium for tree growth. I. Moisture storage capacity. *New Zealand
          Journal of Forestry* 12:189–199.
Wright, A. C. S.
   1964   The "Andosols" or "Humic Allophane" soils of South America. In *FAO-UNESCO
          Meeting on the classification and correlation of soils from volcanic ash* (Tokyo, Japan).
          World Soil Resources Report No. 14:9–22, FAO-UNESCO.
Yamada, Y.
   1968   Relation between 14C and color of humic acid solution from some volcanic-ash soils in
          Japan. *Journal of the Science of Soil and Manure* (Japan) 39:447–451.
Yamamoto, T., and H. W. Hall
   1967   Erodibility indices for wildland soils of Oahu, Hawaii, as related to soil-forming factors.
          *Water Resources Research* 3:758–798.
Yamane, I., I. Ito, K. Sato, and D. Kumada
   1958   On the relationship between vegetation and soil at mountain grassland in north-eastern
          Japan, 3. Growing process and inorganic and organic constituents of predominant plant
          species and some characteristics of soils. *Reports of the Institute of Agricultural Research,
          Tohoku University* 9:1–43.
Yoshida, M.
   1970   Acid properties of montmorillonite and halloysite. *Journal of the Science of Soil and
          Manure* (Japan) 41:483–486.
   1971   Acid properties of kaolinite, allophane, and imogolite. *Journal of the Science of Soil and
          Manure* (Japan) 42:329–332.
Yoshida, Y., N. Obata, and T. Shindo
   1966   [Effects of boron applications on the growth of lucerne on Yatsugatake and acid humic
          volcanic-ash soil. Results of a 5-year field experiment]. *Journal of the Science of Soil and
          Manure* (Japan) 37:516–521. Abstracted in *Soil Science and Plant Nutrition* 12:42–43.
Youngberg, C. T.
   1975   Effects of fertilization and thinning on the growth of Ponderosa pine. *Soil Science Society
          of America Proceedings*. 39:137–139.
Youngberg, C. T., and C. T. Dyrness
   1964   Some physical and chemical properties of pumice soils in Oregon. *Soil Science* 97:391–
          399.
   1965   Biological assay of pumice soil fertility. *Soil Science Society of America Proceedings*
          29:182–187.
Youngberg, C. T., M. E. Harward, G. H. Simonson, and D. Rai
   1975   Nature and cause of stream tubidity in a mountain watershed. In *Forest soils and forest
          land management*. (Proceeding of the Fourth North American Forest Soil Conference),
          edited by B. Bernier and C. H. Winget. Quebec: Les Presses De L'Université Laval.
Yuan, T. L.
   1974   Chemistry and mineralogy of Andepts. *Soil and Crop Science Society of Florida Proceed-
          ings* 33:101–108.
Zachara, J.
   1979   Clay genesis and alteration in two tephritic subalpine podzols of the central Cascades,
          Washington. Master's thesis, College of Forest Resources University of Washington,
          Seattle.

# 5

## On the Damage Caused by Volcanic Eruptions with Special Reference to Tephra and Gases

SIGURDUR THORARINSSON

## THE RANGES OF VOLCANIC DAMAGE

As indicated in the title, this chapter deals in particular with two aspects of eruption damage, that caused by the spreading and fall of tephra, and the pollution by toxic volcanic gases. The chapter is based mainly on experiences gained in Iceland during 1100 years of settlement. At the outset, however, a summary sketch is given of the damage caused by erupting volcanoes in general. For this purpose I have put together in a single table (Table 5.1) the three main types of volcanic eruptions, the chief agents of damage or destruction characteristic of each, and the range of their damaging effects. The table refers to eruptions in historical times.

The division of volcanic eruptions into three main types—effusive, mixed, and explosive—is a usual one. The boundary between explosive eruptions and the other types is easy to draw, if an explosive eruption is defined as one in which no lava is produced at all. Less clear-cut is the borderline between mixed and effusive eruptions, since nearly all lava-producing eruptions also produce some tephra. This borderline ought, however, to be defined somehow. In this chapter those eruptions are regarded as effusive where the volume of freshly fallen tephra is within 5% of the total volume of tephra and lava. When the volume of tephra and lava is recalculated into that of dense rock of similar chemical composition, the corresponding boundary can be placed at 2 volume or weight percent.

Table 5.1 is an improved version of an earlier attempt at such a table

VOLCANIC ACTIVITY AND HUMAN ECOLOGY

**TABLE 5.1**
**The Main Types of Volcanic Eruptions: Their Chief Agents of Damage and the Range of Their Damaging Effects**

| | TYPE OF ERUPTION | DAMAGING AGENT | | | | | | | | |
|---|---|---|---|---|---|---|---|---|---|---|
| | | Lava | Tephra | Gases | Glowing avalanches | Base surges | Mud-flows | Water-floods | Earth-quakes | Tsu-namis |
| **SHORT RANGE** (< 30 km) — EFFUSIVE | EFFUSIVE | ╪ | + | ╪ | | | | ┿ | ┿ | |
| **SHORT RANGE** (< 30 km) — MIXED | MIXED | ╪ | (+) | + | | ┿ | ┿ | (+) | ┿ | (+) |
| **SHORT RANGE** (< 30 km) — EXPLOSIVE | EXPLOSIVE | | (+) | + | ╪ | ┿ | ┿ | (+) | ┿ | (+) |
| **MEDIUM RANGE** (30–500 km) — EFFUSIVE | EFFUSIVE | (+) | (+) | (+) | | | | (+) | (+) | |
| **MEDIUM RANGE** (30–500 km) — MIXED | MIXED | | | + | | | (+) | ┿ | (+) | + |
| **MEDIUM RANGE** (30–500 km) — EXPLOSIVE | EXPLOSIVE | | ╪ | + | (+) | | (+) | ┿ | (+) | + |
| **LONG RANGE** (500–10000 km) — EFFUSIVE | EFFUSIVE | | | [(+)] | | | | | | |
| **LONG RANGE** (500–10000 km) — MIXED | MIXED | | (+) | [(+)] | | | | | | (+) |
| **LONG RANGE** (500–10000 km) — EXPLOSIVE | EXPLOSIVE | | (+) | [(+)] | | | | [(+)] | | + |
| **"GLOBAL"** (> 10000 km) — EFFUSIVE | EFFUSIVE | | | [(+)] | | | | | | |
| **"GLOBAL"** (> 10000 km) — MIXED | MIXED | | | | | | | | | |
| **"GLOBAL"** (> 10000 km) — EXPLOSIVE | EXPLOSIVE | | [(+)] | | | | | | | |

RANGE OF DAMAGING EFFECT

╪ ALWAYS,   ┿ USUALLY,   + COMMONLY,   (+) RARELY,   [(+)] VERY RARELY,        EMPTY SQUARES = NONEXISTENT

presented in a paper (Thorarinsson 1977:3). In the table, which is based on eruptions in historical times, I distinguish among four distance ranges of damaging effects. The boundaries between these ranges are, understandably, somewhat arbitrary but, I hope, not unreasonable.

Before dealing with each of the types of damage listed in Table 5.1, it can be stated that, generally speaking, the amount of damage caused by volcanic eruptions depends, in addition to the type and size of the eruption, on the distance from the volcano to inhabited or habitable areas, and on the type and density of the settlements within the range of the damaging effects (see Figure 5.1). And, as this chapter refers mainly to Iceland, it should also be kept in mind that about 75% of the country is devoid of vegetation and that the population is only 220,000, or, on the average 2.2 inhabitants per square kilometer.

## Lava Flows

The damage caused by lava flows in mixed and effusive eruptions is usually fairly local, or "short-range," as defined in Table 5.1, and exclusively so if the

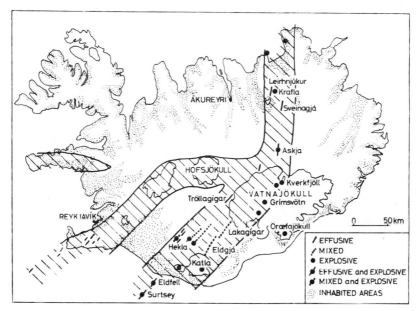

**FIGURE 5.1.** Inhabited areas in Iceland in relation to volcanoes active during the 11 centuries of settlement. Coarse striation: postglacial volcanic zones.

lava is intermediate or acid in composition. In explosive eruptions there are, by definition, no lava flows.

In the last millennium an occasional lava flow in Iceland and Hawaii has reached over 30 km from the crater(s). The longest lava flow in Iceland in historical times, the Lakagígar flow of 1783, stretches 65 km from the crater row. Within their extent, lava flows cause total destruction (Figure 5.2). The range of such destruction can in exceptional cases be shortened somewhat either by diverting the flow by means of earth banks or by increasing its viscosity by chilling it with water pumped on its front and surface. Such an operation was for the first time successfully carried out during the eruption on Heimaey in 1973, on the initiative of the Icelandic physicist Thorbjörn Sigurgeirsson, where the harbor and important factories were saved from being overrun by lava. One cubic meter of water will cool about 0.7 m³ of lava from 1100°C to 100°C when totally converted to steam, and on Heimaey the pumping rate, at the height of the operation, was nearly 1 m³/sec, chilling about 60,000 m³ of lava per day (Jónsson and Matthíasson 1974).

Each lava flow apparently has a natural range, which is mainly dependent on the viscosity of the lava. The viscosity in turn is mainly dependent on the silica content and temperature of the lava. The temperature normally decreases as distance from the crater increases, causing greater viscosity and slowing up the advance. When approaching a certain range, individual for each flow, the lava therefore tends to change course and flow in another direction until it reaches an equal distance from the crater. The success of the water chilling on Heimaey was probably partly due to the fact that the undis-

**FIGURE 5.2.** Lava overflowing the town of Vestmannaeyjar. The lava covered about 200 houses. (Photo: S. Thorarinsson, 28 March 1973.)

turbed natural reach of the Eldfell lava flow was not much longer than the distance to the harbor.

Outside the area covered by lava flows they can also cause damage by setting grasslands or woods on fire.

### Glowing Avalanches

Ever since the destruction of St. Pierre in 1902, the glowing avalanches of the Peléan type have been more feared than other effects of volcanic activity, and within the range of both the avalanches as such and the accompanying glowing clouds (*nuées ardentes*), the destruction is almost total. The range is usually short, but it has recently been pointed out by Walker and Sparks (1977) that along with the formation of the avalanches depositing ignimbrites goes a formation of vitric-enriched, fine-grained tephra (ash), which may have a volume comparable to the associated ignimbrites and can occasionally spread over vast areas.

### Base Surges

According to J. G. Moore (1967), who was the first to study base surges associated with a volcanic eruption, a base surge is "a ring-shaped basal cloud that sweeps outward as a density flow from the base of a vertical eruption column [p. 337]." (See Figure 5.3.) The range of damage of base surges is always, or at least nearly always, a short one. In the powerful phreato-magmatic

**FIGURE 5.3. Base surge, spreading from the base of the eruption column of Surtsey. (Photo: S. Thorarinsson, 16 November 1963.)**

eruption of the Taal volcano in Luzon in 1965, which was studied by Moore (1967), this range was 6 km. Near the eruption center base surges can be highly destructive.

Base surges frequently occurred during the first winter of the Surtsey eruption. The highest initial velocity measured was about 40 m/sec. The initial velocity of a base surge observed during the eruption of Myojin Reef, 420 km south of Tokyo, was 65 m/sec (Morimoto 1960).

## Mudflows

The damaging range of volcanic mudflows is usually less than 25 km but may occasionally exceed that distance, even by a wide margin. A mudflow from the Cotopaxi volcano in 1877 was 160 km long. Volcanic mudflows can have various causes (Macdonald 1972:170–182) and they are frequently very destructive. In Iceland, where volcanic mudflows are not known to have caused any damage to speak of in historic times, they are usually due to hot tephra, mainly bombs and scoria, falling on snow and ice around the vent or on the flanks of the erupting volcano.

The distinction between volcanic mudflows and volcanic water floods is sometimes uncertain, as the floods may be so heavily laden with debris, chiefly tephra, that they could just as well be termed mudflows.

### Water Floods

Volcanic water floods can originate in the same way as mudflows, by melting of snow or ice by tephra or lava. Such floods usually accompany the initial phase of Hekla's eruptions. The flood associated with the beginning of the Hekla eruption on 29 March 1947 reached a peak discharge of 900 m³/sec and the total volume was estimated at $3 \times 10^6$ m³ (Kjartansson 1951).

The catastrophic release of water from a crater lake may cause such floods too. A special type of volcanic water flood that is characteristic of Iceland is the *jökulhlaup*, or glacier burst. Jökulhlaups can be caused by the release of water stored in ice-dammed lakes, but *volcanic jökulhlaups* are floods, sometimes enormous, caused by subglacial eruptions (explosive or mixed), or solfatara activity. Jökulhlaups in historic times are very rare outside Iceland. In Iceland they are common, as some of the most active volcanoes are subglacial, such as Katla and Grímsvötn. The jökulhlaups of Grímsvötn, lasting usually 1–2 weeks, reach a total volume of 6–7 km³ and a peak runoff of about 40,000 m³/sec with about a 10-year interval between them, but since 1938 they have been smaller and more frequent (Thorarinsson 1974). Floods from the Katla volcano, occurring usually twice a century, reach a peak runoff estimated at least at 100,000 m³/sec or more, thus approaching the runoff of the river Amazonas.

The range within which jökulhlaups can cause damage usually exceeds 25 km. The damage is usually caused only by the flood per se and depends both on the size of the flood and the suddenness of its increase. Icebergs carried by the flood (Figure 5.4) can add to the damage. The floods also transport volcanic gases, the effects of which are dealt with later on.

### Earthquakes

The damaging effects of volcanic earthquakes are usually short-range, as defined in Table 5.1. Their magnitude rarely exceeds 5.5 on the Richter scale and is normally much smaller, but their intensity can, in limited areas, be strong enough to bring about serious damage and ruin. In Iceland damaging earthquakes have repeatedly occurred at the beginning of or during eruptions in Hekla.

### Tsunamis

Volcanic tsunamis (huge sea waves) can result from violent submarine explosions, from caldera collapses, and from other volcano-tectonic move-

**FIGURE 5.4. Icebergs standed on Mýrdalssandur after being transported 3 km by the Jökulhlaup from Katla on 12 October 1918. (Photo: K. Gudmundsson.)**

ments. The greatest and most disastrous volcanic tsunamis known in historic times are those caused by the Krakatoa eruption in 1883. Contrary to other damaging volcanic effects, except fluorine contamination, tsunamis are more likely to cause damage outside than inside the 25-km range. In Iceland tsunamis are not known to have occurred.

## TEPHRA FALL

When considering the damage caused by tephra, it is appropriate to distinguish between that directly effected by the spreading and fall of the tephra as such, and that caused more indirectly by the tephra as a transporter of noxious volcanic volatile compounds, especially fluorine. In this chapter the latter are treated under gases, although they are closely linked with the spreading of tephra. They are to a great extent confined to areas where the tephra layer is so thin that without the gases it would have caused little harm.

As seen in Table 5.1, the damage from tephra fall is found within all ranges of distance from the source. In the very rare cases where the effects are termed "global," the distinction between the effects of gases and tephra is vague, as it is only the finest volcanic dust that is spread so far, and that dust is likely to be more or less polluted with volatile compounds, spread as aerosol.

Next to the volcano it is largely the thickness of the tephra layer that decides the degree of damage wrought in an eruption. The tephra fall can destroy or seriously damage buildings by burying them or crushing them by its weight, which may be enhanched by rain (see Figure 5.5). In the Eldfell eruption on Heimaey in 1973 the specific weight of the freshly fallen tephra was 0.6, which means a load of 600 kg/m² on a flat roof if the layer is 1 m thick. Heavy rain would increase the specific weight to 1.0 (Einarsson,Th. 1974). Most roofs were able to withstand 30–40 cm thickness of dry tephra, but if the layer became wet they were likely to collapse. Therefore, one of the most urgent tasks after each heavy tephra fall was to shovel the tephra from the roofs.

Hazard to people and animals may be caused by falling volcanic bombs and xenoliths at various distances from the volcano, depending on the violence of the explosive activity, but this happens only rarely outside the 30-km range.

FIGURE 5.5. Houses in Vestmannaeyjar buried by the Eldfell tephra. In all about 100 houses were buried by tephra. (Photo: S. Thorarinsson.)

During the very violent initial phase of the Hekla eruption that began on the eve of 25 July 1510, a man is reported to have been stunned when hit on the head by a stone as he stood in front of the main door at Skálholt, the episcopal see in South Iceland, 45 km west of Hekla. A bomb, transported 32 km by air during the Plinian phase of the 1947 Hekla eruption and smashed on landing, must have weighed at least 20 kg (Thorarinsson 1954b).

Within short range from the volcano, falling bombs and scoria can be hot enough to set buildings on fire. Roofs of farmhouses 10 km distant from the erupting Hekla craters in the year 1300 were burnt off by falling scoria. Within 1200 m from the Eldfell volcano on Heimaey all windows facing the eruption column had to be covered by sheet iron in order to prevent bombs and stones from going through the window panes and setting the houses on fire.

Suffocation from great quantities of fine dust is likely to have been the cause of, or at least a contributing factor in, the deaths of many inhabitants of Pompeii in A.D. 79.

A particular type of damage effected by the Eldfell tephra of 1973 within a short range from the volcano was that coarse-grained tephra covered the grassland slopes that provided the nesting grounds for a great number of purffins. These birds dig deep holes in the ground for their nests, and they made a deplorable sight in the spring of 1973, trying in vain to clear their old holes or dig new ones, their feet bleeding from the rough and prickly surface of the scoria and lapilli.

The damage of tephra falls in the medium-distance range is mostly confined to more or less severe spoiling of grasslands and cultivated land, which in extreme cases may result in complete devastation. This type of damage only rarely extends outside the medium-distance range, 500 km. From this it follows that away from the volcano, in the medium-distance range, and occasionally in the far range, the fall of tephra mostly affects rural areas and settlements.

What, then, is the relation between the thickness of a tephra layer and the degree and duration of the damage caused to the rural areas affected?

As will be shown below by some selected examples, research in tephro-chronology has made it possible to study the relation between the thickness and distribution of individual tephra layers and the damage they caused to rural settlements, as recorded in written sources or revealed by archaeological studies. It should be kept in mind that until the late eighteenth century almost all settlements in Iceland were entirely rural and, with very few exceptions, the farmsteads were not clustered in hamlets but isolated and rather far from one another. Icelandic farming has depended almost entirely on the raising of livestock: sheep, goats, cattle, and horses.

## The Eruption of Hekla in 1104

After a repose of at least 250 years Hekla erupted in 1104, probably in the autumn. Chronicles from the late thirteenth century, which can be regarded as reliable, state that it was "the first coming up of fire in Mount Hekla," and mention the year 1104. This is all that is known about this eruption from written sources, but comprehensive excavations of farmsteads buried by the tephra, together with detailed mapping of the tephra layer (see Figure 5.6), have added greatly to our knowledge of this eruption and its consequences (Thorarinsson 1967:23–38). The eruption was wholly explosive, producing dacite tephra that was carried mainly to the north. Within the 0.2-cm isopach of the layer in its compressed state the tephra covers more than half of the country, or 55,000 km². The finest ash from this eruption has been identified in peat bogs in Norway and Sweden (Persson 1966, 1967). The total volume of the freshly fallen tephra is estimated at $2.0 \times 10^9$ m³, of which $1.4 \times 10^9$ m³ fell on land (Larsen and Thorarinsson 1978:41). Probably the eruption commenced in late autumn, and the main tephra-producing phase, giving vent to the bulk of the tephra, almost certainly lasted less than 1 day. The tephra fall laid waste three settlements in South Iceland that were situated near the axis of maximum thickness of the tephra layer. These were the valley of Thjórsár-dalur, 13–22 km distant from Hekla, with at least 10 farms; four or five farms on the inland plain 50 km away; and a small holding at Lake Hvítárvatn 70 km away. No farm within approximately the 10-cm isopach of the compressed layer was ever built again. It should be kept in mind that the settlement on the inland plateau was at about 350 m above sea level, which is an unusually high elevation for settlements in Iceland, and would probably have been abandoned later on without the tephra fall of 1104. The Thjórsárdalur valley, on the other hand, with all its farms below 200 m elevation, was certainly well suited for the Icelandic type of farming. The area of that valley that was never resettled is situated mainly within the 25-cm isopach of freshly fallen tephra.

### The Eruption of Hekla in 1693

We possess trustworthy and detailed descriptions of Hekla's eleventh historic eruption, which commenced in the early evening of 13 February 1693 and lasted at least 7 months. These sources have been supplemented with a study of the tephra layer (see Figure 5.7). The eruption was a mixed one, producing besides tephra a considerable volume of lava. The initial phase of the eruption was very violent, and the uprush of tephra during the first hour or so must have averaged about 60,000 m³/sec. The tephra is andesitic and was carried northwestward.

Detailed and reliable written sources are also preserved concerning the damage caused by the eruption. Among these sources is a farm register made in 1695 and a very detailed land register made in 1709. From the contemporary descriptions we can conclude that the 1693 tephra layer was, in the districts near Hekla, almost three times thicker than is shown in the isopach map here presented. The total volume of the 1693 tephra as freshly fallen is roughly calculated at 300 million m³, of which 220 million m³ fell on land. Fine ash was carried as far as Norway. In Iceland the tephra covers about 22,000 km². Five

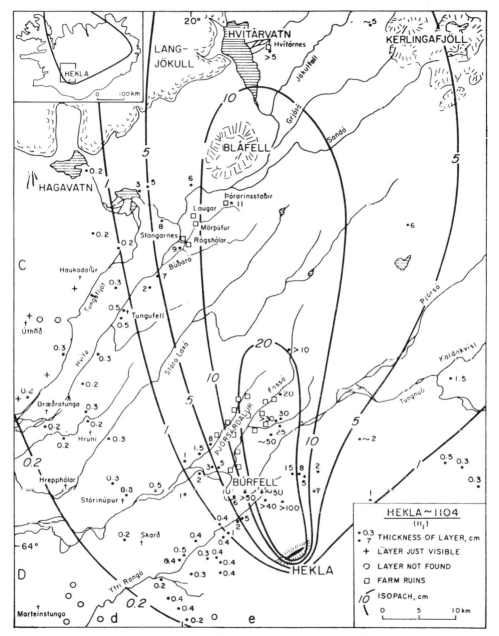

**FIGURE 5.6.** The tephra layer of Hekla in 1104, proximal part. The map in the upper left corner shows the outlines of the entire layer on land. (From Thorarinsson 1967.)

135

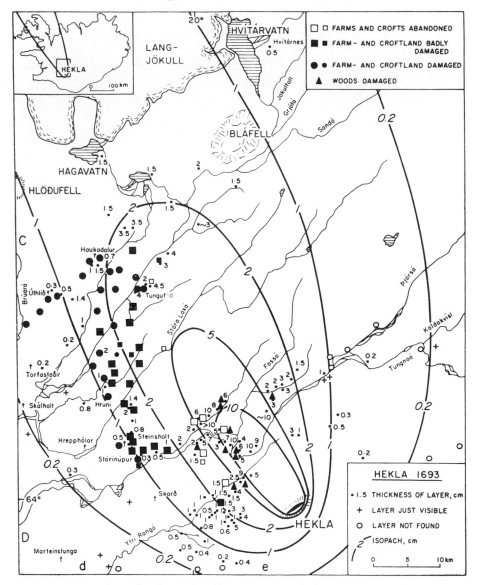

**FIGURE 5.7. The tephra layer of Hekla in 1693, proximal part. The small map shows the outlines of the entire layer on land. (From Thorarinsson 1967.)**

farms and two small holdings were abandoned for a year or more; one of the farms, which was covered by a layer probably about 25 cm thick as freshly fallen, was never resettled. Two of the farms, abandoned for 2 and 4 years, respectively, were covered by a layer initially at least 15 cm thick, and two farms, abandoned for about a year, were covered by a layer somewhat less than 10 cm thick. There was also considerable damage to meadows and cultivated fields of about 50 more farms situated up to 50 km from Hekla and mostly within about the 3-cm isopach of the freshly fallen tephra layer. Outside that area some farmland suffered minor damages, presumably mainly because of fluoride contamination which will be discussed later on in this chapter.

Contemporary chroniclers tell us that ptarmigan not only died in droves in the districts where grassland was damaged, but also lay dead in heaps in the northern part of the district of Borgarfjördur, 120 km away from Hekla. Trout died in large numbers in lakes and rivers 120–180 km from the volcano, where the tephra was sandy and the thickness of the layer may have been 1–3 cm or so (Thorarinsson 1967:88–105).

## The Eruption of Askja in 1875

Above the inland desert plateau of North Iceland rises the volcanic Dyngjufjöll massif with the Askja caldera. No explosive eruption is known to have taken place in the caldera in historic times. In February 1874 solfatara activity seemingly increased in the caldera. Possibly there was a small eruption in its southwest corner in early January 1875, but after that nothing noteworthy is known to have happened until Easter Sunday, 28 March. About 9 p.m. that day a dark cloud was seen rising from Askja and at about 3 a.m. on 29 March a grand scale eruption commenced. During the following 8.5 hours or so a tremendous amount of rhyolite tephra ($SiO_2$ 70%) was hurled up, presumably mainly from fissures that opened up in the south part of the caldera. The tephra was carried eastward and spread over a sector of 9.500 km² in East Iceland. After the tephra fall that sector looked like a grayish-white desert. The volume of the freshly fallen tephra on land was $0.7 \times 10^9 m^3$.

Between 8 and 10 a.m. on 29 March that is, about 15 hours after the eruption began to gain momentum, fine-grained sand began to fall on the west coast of Norway, 1140 km from Askja as the crow flies, and at about 11 a.m. on 30 March dust began to fall in Stockholm, about 1860 km from Askja, to such a degree that light from the gas lamps in the city's streets was obscured by it.

The dust fall caused great interest among natural scientists in Scandinavia, since nothing was known about its origin until a vessel arrived from Iceland many weeks later. Thanks to this interest, we possess a map of the tephra sector (Figure 5.8), which is the first one in the world of an extensive tephra layer. The area of the tephra sector on the map is about 650,000 km².

Figure 5.9 is an isopach map of the freshly fallen tephra on land and shows also the effect of the tephra fall on the rural settlements most seriously hit by it. We find that the 18 farms abandoned for a year or more are situated within the 15-cm isopach of the freshly fallen tephra, and that 7 of the farms that were abandoned for 5 years or more are inside, or only a short distance outside, the 20-cm isopach. In addition to the 18 farms abandoned immediately or shortly after the tephra fall, some farms a little outside the 15-cm isopach had to be given up some years later. This abandonment was caused chiefly by soil erosion by wind and water that was greatly enhanced by the coarse-grained Askja tephra, which blew to and fro and ground away the soil-protecting grass cover

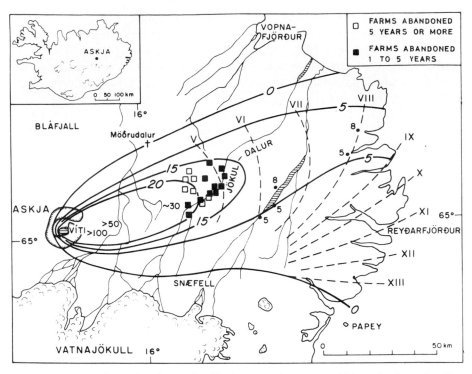

**FIGURE 5.8.** The tephra layer of Askja in 1875, the sector within Iceland. Broken lines: isochrones for the beginning of the tephra fall. (From Thorarinsson 1963.)

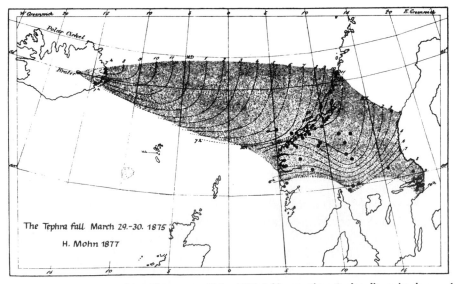

**FIGURE 5.9.** The entire tephra sector of the 1875 Askja eruption. Broken lines: isochrones for the beginning of the tephra fall (Greenwich time). (From Mohn 1877.)

and filled the beds of brooks and rivulets so that they diverted and eroded new ravines and channels.

## Summary

Summing up the experience of the eruptions just described and of other eruptions in Iceland that have been similarly studied, we find that in lowland areas the freshly fallen tephra layer has to be 30–50 cm thick in order to cause abandonment of rural lowland districts lasting decades. In highland districts bordering the uninhabitable interior of the country, 20-cm thick tephra has proved sufficient to cause long-lasting abandonment. A thickness of 15 cm or more was needed to cause 1–5 years abandonment of lowland farms, and, with a few exceptions, 8–10 cm or even more has been required to cause abandonment of shorter duration than a year (Thorarinsson 1971).

When the thickness of the freshly fallen layer is less than 1.0–0.5 cm, the damage to vegetation by the tephra as such is insignificant apart from that caused by toxic, water-soluble gas compounds adhering to the tephra. Taking our entire planet into consideration, we find that very rarely in the last few thousand years has the 0.5 isopach of tephra layers extended outside a 1000-km distance from the source. One of the very biggest explosive eruptions on earth during the last 4000 years was the Minoan eruption of the Santorini volcano in the fifteenth century B.C. According to the most recent isopach maps of the Minoan tephra (Watkins et al. 1978), the 1-cm isopach of the freshly fallen tephra extends about 800 km in the wind direction from the volcano, and the distance to the 0.5-cm isopach may have been little more than 1000 km. When Volcán Quizapú in the Chilean Andes erupted in April 1932, ash fell to a thickness of about 2 cm in the streets of Buenos Aires, about 1150 km downwind from the volcano (Larsson 1937).

Among other effects of tephra falls experienced in Iceland are injuries to the eyes and respiratory organs. It also happens that grazing animals are killed when eating ash-laden vegetation, although the tephra is not contaminated by toxic substances.

Another danger associated with tephra falls is lightning. It can occur very often locally but may in big eruptions occur outside—occasionally far outside—short range. The effects are the same as in severe, purely meteorological thunderstorms: the ignition of buildings and occasional killing of animals and people.

Still another effect may be mentioned. During the Plinian initial phase of the Hekla eruption that commenced on 29 March 1947, the tephra was carried toward the south. After the tephra fall the fishing grounds off the coast within the tephra sector were more or less covered by floating tephra, which subsequently drifted westward with the sea currents. As an immediate consequence of the tephra fall, cod disappeared completely from the area and were not caught there until about 2 days later. It seems that the fish fled temporarily from the tephra sector.

In this connection it is worth mentioning that ecological disturbances caused by tephra or gases can have effects outside the otherwise damaging range. Macdonald (1972:412) mentions a good example of this: In the region west of the Paricutín volcano in Mexico, which erupted in 1943–1952, sugar cane was badly affected, not directly by the tephra but by an infestation of caneborers that developed as a result of destruction by the widespread ash of another insect that normally preyed on the borers and kept them under control.

The damage to vegetation caused by tephra layers as such not only depends on the thickness of the layer but is also greatly influenced by the time of year when the tephra fall takes place and the weather conditions during or following the eruption. A thin layer of tephra deposited during the growth season may do much more harm than a thicker layer deposited in other seasons. The chemical composition of the tephra can influence the duration of the damage. Under equal conditions silicic glass is more resistant to chemical weathering than more basaltic glass, and also poorer in some elements such as calcium. Tephra layers may also lack trace elements vital for grazing animals, such as cobalt.

On the other hand, it is known from not a few eruptions in Iceland and elsewhere (cf. Wilcox 1959), that tephra layers, especially basic ones, such as the layers from the Katla volcano, have improved grass growth because of beneficial chemical and physical effects of the tephra that is deposited on the soil and later worked into it. A thin layer of black tephra increases the absorption of radiation heat and may contain nutrients such as calcium and potassium.

Since this chapter is based almost solely on experience in Iceland, it should be kept in mind that experience in one country is not applicable to a different environment without some reservations. However, the Icelandic experience ought to be applicable without reservations to other cold-temperate areas such as Alaska and Kamchatka.

As an example of the effects of a large tephra fall on quite another environment we can note the large explosive eruption of Santa Maria in Guatemala from 24 October to 6 November 1902, thoroughly described by Sapper (1905: 101–153). The magnitude of that eruption was more than the double of the above-described eruption of Hekla 1104; the calculated volume of tephra, already somewhat compressed, was $5.5 \times 10^9$ m³. The tephra layer covered about 270,000 km² and the bulk of the tephra was produced in the first 2 days. The explosive activity was extremely violent. Fist-size lumps of pumice fell 75 km from the volcano and xenoliths broke through roofs of corrugated iron in a village 14 km distant downwind. Between 500 and 1000 people were killed in houses that collapsed under the weight of the tephra, but many were also killed by lightning and by bombs and stones hurled from the volcano. In one area nearly all big trees were destroyed by lightning. Where the thickness of the tephra layer did not exceed 1 m or so only one rainy season was needed for the coffee bushes to recovery. Grassland was rather seriously damaged where the thickness of the layer exceeded 15 cm but had recovered to a considerable extent within a year, and where the tephra layer was very thin it proved beneficial for the vegetation.

## VOLCANIC GASES

Every magma contains an unspecified amount of volatiles in solution. When separated from the magma during eruptions the dissolved volatiles form volcanic gas. To sample this gas without contamination is extremely difficult. Hitherto only basaltic lavas have been satisfactorily sampled for escaping gases.

The Surtsey eruption of 1963–1967 provided unusually good opportunities to sample gases, and some of the samples obtained there are among the least contaminated ever obtained. Analyses of three samples obtained from the lava crater on 21 February 1965 are presented in Table 5.2.

Not only does the content and percentage of the individual gases vary from one magma to another, there is also a wide range of values for the individual components in most sets of analyses. Furthermore, there appears to be some spontaneous fluctuation in the chemistry of emitted gases. With all these reservations the Surtsey gas analyses presented in Table 5.2 are probably a fairly representative example of the volatiles released from basalt magmas.

Knowing the water content in the magma and how much of that water is retained in the lava, one can estimate from analyses such as those presented in Table 5.2 the amount of the individual gas species that were released. The water content of basalt magmas appears to be between 0.45 and 0.74%, and about 0.1% remains in the magma.

In these analyses fluorine is left out. The content of fluorine has, however, been measured in some Icelandic lavas and tephras. The following figures for the ppm of fluorine were obtained at the Nordic Volcanological Institute in Reykjavík (Sigvaldason pers. comm.): Hekla, 1970: 1200; Eldfell (Heimaey), 1973:700; Katla: 450; Askja, 1961: 400; Surtsey: 370. How much of the fluorine content of the magma is retained in the lava and tephra is not known with certainty, but it is likely to be little below one-third (Bailey 1977).

The volatiles spread in mainly three ways from the volcanic vent: (a) as

**TABLE 5.2**
**Chemical Analyses of Surtsey Gases in Moles Percent (Except S)**[a]

| Sample no. | $H_2O$ | HCl | $SO_2$ | $Co_2$ | $H_2$ | Co | $O_2$ | $H_2 + Ar$ |
|---|---|---|---|---|---|---|---|---|
| 17 | 86.16 | 0.40 | 3.28 | 4.97 | 4.74 | 0.38 | 0.00 | 0.07 |
| 22 | 86.16 | 0.40 | 1.84 | 6.47 | 4.70 | 0.36 | 0.00 | 0.07 |
| 24 | 86.13 | 0.43 | 2.86 | 5.54 | 4.58 | 0.39 | 0.00 | 0.07 |

[a]From Sigvaldason and Elísson (1968:802).

air-transported acid aerosols; (*b*) as compounds adsorbed on tephra; and (*c*) as microscopic salt particles. Óskarsson (1978) found that the adsorption of the water-soluble compounds in general is controlled by the mode of eruption and other environmental factors rather than by the composition of the gas phase. The microscopic salt particles of the halogens and sulfate are spread mainly as a suspended load and so are to a great extent also the acid aerosols of the upper part of the eruption column, whereas from the middle section of the eruption column the acid gases are transported by tephra particles bound to their glassy surfaces by chemiadsorption. The higher the grade of explosivity, the higher is the percentage likely to be transported by the tephra (Óskarsson 1978). Rose (1977) studied the scavenging of volcanic aerosol by the tephra from the 1974 eruption of Fuego in Guatemala. The initial concentration of S (1600 ppm) and Cl (1100 ppm) in the magma was inferred from microprobe analyses of trapped glass inclusions in phenocrysts. During the explosive eruption 33% of the S and 17% of the Cl fell back to earth as acid aerosols on the tephra, 5% of the S and 20% of the Cl were trapped in the tephra, and the remaining S and Cl were released to the atmosphere.

The noxious compounds of volcanic gases are mainly compounds of F, S, and C, the Cl level in the atmosphere being so high that an addition of volcanic Cl does not cause an appreciable change. The oxidation of $SO_2$ to $SO_3$ and $H_2SO_4$ is very rapid, so that aerosol of S is mainly $H_2SO_4$. Cadle *et al.* (1969) found that 95% of the aerosols produced during the 1967 Kilauea summit eruption were $H_2SO_4$ droplets.

The danger from drifting volcanic gases is commonly a short-range one. Farther away damage is mostly caused by the aerosols transported by tephra. The acid sulfur and fluorine compounds affect chiefly the vegetation and through it grazing animals. In the immediate vicinity of the volcano, however, the fumes can easily become so dangerous to inhale that gas masks are needed.

Volcanic gas released by subglacial eruptions and solfatara activity, being unable to escape through the ice, is transported in the hlaupwater to its outlets at the glacier margin and thereafter escapes into the air. The $H_2S$ stench from jökulhlaups in Iceland is occasionally noticed all over the country. Silver turns black in districts bordering the floods. A trustworthy account of a large jökulhlaup from Grímsvötn in late May 1861 states that birds in the vicinity of the flood were killed in masses, mainly "ducks, geese, golden plovers, whimbrels and arctic skuas." The stench from a much smaller jökulhlaup from Grímsvötn in July 1954 was so repulsive that the present writer felt sick in an airplane flying over the flood. Grass along the floodbed withered and mountain ashes on nearby slopes shed their leaves. In a short valley opening out to the flood plains near the outlets from the glacier, all birds were killed, as were small trout in a rivulet in that valley (Thorarinsson 1954 a). No doubt sulfur compounds, mainly $H_2S$, were responsible for the lethal contamination by the jökulhlaup. Some sheep, grazing along the jökulhlaup from Grímsvötn in 1948 got almost completely blind, but their blindness lasted only a few days (Björnsson 1972).

## Fluorine and Fluorosis

Fluorine is the toxic element that most frequently has caused damage after tephra falls in Iceland. The tephra is then contaminated by fluorine, which is transported by the tephra particles. Óskarsson (1978) found that the chemical adsorption of soluble fluorine on the glassy surface of the tephra grains occurs at a temperature below 600°C in the eruption column, and he also found by x-ray diffraction analysis that the fluorine compound on the tephra grains is calcium fluorsilicate.

As the fluorine adheres to the surface of the grains, fine-grained tephra is likely to transport more fluorine than a coarse-grained one. The deposition per unit of tephra-transported fluorine thus depends on both the particle size of the tephra and the thickness of a layer, and reaches its maximum at a distance from the volcano determined by the velocity of the transporting medium (Óskarsson 1978). Most dangerous for the grazing animals is a layer of fluorine-contaminated tephra that is so thin that it does not hinder grazing and so fine-grained that the tephra particles stick easily to the vegetation and thus get into the digestive organs of the animals. Therefore, fluorosis in grazing animals is usually more severe farther away from the erupting volcano.

Experimental feeding of sheep with various amounts of fluorine-contaminated grass showed that a mild degree of chronic fluorosis was produced in the animals receiving 20–40 mg fluorine daily for 6 months. It was also shown that if the fluorine content exceeds 250 ppm of the dry content of grass, it causes fluorine poisoning that can kill the animals in 2–3 days (Sigurdsson and Pálsson 1957). Early symptoms of fluorosis in animals are dental lesions and damage to the membranes of the joints.

The earliest description of dental lesions in Iceland, now known to have been due to fluorine toxication, is to be found in a contemporary account of the aforementioned Hekla eruption that commenced on 13 February 1693. The description was written in 1694 by a farmer and chronicler, Oddur Eiríksson, living in the Borgarfjördur district about 120 km west of Hekla, and probably the chronicler describes his experience of the disease in that district, where the thickness of the tephra layer hardly exceeded 3–4 mm. The farmer writes about the effects of the eruption:

> In the following autumn and winter people noticed that on the teeth of grazing sheep were yellow spots and some black ones; in some animals the teeth were all black; the teeth fell out in some cases, but small, round-pointed teeth came up afresh, like the teeth of a dog or a catfish; where the spots came the tooth turned soft so it could be shaved like wood. In some animals the flesh peeled away from around the front teeth and molars. . . . People thought that this was due to the sand-fall from Hekla [Thorarinsson 1967:103–104].

Another chronicler, the Reverend Jón Halldórsson, adds that young cattle and young horses also got yellow spots on their teeth and that horses and sheep that shed teeth that year got bluish black streaks over any new teeth they had; because of that many called them ash-teeth, and their age could be told as long as they lived (Thorarinsson 1967:103–104).

Hannes Finnsson (1796:143–144) was the first to describe the outgrowth on molars known in Iceland as *gaddur* (spike). These outgrowths make it impossible for the animals to ruminate and masticate. Icelandic farmers have invented a special type of pliers to break off these outgrowths in sheep and horses.

Many sheep and about 40 cows were killed by fluorosis after the Hekla eruption that began on 2 September 1845, and horses, especially yearlings, were affected too (Thorarinsson 1967:147–148).

K. Roholm studied a collection of sheep bones. Both mandibles and extremities, of adult sheep affected by fluorine during the 1845 Hekla eruption. He found that these bones were covered with a characteristic coating of a porous and brittle, osseous tissue. In severe cases the thickness of the bones was twice the normal (Roholm 1937:13). Such deformations, as well as the above-mentioned prominences (*gaddur*) on the molars, ought to be easily detected archaeologically.

After the tephra fall of the initial phase of the 1947 Hekla eruption on 29 March, tephra and water were sampled for testing of fluorine content, as fluorine contamination was feared in the rural districts south of the volcano, which were blanketed by a 1–10-cm-thick layer of coarse-grained (1–10 mm diam.) rhyodacitic tephra. The tephra was found to contain 70–110 ppm fluorine. The fluorine content in the water, normally very low in Iceland, or about 0.2 ppm, was after the tephra fall 1.0–9.5 ppm in surface streams. In late May the content had greatly decreased and no symptoms of fluorosis were observed (Stefánsson and Sigurjónsson 1957). In May and June sandy, andesitic ash fell now and then in settled districts west of Hekla. The aggregate thickness was on the whole less than 1 cm. Not long after the beginning of these ash falls, signs of fluorisis began to appear in sheep. On a few farms near Hekla a high percentage of sheep were affected, the lambs more seriously than the ewes, but most of them recovered. The fluorine content of the grass averaged 25 ppm of dry matter. Heavy rainfalls later in the summer washed most of the fluorine away (Sigurdsson and Pálsson 1957).

The eruption in Iceland that has been studied in greatest detail with regard to its fluorine transport is the Hekla eruption of 1970 (Thorarinsson 1970; Thorarinsson and Sigvaldason 1972; Georgsson and Pétursson 1972; Óskarsson 1978).

This eruption began at 9:23 p.m. on 5 May 1970. Nearly all the tephra produced was expelled during its initial phase, which lasted a good 2 hours. At 10:10 p.m. the tephra vapor column reached its maximum height, 16,000 m.

The tephra production during the initial phase averaged 10,000 $m^3$/sec, and the total production amounted to $7 \times 10^7$ $m^3$, or $4.5 \times 10^7$ tons.

The tephra of the initial phase was carried toward NNW with an average speed of 74 km/hr and reached the north coast of Iceland, 180 km distant from Hekla, at midnight. Figure 5.10 is an isopach map of the tephra layer.

The particle size distribution pattern studied by N. Óskarsson (1978) suggests that the tephra was transported mainly as fallout and that there was only a minor amount of suspended load, namely, material that was so fine-grained that it followed the turbulent motion of the transporting air masses.

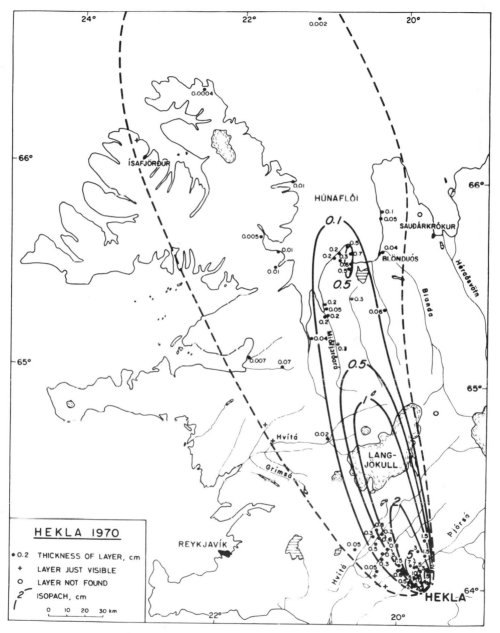

FIGURE 5.10. The tephra layer of Hekla in 1970. (From Thorarinsson 1970.)

The tephra that fell on farmlands near the north coast at first was fine sandy but later mainly silty. Its specific weight was 0.8. Some people in the north thought that they sensed an unpleasant smell associated with the ash-fall, and others felt discomfort in the chest. When the ash fell on quiet pools of water it seemed to form a film on the surface. In South Iceland the grain size was mainly that of coarse sand, specific weight 0.6. Chemically the tephra was basaltic andesite.

This tephra proved very rich in fluorine. Analyses of samples taken on the first days after the tephra fall showed up to 2000 ppm of water-soluble fluorine in South Iceland and 1400 ppm in North Iceland. Within 2 weeks, the fluorine content of the ash had decreased to about 10% of the original value and within 3 weeks to between 1 and 2%. During this period it rained a lot.

On the first days after the tephra fall samples of vegetation on farm land in the south were taken for fluorine analysis. The highest initial value, in a sample taken on 7 May at a place 38 km from Hekla, where the tephra layer was 1 cm thick, was 4300 ppm of dry matter. Samples taken near the north coast on 18 and 19 May revealed a fluorine content of 350–750 ppm of dry matter. The fluorine content of the grass fell exponentially with time, partly because of heavy rainfalls. Within 35–40 days it had dropped to less than 30 ppm fluorine—a value that is generally accepted as innocuous.

Noxious levels of fluorine thus prevailed for a maximum of 5–6 weeks in contaminated districts, but exposure of sheep to the fluorine varied greatly because of differences in sheep management. Some farmers kept their sheep away from the contamination for the first 3 weeks.

Signs of chronic fluorosis were found within a thickness of the layer as low as 1 mm, but acute sickness occurred where the layer was about 0.5 mm and would escape field observations in soil sections. In farming areas within the 1-mm isopach acute fluorosis killed about 3% of the sheep and 8–9% of the lambs, a total of 7000–8000 animals. Some farmers lost more than one-quarter of their lambs. Deaths began to occur in the first days after the tephra fall as a result of convulsive seizures, pulmonary edema, and kidney and liver changes. The morbidity percentage was highest in the districts in the north, 150–180 km from Hekla. In the bones of the lambs fluorine concentrations increased about fourfold the normal value, in adult sheep about 50%.

Dental fluorosis in sheep occurred in 25% of the third incisors, which erupted 9–13 months after the tephra fall. Severe dental lesions were found in sheep that had been exposed to fluorine only in the first 2 days after the tephra fall.

Fluorosis was also considered to have been more or less responsible for the deaths of some mares and foals. The cattle, that spring were not let out in the contaminated areas before July and thus escaped fluorosis.

The Eldfell eruption on Heimaey, 23 January–25 June 1973, produced about $2 \times 10^7$ m³ of Hawaiitic basaltic tephra. In February and March fine-grained ash fell on rural areas on the mainland, forming a 0.5–1.0-mm-thick layer. In Mýrdalur, about 60 km from the volcano, the highest fluorine content measured on 19 February was 3000 ppm. A sample taken on 23 March contained 830 ppm.

Vegetation damages began to appear in late February and increased during March and April. They were first noticed on conifers in gardens, the needles of which gradually became brown all over. Some young pines were gradually killed. Moss was seriously affected and completely killed in some limited areas. Lichens suffered much. Among higher plants heather and crowberry bushes were most seriously affected. Most of the vegetation recovered well during the following summer, and on the whole the vegetation was noticeably luxuriant that summer, presumably to some degree because of the ash.

The highest fluorine content measured in the grass on 27 March was 136 ppm of dry matter. That content had dropped to 7 ppm on 17 June. Since domestic animals were on the whole kept indoors during the period of ash fall and the following months, they were little affected by the pollution (E. H. Einarsson 1974).

It is noteworthy that the eruptions of the subglacial Katla volcano have never been known to cause fluorine pollution. Yet this volcano is one of the most active in Iceland, erupting two times each century, and has in historic times produced more basaltic tephra than any other volcano in Iceland or elsewhere in the world. Furthermore, neither the Surtsey eruption nor the eruptions in the subglacial Grímsvötn caldera are known to have caused fluorine pollution.

The three last-mentioned volcanoes have in common that their eruptions were phreato-magmatic, their explosive activity being due to contact between the magma and water. It is tempting to see some causal relations between this fact and the absence of fluorine pollution, but it should also be kept in mind that the Surtsey lava is low in fluorine (300 ppm) compared with the Eldfell lava (700 ppm), and that the Katla tephra is also rather low in fluorine (450 ppm).

Fluorine spread without the aid of tephra can also affect plants and animals if the emission from the volcano in question is long-lasting. Samples of vegetation taken on the slopes of Etna at between 2450 and 2000 m elevation in the summer of 1976 showed a fluorine concentration of 113–205 ppm $F^-$ of dry matter; the higher concentration was found in the direction of the prevailing wind from the summit crater (Garrec et al. 1977). The accumulation of gaseous hydrofluoric acid by the plants is thought to be affected by the proportion of $SO_2$; it was found by experiments in fumigation chambers (Mandl et al. 1975) that accumulation of fluorine in plant leaves was less in pollution caused by $SO_2 + HF$ than by HF alone, which could be because closure of the stomata was caused by the presence of $SO_2$.

## Emission of Carbon Dioxide

Emission of heavy gases, especially carbon dioxide associated with volcanic activity, can be dangerous and even lethal when they form ponds where animals and people can perish. In Iceland this phenomenon was studied for the first time after the Hekla eruption 1947–1948. This eruption began on 29 March 1947, and lava flowed continuously until 21 April 1948. Two or 3 weeks before it ended, some brooks issuing from underneath old lava flows from the volcano began to precipitate lime. Probably about the same time exhalations of $CO_2$, mixed with atmospheric air, began issuing from old lava fields, and a month later carcasses of suffocated animals began to be found in depressions in these fields, all within an area of about 15 km², 7–12 km from Hekla's lava crater. In all about 100 separate mofettes were gradually found. Their temperature was approximately that of the groundwater.

In completely calm weather, and usually at night, the gas formed ponds up to several meters deep in the depressions. The biggest pond, in Loddavötn, reached a maximum area of about 6000 m². Animals that happened to be in the depressions when the ponds were formed rapidly suffocated (see Figure 5.11), and the grass around the orifices withered. Mice put into the ponds had to be taken out within a minute if they were to survive. Insects such as arthropoda could stand the gas more easily.

The first mammal, a fox, was found killed on 19 May. Nearly a month later the depressions were fenced off and some ponds drained out through ditches, curious work for those unfamiliar with the circumstances to observe. By that time 15 sheep, 19 birds, 1 fox, and a number of insects had been found dead. The mofette activity remained fairly constant during the first 5 months or so. It began to decrease somewhat in September and markedly in October and

FIGURE 5.11. Sheep and redwing suffocated in a pool of carbon dioxide near Hekla in May 1948. (Photo: S. Thorarinsson.)

**TABLE 5.3**
Chemical Analyses of Gas Sampled from Eldfell Lava[a]

| Sample | $H_2O$ | HCl | $SO_2$ | $CO_2$ | $H_2$ | CO | $O_2$ | Na + Ar | $CH_4$ |
|--------|------|-----|------|------|------|------|------|---------|------|
| 1 | 45.0 | — | 5.5 | 9.63 | 19.63 | 0.22 | 4.01 | 19.95 | 0.16 |
| 2 | 63.0 | — | | 17.20 | 18.57 | 0.15 | 0.30 | 0.81 | 0.15 |

[a]Bragi Árnason (personal communication).

November, and ceased altogether in December 1948, but precipitation of lime, although gradually decreasing, still occurred in some brooks a decade later.

The average of analyses of eight samples of the mofette gases gave 24% $CO_2$, 0.85% CO, 13.7% $O_2$, and 62% $N_2$ (Kjartansson 1957:36). The geologist Gudmundur Kjartansson, who studied these phenomena thoroughly, estimated that the aggregate average discharge of the gaseous mixture during the first 6 months was 3 m³/sec as a minimum, and the total volume produced at least 47 million m³, containing 12 million m³, or about 24,000 tons of $CO_2$. Kjartansson interpreted the gas exhalations as having issued from groundwater saturated with $CO_2$ and thus not directly from the magma, although they were primarily magmatic (Kjartansson 1957).

Emission of carbon dioxide with some amount of carbon monoxide was one of the problems that had to be faced by the people who stayed in the town of Vestmannaeyjar on Heimaey during the Eldfell eruption of 23 January–25 June 1973. About the middle of February, gas began to contaminate cellars in the lower part of town. The gas emanation increased rapidly; in calm weather the town was swept in bluish haze, and for the first time the people had reason to praise the windy climate of Vestmannaeyjar. The emanation was very high through March and April, after which it gradually diminished and was practically over in late August.

Sampling of gas from the Eldfell lava proved difficult; Sample 1 in Table 5.3 is obviously contaminated by atmospheric air. Nevertheless, Tables 5.3 and 5.4 prove that the composition of the gas emanating through the ground in town was very different from that of the gas obtained directly from the lava.

**TABLE 5.4**
Examples of the Composition of Gas Collected in Cellars in Vestmannaeyjar in March 1973[a]

| Moles percent | Telegraph station | House No. 4 Grænahlíd |
|---------------|-------------------|------------------------|
| $CO_2$ | 93.3 | 80.0 |
| $H_2$ | 2.6 | 10.6 |
| CO | 0.3 | 0.5 |
| $CH_4$ | 0.3 | 0.6 |
| $N_2$ | 2.5 | 6.8 |
| $O_2$ | 1.0 | 1.7 |

[a]Bragi Árnason (personal communication).

Since CO is rapidly fatal when exceeding 0.4%, and $CO_2$ when exceeding 10%, the gas emanation really involved risk to life. At the end of March, therefore, it was decided to evacuate those parts of town by night where the gas emanation exceeded certain limits and to reduce the day traffic to the minimum possible in those areas. At times it was difficult to drive a car through the gas layer because the motor would hiccup or stall from lack of oxygen. Sometimes the layer of gas was so thin that sparrows lay dead in numbers while gulls and pigeons walked there unconcerned. Occasionally such deep pools formed that lives of pedestrians were endangered. Thanks partly to rigorous precautions and partly to luck, only one person was killed by the gas.

## THE LAKAGÍGAR ERUPTION OF 1783

The Lakagígar eruption in Iceland commenced on Whitsunday, 8 June 1783 and lasted until early February 1784. As is well known, this eruption produced the largest lava flow on earth in historic times (see Figure 5.12). The lava covered 565 km² and its volume was estimated at 12.3 km³. It overran 2 churches and 14 farmsteads, and 30 additional farms were badly damaged. Although the lava flow did not directly cause any loss of life, the effects of the tephra and gases were disastrous.

Thoroddsen (1925) estimates the volume of freshly fallen tephra as $3 \times 10^9$ m³. A more recent estimate, based on measurements on the tephra layer in soil sections, is $8.5 \times 10^8$ m³ (Thorarinsson 1969), and the aggregate volume of lava and tephra calculated as lava is about 12.5 km³. The production of gas must have been enormous. We can get some idea of this by assuming that the percent distribution of volcanic gases was the same as in the Surtsey magma according to the analyses presented earlier. Assuming further that the $H_2O$ content of the Lakagígar magma was 0.5 weight percent and that 0.1% remained in the lava, we get the weight and volume figures for the gas released by the eruption shown in Table 5.5.

These figures are likely to be rather on the low side. Hammer (1977) has concluded from electric conductivity in Greenland ice cores that $H_2SO_4$ aerosols from the eruption in 1783 amounted to $10^8$ (100 million) tons.

Fluorine in the Lakagígar lava has been measured and found to be 320 ppm. Assuming that probably about one-third of the fluorine in the Lakagígar magma was released, the fluorine produced by the eruption amounts to about $5 \times 10^6$ tons.

Most of the tephra produced by the Lakagígar eruption fell on uninhabited areas, and nowhere in rural settlements did the thickness of the freshly fallen layer exceed 3–4 cm—in most settlements it was less than 1 cm. Yet the damaging effect of the tephra and gases was soon seriously felt. Raindrops that fell on the Sída district on 10 June burned holes through dock leaves and dots on the skin of newly shorn sheep. The stench caused by the volcanic gases soon became almost unbearable and birds died in large numbers. Gradually dust

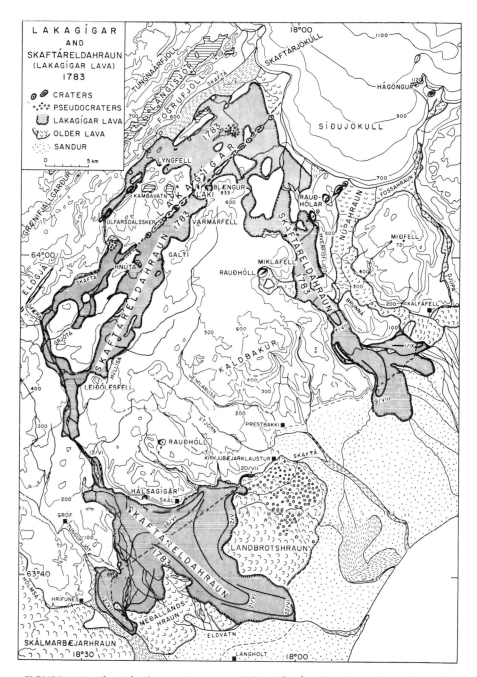

**FIGURE 5.12.** The Lakagígar crater row and the Lakagígar lava flow. (From Thorarinsson 1969.)

TABLE 5.5
Assumed Volume and Weight of Volatiles Released by the Lakagígar Eruption[a]

|        | Volume km³ (0°C 1 atm) | Weight tons |
|--------|------------------------|-------------|
| Lava   | 12.5                   | $3.0 \times 10^{10}$ |
| $H_2O$ | 0.12                   | $1.2 \times 10^{8}$ |
| HCl    | 0.71                   | $1.2 \times 10^{6}$ |
| $SO_2$ | 4.60                   | $1.3 \times 10^{7}$ |
| $CO_2$ | 9.80                   | $1.9 \times 10^{7}$ |
| $H_2$  | 8.20                   | $0.7 \times 10^{6}$ |

[a]From Thorarinsson (1969:12).

and a bluish haze swept over almost the entire country, and the haze remained the whole summer and autumn of 1783. The haze and dust affected vegetation, animals, and people. In remote parts of the country large patches of the cultivated fields turned light yellow and the grass withered down to the roots. Birch trees and shrubs shed their leaves after they had shrunk and turned black, and most of the trees were killed. Some plants were more sensitive to the pollution than others; for instance, clover (*Trifolium repens*) disappeared for 3 years in districts up to 100 km from the volcano. In the greater part of the country, the Iceland moss (*Lichen islandicus*), which at that time was part of the people's diet in many districts, also disappeared. All over the country the grass growth was stunted and large groups of grazing livestock sickened and perished. No fish were caught in the usual fishing grounds between the summer of 1783 and May 1784 (Finnsson 1796:133–140).

A very important account of the eruption and the horrors people and animals suffered, mainly because of the pollution from the tephra and gases, was written by Jón Steingrímsson, a remarkable character and a keen observer of nature, who lived through this terrible eruption as a parson at Prestbakki in the Sída district, one of the districts that became partly covered by the Lakagígar lava. This unique document was written in Icelandic and has never been translated in its entirety, but a portion of it (from Steingrímsson 1915–1917) is here quoted in translation by Jóhann S. Hannesson. Steingrímsson writes:

> The pestilential effect of the fire caused death and damage to horses, sheep, and cattle after the following fashion. The horses lost all flesh; on some the hide rotted all along the back; manes and tails decayed and came off at a sharp pull. Knotty growths appeared about the joints, particularly around the fetlock. The head swelled inordinately, whereupon followed a paralysis of the jaw, so that the beasts could not graze or feed, for what they were able to chew dropped from their mouths again. The entrails corrupted; the bones withered, quite drained of marrow. Some survived through being in good time looped[1] all around the head and as far back as the shoulders.

[1]"Looping" consisted in threading a seton (twist of horsehair or woolen yarn) through a fold pinched in the hide and tying the ends together to form a loop. The purpose was to promote drainage of morbid humors.

Sheep suffered yet more grievous harm; there was hardly a member but knotted, particularly the jaws, so that the knots pushed out through the skin close to the bone. Brisket, hips, and legs—around those parts large bony excrescences developed, which caused the leg bones to curve or else deformed them in various ways. Bones and knuckles, as soft as if chewed, lungs, liver and heart, in some swollen, in some shrivelled, the entrails rotten and soft, full of sand and worms. The shred of flesh that remained was after the same fashion. What passed for meat was both rank and bitter, and thereto full of strong poison, wherefore the eating of it proved the death of many a man, notwithstanding that people tried to cure it, clean it and salt it, according as their skill and means permitted.

Cattle, too, were subjected to this same plague, large growths formed on jaws and collarbones; the leg-bones sometimes split, sometimes knotted variously, with knobs as large as a man could grip with both hands. Of the hips and other joints the same was true; they became deformed, and then knitted, and so lost all pliancy. The tail fell off with the tuft of hair, sometimes half of it, sometimes less. The hooves loosened and slipped off, or sometimes split in the middle (a beast becoming footsore was the first sign of the pestilence). The ribs were deformed all along the side and then broke apart in the middle, being unable to support the weight of the beast when it must needs lie on its side. No knucklebone was so hard that it might not easily be shaved with a knife. The hair of the hide fell off in patches; the inward parts were soft, as already said of the sheep, and in many respects unnatural. A few cows that were not overly crippled were saved by pouring back down their throats the milk that could be pulled from them. In this matter one thing was noteworthy, namely that calves thrown during these hard times had excellent marrow in their bones, with but little withering, although the dams had famine-withered marrow in every bone.

Those people who did not have sufficient old and wholesome food throughout this time of pestilence also suffered grievous distress. On their breastbones and ribs, the back of their hands, their insteps, legs and joints, bumps and knots and hard knobs appeared. Their bodies puffed up; the gums swelled and cracked, with sore pains and toothache. The sinews contracted, especially at the hough, which disease is called scorbutus, that is to say, curd- or water-dropsy in its extreme stages. Of this pestilential disease having so severely afflicted any man that his tongue rotted and fell out, I know no instance here or elsewhere, unless there be truth in a certain report to that effect, concerning a man of this parish, who died in the district of Sudurnes, he having often in the past been troubled with a sore throat.

The pestilence affected the inward powers and parts, causing weakness, shortness of breath, throbbing of the heart, excessive discharge of urine and enfeeblement of the parts thereto pertaining, from which ensued diarrhoea, dysentery and worms in the intestines, sore boils on neck and thighs, and particular loss of hair in many both young and old. This sickness and the mortality that followed in its train proceeded, in my judgment, from insalubrious air, overmuch drinking of water, and unwholesome food, such as was the flesh of the plague-stricken beasts and also the barley which the person then in authority was so forward in proffering; in which matter the observed facts speak for themselves, that people died having store of barley, but survived wondrously if they had plenty of rye [pp. 37–38].

The Lakagígar eruption proved truly disastrous for the Icelanders. Table 5.6 speaks for itself on this, although it should be kept in mind that severe weather conditions in North Iceland are likely to have had a share in the decimation of the livestock. Since the population at that time was entirely

**TABLE 5.6**
**Livestock in Iceland before and after the Lakagígar Eruption**[a]

|          | Number 1783 | Number 1784 | Perished 1783–1784 | Percentage |
|----------|-------------|-------------|--------------------|------------|
| Cattle   | 20,067      | 9804        | 10,263             | 50.1       |
| Sheep    | 236,251     | 49,613      | 186,638            | 79.0       |
| Horses   | 35,939      | 8683        | 27,256             | 75.9       |

[a]From Ólafsson (1861).

rural, basing its livelihood on the livestock, the eruption resulted in a famine that, together with the various diseases that afflicted the people, decimated the population from 48,884 in 1783 to 38,363 in 1786, a reduction of 10,521, or 21.6%. In the rural districts near the volcano the death toll was 37.4% (Ólafsson 1861). The famine following the 1783 eruption is still referred to as the Haze Famine.

## Long-range and Global Effects of the Lakagígar Eruption

The effects of the tephra and gases emitted during the Lakagígar eruption were not limited to Iceland. So much ash fell on the Faroe Islands that grasslands were blackened. In Caithness in Scotland the crop was spoiled by ash-fall and long afterward people talked about the year 1783 as "the year of the ashie" (Geikie 1893:217). In Kent in England it was observed that rain from dust clouds killed insects and plants and visibly affected shining surfaces of copper vessels. In some places in Norway rain blackened the grass and scorched leaves on the trees. Some dust fell in Holland and in northern Italy.

During the latter half of June 1783 a bluish gray haze or mist spread over nearly the whole of continental Europe. In many places the air was hazy for up to 4 months and the sun was red until January 1784. In England the sun remained copper-colored until it had risen 30° above the horizon, and in Languedoc in southern France the haze was at times so dense that the sun was invisible below 17° above the horizon. In Pavia, northern Italy, it was so dim at day that the sun could easily be looked at. Sulfuric stench pervaded many of the areas where the haze was observed.

Benjamin Franklin, then living in Paris as the first plenipotentiary of the United States, took great interest in this haze. He wrote the following (here quoted from Lamb 1970):

> During several of the summer months of the year 1783, when the effect of the Sun's rays to heat the Earth in these northern regions should have been the greatest, there existed a constant fog over all Europe and great part of North America. This fog was of a permanent nature; it was dry and the rays of the sun seemed to have little effect towards dissipating it, as they easily do to a moist fog. . . . They were indeed rendered so faint in passing through it that, when collected in the focus of a burning glass, they would scarce kindle

brown paper. Of course, their summer effect in heating the Earth was exceedingly diminished. Hence, the surface was early frozen. Hence the first snows remained on it unmelted and. . . . Hence, perhaps the winter of 1783–84 was more severe than any that happened for many years [p. 433].

Franklin thought that this haze was either due to a smoke from consumption by fire, or caused by volcanic activity in Hekla and the new volcanic island that was erupting in Iceland in the spring of 1783. He concludes wisely that "it seems, however, worth the inquiry whether other hard winters recorded in history were preceded by similar permanent and widely extended summer fogs."

About 25 June the haze had reached Moscow. At the end of that month it reached Syria and the Altai Mountains in western Siberia. Haze was also observed in North Africa (Hólm 1784:91–92; Traumüller 1885; Thoroddsen 1914, 1925:62–66).

In 1783, some part, and even the main part, of the haze was attributed to the eruption in Asama on Honshu that same summer. The Asama eruption, however, which began on 9 May and lasted until 5 August, was on so small a scale until 6 July that it could not possibly have caused haze in the western part of the Old World. The total production of lava in the Asama eruption amounted to only about 4% of the Lakagígar lava, and the tephra production of Asama, tephra flows included, was about 20% of the Lakagígar tephra (Aramaki 1957). Even if the figures for the tephra production are probably on the low side (cf. Walker 1977), there is hardly any doubt that the haze in Europe and adjacent parts of Africa and Asia was caused by the Lakagígar eruption, and probably it contributed considerably to the haze in North America as well.

The Lakagígar eruption provides an example of the global effect of an eruption's tephra and gases on weather and climate. That volcanic activity can locally affect weather by causing heavy rainfalls and hailstorms has long been known, and in the period since Benjamin Franklin put the blame for the cold winter of 1783–1784 on volcanic activity in Iceland the previous summer, much has been written about the influence of volcanic eruptions on climate, mainly on temperature. Opinions still differ widely, from those that regard this influence as of little importance to those that see it as a possible cause of the Quaternary glacials.

Volcanic eruptions can influence the temperature mainly in two counteracting ways. One is by the emission of $CO_2$, the increase of which has a warming effect, the so-called greenhouse effect, on the lower atmosphere, since $CO_2$ is more transparent to incoming short-wave radiation than to outgoing long-wave one. This factor, which can hardly be called a damaging one except where the climate is very hot, is not likely to have affected the temperature appreciably in historic times, since the annual output of $CO_2$ by volcanic activity is almost certainly a few hundred times smaller than the present annual contribution of $CO_2$ to the atmosphere due to combustion of fossil fuels, which in 1956 was calculated to be $6 \times 10^9$ tons (Plass 1956, cf. also Cadle 1975 and Cadle et al. 1976). This is about 300 times the amount likely to have been produced by the Lakagígar eruption. And in spite of the increase of $CO_2$ in the atmosphere through human activities since the beginning of the industrial era, and especially in the twentieth century, there appears to have been a global lowering of temperature the last 2–3 decades.

The other, and presumably more important, effect is the temperature-lowering effect of volcanic dust and aerosol forming dust veils in the atmosphere that reduce the direct sun radiation. H.H. Lamb, a great authority on these problems, estimated that 1.3°C lowering of temperature occurred after the Lakagígar and Asama eruptions in 1783, and he maintains that temperature lowering in middle latitudes is 0.5–1.9°C in the year after a great eruption (Lamb 1972:421).

Far from denying that big eruptions influence not only the local weather by causing heavy rainfalls, hailstorms, etc., but also the climate on a global scale, the present writer feels that a more reliable quantitative knowledge of volcanic productivity through at least the last millennium is needed in order to evaluate with satisfactory exactness its influence on temperature variations. Many old estimates and calculations of the volume of tephra produced by various historical eruptions need a thorough revision.

Cores from the ice fields of Greenland and Antarctica provide the most exact dating of temperature variations through the last millennia, since the annual layers can be identified and the temperatures calculated from the $^{16}O$ : $^{18}O$ ratio. Furthermore, the production of volcanic aerosols is registered in these cores by variations in their electrical conductivity (Hammer 1977). Further studies of these ice cores seem at the moment to be the most promising way to increase our knowledge about the relation between volcanic activity and temperature variations. More tephrochronological studies, not only on land but also in sediment cores from the ocean floors, are also needed.

## REFERENCES

Aramaki, S.
    1957   The 1783 activity of Asama Volcano. Part II. *Japanese Journal of Geology and Geography* 28:11–33.
Bailey, J. C.
    1977   Fluorine in granitic rocks and melts: A review. *Chemical Geology* 19:1–42.
Björnsson, S.
    1972   Blinda í fé af völdum Skeidarárhlaups [Blindness in sheep caused by a jökulhlaup in Skeidará]. *Jökull* 22:95.
Cadle, R. D., A. L. Lazrus and J. P. Shedlovsky
    1969   Comparison of particles in the fume from eruptions of Kilauea, Mayon, and Arenal volcanoes. *Journal of Geophysical Research* 74:3372–3378.
    1975   Volcanic emission of halides and sulfur compounds to the troposphere and stratosphere. *Journal of Geophysical Research* 80:1650–1652.
Cadle, R. D., C. S. Kiang, and J.-F. Louis
    1976   The global scale dispersion of the eruption clouds from major volcanic eruptions. *Journal of Geophysical Research* 81:3125–3132.

Einarsson, E. H.
    1974   Áhrif flúors frá Heimaeyjargosinu 1973 [Influence of fluorine from the Heimaey eruption
           in 1973]. *Ársrit Ræktunarfélags Nordurlands*: 96–163.
Einarsson, T.
    1974   *The eruption on Heimaey*. Reykjavík: Heimskringla.
Finnsson, H.
    1796   Um mannfækkun af hallærum á Íslandi [On the decrease in the population of Iceland
           because of famine years]. *Rit thess Konunglega Íslenzka Lærdómslista-Félags* 14:30–226.
Garrec, J. P., A. Lounowski, and R. Plebin
    1977   Study of the influence of volcanic fluoride emissions on the surrounding vegetation.
           *Fluoride* 10:152–156.
Geikie, A.
    1893   *Textbook of geology* (3rd ed.). London: Macmillan.
Georgsson, G., and G. Pétursson
    1972   Fluorosis of sheep caused by the Hekla eruption in 1970. *Fluoride* 2(2):58–66.
Hammer, C. U.
    1977   Past volcanism revealed by Greenland ice sheet impurities. *Nature* 270(5637):482–486.
Hólm, S. M.
    1784   *Vom Erdbrande in Island im Jahre 1783*. Copenhagen: E.G. Prost.
Jónsson, V. Kr., and M. Matthíasson
    1974   Hraunkæling á Heimaey [Chilling of lava on Heimaey]. *Tímarit Verkfrædingafélags
           Íslands* 59:70–83.
Kjartansson, G.
    1951   Water flood and mud flows. In *The eruption of Hekla 1947–1948*, Vol. 2(4), edited by
           Trausti Einarsson, Gudmundur Kjartansson and Sigurdur Thorarinsson. Reykjavík:
           Societas Scientiarum Islandica.
    1957   Some secondary effects of the Hekla eruption. In *The eruption of Hekla 1947–1948*, Vol.
           3(1), edited by Trausti Einarsson, Gudmundur Kjartansson and Sigurdur Thorarinsson.
           Reykjavík: Societas Scientiarum Islandica.
Lamb, H. H.
    1970   Volcanic dust in the atmosphere; with a chronology and assessment of its meteorological
           significance. *Philosophical Transactions of the Royal Society of London. A. Mathematical
           and physical series*. 266 (1178):425–533.
    1972   *Climate. Present, past and future*. London: Methuen.
Larsen, G., and S. Thorarinsson
    1976   $H_4$ and other acid Hekla tephra layers. *Jökull* 27:28–46.
Larsson, W.
    1937   Vulkanische Asche vom Ausbruch des chilenischen Vulkans Quizapu (1932) in Argentina
           gesammelt. *Bulletin of the Geological Institute of Uppsala* 26:27–52.
Macdonald, G. A.
    1972   *Volcanoes*. Englewood Cliffs: Prentice-Hall.
Mandl, R.H., L.H. Weinstein, and M. Keveny
    1975   Effects of hydrogen fluoride and sulphur dioxide alone and in combination, on several
           species of plants. Environmental Pollution 9:133–143.
Mohn, M.
    1877   Askeregnen den 29de-30te Marts 1875. *Videnskabernes Selskabs Forhandlinger* 1877 No.
           10.
Moore, J. G.
    1967   Base surge in recent volcanic eruptions. *Bulletin Volcanologique* 30:337–363.
Morimoto, R.
    1960   Submarine eruption of the Myojin reef. *Bulletin Volcanologique* 23:151–160.
Ólafsson, A.
    1861   Um búnadarhagi Íslendinga [On farming conditions in Iceland]. In *Skýrslur um landshagi
           á Íslandi*, Volume 2. Reykjavik: Hid íslenzka bókmenntafélag.

Óskarsson, N.
  1978   Notes on fluorine adhering to tephra of the 1970 Hekla eruption. Preprint issued by the
         Nordic Volcanological Institute, Reykjavík.
Persson, Chr.
  1966   Försök till tefrokronologisk datering av några svenska torvmossar. Geologiska Föreningens
         i Stockholm Förhandlingar 89:181–197.
  1967   Försök till tefrokronologisk datering i tre norska myrar. Geologiska Föreningens i Stock-
         holm Förhandlingar 89:181–197.
Plass, G. N.
  1956   The carbon dioxide theory of climatic change. Tellus 8:140–154.
Roholm, K.
  1937   Fluorine intoxication. A clinical-hygienic study. Copenhagen: Nyt Nordisk Forlag, Ar-
         nold Busch.
Rose, W. I., Jr.
  1977   Scavenging of volcanic aerosol by ash: Atmospheric and volcanological implications.
         Geology 5:621-624.
Sapper, K.
  1905   In den Vulkangebieten Mittelamerikas und Westindiens. Stuttgart: Verlag der E.
         Schweizerbartschen Verlagsbuchhandlung.
Sigurdsson, B. and P. A. Pálsson
  1957   Fluorosis of farm animals during the Hekla eruption of 1947–1948. In The eruption of
         Hekla 1947–1948, Part 3(3), edited by Trausti Einarsson, Gudmundur Kjartansson and
         Sigurdur Thorarinsson. Reykjavík: Societas Scientiarum Islandica.
Sigvaldason, G. E.
  1974   Chemical composition of volcanic gases. In Physical volcanology. Editors L. Civetta et al.
         Elsevier.
Sigvaldason, G. E., and G. Elísson
  1968   Collection and anslysis of volcanic gases at Surtsey, Iceland. Geochimica et Cos-
         mochimica Acta 32:797–805.
Stefánsson, K. and J. Sigurjónsson
  1957   Temporary increase in fluorine content of water following the eruption. In The eruption
         of Hekla 1947–1948, Part 3(2), editors Trausti Einarsson, Gudmundur Kjartansson and
         Sigurdur Thorarinsson. Reykjavík: Societas Scientiarum Islandica.
Steingrímsson, J.
  1915–1917   Fullkomid skrif um Sídueld [A complete description of the Lakagígar eruption]. In
         Safn til sögu Íslands 4. Reykjavík: Hid íslenzka bókmenntafélag.
Stephensen, M.
  1785   Kort Beskrivelse over den nye Vulkans Ildsprudning i Vester-Skaptefields-Syssel paa
         Island i Aaret 1783 Short description of the eruption in the new volcano in the district
         Vestur-Skaftafellssýsla in 1783 Copenhagen: Nicolaus Möller.
Thorarinsson, S.
  1954a  Athuganir á Skeidarárhlaupi og Grímsvötnum 1954 [Observations of Grímsvötn and the
         Skeidará jökulhlaup in 1954]. Jökull 4:34–37.
  1954b  The tephra fall from Hekla on March 29th 1947. In The eruption of Hekla 1947–1948 Vol.
         2(3), editors Trausti Einarsson, Gudmundur Kjartansson and Sigurdur Thorarinsson.
         Reykjavík: Societas Scientiarum Islandica.
  1963   Eldur í Öskju-Askja on Fire Reykjavík: Almenna bókafélagid.
  1967   The eruptions of Hekla in historical times. In The eruption of Hekla 1947–1948 Vol. 1,
         edited by Trausti Einarsson, Gudmundur Kjartansson and Sigurdur Thorarinsson. Reyk-
         javík: Societas Scientiarum Islandica.
  1969   The Lakagígar eruption of 1783. Bulletin Volcanologique 33(3):910–929.
  1970   Hekla. A notorious volcano. Reykjavík: Almenna bókafélagid.
  1971   Damage caused by tephra fall in some big Icelandic eruptions and its relation to the
         thickness of the tephra layers. In Acta of the 1st International Scientific Congress on the

*Volcano of Thera*, edited by A. Kaloyeropoyloy. Athens: Archaeological Services of Greece. General Direction of Antiquities and Restoration.

1974 *Vötnin stríd. Saga Skeidarárhlaupa og Grímsvatna gosa* [The swift flowing rivers. The history of Skeidará jökulhlaups and eruptions in Grímsvötn]. Reykjavík: Bókaútgáfa Menningarsjóds.

1977 At leve på en vulkan. *Geografisk Tidsskrift* 76:1–14. Copenhagen.

Thorarinsson, S. and G. E. Sigvaldason

1972 The Hekla eruption of 1970. *Bulletin Volcanologique* 36(2):269–288.

Thoroddsen, T.

1914 *Eldreykjarmódan 1783* [The volcanic haze in 1783]. In *Afmælisrit til Dr. Phil. Kr. Kaalunds*. Copenhagen: Hid íslenzka frædafélag.

1925 *Die Geschichte der islandischen Vulkane*. Det Kongelige Danske Videnskabernes Selskabs Skrifter, Naturvidenskabelig og Mathematisk Afdeling, 8. Række, IX.

Traumüller, E.

1885 Die trockenen Nebel, Dämmerungen und vulkanische Ausbrüche des Jahres 1783. *Meteorologische Zeitschrift*: 138–140.

Walker, G. P. L., and R. S. J. Sparks

1977 The significance of vitric-enriched air-fall ashes with crystal-enriched ignimbrites. X. *INQUA Congress Birmingham Abstracts*, edited by Keith Clayton.

Watkins, N. D., R. S. J. Sparks, H. Sigurdsson, and T. C. Huang

1978 Volume and extent of the Minoan tephra from Santorini volcano; new evidence from deep-sea sediment cores. *Nature* 271(5041):122–126.

Wilcox, Ray F.

1959 Some effects of recent volcanic ash falls with especial reference to Alaska. *Geological Survey Bulletin* 1028–N:409–476. Washington; United States Government Printing Office

# 6

# Volcanoes as Hazard:
# An Overview

RICHARD A. WARRICK

## INTRODUCTION

Volcanoes can be hazardous to your health. The likelihood of eruptions is often very small, but in many cases the social consequences are high. Vulnerability to volcanic events is largely a matter of human behavior, not volcanic behavior. Individual or societal response to volcanic threat is largely a function of hazard perception, of risk estimation, and of adjustment choice evaluation. To improve response, interdisciplinary research can play a central role in volcanic risk perception and assessment. As we look toward the future, three critical issues to be faced in the United States are

1. Is vulnerability on the increase?
2. How do we effectively use volcanic risk information in hazard decision making?
3. What is a socially acceptable level of volcanic risk?

To address issues like these, we must adopt a more balanced approach to volcanic hazard that includes the social dimensions of risk as well as the environmental, the behavior of humans as well as the behavior of volcanoes.

These are the central themes and conclusions of this chapter on volcanic hazard. The principal aim of the overview is to sensitize the reader to the problems, concepts, and issues surrounding volcanic hazards from a broad, interdisciplinary perspective. The objective is to link together in a meaningful fashion the notions of extreme events, individual behavior, and social re-

161

sponse. What is the spectrum of hazardous events? How do we identify the hazards and estimate and evaluate the risks? What are the options available for reducing the risks? And how do we weigh and balance them? In exploring these questions, volcanic hazard is placed in a framework of risk assessment, which provides the organizing structure for the chapter. The areal emphasis is on the United States.

## THREE VOLCANIC HAZARD SITUATIONS IN THE UNITED STATES

In order to set the stage for further discussion, it may be helpful to the reader to first get a feeling for the hazard context with which we are dealing. The following three cases serve this purpose. Although unquestionably anecdotal, the examples nevertheless are reflective of the three major volcanic situations in the United States—the Aleutian Arc and Cook Inlet region of Alaska; the Hawaiian Islands; and the Cascade Range of the Pacific Northwest (Figure 6.1).

### Mount Katmai, Alaska, 1912

In June 1912, a rare geophysical event happened that was to become one of the most powerful, violent displays of nature to occur in the twentieth century. Several vents of Mount Katmai, located about 480 km southwest of Anchorage, Alaska, erupted with forceful vengeance, hurling 25 km³ of ash and pumice aloft. The ensuing fallout of tephra accumulated to depths of over 3 m near the mountain's flanks. An adjacent valley was devastated for 22 km by a hot pyroclastic flow, which filled the valley floor to an average depth of about 30 m (Gedney, et al. 1970). This same valley was later to be heralded as a unique national landscape and called the Valley of Ten Thousand Smokes, so named because of the numerous steaming fumaroles left by the cooling flow deposit. As to Mount Katmai itself, the active volcano was reduced drastically by over 300 m in height as the drained magma chamber deep below could no longer support the entire mountain, causing a sudden collapse that gave Katmai its new appearance as a caldera.

The few residents of Kodiak—a small, isolated coastal settlement located approximately 160 km downwind of the eruption—were forced to tolerate a 25-cm accumulation of volcanic ash. Places as far away as Seward (400 km) and Cordova (560 km) gained firsthand knowledge of Katmai's temper by way of light ashfall and highly acidic rainfall. Traces of ash were recorded as far away as 1200 km, at Juneau (Wilcox 1959). Relatively mobile peoples were displaced for about 2 decades (see Workman, Chapter 11 of this volume). However, for all its ferocity, the Mount Katmai eruption of 1912 had relatively little impact on human well-being.

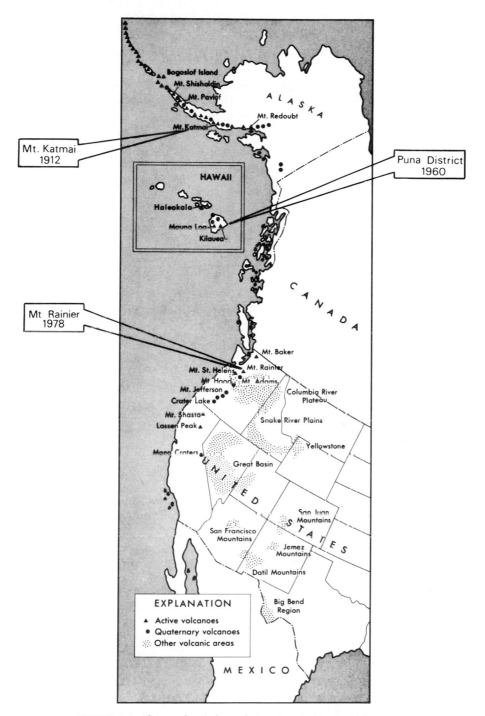

**FIGURE 6.1.** Three volcanic-hazard situations in the United States.
(Adapted from U.S. Geological Survey 1966.)

## Puna District, Hawaii, 1960

To the inhabitants of Puna District on the Big Island of Hawaii, the lava flow eruption of January 1960 came as no real surprise. Indeed, it had been less than 1 month since Kilauea Volcano ceased its last eruption. Kilauea and Mauna Loa, the island's two active volcanoes, frequently remind the islanders of their potency by releasing their pent-up forces every several years on the average. In comparison with the explosiveness of Mount Katmai and its neighboring volcanoes, the eruptions of the Hawaiian volcanoes are mild and effusive. For the islanders, the news of eruption elicits concern mixed with excitement and anticipation, as visitors flock to the volcanoes to witness nature's fireworks.

But in January 1960, the flow of lava that poured from a fissure along the East Rift Zone of Kilauea came in direct conflict with human activity. Although as a rule damage from lava flows is commonplace, the tolls are usually small, being limited for the most part to damage to agriculture. This time, agricultural damage was added to the near destruction of the villages of Kapoho and Koae. Seventy buildings were destroyed and thousands of acres were covered with lava. Kilauea's activity for the year resulted in damages totaling about $6 million (Belt *et al.* 1962). To an island of about 50,000 people, this was a disaster. The homeless were provided for, and the burden of loss was absorbed or shared, either through formal mechanisms (as through the Red Cross) or informally (as through family, friends, and neighbors). The villagers of Kapoho would return to rebuild, to live, and to cope with future eruptions.

## Mount Rainier, Washington, 1978

To the people living in the shadow of Mount Rainier, Washington, the mountain is much more than just a prominent feature of the landscape. It is a means of life support and is inextricably woven into the lifestyles, the settlement patterns, and the behavior of residents in work and play. Expressions of human activity can be found concentrated in the narrow river valleys emanating from Mount Rainier's glacier-capped peak—valleys that because of their natural corridors and land and water resources are magnetic forces in human settlement. The White River Valley is one such valley. So is the Puyallup. Downstream the terrain flattens and the rivers broaden as they enter the Puget Sound lowland, where settlements are more numerous and the population density greater.

Unlike their Hawaiian counterparts, none of the Washington State residents have had firsthand experience with the volcanic hazards associated with Mount Rainier. The last reported eruption of Mount Rainier was in 1882 (International Volcanological Association, 1960). In general, the eruptive characteristics of Mount Rainier and its neighboring volcanoes in the Cascade Range of the Pacific Northwest have a much closer resemblance to Mount Katmai than to Kilauea—infrequent, explosive pyroclastic-type eruptions.

Although it is beyond their direct experience, some of the residents may know something of Mount Rainier's geologic history. For example, some may know that parts of those same valleys containing farms, houses, industries, hydroelectric dams, campgrounds, schools, and other human artifacts have been affected by at least 55 large volcanic mudflows during the last 10,000 years (Crandell, 1971). They might also know that 5000 years ago one exceptionally large volcanic mudflow, the Osceola, traveled down their White River and spread out over 325 km² of Puget Sound lowlands where today two sizable towns rest on 21 m of mud from that event (Crandell and Waldron, 1969).

But it is very likely that most inhabitants are blissfully unaware of the violent potential of the presently dormant Mount Rainier. It is also highly unlikely that the risk and uncertainty of volcanic events, as rare as they are, have entered into individual and collective decisions about location, livelihood, or land use. As more scientific information becomes available on volcanic risks and as critical decisions are faced concerning growth in their region, the inhabitants may be forced to consider the volcanic hazards to their well-being.

## RISK ASSESSMENT OF VOLCANIC HAZARD

Although divergent in time and space, all of the three cases described above—Mount Katmai, Kilauea Volcano, and Mount Rainier contain similarities. Broadly, each describes a situation of hazard and each includes elements of risk and uncertainty. First, let us make explicit the common components of "hazards," and then let us blend our hazard concepts with the more purposeful structure of "risk assessment." This, it is hoped, will get us all thinking on the same plane and will provide a framework for systematic discussion.

### Volcanoes as Hazard: A Conceptual Approach

The notion of volcanic hazard can be conceptualized simply as a system whose major interactive elements bridge the gap between the physical and human realms. Figure 6.2 displays a simplified but useful way of portraying the major relationships. With regard to Figure 6.2, five points of elaboration are warranted.[1]

First and most important is the fundamental distinction between volcanic events and volcanic hazards. Volcanic *hazards* arise from an interaction of human use systems and natural event systems (i.e., volcanic processes and products). In and by itself, nature is neutral. This is a simple but often

---

[1] These five points summarize a long record of investigation of human response to natural hazards. See Burton and Kates (1964); Burton, *et al.* (1968, 1978); White (1973, 1974); Mitchell (1974).

**HAZARDS FROM VOLCANOES AND MAN**

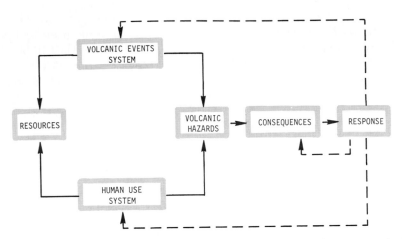

**FIGURE 6.2. Hazards from volcanoes and man. Volcanic hazards (and resources) arise from an interaction of natural and human systems. In responding to the threat of adverse consequences, society may attempt to modify volcanic events, alter loss potential, or redistribute losses. (Adapted from Kates 1970; Burton et al. 1978.)**

overlooked point. Metaphorically, the cataclysmic eruption of Mount Katmai in 1912 is the towering tree falling in the uninhabited forest—does it make a sound if no one is there to listen? Although nature provides the volcanic event, in the last analysis it is the actions of people that are responsible for the resultant hazards.

Second, in part people risk exposure to extreme volcanic events in order to reap the benefits—the *resources*—that the environment has to offer. Kilauea's fertile flanks entice farmers, and the scenic, rugged, lava-shaped coastline of Puna District attracts residents and visitors alike. In this light, hazards can be viewed as the "negative resources," elements of nature to be accepted, prevented, mitigated, or avoided in pursuing the perceived environmental opportunities (Burton *et al.* 1978). In short, the notion of hazard inextricably includes "resource," and thus implies a tradeoff of risk and benefit.

Third, the threat of adverse consequences may initiate individual and/or collective *response*. Consciously, societies may select from a range of hazard adjustments, mechanisms for coping with volcanic hazards: by simple loss bearing or sharing, by modifying events or preventing effects, or by change in land use or location (Burton *et al.* 1978). The first set of adjustments (loss bearing or sharing) is exemplified in the behavior of most residents of Puna District in the face of periodic lava flow eruptions.

Fourth, as implied in Figure 6.2, the entire system of volcanic hazard is *dynamic*. The causes of shifting vulnerability to volcanic events are to be found in some mix of natural, social, or technological change over time. Understanding the dynamic process—the evolving character of volcanic risks and the patterns of human adjustment to them—is central to the development of sound public policies designed to cope with the possible consequences of future volcanic events. For example, in what ways have the patterns of human susceptibility to volcanic events in Alaska been changing since the 1912 eruption of Mount Katmai? And what implications do those patterns of change have for the development of volcanic hazard policies in Alaska?

Finally, let us emphasize that it is the myriad individual and collective decisions concerning land use, location, and hazard adjustment that, in the aggregate, account for the disaster potential. For instance, current decisions regarding purchase of subdivision lots for vacation homes in Puna District, Hawaii, are helping to shape the character of volcanic risk in the future. However, the process of choice under situations of risk and uncertainty is difficult to comprehend fully, being guided by a bewildering array of cultural, social, economic, and psychological factors. If we had more complete understanding of how such decisions are made, we would be better able to explain and predict human behavior under varying conditions of volcanic risk.

## Risk Assessment

In part, decisions made about volcanic hazard are influenced by the ways in which the threats are recognized, the probabilities and consequences of damaging events are estimated, and the risks are evaluated, in light of opportunities available for coping with the hazard. In general, this appraisal of the kinds and degrees of environmental threat is referred to as *risk assessment* (Lowrance 1976; Otway and Pahner 1976; Kates 1977, 1978).

As conceived by Kates (1978), risk assessment is composed of three overlapping but distinguishable elements: hazard identification, risk estimation, and social evaluation. In terms of volcanic hazard, as depicted in Figure 6.3, *identification* involves the recognition of those processes or products of volcanoes which are threatening to human well-being. Volcanic risk *estimation* is the measurement of the threat, a determination of the frequencies of the events and the magnitude of their consequences. Social *evaluation* places meaning on the estimates—that is, how important are they? Thus, evaluation allows the consequences of volcanic events to be weighed and balanced vis a vis the alternative adjustments available for coping with the hazard.

Risk assessment is practiced by the layman as well as the expert, and its methods range from the informal and intuitive to the formal and scientific.

Within the risk assessment framework portrayed in Figure 6.3, the remainder of the chapter will touch upon the major physical, social, and behavioral components that interact to comprise volcanic hazards.

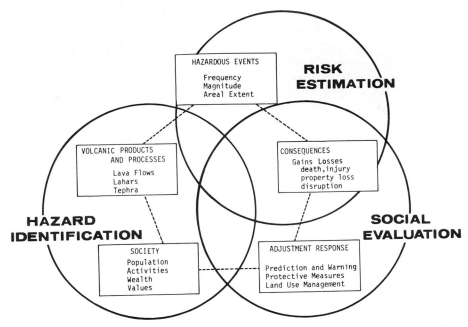

**FIGURE 6.3. Risk assessment in coping with volcanic hazard. (Adapted from Kates 1976.)**

## VOLCANIC HAZARD IDENTIFICATION

What are the specific hazardous phenomena that are associated with volcanoes? How are threats recognized? What role does research play in volcanic hazard identification?

### Specific Volcanic Hazards

The primary hazardous products or processes directly related to volcanic activity include lava flows, pyroclastic flows, tephra, and volcanic mudflows. Indirectly, tsunamis, debris avalanches, forest fires, and lightning also pose threats to people and their environment.[2]

Most inhabitants of Puna District, Hawaii, surely would have no difficulty in identifying the frequently occurring *lava flows* as a hazardous phenomenon of volcanic activity. On their island, lava flows have traveled as much as 55 km from their vent and attained depths of 5 m or more (Macdonald 1972). The lavas of Hawaii's volcanoes, poor in silica, are normally very fluid, allowing easy escape of volatile gases and rapid flow of molten material over the ground surface. However, the flow is typically slow enough to allow people to move out of the way. Structures and their contents, highways, cultivated fields—

---

[2]For more detailed discussions of specific volcanic hazards, see Macdonald 1972; Bolt *et al.* 1975; or Warrick 1975.

anything stationary or unmoved—in the path of lava flows are subject to destruction. Although lava flows also occur in Alaska and the Pacific Northwest, they tend to be overshadowed by the threat of other extreme volcanic events.

*Pyroclastic flows* are rare phenomena often associated with volcanoes displaying explosive characteristics. Pyroclastic flows are turbulent mixtures of rock fragments suspended in gas that move rapidly over the ground surface. Known to reach temperatures of hundreds of degrees centigrade and to travel at speeds in excess of 100 km/hr (Office of Emergency Preparedness 1972), pyroclastic flows are incredibly destructive. A pyroclastic flow from the eruption of Mount Pelée totally annihilated the city of St. Pierre with its 30,000 inhabitants on the island of Martinique in 1902 (the largest volcanic disaster of this century). The speed, direction, and path of the flows are determined in part by the force and nature of emission from the volcano and by gravity and topography. Thus, pyroclastic flows commonly follow valleys and drainage ways.

Violent volcanic eruptions tend to throw thousands of tons of rock fragments, or *tephra*, of varying sizes into the atmosphere. Fragments of less than 4 mm, termed *volcanic ash*, remain airborne for long periods of time and can be carried great distances by wind before being deposited, as in the case of the massive deposits of ash from Mount Katmai in 1912. That tephra can be deadly is evidenced by the famous burial of Pompeii, with its 16,000 victims, under 5–8 m of lapilli and ash (Bullard 1962). The range of adverse human consequences from ash fall includes structural damage to buildings, severe health effects from inhalation of ash and toxic gases released by the cooling particles, contamination of water supplies, reduced visibility, clogged transportation routes, and destruction of vegetation, crops, and livestock (see Wilcox 1959).

A controversial but highly relevant issue is the effect of volcanic ash on climate, particularly as it may disrupt agriculture and the delicate food supply balance in many parts of the world. Several major eruptions are believed to have precipitated mid-latitude drops in mean annual temperature with deleterious impacts on agriculture, as for example the cataclysmic eruption of Tambora in 1815 (Lamb 1972; Watt 1973). The Mount Katmai eruption reportedly reduced solar radiation reaching the Earth's surface by about 20% at Mount Wilson, California, located 3200 km away (Macdonald 1972).

*Volcanic mudflows* (or *lahars*) "represent a form of mass movement of sediment that is intermediate between a flood and an essentially dry landslide . . ." that is generated by, or associated with, volcanoes (Waldron 1967). Mudflows can be initiated by rapid melting of snow or ice on the volcano's slope, by descent of pyroclastic flows into streams, or by ejection of water from a crater lake—as, for example, in the eruption of Kelut, Java, in 1919, which resulted in the death of 5500 persons (Rittmann 1962)—as well as by other causes. Speeds of 30–50 km/hr and lengths of 8–16 km are frequent, although rare mudflows of over 100 km/hr or 160 km in length have been recorded.

Though the residents of Mount Rainier's mountain valleys may not suspect it, the glistening sheath of glacier ice on the dormant volcano could be melted suddenly by renewed volcanic activity, with the possibility of devastating mudflows. The amount of water locked in that ice may be about 15 km$^3$, or roughly equivalent to the storage capacity of Grand Coulee Dam (Warrick 1975). All drainage from Mount Rainier (as with most of the volcanoes in Washington) is west to the Puget Sound–Cowlitz River lowland, where population is densest—the direction that potentially destructive mudflows would take.

Eruptions of coastal or island volcanoes can produce *tsunamis* (or seismic sea waves) that threaten shorelines far from the volcano. The catastrophic eruption of Krakatau in 1883 produced an extremely large tsunami that attained a height of over 18 m as it reached the coasts of Java and Sumatra, killing 36,000 persons (see Furneaux 1964 for an interesting literary account of the tragedy). One of Mount Katmai's neighboring volcanoes, St. Augustine, located in Cook Inlet, has the potential for generating tsunamis; it is monitored closely partly for this reason and partly because of its recent activity (University of Alaska Geophysical Institute 1972; Hobbs, *et al*. 1977).[3]

Other indirect hazards from volcanic activity include *forest fires*, *debris avalanches* and *landslides*, and *lightning*.[4] For example, thousands of burned acres frequently accompany lava flow eruptions in Hawaii (Peterson 1973), and the acute threat of widespread fires on the forest-covered slopes of the Cascade Range in the event of renewed volcanic activity is a real concern.

### Crossing the Awareness Threshold

As individuals depending upon our own senses, we are limited in our abilities to identify the full range of hazards emanating from the environment. We are particularly hampered in our recognition of those events which, because of their infrequence, are beyond our realm of direct experience. Indeed, experience—along with material wealth, perceived role, and personality—is a dominant factor in determining the way in which people identify risk, estimate the likelihood of occurrence and the consequences, and choose among a range of alternatives for coping with the threat (White 1973; Burton *et al*. 1978).

In the absence of direct individual experience, aggregate experience—that of groups or society as a whole—is a source of risk information. Our collective memories transcend the limitations of the individual as a sensor of the environment and, if tapped, substitute for direct experience. Thus, the newcomer to Alaska may read of the day that Mount Katmai blew its top and of the effects that ashfall had on the site he now occupies. Or he may hear from his neighbor about the time in 1953 that ash fall from the surprise eruption of Mount Spurr—a little-known volcano—suddenly jarred the residents of An-

---

[3]In fact, in the 1880s minor damage was reported from severe sea waves generated by one of St. Augustine's eruptions (Davidson 1884, as reported in Forbes 1972).

[4]Lightning occurs from severe convectional activity during violent eruptions. It is interesting to note that the only recorded death from the 9 years of activity of Parícutin Volcano, Mexico, was from a lightning strike (Gutierrez 1972; also see Nolan, Chapter 10 in this volume).

chorage across the threshold of hazard awareness. By word of mouth, from written account to story and legend, individuals may learn of volcanic threats to their well-being.

In populated areas of the United States where volcanoes exist, the kinds of specific volcanic hazards mentioned above (with the exception of lava flows in Hawaii) occur very rarely. Along the Cascade Range of the Pacific Northwest, for example, the 12 prominent volcanoes have been strangely silent for the last 96 years (excluding the activity of Lassen Peak in the 1910s and the recent steaming of Mount Baker). In such circumstances of rarity, the knowledge of the specific hazards either may never be gained or may fade from our collective consciousness. Identification of volcanic hazards is then more difficult. Although the evidence is lacking, one could speculate that only a small percentage of the population at risk knows of the possibility—however remote—of ashfalls or mudflows, or even that the same mountain upon which they ski, hike, or pursue livelihoods is actually a potentially explosive volcano! In such circumstances, the process of hazard identification is complicated and involves searching the environment for clues of past activity—the job of research.

## Identifying Hazards through Research:
## Volcanic Hazard Mapping

In identifying specific volcanic hazards, a critical parameter is their areal extent, especially in relation to existing or future spatial patterns of human activity. From a public policy point of view, a prerequisite to sound decision making regarding adjustment to volcanic risk is the mapping of the hazards.

In the United States, hazard mapping of volcanic phenomena has, for the most part, come about only recently. In Alaska, the first systematic identification of the chain of volcanoes extending along the Aleutian Arc (there are at least 76 Alaskan volcanoes, 39 of which are believed to have been active during the last 200 years [Wilcox 1959]) was performed around World War II.[5] Information was compiled on individual volcanoes—their activity, eruptive products, and other characteristics—through the efforts of the U.S. Geological Survey (see Coats 1950). Several maps that show the actual areal extents of specific eruptive phenomena have been made, like the ashfall of the eruptions of Mount Spurr and Mount Katmai (as reported in Wilcox 1959). Computer models allow rough judgments to be made of the areal extent of possible volcanically induced sea waves from St. Augustine (University of Alaska Geophysical Institute 1972; also see Detterman 1974). But little has been done in the way of comprehensive volcanic hazard mapping in Alaska beyond some rough, small-scale estimates of hazard zones and populations at risk (Warrick 1975). On a broad scale, the only volcanic hazard to most Alaskans is ash fall,

---

[5]It is interesting to note that the fundamental distinction of whether a volcano is extinct or just dormant is not easily made. Prior to 1912, for example, Mount Katmai was thought to be extinct (Tazieff 1967).

essentially because of the current spatial patterns of population distribution within the state.

On the Island of Hawaii, the frequent lava flow eruptions from Kilauea and Mauna Loa leave an unmistakable record on the landscape. The spatial extent of past flows is well documented (Macdonald and Abbott 1970). With data on past activity, places of extrusion (whether along rift zones on the volcano's flank or within the caldera), and slope and terrain, hazard maps have been constructed for the Big Island (Mullineaux and Peterson 1974). There is some question, however, as to whether these maps possess the precision necessary for detailed land use planning.

Within the last 200 years, there also has been activity at Hualalai Volcano, and at Haleakala on the Island of Maui. Though most Islanders are unaware of it, some parts of the islands also are subject to other specific hazards—like ashfalls or pyroclastic flows—resulting from more violent forms of volcanic activity, as indicated by the geologic record. In this regard, hazard maps have been prepared recently for the island of Oahu (Crandell 1975).

In the Pacific Northwest, the recent series of investigations of Cascade volcanoes by the U.S. Geological Survey are concerned expressly with volcanic hazard mapping. Areas studied include Mount Rainier (Crandell 1967; 1971; 1973), Mount Baker (Hyde and Crandell 1978), Mount St. Helens (Crandell and Mullineaux 1976), and others (see Crandell, Mullineaux, and Miller, Chapter 8 of this volume). Thus, for example, with the aid of available hazard maps, Washingtonians living adjacent to Mount Rainier now might be able to identify their location as being subject to mudflow and flood hazards. In addition, the state of California has made rough identifications of volcanic hazard areas for northern California (California Division of Mines and Geology 1973).

In sum, knowledge of the type of specific volcanic hazards and where they occur is a fundamental first step in volcanic risk assessment. The next related questions that the residents of Alaska, Puna District, or Mount Rainier may well ask pertain to timing and effect: How often do these events occur? What are the consequences that can be expected? These questions are part and parcel of risk estimation.

## RISK ESTIMATION

### Estimations by Individuals

How do persons at risk from volcanic hazards estimate the likelihood of occurrence and the losses that might result? The sources of information upon which individuals make risk estimates can be quite diverse, and the process by which information is synthesized to make risk judgments is understood only in a rudimentary sense.

For example, in attemtping to understand the lava flows that occasionally disrupt him and his fellow islanders, the resident of Puna District may rely partly on his personal experience with lava flow occurrence, partly on information supplied by the Hawaiian Volcano Observatory, and partly on his cultural heritage. The resulting estimates may vary greatly among individuals using the same information and may show either congruence or discrepancy with the estimates of the technical experts. Thus, one study (Murton and Shimabukuro 1974) found that of a sample of residents interviewed in Puna, nearly one-quarter either assigned a deterministic explanation to eruptive occurrences (that the events occur in groups or with definable regularity) or could not make judgment at all; the remaining individuals expressed a probabilistic explanation (that the events could occur in any year). Among the experts, the question of periodicity is not clear-cut, despite the array of statistical techniques, years of data accumulation, and competent personnel. Much depends upon the time and space scales within which one is working. Whereas the technical expert develops sophisticated geophysical explanations of occurrence, many Hawaiians rely on a supernatural explanation—the goddess Pele determines occurrence. In such circumstances, the notion of "objective" risk is blurred: For both the technical expert and the layman, estimation of risk is a matter of perception.

Though we have very few studies of individual assessment of volcanic risk per se, we have some idea of the factors that appear to influence risk estimation in analogous situations of environmental hazard.[6] In general, the more frequent the individual's *experience* with the event, the greater the acuity of estimation of its magnitude and recurrence. Mobile *urban* residents are often less accurate than their rural neighbors, who maintain a closer relationship to the natural environment. In situations where there are other *competing social or natural risks*, the perceptual accuracy of the risk in question may be diminished (Golant and Burton 1969). Furthermore, in making locational decisions, an individual's *time horizon*—short-term versus long-term—may influence the way in which risk is estimated. In light of the above factors, one might ask, How would a developer from the U.S. mainland who has had no experience with lava flow eruptions, who resides in an urban environment, and who intends to sell off his investment within 10 years, estimate the risk and consequences of lava flow hazard in making a decision to build a condominium in Puna District, Hawaii?

It has been shown that certain *cognitive limitations* intervene in the risk estimation process (Slovic *et al.* 1974). So, for example, many people fall victim to the "gambler's fallacy"—that if an event occurs in one year it is less likely to occur the next year—in their interpretation of probabilistic events. Almost universally, people, even professionals, have difficulty in making valid conclusions from small samples (Tversky and Kahneman 1971). People may proceed with an "anchoring" process (Slovic *et al.* 1974) in which

---

[6]The following factors are extracted mainly from reviews by White 1974; and Mitchell 1974; and Burton *et al.* 1978.

undue importance is attached to their first rough estimate, such that further information is adjusted to that estimate. Attempts are underway to learn more about these kinds of cognitive obstacles to risk estimation (Slovic *et al.* 1976).

The above factors act in a complex fashion to guide the estimates of risk that people make about extreme events. The relative importance of each is not at all clear. The problem is clouded further by an array of social, cultural, and economic situations that are superimposed upon those variables. At this time, one can do little more than speculate on how the process operates in the context of volcanic hazard. It is hoped that studies like that of Hodge, Sharp, and Marts (Chapter 7 of this volume) will help to piece together the puzzle.

The point is that the way in which individuals appraise hazards influences the decisions they make about them. Therefore, a key to understanding society's vulnerability to low-probability events like volcanic eruptions lies, in part, in understanding individual decision processes and hazard perceptions. Recognition of this perspective is gaining at the national level, as evidenced by the attention being directed toward individual and social perceptions of risk within the Assembly for the Behavioral and Social Sciences of the National Research Council (Holden 1978). With respect to volcanic hazard, such understanding will become more critical in the future as improved scientific information about volcanic hazards becomes available to individuals and public policy decision makers through research.

## Estimating Risks through Research

The estimation of volcanic risks involves both physical and social components: the estimation of the frequencies of events of given magnitudes and the measure of the consequences to human well-being. To date, systematic research has taken substantive steps in the first direction, and relatively few in the second.

### ESTIMATING THE FREQUENCIES

Like the layman who faces tough obstacles in risk estimation, the scientist encounters formidable difficulties in establishing probabilities for most volcanic events. Despite a battery of geologic tools for probing the prehistoric past, in many places the frequencies of volcanic events are so low that meaningful probabilistic statements cannot be made confidently. This is a problem in specifying the degree of risk. Often, only nominal risk statements are possible.

For example, along the river valleys of Mount Rainier where lahars have been identified as a rare but potential threat, the data allow only broad statements of risk—high, moderate, and low (Crandell 1973). For the low probability, catastrophic events, the designation is based only on several occurrences during the last 10,000 years or so—hardly a sound statistical base, but the only one we have. It takes a combination of genetic, stratigraphic, chronologic, and cartographic methods, as applied to the objectives of volcanic hazard appraisal, to uncover the clues (Crandell and Mullineaux 1974). Even in Hawaii, for which there exists a well-documented historical and geologic record of lava flow eruptions, careful attempts to quantify risk at a scale useful for detailed land use planning are frustrated by wide variations in risk at a microscale due to topographical and other factors (Mullineaux and Peterson 1974).

Besides geological studies of a statistical nature, another line of research seeks to reveal the eruptive characteristics of volcanoes by understanding the physical mechanisms involved. Geophysical and volcanological studies, such as those carried out by the Hawaiian Volcano Observatory, are conducted in part with the objective of being able to specify the timing of volcanic events more precisely.

Despite the fact that considerable progress has been made recently in establishing the degree of volcanic risk through both concerted geological and geophysical investigations (particularly in the Pacific Northwest), the ability to state more precisely the probabilities of specific volcanic hazards seems inherently limited. A very large component of uncertainty will remain regardless. We may be relatively confident about a broad estimate of 1 in 1000 per year for a major eruption of Mount Rainier, or 1 in 400 for its neighbor Mount St. Helens (Crandell et al. 1975), but sobered by the knowledge that similar estimations at scales conducive to local hazard planning may never be able to shed entirely their wide confidence bands.

This limitation raises an important issue concerning the interplay between research and hazard response: How precise do probabilistic estimates have to be in order to contribute effectively to hazard planning? At what point do the added costs of refining the estimates outweigh the added benefits? Unfortunately, there are no easy answers. To date, geological and geophysical research has concentrated largely on obtaining first-cut estimates for events about which we know little. It is reasonable to assume that the social value of these endeavors is great relative to the costs. Nevertheless, at some point judgments must be made as to whether further efforts could be expended more gainfully in directions other than this one, as for example in measuring the social consequences of volcanic events.

MEASURING THE CONSEQUENCES

Though research efforts in volcanic risk estimation have come to focus on the questions of what events, where, and how often, surprisingly little attention has been directed toward determining the human consequences. It is this complementary element of volcanic risk estimation that is central to volcanic hazard decision making.

Although a number of studies have examined social consequences of, and human response to, particular volcanic disasters (e.g., Nolan 1972; Rees 1970; Lachman and Bonk 1960; Keesing 1952;), we know very little about the magnitude of loss potential—that is, vulnerability.[7] Thus, though we have a rough guess as to the number of persons currently at risk from a major eruption of

---

[7]The difference here is really one of disaster versus hazard: Hazard vulnerability takes a look further down the road and includes threat as well as tragedy.

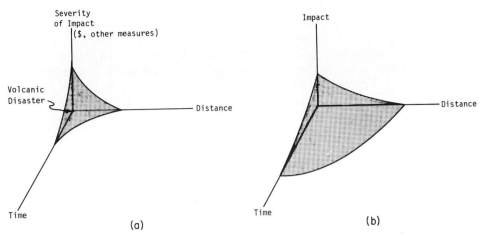

**FIGURE 6.4.** Systemic disruption from volcanic disaster (hypothetical). **(a) Potential systemic disruption.** Consequences are greatest at time and point of disaster. Through social, technological, and economic linkages, impacts diffuse over space and persist over time. Total disruptive impact is represented by the area under the surface, an aggregation of effects over time and space. **(b) An increase in vulnerability.** Although initial consequences at point of impact may not necessarily rise, total disruption increases as effects are felt over greater areas and as social systems take longer to recover fully.

Mount Rainier, we have only vague ideas of the number and kinds of structures subject to damage, the likely number of deaths and injuries, the potential number of dislocated persons, the economic impacts, and so on. Methods for determination of vulnerability are hampered by the lack of historical loss data.

One approach to the problem is to borrow methods from other natural hazard loss assessments like floods, earthquakes, or hurricanes. One method is the creation of *synthetic loss data* from generalized event magnitude-damage relationships (like flood stage-damage curves) to estimate consequences from a distribution of hazard events.[8] This tactic is particularly useful in situations where historical loss data are unavailable, or in highly dynamic social situations where changes in loss potential are evident. Crude projections of loss from volcanic events in Northern California were derived from a rough application of this method (California Division of Mines and Geology 1973).

*Simulation* techniques take the approach a step further, incorporating elements of the physical event (e.g., frequency, severity), the population-at-risk (e.g. persons, buildings), and the vulnerability of the population-at-risk into complex computer models (White and Haas 1975; Friedman 1975). For given geographical areas, estimates of losses and their distributions can be obtained by simulating the occurrence of hazard events over time. One major advantage of simulation techniques is that they allow for the systematic assessment of policies or strategies designed to reduce hazard losses. Simulation

---

[8]For explanation in the flood context, see White 1964.

models have been constructed for inland flooding, coastal flooding, hurricane wind hazard, earthquakes, and others (see Friedman 1975 for specific references).

Another, more qualitative approach asks, What if? This kind of *scenario* building can be illuminating, as, for example, in speculating upon the consequences of a major eruption of Mount Rainier (Cullen 1978). Scenario applications often suffer from a high degree of subjectivity and from confinement to unique and static situations. However, Ericksen (1975) has shown how, in the natural hazard context, scenarios can be developed that combine analytic rigor with intuitive judgment. The application of both simulations and scenarios holds promise for volcanic risk assessment.

The point is this: If society is to make decisions regarding the acceptability of volcanic risks, then sound risk estimations that include both elements of risk—physical characteristics and social consequences—are required. Although in the United States the role of geologic investigations in volcanic risk estimation has been recognized and expanded, the complementary need to assess the social consequences has gone largely unattended. The need for such assessment may take on increased importance in light of the possibility of expanding vulnerability to volcanic hazard.

## Are We Becoming More Vulnerable to Volcanic Hazard?

As we begin to pay closer attention to volcanic hazards, a critical issue is whether, unsuspectingly, we are becoming more and more vulnerable to their occurrence. Of course, as population growth occurs in volcanic areas, the direct potential for loss of life and property expands. This appears to be the case in portions of Hawaii, the Pacific Northwest, and Alaska, where growth is comparatively rapid in relation to other parts of the United States (Warrick 1975).

Beyond the absolute numbers of people and property, however, is the issue of whether vulnerability is increasing disproportionately through the increasingly complex social, technological, and economic linkages that pervade our society. These linkages, for example, take the form of specialized regional industries; complex, interconnected energy networks; major transportation arteries—and our social systems appear to place greater reliance on their normal functioning. When they are disrupted, the effects reverberate throughout social systems. These linkages have temporal and/or spatial components, transporting risks in space to nonvolcanic areas and prolonging impacts over time as social system recovery time lengthens. Imagine total negative consequences of a volcanic event as being the volume under the surface in Figure 6.4a. A major concern is whether total future consequences are mounting as a result of an expansion of the surface along the time and space axes, as in Figure 6.4b. Such impacts can be labeled broadly as *systemic disruptions*. Let us illustrate this notion in both a historical and a contemporary sense.

With respect to systemic disruptions, it can be argued that very rare, extreme events for which societies have little experience—and therefore few adaptive mechanisms for coping with risk—pose a greater threat to social system stability than more frequently occurring events of similar intensity and areal extent. In very extreme conditions, severe systemic disruption may lead to social system breakdown. For instance, a major volcanic eruption at Santorini may have marked the end of the Minoan civilization on Crete (Latter 1969). The volcanic eruptions on Heimaey Island, Iceland, in 1973 or at Trista da Cunha in 1961 precipitated dislocations that surely overshadowed direct property loss (for descriptions, see Blair 1964; Grove 1973). Closer to home, the cultural impacts from the eruption of Parícutin, Mexico, were severe in the local setting and persisted for decades following the eruption (Rees 1970). In contrast, the lava flow eruption of 1960 in Puna District, Hawaii, while inflicting damage in the millions of dollars, created little social disruption; impacts were locally confined and normal patterns of life were resumed quickly. In short, vulnerability appears to be partly a function of both the magnitude of the physical event and society's experience with it.

In considering the risk of volcanic hazard today, one wonders whether the dynamics of social change are so great that, unwittingly, the potential for major social disruptions from volcanic events is building dangerously. Especially at places like Mount Rainier and other Pacific Northwest locations where the threat of volcanic eruption only recently has begun to permeate the social consciousness, the actions taken in local situations may spread vulnerability to other areas and act to retard the recovery process over time. To illustrate, in nearly every river valley practicable, hydroelectric reservoirs have been constructed in the Pacific Northwest. Plans to meet future energy needs include thermal (including nuclear) as well as hydroelectric. What would be the reverberating social and economic effects of a loss of electrical generating capacity from a large mudflow impacting a major power facility? Given that the power grid connects areas distances apart, would the disruptive impact of the volcanic event be transported along similar spatial lines? Over what length of time would the consequences continue to accrue before the system recovered? One could pose similar questions about, say, critical transportation or communication system linkages subject to disruption. Whether we are unknowingly increasing vulnerability by subjecting tender parts of our social organization to rare volcanic events is the central issue.

In terms of volcanic risk estimation, the point to be emphasized is that systemic disruptions may be a critical dimension of loss from future volcanic events. But the potential for such disruptions may be overlooked if attention is focused too narrowly in time and space. In an analogous hazard situation, for instance, Cochrane (1974) calculated conservatively that indirect effects (as manifested in economic losses) equaled direct property damage for simulated recurrence of the 1906 earthquake in present-day San Francisco. Similarly, for many volcanic hazards, risk estimations that do not include careful consideration of disruptions to social and economic systems run the danger of seriously underestimating the risk.

In terms of volcanic hazard planning, the issue of increasing vulnerability suggests that we look discriminately at uses of volcanically hazardous locations. Are there particularly sensitive industries upon which the economy of Washington relies heavily that should be steered away from certain risk zones downvalley or downwind from Mount Rainier, Mount Baker, or Mount St. Helens? Should local, regional, or state growth plans in Alaska guide growth patterns in volcanic areas so as to assure a reduction in possible future disasters? Although resting upon risk estimations, judgments of this sort inherently require social evaluation of volcanic risks in light of choices available for adjusting to the hazards.

## SOCIAL EVALUATION

The final component in the risk assessment of volcanic hazard is social evaluation. What values or meanings are attached to volcanic risks by individuals or groups, and how? Under what circumstances are the risks acceptable or unacceptable? When unacceptable, what alternative actions—or adjustments—exist for coping with the hazard? Let us take the last question first.

### Range of Adjustments to Volcanic Hazard

As you might expect, the range of adjustments that groups and especially individuals can adopt to reduce the risk of volcanic events is rather limited. Purposive adjustments to volcanic hazard fall largely into the categories of prediction and warning, disaster preparedness and emergency action, protective measures, loss sharing, and land use management.

The ability to *predict* volcanic eruptions is highly variable, depending on the location and type of volcano in question. The residents of Puna, for example, are kept posted on Kilauea's behavior by scientists at the Hawaiian Volcano Observatory (HVO). The HVO has been monitoring the Hawaiian volcanoes since 1912. By observing Mauna Loa and Kilauea volcanoes through many cycles of activity, it was discovered that particular patterns of seismic activity, in conjunction with land deformations (like swelling, or "inflation," of the summit) portend lava flow eruptions (Waesche and Peck 1966; Fiske and Kinoshita 1969; Ollier 1969). Seismometers, tiltmeters, and geodimeters are employed extensively to detect these phenomena. It is often possible to pinpoint where the lava will break out. The advance warning of the 1960 eruption of Kilauea, for instance, provided enough time to evacuate people, household contents, and even some structures from the town of Kapoho before it was covered with lava. One major goal of the HVO is to develop the ability to issue reliable, longer-term warning of eruption.

Unfortunately, the infrequence of eruptions in the Cascade Range, Alaska, or other areas of comparable volcanic activity, is not conducive to the development of predictive methods. Whereas continuous monitoring of all these volcanoes at a scale similar to Hawaii is impracticable (McBirney 1966), several restless volcanoes have warranted close attention in light of recent activity—for example, Mount Baker, Washington (Frank *et al.* 1977) and St. Augustine, Alaska. Others have been monitored on an experimental basis (Crandell and Mullineaux 1974). The use of thermal infrared surveys as an aid to prediction has been explored (Lange and Avent 1973). A reliable predictive capability for the U.S. mainland volcanoes will have to await further geophysical research.

Other phenomena that may prove to be precursors of eruptions in certain situations are geomagnetic variations; changes in temperature or gas composition of crater lakes, hot springs, or fumaroles; changes in subsurface temperatures; or even unusual behavior of animals (McBirney 1966; Ollier 1969; Macdonald 1972).

*Disaster preparedness* entails advance planning to cope effectively when disaster strikes. An integral part of disaster preparedness is emergency evacuation plans, which include an established procedure for communication flow, decision making, and orderly evacuation procedures. Typically, where hazard event occurrence is relatively frequent, the reinforcing experience promotes disaster preparedness planning and readiness (Mileti 1975). We can observe this to be the case in Hawaii. When Puna District is threatened by a possible eruption, the HVO issues a warning, and the Civil Defense Agency along with the Red Cross, police and fire departments, state and country governments, and the National Guard coordinate in further disseminating the warning and in evacuating people and valuables from the threatened area (see Hawaiian Civil Defense Division 1971).

Several means of *protection* against volcanic hazards have been tried or suggested, but the effectiveness of any of them is uncertain. The use of explosives on lava flows to alter direction and movement by breaching the hardened crust at a critical spot is one such means. In Hawaii, lava flows were bombed twice, in 1935 and 1942, but the effectiveness of those actions is difficult to evaluate (see Jaggar 1945; Macdonald and Abbott 1970). The construction of permanent barriers to divert lava flows away from inhabited areas has long been proposed for Hilo, Hawaii, but rejected on the grounds of unfavorable benefit–cost ratios (Corps of Engineers 1966); recent activity at Mauna Loa has stimulated a reappraisal. Emergency diversion barriers were hastily erected by bulldozers during the 1960 eruption of Kilauea in an effort to halt advancing flows at Kapoho but failed to divert the flows more than temporarily (Macdonald and Abbott 1970). Watering the flow margin to retard forward movement was also tried at Kapoho. Macdonald (1962) indicates that the action was effective in reducing radiant heat and in temporarily (for several hours) checking the advance. Similar results were experienced on Heimaey Island, where vast quantities of salt water were sprayed on the leading edges of lava flows (Grove 1973; U.S. Geological Survey 1976).

For protection against volcanic mudflows and floods, existing dams—as in the river valleys of Mount Rainier and other Cascade volcanoes—might well hold back the onslaught if drawdown were initiated quickly at the first signs of danger (Crandell 1973). Waldron (1967) has suggested that rehabilitation of headwater areas after initial mudflows may reduce the frequencies of further mudflows, particularly in cases where watersheds have been altered hydrologically by accompanying ashfalls.

To prevent damage or injury from ashfall, protective actions that could be taken by individuals include shoveling ash from roofs to prevent structural damage, wearing protective goggles and breathing masks, and storing water as a hedge against contaminated supplies (see Wilcox 1959; Crandell and Waldron 1969).

Two ways of spreading the burden of loss are *insurance* and *relief and rehabilitation*. Insurance against the risk of loss from volcanic hazards is available only on a limited scale in the United States. Although not normally covered under standard forms of property insurance, volcanic risk may be included in certain forms of supplemental all-risk coverage, at the discretion of individual insurers (Warrick 1975). At the other extreme, in New Zealand all specific volcanic hazards are covered under a universal national disaster insurance (O'Riordan 1971).

Although there has never been a national disaster declaration from volcanic eruption within the United States, the maze of federal, state, and local relief and rehabilitation organizations would come alive in the event of a major volcanic disaster. The residents of Puna, for example, benefited from about $57,000 expended by the local American Red Cross to aid victims of the 1960 eruption (Hawaiian Chapter of the American Red Cross, no date). At the international level, the United Nations Disaster Relief Organization oversees international aid to volcanic disaster victims.

Finally, *land use management* refers to the control of the type and extent of occupancy of hazardous areas in order to reduce loss potential. Included are such measures as relocation, density requirements, structural requirements, and zoning. Given that the rarity and intensity of volcanic events make most of them uncontrollable, land use management is really the only broad, practicable alternative for volcanic hazard reduction over the long term (aside from possible predictive capabilities). However, the infrequency of most volcanic events works against its adoption. Moreover, difficult choices are required since many land uses in volcanically hazardous areas are tied directly to volcanoes or their eruptive products: for example, the agricultural benefits derived from the added soil fertility of ashfalls, or the recreational attractiveness of volcanic mountains. If land use management is to be seriously considered as a viable tool for reducing volcanic risk, then demanding judgments will be necessary as to the level of risk to be accepted (and by whom) in relation to the benefits to be derived from volcanic locations. In fact, this is a problem inherent in the whole adjustment process.

In general, the choice of adjustment to volcanic hazard—whether it be to do nothing at all, to develop a predictive capability, to formulate volcanic-specific disaster preparedness plans, to adopt protective measures or insurance provisions, to implement land use management, or some combination thereof—is one of evaluation. This involves weighing and balancing risks, costs, and benefits of alternative options. It depends upon how, as individuals or society, we appraise the hazard, perceive the range of adjustments and their consequences, and employ methods of choice.

## The Process of Adjustment Choice

As individual decision makers, how do we make choices about low-probability events like volcanic eruptions? As a society, which adjustments, or mix of adjustments, **should** we adopt to lessen volcanic risk? The two questions are linked: The understanding of individual choice evaluation—of value and preference—has implications for action we take as a society in adjusting volcanic risks to more socially acceptable levels. Some insights can be gained from studies of similar natural hazards.

In pulling together studies conducted on a large range of natural hazards, Burton et al. (1978) observed broadly that adjustment choices made by individuals are the result of a process of (a) appraisal of the probability and magnitude of the event; (b) canvassing the range of possible alternative actions; (c) evaluating the consequences of selected actions; and (d) choosing one or more actions. Thus, the way in which people estimate rare events and their consequences (as discussed above) can influence this process considerably. Likewise, the way in which individuals evaluate (or place meaning or value on) events and their outcomes in light of perceived alternative adjustments will help guide final choice, whether it be to rely on evacuation and relief in Puna District, or to move to a less risky location downvalley from Mount Rainier. How do individuals evaluate? Several models attempt to explain choice.

A *rational model* of decision making under risk and uncertainty portrays individuals as omniscient beings who are able to calculate the expected values arising from a range of probabilistic events and their associated consequences. The final choice of action reflects the individual's ability to maximize expected utility among all possible available alternatives.[9] Thus, according to this model of man, our economically rational resident of Puna District would accurately assimilate all relevant technical information on Kilauea's eruptive characteristics, comparatively assess the full costs and benefits of all possible alternatives, and systematically select the most favorable course of action on the basis of best average returns. In short, he is an *optimizer* (Slovic et al. 1974).

However, a number of empirical studies of natural hazards have demonstrated that such a model of choice does not always adequately explain behavior. For example, it was discovered that many farmers on U.S. flood plains failed to make intensive uses of newly protected land, despite the fact that the Soil Conservation Service's justifications assumed that farmers would attempt to reap optimum average returns (Burton 1962). Some people inexplicably

---

[9] The theoretical foundation was laid by von Neumann and Morganstern (1947).

refuse to evacuate their seacoast homes upon receiving news from an elaborate warning system that hurricane landfall is extremely probable (Burton *et al.* 1969). Most flood plain dwellers fail to purchase flood hazard insurance policies, although the policies are heavily subsidized by the U.S. government (Kunreuther 1977). In terms of expected utility, often these are not optimal decisions.[10]

Alternatively, a model of *bounded rationality* supposes that people make decisions in a boundedly rational manner, content to be *satisficers* rather than optimizers (Simon 1957; Kates 1962; Slovic *et al.* 1974). According to this model of decision making, choice among a range of alternatives in dealing with hazard is based on the individual's perception of them, which is conditioned by environmental, social, and psychological factors. In effect, a model of bounded rationality leaves room for the influence of cultural traits, social norms, cognitive limitations, and personality differences on the individual's perception of the states of nature and alternative actions, and on his evaluative processes. Thus, although seemingly irrational to the Western eye, the decision of the Hawaiian living in Puna District to refuse to evacuate the contents of his home lying in the path of an advancing lava flow is consistent with his beliefs in the goddess Pele, who determines such dramatic events and, consequently, peoples' fates (Murton and Shimabukuro 1974).

It may well be asked, Why is it important to understand individuals' choice processes under risk and uncertainty of volcanic hazards when it is demonstrably difficult to change human nature? There are two responses to this question. First, though it can be argued that boundedly rational decisions may be adequately adaptive in many contexts, like business (Cyert and March 1963), the evidence from natural hazards research indicates that such decisions often may lead to results contrary to actual value preferences, and in the long run exacerbate hazard risks instead of mitigating them. This is particularly true for infrequent events, like volcanic eruptions of Mount Rainier, which deny feedback necessary for trial-and-error learning.

And, second, boundedly rational people should not be viewed as incorrigible. The central premise underlying studies of decision making under environmental risk and uncertainty is that better understanding of the decision process will allow exploration of ways in which alteration of the environmental, social, and informational conditions of choice might lead to better decisions. It is hoped that this would achieve volcanic risk levels that are more consistent with individual and societal goals of hazard reduction.

It is this last point that leads us to address two closely related issues surrounding societal response to volcanic hazards, namely: How can scientific information on volcanic hazards best be used? What is an acceptable level of volcanic risk?

---

[10]*Subjective* models of decision making substitute subjective (or personally derived) probabilities for objective probabilities but similarly depend on expected utility theory. These, too, tend to fall short in explanatory power in real risk situations (Savage 1954; in hazard context, see Slovic *et al.* 1974).

## How Can Emerging Volcanic Risk Information Best Be Utilized?

In March 1975, a volcanic event began at Mount Baker, Washington, that was to highlight several critical needs in our social capability for handling not only rare volcanic events but numerous other environmental and technological risks as well. For the first time since the eruption of 1870, Mount Baker showed signs of restlessness. Thermal activity increased, and subsequently major fumarolic disturbances were recorded. Among scientists there was speculation as to a renewed period of volcanic activity. As a result, there was a reassessment of the short-term risk to lives and property in the area, particularly from landslides and mudflows. The calculated paths of such flows cut through the forested recreational slopes of Mount Baker and terminate in Baker Lake, a reservoir used for both recreation and hydroelectric power. A mass movement into the lake could create waves 1–10 m high, depending on the characteristics of the event. The U.S. Geological Survey issued a press release that estimated the likely magnitude of such events and their rough probabilities of occurrence.

Responding to the new risk information, the U.S. Forest Service and other agencies took precautionary measures to reduce possible losses, including the closure of a number of campgrounds and other recreational areas to the public. The water level of Baker Lake was purposefully kept low in order to accommodate the volume of a mass movement should it occur.

The reaction to these measures was sharp and controversial. Many of the residents of the nearby town of Concrete and the Upper Skagit River Valley felt that the risk was small and certainly not worth the economic loss projected from decreased tourism as a result of actions on the part of the Forest Service. The Puget Power Company faced losses in hydroelectric generating capacity amounting to hundreds of thousands of dollars as a consequence of keeping the reservoir level low; there was concern over power shortage. The owner of a resort on Baker Lake faced a substantial financial loss. With recreational facilities closed, 2000–3000 campers would be deprived of recreational opportunities. There was strong sentiment that this was just another example of meddling by the federal government.[11]

The case of Mount Baker illustrates the gap that exists between scientific information on risk and the subsequent use of that information in affecting sound policy decisions.[12] Increasingly, the public will rely upon technical information—like the U.S.G.S. risk estimation for Mount Baker—for decisions concerning adjustment to volcanic hazards. Yet, one cannot expect the translation of the information to be either simple or straightforward, for three reasons. First, it would be naive to believe that such information is interpreted consistently by all persons involved. What does a .01 chance in any given year

---

[11]See Hodge et al., Chapter 7 in this volume.
[12]See SCOPE (1973) for a broad discussion of this subject.

of a mudflow $30 \times 10^6 m^3$ traveling at speeds up to 30–50 km/hr mean to a resort owner on Baker Lake who has never in his life witnessed such an event? If we have learned anything from a multitude of studies of natural and technological hazards, it is this: It cannot be assumed that decision makers, from the individual at risk to the national policy formulator, will interpret risk information in the same way as the scientist—the geologist, the hydrologist, the seismologist. Relatively little systematic attention has been paid to the question of what mode of presentation of risk information would ease accurate interpretation by all decision makers involved.

Second, it cannot be assumed, in a normative fashion, that all persons will evaluate and make their decisions in a "rational" way. Thus, like the nuclear engineer who cannot understand why a large segment of the public rejects the option of nuclear power on the basis of exceedingly small risks, the geologist and volcanologist cannot expect evaluations of risk by the public to be identical to their own. The model of "bounded rationality" appears to apply at all levels, from expert to layman. To the Forest Service at Mount Baker the preventive actions taken were reasonable in light of the risks; to the citizens of Concrete, the actions were an overreaction. Better understanding of how people actually evaluate risk and make choices would help bridge the gap.

Third, the way in which information flows—through which channels and by whose direction—can significantly shape social response. As information is disseminated, it gets modified in various ways and supplemented with informal interpretations along the route. If a more precise a priori knowledge of the information flow process were at hand, risk information could be channeled accordingly. In the case of Mount Baker, the possibility of promoting a smoother decision-making process by knowing **how** to disseminate new volcanic risk information underscores the need to learn more about the process. A recent, similar situation at Hilo, Hawaii, in which a prediction was issued for a large lava flow eruption of Mauna Loa, prompted a research proposal to trace the flows of information and its use in what amounted to a rare social experiment (Murton and Sorensen 1977). It is knowledge of this sort that would allow better use to be made of information concerning future volcanic events.

In short, our ability to generate information on volcanic risk is outstripping our knowledge of how to make the best use of it. How and to whom information is presented and its mode of presentation can be influential in shaping public response. A fuller understanding of the ways in which scientific information can be conveyed to promote desirable decisions by users becomes crucial if the labors from research on volcanic hazards are to be realized fully. This applies as much to risk information from geophysical monitoring of volcanoes in the Cook Inlet Region of Alaska, from risk mapping of lava flow zones in Hawaii, and from geologic investigation of mudflows at Mount Rainier as it does to renewed volcanic activity at Mount Baker, Washington. Similarly, the complementary issue of acceptable risk levels applies to all volcanic areas.

## What Is an Acceptable Level of Volcanic Risk?

To the Puna resident, is it worth the risk to build a home where someday a lava flow may obliterate it? To a community in the Puyallup River Valley near Mount Rainier, is it worth the trouble and expense to establish a warning system to save lives in the event of a rare mudflow? To the state of Alaska, is it worth establishing emergency contingency plans to cope with an infrequent Mount Katmai-type ashfall on Anchorage, an event that could seriously strain the state's economy? To the federal government, is it worth supporting research directed at gaining understanding of the physical and social aspects of seemingly remote volcanic hazards? At all levels of social organization, the question of acceptable level of risk becomes a major issue as we increasingly cross over the threshold of volcanic hazard awareness and are forced to make conscious risk decisions—even if the choice is to do nothing.

In this regard, volcanic risks are similar to many low-probability technological risks we now have to confront head-on, like nuclear power plants, transportation of liquid natural gas (LNG), or atmospheric impact of fluorocarbons. Resulting controversies over these technologies revolve around the concern for risk and uncertainty in their use and around differing perceptions of what is and is not acceptable. Social choice is constrained by imprecise knowledge of individual and societal preference for risk under varying social, economic, and environmental conditions. At a fundamental level, many of these problems of risk are the same across natural and technological hazards (Lowrance 1976).

The potential gravity of a number of such hazards has drawn considerable attention to finding ways of helping to make rational, socially desirable risk decisions (Kates 1977). Efforts have been directed largely toward normative decision tools like comparative risk analysis and its cousin, benefit–risk analysis, which presumably indicate what society should be willing to accept in the way of risk (see Starr 1969, 1972; Otway 1972; Okrent 1975).

Methods of *comparative risk analysis* assume that socially acceptable levels of risk can be discerned from investigations of historical trends in societal risk acceptance, as in transportation, occupational, and natural hazard risks. Furthermore, it is assumed that these risk patterns are sufficiently enduring to serve as guidelines for public policy decisions. Thus, for example, a local official at Mount Baker, in trying to gauge the restrictiveness of recreation closures around a newly disclosed risk area, might well ask, How does the risk to life and limb from a landslide or volcanic mudflow compare with everyday risks faced (and presumably accepted) by the Washington state populace, like automobile accidents, fires, skiing, or cigarette smoking? On this basis, our local official might initially dismiss volcanic risk as being too insignificant. Yet, upon closer inspection of his method, one finds considerable ambiguity in the risk data upon which it is based. For instance, the risk of nuclear reactor core meltdown, though probably lower than many volcanic risks (according to engineering estimates), creates nationwide controversy; on the other hand, the high risk of death from automobile accidents is seemingly well tolerated. Clearly, other factors enter the risk acceptance equation.

As developed by Starr (1972), *benefit–risk analysis* explicitly takes into account benefits derived in relation to risks accepted. Risks are seen to vary with benefits gained, but within two discernible risk patterns, a voluntary and an involuntary. Thus, voluntary risks with high benefits are most tolerated, and involuntary risks with little return are least tolerated. In attempting to make a decision on recreational area closure, our local official at Mount Baker might estimate the benefits of the recreation area and then compare the volcanic risk with these social benefit–risk patterns to see what society **should** be willing to accept. However, our official should be cautioned that as a formal decision tool, the benefit–risk method has its drawbacks (see Otway and Cohen 1975).

Operationally, in a less formal way, benefit–risk methods may prove useful in forcing explicit recognition of the tradeoffs involved in the use of volcanically hazardous locations (Warrick 1975). Presently, even rough benefit–risk comparisons are rarely made. But changing awareness and perceptions of volcanic hazards may well alter the situation in the future as the results of volcanic hazard research become known. Benefit–risk comparisons may find considerable applicability in local situations where issues of growth and land use management become pressing, as in Puna District, where expensive residential and recreational developments multiply in notoriously hazardous lava flow zones.

In general, normative decision tools, though holding promise for applications in volcanic hazard decision making, come up against the problem of discrepancies between what people **should** accept and what they **do** accept in the way of risk. Obviously, the social evaluation of risk is a pivotal point in decision making but one that is not yet fully comprehended. Much more needs to be known about the risk acceptance process in general—about the role of physical, economic, psychological, and social factors in choice—before risk comparison methods can be calibrated finely enough so that judgments as to which levels of volcanic risk are acceptable can be made with confidence.

## SUMMARY AND CONCLUSION: TOWARD A MORE BALANCED APPROACH TO VOLCANIC HAZARD

So far we have touched upon a wide variety of concepts and issues that we hope have led the reader to view the problem of volcanic hazard from a slightly different perspective. From the approach taken herein, it is emphasized that hazards arise fundamentally from an interaction of natural and human systems. Ultimately, the responsibility for hazard rests on the human side of the equation. It is largely the actions of people that create or diminish vulnerability to one of nature's most spectacular phenomena—volcanoes.

To recapitulate, we found it convenient to view volcanic hazard from the vantage point of risk. In setting the hazard in a framework of risk assessment, it was possible to disaggregate the volcanic hazard problem into components: hazard identification, risk estimation, and social evaluation.

Hazard identification involves the recognition of harmful events or processes. Among those major volcanic phenomena potentially detrimental to people are lava flows, pyroclastic flows, ashfalls, volcanic mudflows, and volcanically generated tsunamis. Awareness of them, however, is another matter, since in most cases these phenomena occur only rarely. Geologic research helps to identify such hazards and where they might occur, partly with the intention of easing people across the threshold of awareness in the absence of disaster. In the United States, research of this nature has made notable headway during the last decade; volcanic hazard maps in various forms have begun to appear.

Volcanic risk estimation attaches a likelihood of occurrence to the events and measures their consequences. Studies of similar extreme, low-probability events suggest that individuals may run up against formidable obstacles in estimating volcanic risks. These range from sheer lack of experience to systematic cognitive limitations in handling probabilistic information. Although geologic investigations make progress in filling the knowledge gap by attempting to establish the probabilities of volcanic events, relatively little attention is paid to the estimation of the likely social consequences accompanying those extreme events. Yet both the elements of risk estimation—likelihood and consequences—are central to hazard decision making.

Social evaluation places value on volcanic risk estimates in relation to alternatives available for coping with the hazards. The range of possible human adjustments to volcanic hazards is rather limited and includes methods of prediction and warning, disaster preparedness, protective measures, loss sharing, and land use management of hazard areas. If we extrapolate from our rudimentary understanding of similar natural hazards, the processes by which people perceive, evaluate, and choose among alternatives for coping with volcanic hazards may be best described by a model of "bounded rationality." Understanding the evaluation process is critical if, as a society, we are to formulate sound volcanic hazard policy that is consistent with underlying social values and preferences and that can be implemented effectively.

Within the above framework, three major issues were highlighted. First, are we becoming more vulnerable to extreme volcanic events? The growing complexity of society suggests that we may be creating situations of increased social disruption in which debilitating social impacts may be extended, both spatially and temporally, in times of disaster. Second, how can emerging volcanic risk information best be utilized? The recent events at Mount Baker, Washington, serve to illustrate the widening gap that exists between the proliferating scientific information on risk and the subsequent use of that information in affecting sound policy decisions. Third, what is an acceptable level of volcanic risk? This issue lies at the crux of decisions to cope with volcanic threat, whether by individuals, communities, states, or the nation. Although there are no ready answers, the increasing attention being brought to bear on this issue in the fields of risk and technology assessment is encouraging.

In conclusion, this broad overview points to the need to consider fully the interdisciplinary nature of volcanic hazard. Encouragingly, there are definite signs that the trends are in that direction. In the past, volcanic research pursued a rather narrow line in which efforts were concentrated on the physical character of volcanoes. Geophysical and volcanologic research sought to achieve a greater understanding of the processes and mechanisms of volcanism, and geologic studies worked toward unraveling the mysteries of past volcanic activity. The primary objective of this research was pure scientific understanding; the quest for knowledge to help man better adjust to the threat of volcanic eruption was, for the most part, a secondary or incidental outcome of the research. Recently, however, there has been an apparent shift in emphasis; that is, there seems to be an increase in research directed specifically toward providing information relevant to volcanoes **as a hazard**. For instance, the recent series of investigations by the U.S. Geological Survey in the Cascade Range and in Hawaii are expressly concerned with volcanic hazards. Apparently, a growing realization of the latent threat of volcanic events, in light of continued human encroachment into volcanically hazardous areas, has stimulated the perceived need for research of this kind.

As we look to the future, attention paid to volcanic hazard could benefit from yet a broader-based approach. In order to address the sorts of issues raised in this chapter, there needs to be a more finely balanced blend of perspectives. Research inputs from the social and behavioral sciences are required to complement the physical and technological. Thus, we need not only refined estimates of probabilities and magnitudes of rare volcanic events, but also estimates of the accompanying loss of life and property and social disruption. We need not only research directed toward developing a predictive capability (as for the volcanoes in the Cascade Range), but also knowledge about how people would respond to the warning. We need not only better estimates of volcanic risk, but also greater understanding of how that information may be communicated to facilitate effective decision making and social response.

In short, whereas volcanic threat at one time may have been largely beyond our realm of consciousness, there is every indication that, as a society, we are awakening to the hazard. It is possible that many future public policy decisions may be forced to treat volcanoes explicitly. Presently, our arsenal of knowledge and decision tools appears inadequate. As this chapter attempted to illustrate, it is a more balanced approach to the problem that will allow informed, well-reasoned judgments to be made concerning volcanoes as hazards.

## REFERENCES

Belt, Collins, and Associates
   1962   A plan for the metropolitan area of Hilo. Report to the County of Hawaii. Hilo, Hawaii.
Blair, J. P.
   1964   Home to Tristan da Cunha. National Geographic 125 (January):60–81.

Bolt, B. A., W. L. Horn, G. A. Macdonald, and R. F. Scott
  1975  *Geological hazards*. New York: Springer-Verlag.
Bullard, F. M.
  1962  *Volcanoes in history, in theory, in eruption*. Austin: University of Texas Press.
Burton, I.
  1962  *Types of agricultural occupance of flood plains in the United States*. Chicago: University
         of Chicago, Department of Geography Research Paper No. 75.
Burton, I., and R. W. Kates
  1964  The perception of natural hazards in resource management. *Natural Resources Journal*
         3:412–441.
Burton, I., R. W. Kates, and R. E. Snead
  1969  *The human ecology of coastal flood hazard in Megalopolis*. Chicago: University of
         Chicago, Department of Geography Research Paper No. 115.
Burton, I., R. W. Kates, and G. F. White
  1968  *The human ecology of extreme geophysical events*. Natural Hazards Research Working
         Paper #1. Toronto: University of Toronto.
  1978  *The environment as hazard*. New York: Oxford University Press.
California Division of Mines and Geology
  1973  *Urban Geology: Master Plan for California*. Bulletin #198. Sacramento, Calif.
Coats, R. R.
  1950  *Volcanic activity in the Aleutian Arc*. U.S. Geologic Survey Bulletin #947-B. U.S.
         Department of the Interior. Washington, D.C.: U.S. Government Printing Office.
Cochrane, H. C.
  1974  Predicting the economic impact of earthquakes. In *Social science perspectives on the
         coming San Francisco earthquake: Economic impact, prediction, and reconstruction*,
         edited by H. C. Cochrane, J. E. Haas, M. J. Bowden, and R. W. Kates, Natural Hazard
         Research Working Paper #25. Boulder: University of Colorado.
Corps of Engineers
  1966  *Review report on survey for lava flow control, island of Hawaii*, State of Hawaii. U.S.
         Department of the Army. Honolulu, Hawaii: Corps of Engineers, Ford Armstrong.
Crandell, Dwight R.
  1967  *Volcanic hazards at Mount Rainier, Washington*. U.S. Geological Survey, U.S. Depart-
         ment of the Interior. Washington: U.S. Department Printing Office.
  1971  *Postglacial lahars from Mount Rainier Volcano, Washington*. Geological Survey Profes-
         sional Paper #677. U. S. Department of the Interior. Washington: U. S. Government
         Printing Office.
  1973  *Potential hazards from future eruptions of Mt. Rainier, Washington*. U. S. Geological
         Survey Misc. Geological Investigations Map #1-836. U. S. Department of the Interior.
         Washington: U. S. Government Printing Office.
  1975  *Assessment of volcanic risk on the island of Oahu, Hawaii*. U. S. Geological Survey
         Open-File Report 75–287.
Crandell, D. R., and D. R. Mullineaux
  1974  Appraising volcanic hazards of the Cascade Range of the northwestern United States.
         *Earthquake Information Bulletin* 6 (No. 5, September–October):3–10. Reston, Va.: U. S.
         Geological Survey.
  1976  *Potential hazards from future eruptions of Mount St. Helens Volcano, Washington*. U. S.
         Geological Survey Open-File Report 76–491.
Crandell, D. R., D. R. Mullineaux, and M. Rubin
  1975  Mount St. Helens Volcano: Recent and future behavior. *Science* 187:438–441.
Crandell, D. R., and H. H. Waldron
  1969  Volcanic hazards in the Cascade Range. In *Geologic Hazards and Public Problems:
         Conference Proceedings*, edited by Robert A. Olson and Mildred Wallace. Office of
         Emergency Preparedness. Region Seven. Santa Rosa, Calif.

Cullen, J.
  1978   Impact of a major eruption of Mount Rainier on public service delivery systems in the Puyallup Valley, Washington. Unpublished master's thesis, Department of Urban Planning, University of Washington.
Cyert, R. M., and J. G. March
  1963   A behavioral theory of the firm. Englewood Cliffs, N. J.: Prentice-Hall.
Davidson, G.
  1884   Notes on the volcanic eruption of Mount St. Augustin, Alaska, October 6, 1883. Science 3:186–189.
Detterman, R.
  1974   U. S. geological survey map GQ-1068. U. S. Department of the Interior, Washington, D. C.: U. S. Government Printing Office.
Ericksen, N. J.
  1975   Scenario methodology in natural hazards research. Boulder: University of Colorado Institute of Behavioral Science.
Fiske, Richard S., and Willie T. Kinoshita
  1969   Inflation of Kilauea Volcano prior to its 1967–68 eruption. Science 165:341–349.
Forbes, R. B., J. Kienle, and D. Harlow
  1972   Eruptive history and microseismicity of Augustine Volcano, Alaska. College, Alaska: University of Alaska Geophysical Institute.
Frank, D., M. F. Meier, and D. A. Swanson
  1977   Assessment of increased thermal activity at Mount Baker, Washington, March 1975–March 1976. U. S. Geological Survey Professional Paper 1022-A.
Friedman, D. G.
  1975   Computer simulation in natural hazard assessment. Boulder: University of Colorado Institute of Behavioral Science.
Furneaux, R.
  1964   Krakatoa. Englewood Cliffs, N. J.: Prentice-Hall.
Gedney, L., C. Matteson, and R. B. Forbes
  1970   Seismic refraction profiles of the ash flow in the Valley of Ten Thousand Smokes, Katmai National Monument, Alaska. Journal of Geophysical Research 75:2619–2624.
Golant, S., and I. Burton
  1969   The meaning of a hazard: Application of the semantic differential. Natural Hazards Research Working Paper No. 7. Boulder: University of Colorado Institute of Behavioral Science.
Grove, N.
  1973   Volcano overwhelms an Icelandic village. National Geographic 144 (July):40–67.
Gutierrez, C.
  1972   A narrative of human response to natural disaster: The eruption of Paricutin. College Station: Texas A & M University, Environmental Quality Program.
Hawaiian Chapter of the American Red Cross
       Report of Red Cross Involvement in Disasters in Hawaii. Honolulu, Hawaii.
Hawaiian Civil Defense, Department of Defense
  1971   The State of Hawaii Plan for Emergency Preparedness: Disaster Assistance, Volume 3. Honolulu, Hawaii.
Hobbs, P. V., L. F. Radke, and J. L. Stith
  1977   Eruptions of the St. Augustine Volcano: Airborne measurements and observations. Science, 195 (4281):871–872.
Holden, C.
  1978   ABASS: Social sciences carving a niche at the academy. Science 199 (434):1183–1187.
Hyde, J. H., and D. R. Crandell
  1978   Postglacial volcanic deposits at Mount Baker, Washington, and potential hazards from future eruptions. U. S. Geological Survey Professional Paper 1022-C.

International Volcanological Association
    1960   *Catalogue of the active volcanoes of the world*, Parts 3 and 9. Naples, Italy: International Volcanological Association.

Jaggar, T. A.
    1945   *Volcanoes declare war*. Honolulu: Paradise of the Pacific, Ltd.

Kates, R. W.
    1962   *Hazard and choice perception in flood plain management*. Chicago: University of Chicago Department of Geography, Research Paper #78.
    1970   *Natural hazard in human ecological perspective: Hypotheses and models*. Natural Hazards Research Working Paper #14. Boulder: University of Colorado.
    1976   *Risk assessment of environmental hazard*. SCOPE 8. International Council of Scientific Unions, Scientific Committee on Problems of the Environment. Paris, France.
    1978   *Risk assessment of environmental hazard*. New York: Wiley.

Kates, R. W. (Editor)
    1977   *Managing Technological Hazard: Research Needs and Opportunities*. Boulder: Institute of Behavioral Science, University of Colorado.

Keesing, F. M.
    1952   The Papuan Orokaiva vs. Mt. Lamington: Cultural shock and aftermath. *Human Organization* 11:16–22.

Kunreuther, H. R. Ginsberg, L. Miller, P. Sagi, P. Slovic, B. Borkan, N. Katz
    1977   Limited knowledge and insurance protection: Implications for natural hazard policy. Executive Summary. Wharton School, University of Pennsylvania.

Lachman, R. and W. J. Bonk
    1960   Behavior and beliefs during the recent volcanic eruption at Kapoho, Hawaii. *Science* 131:1095–1096.

Lamb, H. H.
    1972   *Climate: Present, past, and future, Volume 1: Fundamentals and climate now*. London: Methuen.

Lange, I. M., and Jon C. Avent
    1973   Ground-based thermal infrared surveys as an aid in predicting volcanic eruptions in the Cascade Range. *Science* 182 (October 19):279–281.

Latter, J. H.
    1969   Natural disasters. *The Advancement of Science* 25:362–380.

Lowrance, W. W.
    1976   *Of acceptable risk*. Los Altos, Calif.: William Kaufmann, Inc.

Macdonald, G. A.
    1962   The 1959 and 1960 eruptions of Kilauea Volcano, Hawaii, and the construction of walls to restrict the spread of the lava flows. *Bulletin Volcanologique* 24:248–294.
    1972   *Volcanoes*. Englewood Cliffs, N. J.: Prentice-Hall.

Macdonald, G. A., and A. T. Abbott
    1970   *Volcanoes in the sea: The geology of Hawaii*. Honolulu: University of Hawaii Press.

McBirney, A. R.
    1966   Predicting volcanic eruptions. *Discovery* 27 (April):20–25.

Mileti, D.
    1975   *Natural hazards warning systems in the United States: A research assessment*. Boulder: University of Colorado Institute of Behavioral Science.

Mitchell, J. K.
    1974   Natural hazards research. In *Perspectives on environment*, edited by I. Manners and M. Mikesell, Commission on College Geography, Publication 13. Association of American Geographers.

Mullineaux, D. R., and D. W. Peterson
    1974   *Volcanic hazards on the Island of Hawaii*. U. S. Geological Survey Open-File Report 74-239.

Murton B., and S. Shimabukuro
   1974   Human adjustment to volcanic hazard in Puna District, Hawaii. *Natural hazards: Local, national, global*, edited by G. F. White. New York: Oxford University Press.
Murton, B., and J. Sorensen
   1977   Draft of research proposal on information flow in volcanic hazard warning. Hawaii: Department of Geography, University of Hawaii. (Unpublished)
Nolan, M. L.
   1972   The towns of the volcano: A study of the human consequences of the eruption of Paricutín Volcano. Unpublished doctoral dissertation, Department of Geography, Texas A & M University.
Office of Emergency Preparedness
   1972   *Disaster preparedness*, Volumes 1 and 3. Executive Office of the President. Washington: U. S. Government Printing Office.
Okrent, D.
   1975   *Risk-benefit methodology and application: Some papers presented at the Engineering Foundation Workshop, Asilomar.* Los Angeles: University of California, Dept of Energy and Kinetics, UCLA-ENG 7598.
Ollier, C.
   1969   *Volcanoes*. Cambridge, Mass.: MIT Press.
O'Riordan, T.
   1971   *The New Zealand Earthquake and War Damage Commission—A study of a national natural hazard insurance scheme*. Natural Hazards Research Working Paper No. 20. Boulder: University of Colorado Institute of Behavioral Science.
Otway, H. J.
   1972   The quantification of social values. In *Risk vs benefit: Solution or dream*, edited by H. J. Otway. Los Almos, New Mexico: Los Alamos Scientific Laboratory of the University of California.
Otway, H. J., and J. J. Cohen
   1975   *Revealed preferences: Comments on the Starr benefit–risk relationship*. Laxenburg, Austria: International Institute for Applied Systems Analysis, RM 75 7.
Otway, H. J., and P. D. Pahner
   1976   Risk assessment. *Futures* 8:122–134.
Peterson, D. W.
   1973   Personal communication. Hawaii National Park: U. S. Department of the Interior.
Rees, J. D.
   1970   Paricutín revisited: A review of man's attempts to adapt to ecological changes resulting from volcanic catastrophe. *Geoforum* (April):7–26.
Rittmann, A.
   1962   *Volcanoes and their activity*. E. A. Vincent (trans.). New York: Wiley.
Savage, L. J.
   1954   *The Foundations of Statistics*. New York: Wiley.
Scientific Committee on Problems of the Environment (SCOPE)
   1973   *International research on societal response to scientific information about man-made environmental hazards*. A Report of a SCOPE Seminar, Holcomb Research Institute, Butler University, Indianapolis, Indiana, August 21–24, 1973. SCOPE/EPRI/UNEP Risk Assessment Workshop Background Paper #3.
Simon, H. A.
   1957   *Models of man*. New York:Wiley.
Slovic, P., B. Fischhoff, and S. Lichtenstein
   1976   Cognitive processes and societal risk-taking. In *Cognition and social behavior*, edited by J. S. Carroll and J. W. Payne. Potomac, Md.: Lawrence Erlbaum Associates.
Slovic, P., H. Kunreuther, and G. F. White
   1974   Decision processes, rationality, and adjustment to natural hazards. In *Natural hazards: Local, national, global*, edited by G. F. White. New York: Oxford University Press.

Starr C.
   1969   Social benefit versus technological risk. *Science* 166:1232–1238.
   1972   "Benefit–cost studies in sociotechnical systems. In Committee on Public Engineering Policy, *Perspectives on benefit–risk decision making*, edited by C. Starr. Washington: National Academy of Engineering.
Tazieff, H.
   1967   The menace of extinct volcanoes. *Impact* 17 (2):135–148.
Tversky, A., and D. Kahneman
   1971   Belief in the law of small numbers. *Psychological Bulletin* 76:105–110.
U. S. Geological Survey
   1966   *Volcanoes of the United States*. U. S. Department of the Interior. Washington: U. S. Government Printing Office.
   1976   *Man against volcano: The eruption on Heimaey, Vestmann Islands, Iceland*. United States Department of the Interior. Washington, D.C.: U.S. Government Printing Office.
University of Alaska Geophysical Institute
   1972   *Volcanic risk in the Cook Inlet region*. College, Alaska: Author.
von Neumann, J., and O. Morgenstern
   1947   *Theory of games and economic behavior*. Princeton, N.J.: Princeton University Press.
Waesche, H. H., and D. L. Peck
   1966   Volcanoes tell secrets in Hawaii. *Natural History* 65 (March):20–29.
Waldron, H. H.
   1967   *Debris flow and erosion control problems caused by the ash eruptions of Irazu Volcano, Costa Rica*. U. S. Geological Survey Bulletin #1241-1. U. S. Department of the Interior. Washington, D.C.: U. S. Government Printing Office.
Warrick, R. A.
   1975   *Volcano hazard in the United States: A research assessment*. Boulder: Institute of Behavioral Science, University of Colorado.
Watt, K. E. F.
   1973   Tambora and Krakatoa: Volcanoes and the cooling of the world. *Saturday Review of Science* 4 (January):43–44.
White, G. F.
   1964   *Choice of adjustment to floods*. Research Paper #93. Chicago: University of Chicago Department of Geography.
   1973   Natural hazards research. In *Directions in geography*, edited by R. J. Chorley. London: Methuen.
White, G. F. (Editor)
   1974   *Natural hazards: Local, national, global*. New York: Oxford University Press.
White, G. F., and J. E. Haas
   1975   *Assessment of research on natural hazards*. Cambridge, Mass.: MIT Press.
Wilcox, R. E.
   1959   *Some effects of recent volcanic ash falls, with especial reference to Alaska*. U. S. Geological Survey Bulletin #1028-n. U. S. Department of the Interior. Washington, D.C.: U. S. Government Printing Office.

# 7

# Volcanic-Hazards Studies in the Cascade Range of the Western United States

DWIGHT R. CRANDELL
DONAL R. MULLINEAUX
C. DAN MILLER

## INTRODUCTION

The Cascade Range stretches from northern California through Oregon and Washington into southern British Columbia (Figure 7.1). Scattered along the Cascades south of Canada are scores of volcanoes of geologically recent age. They range in size from small cinder cones to the massive complex cones of Mount Shasta and glacier-covered Mount Rainier, whose summits are thousands of meters above their bases. Eruptions of some of these volcanoes in the past must have significantly affected the lives of nearby inhabitants; indeed, accounts of volcanic activity are common in legends of Indians who lived in the Pacific Northwest, and similar accounts appear in the written records of early settlers.

Because they are situated within a broad mountain range, most Cascade volcanoes are relatively distant from major centers of population. Furthermore, the largest population centers lie west of the Cascades; thus, the prevailing westerly winds in this region have carried most ash erupted at the volcanoes away from the sites of such cities as Portland and Seattle. Consequently, few ash eruptions in the recent past have affected areas that are densely populated today.

Only a few Cascade volcanoes have erupted during the time of written records, which in this region is the time since shortly after 1800. Furthermore, historic eruptions of most of these volcanoes have been infrequent compared with a person's lifetime and on a relatively small scale in terms of volume of

195

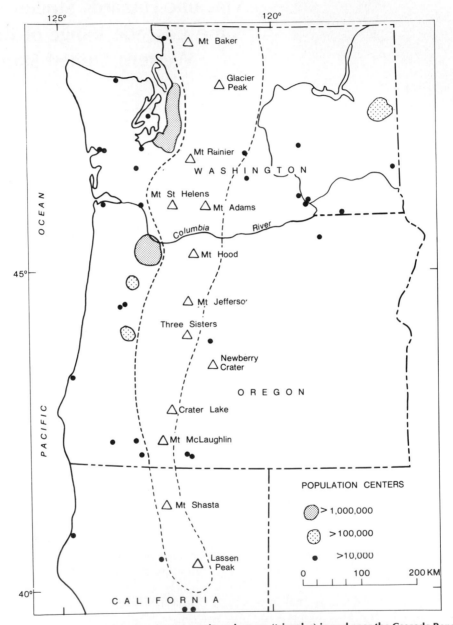

FIGURE 7.1. Location of major composite volcanoes (triangles) in and near the Cascade Range (dashed line), and population centers in Washington, Oregon, and northern California.

rock erupted and area affected. Thus, the sparse record of historic activity discloses neither the probable frequency of future volcanism nor the full range in kinds and scales of future eruptive activity possible at even these few volcanoes. This kind of information can be acquired, however, by studying the eruptive history of the volcanoes during geologically recent time and thereby extending our knowledge of their behavior back thousands of years.

The lack of an extensive historic record of eruptions in the Cascade Range has resulted in a general public disregard for the possibility of future eruptions, much less their likely effects. A disregard for potential volcanic hazards is understandable because the public generally is not aware that many volcanoes in the Cascade Range are only dormant, not extinct. Moreover, the infrequency of eruptions during the last hundred years hardly encourages the perception of volcanic eruptions as a direct and immediate danger comparable to others encountered in everyday living. Nevertheless, future volcanic events like some in the recent past could affect vast areas and tens of thousands of people.

Detailed studies by the U.S. Geological Survey have been in progress for about 10 years for the express purpose of assessing hazards that could result from future eruptions in the Cascade Range. These studies have led to the recognition of the kinds, scales, and areas affected by geologically recent eruptive events at six potentially dangerous volcanoes: Mounts Baker, Rainier, and St. Helens in Washington; Hood in Oregon; and Mount Shasta and the Lassen Peak area in California. Assessments of potential hazards based on these data can be used by land-development planners, engineers charged with selecting sites for dams and nuclear power plants, governmental agencies responsible for disaster response and mitigation, and the general public in making informed decisions regarding areas in which to invest and live.

These investigations are notably different from most studies concerned with future eruptions of dormant volcanoes. Such studies attempt to recognize an impending eruption after molten rock has already begun to move into a volcano but has not reached the surface, and they emphasize geophysical and geochemical monitoring techniques that could permit an eruption to be predicted by detection of precursory events. Such techniques include monitoring of earthquakes, ground-tilt measurements, and changes in heat flow, as well as many others. In contrast, the studies described in this chapter are concerned chiefly with forecasting the kinds, scales, and distribution of hazardous volcanic phenomena that are possible if an eruption does occur, as well as the likelihood of such events.

Maps that show areas of potential danger from future eruptions have now been compiled for several volcanoes in the Cascade Range (Crandell 1973; Crandell and Mullineaux 1978; Hyde and Crandell 1978) and Hawaii (Mullineaux and Peterson 1974; Crandell 1975). Similar maps have been prepared for volcanoes in other countries, such as Kelud Volcano in Indonesia (Neumann van Padang 1960), Mount Vesuvius (Pinna and Scandone in Barberi and Gasparini 1976), La Soufrière on Guadaloupe in the French West Indies

(Denis Westercamp, written commun. 1977), Tenerife in the Canary Islands (Booth 1977), some volcanoes in Kamchatka, U.S.S.R. (Bolt *et al.* 1975: 121–122), and Cotopaxi Volcano in Ecuador (Miller *et al.* 1978). Volcanic-risk maps like these are desirable for recently active volcanoes throughout the world, and especially in areas where eruptions could affect large numbers of people. Such maps, along with nontechnical descriptions of the potential hazards, are immediately useful in pointing out areas of danger when the volcano does erupt and can also be used for long-term land-utilization planning for regions around volcanoes.

Although the primary purpose of our studies is to assess potential hazards, detailed information concerning the deposits from past eruptions is useful in other kinds of studies. For example, identifiable deposits of widespread volcanic ash can be used as time-stratigraphic markers in geological, archaeological, and anthropological investigations once their age and distribution have been determined. Knowledge of the chronology and character of eruptions is also a prerequisite to the investigation of such fundamental questions as the origin and evolution of magmas and the possible periodicity or cyclicity of eruptions at a volcano.

## METHOD OF VOLCANIC-HAZARDS ASSESSMENTS

In order to reconstruct the history of a volcano during a given period of time, we study the rocks and unconsolidated deposits produced by volcanic activity during that period. Specifically, we conduct detailed studies of the stratigraphic successions of volcanic rocks and other deposits on all sides of a volcano, determining insofar as possible the kinds of volcanic events they represent as well as the extents, volumes, and chronologic ages of the deposits. For example, we may recognize the deposit of a hot pyroclastic flow by the presence of charred wood and by the emplacement temperature of the deposit as indicated by the magnetic properties of rock clasts in it (Hoblitt and Kellogg 1976). Radiocarbon dating of the wood will provide the approximate age of the pyroclastic flow as well as the age of the eruption that produced it. The areal distribution of pyroclastic-flow deposits provides a clue to areas that could be affected by activity of the same kind in the future, if the topography of the volcano and adjacent areas has not significantly changed.

The age and areal extent are determined for all the products of volcanism—lava flows, domes, pyroclastic flows, tephra,[1] and mudflows—around the volcano. When these studies are completed, a composite record of volcanic events can be assembled for whatever time period seems adequate to illustrate a wide range of past activity for that volcano. The frequency of past eruptive activity can also be inferred from this record, but this is only a minimum frequency because of the possibility that some products of volcanism

---

[1]In this chapter the term *tephra* is restricted to airfall deposits.

were removed or buried by later events and deposits. Moreover, the records of historic eruptions in the Cascade Range indicate that not all eruptive events produce recognizable deposits.

After a composite record of volcanic events has been compiled, we can prepare volcanic-risk maps that show the possible areal extent of volcanic phenomena in the future as inferred from the extent of similar phenomena in the past. However, there are limitations on the use of past events to forecast future hazards. Inasmuch as the volume of a future eruption seldom, if ever, can be predicted, maps that show possible extents of various volcanic phenomena obviously cannot be used to predict the exact degree of hazard at a specific location. A second limitation relates to extrapolating from one event or area to another. For example, the Electron Mudflow extended about 55 km from Mount Rainier down one valley about 500 radiocarbon years ago (see Figure 7.5, p. 206). This mudflow was used by Crandell (1973) as a model to portray the potential extent of mudflow hazard in certain other valleys that drain Mount Rainier. The Electron Mudflow evidently was caused by a massive clay-rich avalanche of hydrothermally altered rock from the volcano. The same model should not be used, therefore, for a valley that does not head in an area of altered rock on the volcano. The Electron Mudflow was one of the four largest mudflows from Mount Rainier during Holocene time. Future mudflows as large as these are far less likely than mudflows of much smaller volume, and the smaller mudflows do not extend as far, or as high on valley walls. Because large mudflows are relatively infrequent, and small mudflows travel relatively short distances, the degree of mudflow risk decreases progressively away from the volcano along any valley. The risk also decreases with increasing height above the present river at any place in a valley. These concepts are shown diagrammatically in Figure 7.2.

The degree of hazard from future eruptions of tephra is assessed by determining the areal extent, thickness, and approximate volume of tephra deposits that have resulted from the past eruptions of a given volcano. These data can be used in conjunction with records of wind directions and speeds to construct a model of probable fallout areas and distance–thickness relations for future tephra eruptions of roughly the same character and volume as those of the past.

For example, models based on selected tephra eruptions of the past were used at Mount St. Helens to indicate potential hazards from tephra of three volumes, on the order of 0.01, 0.1, and 1 km³ (Crandell and Mullineaux 1978). The areas of most probable tephra fallout (Figure 7.3a) were based on the frequency of wind directions and speeds at the altitudes that probably would be reached by the eruption column. Thus, a 155° sector extending away from the volcano from about north–northeast to south–southeast is the area in which the fall of tephra is most likely. Wind speeds, which determine how far tephra is carried, are also relatively high toward that sector. The overall risk is proportional to tephra thickness, which decreases progressively away from the volcano (Figure 7.3b). Although all of the 155° sector east of the volcano has a

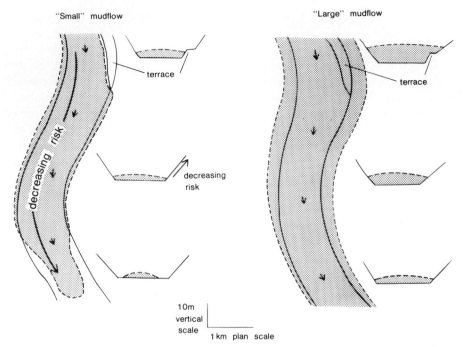

FIGURE 7.2. Diagrammatic maps and cross sections of hypothetical "small" and "large" mudflows, showing relation of potential risk to the length and thickness of the mudflow in a valley. The shaded portions show extent and maximum height reached by mudflow, but thicknesses that remain after deposition may be substantially less (Crandell 1971: 7). Short arrows point downvalley. The frequency of mudflows is inversely proportional to their size.

similar degree of risk at the same distance from Mount St. Helens, only a small part of that sector would be threatened by a single tephra eruption at any time of moderate to strong winds. This forecast is based on fallout patterns of several single tephra deposits from Mount St. Helens in the past, which cover sectors much narrower than 155° (Figure 7.4).

We are still in a developmental stage with respect to volcanic-hazards appraisals in the United States. Techniques are still being sought by which we can identify and assess a potential hazard from the character, distribution, and age of a particular volcanic rock unit or succession of rock strata formed by past eruptions. An important question for which we do not yet have an answer is, Have the volcanoes under investigation exhibited any cyclic or evolutionary behavior with respect to the timing of eruptions or the character of eruptive activity? Recognition of such behavior, if it has occurred, and recognition of the stage a volcano is now in, would enhance our ability to identify the most dangerous volcanoes and to gauge the possible consequences of the next eruption.

One difficult problem in assessing potential volcanic hazards is in determining the relative degree of risk that will result from various kinds of phenomena that may occur in the future. We are not, however, attempting to

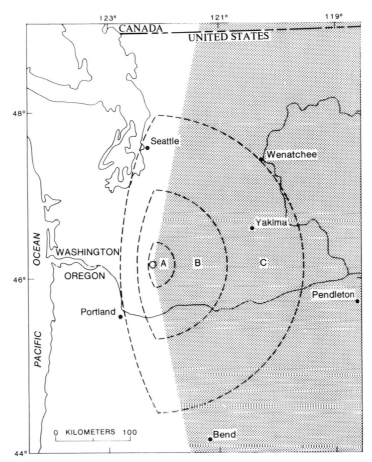

**FIGURE 7.3a. Map of tephra-hazard zones around Mount St. Helens, Washington. Potential tephra thickness is greatest in Zone A, and progressively less in Zones B and C. Winds blow into sector marked by pattern about 80% of the time; thus, tephra from most future eruptions will fall in that sector.**

determine the levels of risk that are acceptable or unacceptable; this is largely a socioeconomic problem to which we can contribute only one part of the data necessary to make decisions (see Warrick 1975:73–104, and Warrick, Chapter 6 of this volume). In addition, there is a problem in finding ways to communicate the nature of volcanic hazards clearly and effectively to a public that lacks technical knowledge of volcanic phenomena.

## RECENT ERUPTIVE HISTORIES OF FIVE CASCADE VOLCANOES

Detailed studies of the Holocene histories of five Cascade volcanoes show differences in style and frequency of eruptive behavior (Table 7.1). The recent

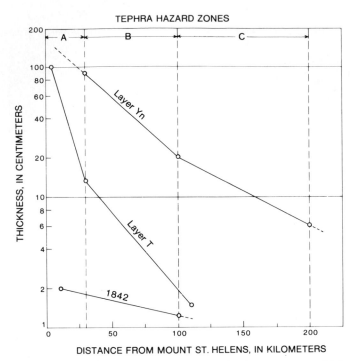

**FIGURE 7.3b. Graph showing progressive decrease of estimated average thickness of tephra with increase of distance along thickest parts of distribution lobes for three tephra layers of markedly different volumes from Mount St. Helens. Tephra layers 1842, T, and Yn have estimated volumes on the order of 0.01, 0.1, and 1 km³, respectively (Crandell and Mullineaux 1978).**

eruptive histories of Mount Rainier and Mount Baker have some similarities to each other as well as some important differences; both differ greatly from Mount St. Helens and Mount Hood. During the last few thousand years Mount St. Helens has exhibited a great diversity of eruptive phenomena and has erupted more frequently than any other volcano in the Cascade Range. The differences that are apparent from the following brief summaries of recent activity clearly show that the future behavior of one volcano cannot be anticipated from the past eruptive record of another. Ideally, in order to assess potential volcanic hazards adequately, detailed studies should eventually be made of all the major Cascade volcanoes that have been active during Holocene time.

## Mount Baker

Mount Baker is a moderately large andesitic stratovolcano near the northern boundary of Washington (Figure 7.1). It was built mostly during Pleistocene time, and probably had attained its present size before the last major glaciation.

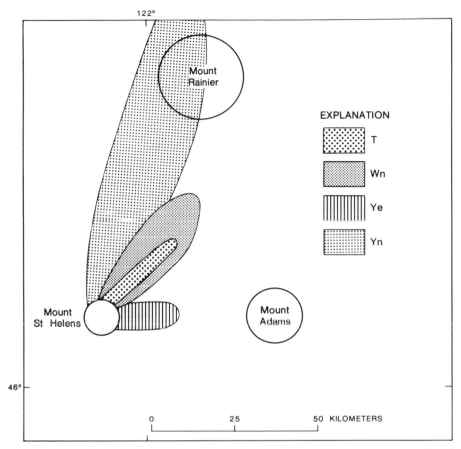

FIGURE 7.4. Orientation and width of four tephra lobes. Each lobe consists of air-fall tephra 20 cm or more thick from a single eruption. Layer T was erupted about 1800, layer Wn about 1500, and layers Yn and Ye about 4000 years ago. Lesser thicknesses of each tephra layer have been traced farther downwind to distances of more than 300 km from Mount St. Helens.

The first known eruption during postglacial time occurred about 10,300 years ago, when a small volume of black lithic ash was deposited on and beyond the east and northeast flanks of the cone (Hyde and Crandell 1978). It was followed by a major eruptive period that probably occurred sometime between 10,000 and 8000 years ago. At that time, lava flows, hot pyroclastic flows, and mudflows originating at Sherman Crater, a vent lower than and 800 m south of the summit of Mount Baker, formed a thick fill in a valley on the east side of the volcano. A little later, pumice was erupted at a vent near the south base of the volcano, and andesite lava from that vent subsequently flowed 12 km downvalley to the east. During the last 6000 years lithic ash was erupted several times, but even the largest ash deposit probably has a volume of less than 0.2 km³. Large avalanches of hydrothermally altered rock debris occurred at least eight times during the last 10,000 years and formed mudflows as long as 30 km

**TABLE 7.1**

**Types of Eruptive Activity at Five Cascade Range Volcanoes during the Last 11,000 Years**[a]

| $^{14}$C yr B. P. | Mount Baker | Mount Rainier | Mount St Helens | Mount Hood | Mount Shasta Black Butte |
|---|---|---|---|---|---|
| 0 | A | A | L L D A; D A P P P M | D A P M | P M |
| | | | | | M |
| 1000 | | | D A | | ? ? P |
| 2000 | | L A M P | L A A P; L D A P; L D A A P M; D A P M M; D A P M M; A A P M; A P M; A; A | D P M | L D P |
| 3000 | A | | | | ? ? P M P M |
| 4000 | | A A | A; A | | ? ? M M |
| 5000 | | A A M; A A; A | | | L P |
| 6000 | | A; A; A | | | ? M; P |
| 7000 | | | | | ? ? P |
| 8000 | | M | A | | ? ? ? |
| 9000 | L L A P M | | | | L ? D D P P M |
| 10,000 | | A; A | A | | A A P P |
| 11,000 | | ? ? | A | | |

[a]L = lava flow; D = dome; A = air-fall tephra; P = pyroclastic flow; M = mudflow known or inferred to have been caused by eruptive activity. Events of unknown age are arbitrarily placed midway between known limits (indicated by heavy horizontal dash), or between inferred limits (indicated by query). Domes include those whose former presence is inferred from the character of pyroclastic-flow deposits.

in valleys that drain the volcano (Hyde and Crandell 1978). During the nineteenth century, Mount Baker erupted on a small scale several times—the most recent eruption apparently occurred in 1870 (Harris 1976, and references therein).

Hydrothermal activity at Sherman Crater increased significantly in March 1975 and caused concern that an eruption might be imminent (Malone and Frank 1975; Frank et al. 1977). The relatively high level of activity continued through 1977, but no other phenomena indicative of a forthcoming eruption had yet occurred at the time of this writing (April 1979).

## Mount Rainier

Mount Rainier, a large andesitic stratovolcano in the central part of the Cascade Range of Washington, was, like Mount Baker, constructed chiefly during Pleistocene time, but repeated eruptions have occurred during the last 10,000 years. Mudflows have exceeded all other products of the volcano in both frequency and volume during this time, although most of these consist largely of debris derived from older, hydrothermally altered rocks of the volcano rather than from newly erupted material (Crandell 1971).

The two most active periods of Mount Rainier's recent eruptive history were between about 6500 and 4500 years ago and between about 2500 and 2000 years ago (Table 7.1). The earlier period included the formation of at least eight tephra deposits, two of which consist partly or wholly of older rock debris possibly ejected during steam explosions (Mullineaux 1974). The other six tephra deposits consist of newly erupted magma of andesitic composition. During this period tremendous avalanches of altered rock from the volcano— probably triggered by volcanic explosions—produced three mudflows that range in individual volume from perhaps 0.1 to 2 km³. The largest mudflow of this period, the Osceola Mudflow, occurred some 5000 years ago and covered an area of more than 300 km² in the southern Puget Sound lowland, an area in which tens of thousands of people now live (Figure 7.5). Evidence of human occupation of that area before and possibly at the time the mudflow occurred has been found at one site, where knives, projectile points, scrapers, a drill, and other artifacts were obtained from horizons stratigraphically beneath the mudflow deposit (Hedlund 1976).

The avalanches of altered rock that produced the Osceola Mudflow left a large depression at the summit of the volcano. The depression was partly filled about 2200 years ago by a lava cone at least 300 m high and 1.5–2 km in diameter. Melting of snow and ice by lava flows during building of the cone produced floods and mudflows that carried fresh rock debris down some valleys. A tephra layer that has an estimated volume of about 0.3 km³ was also formed during this period and is the most voluminous tephra erupted at Mount Rainier during Holocene time.

The youngest large mudflow from Mount Rainier, which is known as the Electron Mudflow, occurred about 500 radiocarbon years ago and reached a point about 55 km downvalley from the volcano (Figure 7.5). Because the mudflow contains a high proportion of altered rock debris, it evidently originated in an area of hydrothermally altered rock on the west side of the volcano; however, no independent evidence has been found to suggest that the mudflow was caused by eruptive activity. The last significant eruption of Mount Rainier

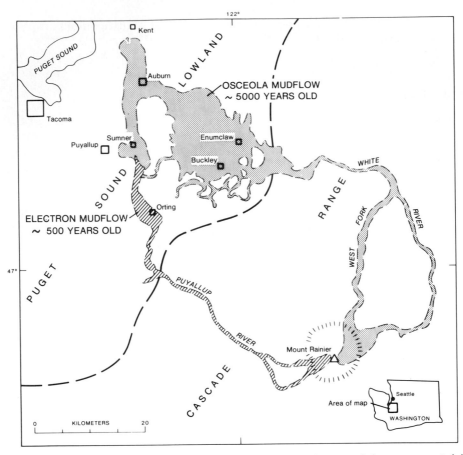

**FIGURE 7.5. Map showing extent of two large mudflows that extended trom Mount Rainier into the Puget Sound lowland during Holocene time.**

occurred about 150 years ago when tephra of very small volume was deposited on the east side of the volcano (Mullineaux *et al.* 1969; Mullineaux 1974).

## Mount St. Helens

Mount St. Helens is the youngest, one of the smallest, and compositionally one of the most diverse of the major volcanoes of the Cascade Range. It is situated in southern Washington (Figure 7.1), and its first activity occurred between 40,000 and 50,000 years ago, prior to the last major glaciation (Fraser Glaciation) of the Cascade Range. Mount St. Helens has been more active and more explosive during the last 4500 years than any other volcano in the Cascade Range (Crandell *et al.* 1975); with few exceptions, lava flows evidently were produced only during the last 2500 years (Table 7.1).

There have been four major eruptive periods during the last 4000 years, each of which extended over hundreds of years. The first period was characterized by repeated eruptions of pumiceous tephra, at least three of which had enough volume, perhaps 1 km³ or more, to form recognizable deposits many hundreds of kilometers downwind from the volcano. Pyroclastic flows alternated with tephra eruptions, and large mudflows, possibly generated by hot pyroclastic flows moving across snowfields, extended down valleys west of the volcano to distances of at least 50 km. This period probably ended about 3300 years ago and was followed by a dormant interval of perhaps as long as 300 years.

During the second eruptive period, relatively small volumes of tephra— less than 1 km³—were erupted several times. In addition, large pyroclastic flows moved away from the volcano in many directions. Some of the resulting pyroclastic-flow deposits consist wholly of nonvesicular dacite and probably originated from the explosive disruption of domes being extruded at the volcanic center. Other pyroclastic flows were made up largely or wholly of pumice and may have originated during eruptions that produced columns of gas and pumice high above the volcano. Mudflows extended down some valleys as far as 70 km during this period.

The behavior of the volcano became more varied during the third eruptive period, which began about 2500 years ago. During the last 2500 years lava flows have been a major product of the volcano, even though tephra, domes, and pyroclastic flows like those of earlier periods continued to be formed. The third period was characterized by the eruption of different kinds of rock during short periods of time. Andesitic or dacitic tephra was erupted at least four times, basaltic tephra and basalt lava flows at least twice, dacitic or andesitic pyroclastic flows at least twice, and andesitic lava flows at least twice. The eruptive period ended about 1700 years ago.

The last major period of eruptive activity began about 500 years ago, and eruptions of different rock types in rapid succession continued intermittently until the mid-1850s. During this period at least three dacite domes were erupted, one of which forms the present summit of the volcano (Figure 7.6). Pyroclastic flows and mudflows were caused by avalanches of hot rock debris during emplacement of the summit dome 350–400 years ago, and later as a dacite dome was formed on the north flank of the volcano during the first half of the nineteenth century. Eruptions of dacitic tephra occurred about 450 and 175 years ago; the volume of the older tephra probably approached 1 km³. Andesite lava flows of this period moved down almost every flank of the volcano probably between 400 and 450 years ago, and perhaps down the southwest side sometime during the first half of the nineteenth century. The last recorded eruption evidently occurred in 1857 (Harris 1976: 186–187).

## Mount Hood

Mount Hood is the northernmost volcano in the Cascade Range of Oregon and lies about 75 km east–southeast of Portland. Although Wise (1969) indicated that the volcano is of late Pleistocene age, early to middle Pleistocene

FIGURE 7.6. West side of Mount St. Helens, Washington. Pyroclastic-flow deposits on south-west flank of volcano were formed 400–450 years ago. The summit of the volcano is a dacite dome 350–400 years old, and Goat Rocks is a dacite dome that was formed between 1800 and 1850. The volcano in the background is Mount Adams. (Photograph by Delano Photographics, Inc., Portland, Oregon.)

activity also can be inferred from deposits of volcanic mudflows in the eastern part of the Portland metropolitan area (Trimble 1963: 97). Three major erup-tive periods have occurred at Mount Hood since the maximum of the last major glaciation (Crandell and Rubin 1977). Each period was characterized by the extrusion of one or more dacite domes, and avalanches of hot rock debris from the flanks of the domes produced lithic pyroclastic flows.

The first eruptive period began during a late stage of the last glaciation; domes were formed at the summit of Mount Hood while large glaciers still occupied cirques and valleys on the volcano. Hot avalanches of debris from the domes, as well as mudflows, built large fans on the flanks of the volcano between glaciers. As glaciers receded, continued eruptions caused pyroclastic flows and mudflows to be deposited between ice margins and valley sides. During this period, floods and mudflows formed deposits across the floors of major valleys on all sides of Mount Hood. This eruptive period has not been radiometrically dated, but the relation inferred between the volcanic deposits and glaciers substantially larger than those of the present suggests that the eruptions occurred between 12,000 and 15,000 years ago. This suggestion is supported by soil profiles on the volcanic deposits, which resemble soils on glacial deposits formed during the last major glaciation.

About 1800–1500 years ago, after a long dormant period (Table 7.1), avalanches of hot rock from a growing dome high on the south side of the volcano formed a broad fan of pyroclastic-flow deposits on the south flank of the cone. Mudflows derived from the pyroclastic-flow deposits blanketed the floor of a valley west of the volcano and extended to the Columbia River, 75 km from Mount Hood (see Figure 7.11, p. 215).

The emplacement of another dome (Figure 7.7) on the south side of the volcano during the third eruptive period, about 200–250 years ago, resulted in hot pyroclastic flows and mudflows that were limited to valleys on the south and west sides of the volcano. An eruption of very small volume sometime during the last 200 years resulted in a tephra deposit that consists of widely scattered pumice lapilli on the eastern and southern flanks of the volcano. This deposit is unusual in that Mount Hood, unlike many other Cascade volcanoes, erupted virtually no other pumiceous tephra during Holocene time. Deposits of fine lithic tephra a few centimeters to several meters thick blanket much of the lower slopes of Mount Hood. These deposits are believed to be chiefly the

FIGURE 7.7. South side of Mount Hood, Oregon. Dashed line outlines the fan of pyroclastic-flow deposits and lahars formed about 1600 years ago. Dome (arrow) in summit crater was erupted 200–250 years ago.

products of clouds of ash that accompanied the hot pyroclastic flows of the last two eruptive periods (Crandell and Rubin 1977).

## Mount Shasta

Mount Shasta is a massive andesitic and dacitic stratovolcano located in northern California (Figure 7.1). The cone consists of lava flows, domes, and pyroclastic deposits erupted at four main volcanic centers (Christiansen and Miller 1976). Although the volcano was constructed chiefly during Pleistocene time, two of the four centers were active during Holocene time. Holocene eruptions also occurred at Black Butte (Figure 7.8), a group of overlapping dacite domes about 13 km west of Mount Shasta. The bulk of the fragmental volcanic debris produced at Mount Shasta and Black Butte during Holocene time consists of pyroclastic flows and associated mudflows (Table 7.1).

In contrast to many other Cascade Range volcanoes, Mount Shasta is not drained by a radial system of deep river valleys but is nearly surrounded by broad fans of volcanic debris (Figure 7.8). As a result, pyroclastic flows and mudflows spread out on the fans, traveled relatively short distances, and

FIGURE 7.8. View eastward toward Mount Shasta, Shastina, and Black Butte, California. Smooth fans composed of pyroclastic-flow deposits lead down from Mount Shasta and Shastina toward the communities of Mount Shasta (north edge of town shown in lower right corner) and Weed (just out of picture, lower left corner).

covered broad areas at the base of the volcano, whereas similar deposits at other volcanoes in the Cascades are confined to deep valleys leading away from the volcano.

During Holocene time Mount Shasta was most active between about 10,500 and 9000 years ago. Early in this eruptive period a series of pyroclastic flows originated at a vent or vents near the present summit and traveled more than 20 km in all directions except to the west and southwest. These pyroclastic flows were followed by two eruptions of pumiceous tephra, one of which forms a widespread blanket of pumice on all flanks of Mount Shasta except the west. These two eruptions are the only ones known of pumiceous tephra at Mount Shasta during Holocene time.

About 10,000 years ago, andesite lava flows were erupted from a vent near Mount Shasta's present summit and from vents on the flanks and at the summit of Shastina (Figure 7.8). At about the same time dacite domes were formed at Black Butte and at the summit of Shastina. Collapse of these domes, possibly caused by explosions, produced pyroclastic flows whose deposits cover an area of more than 110 km² west of Mount Shasta (Miller 1978a).

More recent activity at Mount Shasta includes additional cone-building eruptions of andesite lava near the present summit of the volcano (Table 7.1). These were followed by the eruption of a dacite dome sometime during the last several thousand years; this dome forms the present summit. During the last 8000 years there were eight or nine eruptions of andesitic or dacitic fragmental material at the summit vent. These eruptions were relatively small in volume and were spaced from a few hundred to about a thousand years apart. Most of them produced hot mudflows that traveled as far as 30 km from the summit of the volcano as well as hot pyroclastic flows (Table 7.1).

The most recent eruption that produced recognizable deposits occurred about 200 years ago, when a pyroclastic flow traveled at least 12 km down the southeast side of Mount Shasta and hot mudflows extended more than 20 km south of the volcano.

Three communities, each with populations of several thousand, are situated on broad fans at the base of Mount Shasta. Future pyroclastic flows and mudflows could cover almost any part of these fans; thus, it is possible that pyroclastic flows and mudflows resulting from even relatively small future eruptions could destroy parts of one or more of these communities (Miller 1978b).


## EXAMPLES OF VOLCANIC-HAZARDS ASSESSMENTS

Potential hazards from future eruptions of Mounts Rainier, St. Helens, and Hood are summarized in the following pages as examples to show how volcanic-hazards assessments differ in some respects from one volcano to another.

## Mount Rainier

The recent history of Mount Rainier indicates that mudflows and tephra are the principal hazards, although tephra eruptions like those of the recent past would be of relatively small volume and would not seriously affect areas very far from the volcano. The most catastrophic events at Mount Rainier during Holocene time were massive avalanches of hydrothermally altered and water-saturated rock debris that produced enormous mudflows. Mudflows like these today could be disastrous if they extended into densely populated areas. The largest area of altered rock that can be recognized today as a potential source of mudflows is high on the west flank of the volcano (Crandell 1971: 69); thus, valleys that head on that flank are likely to be affected by large mudflows in the future (Figure 7.9). It is also possible that large masses of altered rock exist elsewhere, unseen below glaciers or unaltered rock; such masses could

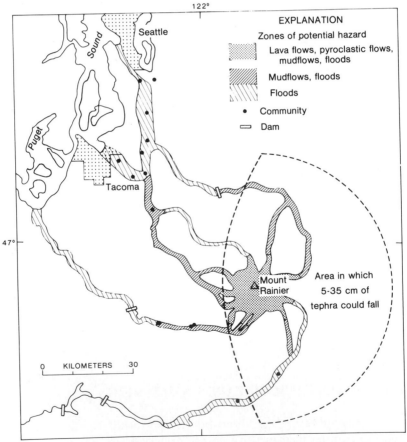

**FIGURE 7.9. Map of the region near Mount Rainier, Washington, showing zones of potential hazard from future eruptions (modified from Crandell 1976). Thickness and extent of tephra are based on the largest single deposit of Holocene age, which covered only part of the area shown.**

provide a source of large mudflows in other valleys if a flank of Mount Rainier were disrupted by explosions or earthquakes. In addition, mudflows of unaltered debris may be expected if renewed activity were to cause large-scale melting of snow and ice within and adjacent to the summit craters, or if lava were to be erupted beneath or onto one of the large glaciers. Resulting floods and mudflows could affect valley floors for distances of many tens of kilometers downstream from the volcano.

Tephra eruptions like those of the recent past probably would be troublesome but not catastrophic. The fall of tephra within Mount Rainier National Park could be frightening and have locally severe effects but probably would not constitute a serious threat to life. Beyond the park and within about a 155° sector from north–northeast to south–southeast of the volcano's summit, a fall of tephra could disrupt transportation, recreation, and agriculture, and create clean-up problems as far away as central Washington. The effects of tephra at that distance, however, should be minor and not cause significant economic loss.

## Mount St. Helens

Because of its long record of spasmodic explosive activity, Mount St. Helens probably is the volcano most likely to endanger people and property in the western United States (Crandell and Mullineaux 1978). Future eruptions are likely to produce lava flows, domes, tephra, and pyroclastic flows; most of these eruptions will also produce mudflows (Figure 7.10). Domes and most lava flows will not directly affect areas that are more than a few kilometers beyond the base of the volcano. Future pyroclastic flows like those of the past will affect the flanks of the volcano, broad areas beyond the base of the volcano to a distance of at least 5 km, and valley floors to a distance of perhaps 15 km. Mudflows and floods generated by hot pyroclastic flows or lava flows moving across snow could inundate valley floors at distances of many tens of kilometers from the volcano.

Hazard zones associated with future tephra eruptions at Mount St. Helens are inferred from records of modern wind directions and speeds as well as from the extents of tephra deposits formed about 1842, about 1800, and about 4000 years ago (Figure 7.3). These three tephra deposits represent volumes of material on the order of 0.01, 0.1, and 1 km³, respectively, and serve as examples of a range of likely future conditions. Judging by the record of the past, we estimate that small eruptions will occur about once per 100 years, on the average. Eruptions that produce as much as 0.1 km³ of tephra may occur as often as once per 500 years, and significantly larger eruptions may occur once every 3000–4000 years.

An eruption of Mount St. Helens could present a significant potential hazard to several reservoirs; one reservoir is situated in the Lewis River valley about 10 km south of the volcano, and two others are present farther downstream (Figure 7.10). Crandell and Mullineaux (1978) estimate that a

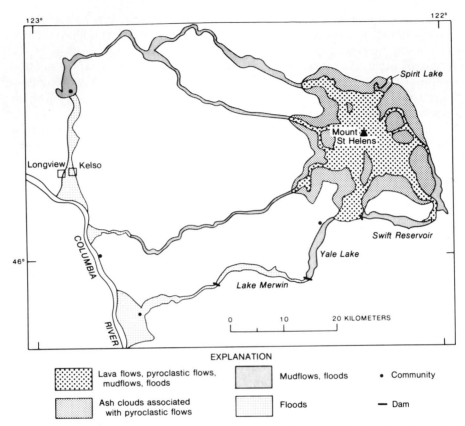

**FIGURE 7.10. Map of the region near Mount St. Helens, Washington, showing zones of potential hazard from future lava flows, pyroclastic flows, mudflows, and floods. (Modified from Crandell and Mullineaux 1978.)**

mudflow with a volume of as much as 110 million m³ could move into the reservoir, and recommend that if an eruption begins, the volume of the reservoir should be reduced as quickly as possible by 125 million m³ to provide reasonable assurance against overtopping and failure of the dam.

### Mount Hood

The potential hazard from Mount Hood is regarded as less than that from either Mount Rainier or Mount St. Helens because Mount Hood has erupted only infrequently during the last 15,000 years. The extrusion of domes was the most common type of eruption during this period and future activity probably will be of the same kind. The next dome probably will be erupted at the summit crater, which is open to the south; thus, the south slope of the volcano will be endangered. However, a dome conceivably could be extruded almost anywhere on the volcano. The formation of a dome almost surely will be accompanied by avalanches of hot rock debris; if these avalanches are large enough, they will form hot pyroclastic flows that will move far downslope. If an eruption occurs when snow blankets the volcano, pyroclastic flows probably will give rise to floods and mudflows that could move far down the valleys downslope from the new dome. Mount Hood's recent eruptive history suggests that the likelihood is very low that a large volume of pumiceous tephra will be erupted either alone or in conjunction with the extrusion of a dome. Thus, it is anticipated that hazards accompanying future eruptions of Mount Hood will, for the most part, be confined to valley floors (Figure 7.11).

## Regional Assessments

Examinations of individual volcanoes can provide the data necessary to compile maps that show areas of potential volcanic hazards for entire states or regions. A preliminary map (scale 1:1 million) of potential volcanic hazards in Washington has been published (Crandell 1976), and similar maps are being prepared for Oregon and California. In addition, a map (scale 1:7 ½ million) has been prepared for volcanic hazards throughout the western conterminous United States (Mullineaux 1976). Such maps can portray the nature and degree of hazard at a specific location not only from the nearest volcano but from more distant volcanoes that may be more dangerous because they are more explosive. For example, the principal hazard from tephra in north-central Oregon is not from the closest volcano, Mount Hood, but from Mount St. Helens in southern Washington. In fact, about a centimeter of tephra from

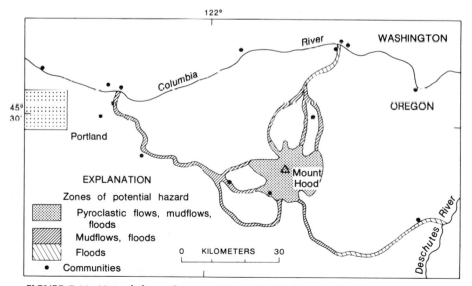

**FIGURE 7.11.** Map of the region near Mount Hood, Oregon, showing zones of potential hazard from future eruptions. The volcano has not erupted an appreciable amount of tephra during the last 10,000 years or more. (Crandell and Rubin 1977.)

Mount St. Helens fell along the Columbia River Valley near The Dalles as recently as 1842. One limitation of compilations like these is that basic information rarely is adequate for every volcanic center, and the available information may vary widely in completeness and reliability. Thus, regional assessments ideally should be updated from time to time as more information about specific volcanoes becomes available.

## DISCUSSION

For any type of eruption there is a direct relation between the volume of material erupted and the size of the area affected by this material. Most long-lived stratovolcanoes seem to erupt small volumes of material more often than large volumes. Consequently, the greater the distance from a volcano, the less likely an area is to be affected by eruptive phenomena. Likewise, topographically high areas are affected less frequently than are low areas by products of volcanism that move along the ground, and areas upwind from a volcano are affected by tephra less often and less severely than are downwind areas.

These generalizations can be applied to evaluations of long-term safety for communities in and near the Cascade Range. Except for those near Mount Shasta, most communities are so far from any volcano that only relatively thin tephra deposits are likely to fall on them, and only a mudflow or flood of unusual size could reach them. Nevertheless, parts of the western United States face a progressive increase in the potential impact of future eruptions because of a growing population, increasing recreational use of areas around volcanoes, and accelerating energy needs that may be met in part by new hydroelectric and nuclear power plants downvalley and downwind from volcanoes. Even though the recent histories of most volcanoes in the Cascade Range do not imply that catastrophic eruptions are imminent, the kinds of land use that develop today around volcanoes probably will persist for many decades if not centuries. In the perspective of time periods like these, we believe that many of the volcanoes in the western United States constitute a real threat to life and property. That threat can be partly mitigated by land-use zoning in relatively high-risk areas, careful selection of safe sites for critical installations, preparation of contingency plans for warning and evacuation when eruptions occur, and monitoring of volcanoes believed to be especially dangerous.

Land-use zoning and selection of safe sites relative to volcanic hazards require long-range forecasts of the probability of volcanic activity and identification of areas of potential risk. Forecasts of the likelihood of future tephra eruptions and their volume can be based on the recent eruptive history of a volcano, and records of wind-direction frequencies allow the distribution of future tephra deposits to be estimated. For example, the frequency of relatively large volume (on the order of 1–3 km$^3$) tephra eruptions at Mount St. Helens is about one per 3000 years, and the wind blows from the volcano toward a 45°

sector centered on Portland only about 3% of the time; thus, the resultant probability that Portland will be affected in any one year by tephra from an eruption of this size is about 1 in 100,000. Such an eruption could result in tephra as thick as 6 cm at Portland if the thickest part of the lobe fell on the city. This low level of risk from future tephra eruptions seemingly is confirmed by the volcano's past behavior—during the 40,000–50,000 years that Mount St. Helens has been in existence, tephra of relatively large volume evidently was carried to the southwest only once. In contrast, meteorological records show that winds above the volcano blow toward the east–northeast into a 45° sector that includes Yakima about 30% of the time. This results in a probability of about 1 in 10,000 in any one year that Yakima would be significantly affected by a large-volume eruption of Mount St. Helens. According to Figure 7.3, the thickness of tephra that can be expected at Yakima from such an eruption could be slightly more than 10 cm.

In contrast, only limited areas near a volcano are vulnerable to other products of eruptions. These areas are mainly the floors of valleys that head at the volcano. Although the frequency of future eruptive activity can be estimated, there is as yet no fully satisfactory way of estimating the probability that a future mudflow or pyroclastic flow will affect a specific valley or site within a valley.

We are uncertain as to the kinds of socioeconomic actions that are warranted in the near future to avoid or mitigate the effects of future volcanic eruptions in the Cascade Range. Such eruptions clearly have a low probability of occurrence in any one year. Fortunately, it is much more likely that the next tephra eruption in the Cascades will be of small volume (less than 0.1 km³) than of large volume. The effects of a small eruption should be areally limited, should cause significant economic loss only locally, if at all, and probably would endanger few people. The chance in any one year that such an eruption will occur somewhere in the Cascade Range probably is a little more than 1 in 100. The probability of an eruption with a volume on the order of one or several cubic kilometers is substantially less, perhaps no greater than 1 in 1000 in any one year. However, such an eruption could have regional effects whose severity would depend on the location of the active volcano, the character and duration of the eruption, and wind directions. The probability in any one year of an even larger tephra eruption, with a volume on the order of 10 or several tens of cubic kilometers, is probably no greater than 1 in 10,000 and may be no more than 1 in 100,000. Accompanying this low probability, however, is the virtual certainty that such an eruption would be a regional disaster.

The relative probabilities of eruptions of various volumes suggest that, for most purposes, preparation can be justified now only for those of small scale. On the other hand, severe consequences could result if a volcanic mudflow were to cause a dam to fail, or if fall of tephra interfered with the operation of a nuclear power plant. Even though volcanic phenomena like these have only a small probability of occurrence in any one year, their potential consequences demand consideration in the planning and operation of such installations.

Studies that emphasize the areal distribution of potential volcanic hazards and that can be applied to land-use planning are currently on a very small scale in the United States by comparison with research being directed toward such fundamental problems as the place and origin of magmas and the evolution of volcanoes. The modest effort that is directly applicable to volcanic-risk assessment, however, probably is commensurate with the relatively low degree of potential risk that can be inferred from the infrequency of eruptions in the western conterminous United States.

A comparably low level of effort, however, may not be appropriate for countries where frequently active volcanoes endanger people or disrupt their lives and seriously affect the economy. In such countries, intensive research efforts could lead to the preparation of volcanic-risk maps that would have lasting value. A new emphasis on such hazard-related studies of volcanoes may now be developing outside the United States; some of these studies are listed in an earlier part of this chapter (p. 197). Similar programs would be useful in many other countries, such as in some parts of Central and South America. It should be emphasized, however, that the data necessary for reliable assessments may not be easy to acquire and ordinarily will require specialized studies of each volcano by qualified scientists.

# REFERENCES

Barberi, F., and P. Gasparini
   1976  *Volcanic hazards: International Assoc. Engineering Geology Bulletin,* No. 14:217–232.
Bolt, Bruce A., W. L. Horn, G. A. Macdonald, and R. F. Scott
   1975  *Geological hazards.* Springer-Verlag, New York.
Booth, Basil
   1977  Mapping volcanic risk. *New Scientist* 75:743–745.
Christiansen, Robert L., and C. Dan Miller
   1976  Volcanic evolution of Mt. Shasta, California. *Geological Society of America Abstracts with Programs* 8:360–361.
Crandell, Dwight R.
   1971  Postglacial Lahars from Mount Rainier Volcano, Washington. U.S. Geological Survey Professional Paper 677.
Crandell, Dwight R.
   1973  Map showing potential hazards from future eruptions of Mount Rainier, Washington. U.S. Geological Survey Misc. Geologic Investigations Map I-836.
Crandell, Dwight R.
   1975  Assessment of volcanic risk on the Island of Oahu, Hawaii. U.S. Geological Survey Open-File Rept. 75–287.
Crandell, Dwight R.
   1976  Preliminary assessment of potential hazards from future volcanic eruptions in Washington. U.S. Geological Survey Misc. Field Studies Map MF-774.
Crandell, Dwight R., and Donal R. Mullineaux
   1978  Potential hazards from future eruptions of Mount St. Helens Volcano, Washington. U.S. Geological Survey Bulletin 1383-C.
Crandell, Dwight R., Donal R. Mullineaux, and Meyer Rubin
   1975  Mount St. Helens Volcano—Recent and future behavior. *Science* 187:438–441.

Crandell, Dwight R., and Meyer Rubin
  1977   Late-glacial and postglacial eruptions at Mt. Hood, Oregon. *Geological Society of America Abstracts with Programs* 9:406.
Frank, David, Mark F. Meier, and Donald A. Swanson
  1977   Assessment of increased thermal activity at Mount Baker, Washington, March 1975–March 1976. U.S. Geological Survey Professional Paper 1022-A.
Harris, Stephen L.
  1976   *Fire and ice, the Cascade volcanoes.* Seattle: The Mountaineers and Pacific Search Books.
Hedlund, Gerald C.
  1976   Mudflow disaster. *Northwest Anthropological Research Notes.* 10:77–89.
Hoblitt, Richard P., and Karl S. Kellogg
  1976   Emplacement temperatures of unsorted and unstratified deposits of volcanic rock debris as determined by paleomagnetic techniques. *Geological Society of America Abstracts with Programs* 8:919–920.
Hyde, Jack H., and Dwight R. Crandell
  1978   Postglacial volcanic deposits at Mount Baker, Washington, and potential hazards from future eruptions. U.S. Geological Survey Professional Paper 1022-C.
Malone, Stephen D., and David Frank
  1975   Increased heat emission from Mount Baker, Washington. EOS (American Geophysical Union Transactions) 56:679–685.
Miller, C. Dan
  1978a  Holocene pyroclastic-flow deposits from Shastina and Black Butte, west of Mt. Shasta, California. *U.S. Geological Survey Journal of Research,* 6, 611–624.
Miller, C. Dan
  1978b  Potential hazards from future eruptions in the vicinity of Mount Shasta Volcano, northern California. U.S. Geological Survey Open-File Rept. 78–827.
Miller, C. Dan, Donal R. Mullineaux, and Minard Hall
  1978   Reconnaissance map of potential volcanic hazards from Cotopaxi Volcano, Ecuador. U.S. Geological Survey Miscellaneous Investigations Map I-1072.
Mullineaux, Donal R.
  1974   Pumice and other pyroclastic deposits in Mount Rainier National Park, Washington. U.S. Geological Survey Bulletin 1326.
Mullineaux, Donal R.
  1976   Preliminary overview map of volcanic hazards in the 48 conterminous United States: U.S. Geological Survey Miscellaneous Field Studies Map MF-786.
Mullineaux, Donal R., and Donald W. Peterson
  1974   Volcanic hazards on the Island of Hawaii. U.S. Geological Survey Open-File Rept. 74-239
Mullineaux, Donal R., Robert S. Sigafoos, and Elroy L. Hendricks
  1969   A historic eruption of Mount Rainier, Washington. *In Geological Survey Research 1969.* U.S. Geological Survey Professional Paper 650-B.
Neumann van Padang, M.
  1960   Measures taken by the authorities of the volcanological survey to safeguard the population from the consequences of volcanic outbursts. *Bulletin Volcanologique* 23:181–192.
Trimble, Donald E.
  1963   Geology of Portland, Oregon and Adjacent Areas. U.S. Geological Survey Bulletin 1119.
Warrick, Richard A.
  1975   Volcano hazard in the United States—A research assessment. Program on Technology, Environment and Man, Monograph #NSF-RA-E-75-012. Boulder: Institute of Behavioral Science, University of Colorado.
Wise, William S.
  1969   Geology and petrology of the Mt. Hood Area—A study of high Cascade volcanism. *Geological Society of America Bulletin* 80:969–1006.

# 8

# Contemporary Responses to Volcanism: Case Studies from the Cascades and Hawaii[1]

DAVID HODGE
VIRGINIA SHARP
MARION MARTS

Although volcanoes are recognized as potentially serious hazards, those of the lower 48 states have been dormant for decades and the volcano hazard has been outside the experience of most mainland Americans. Volcanoes in Hawaii and Alaska have been more active, but the lava flows characteristic of the Hawaiian volcanoes have not posed a significant threat to human life, and the more explosive Alaskan eruptions have been remote from most settlement (Warrick 1975; see also Warrick, Chapter 6 of this volume).

The unusual steaming activity of Mount Baker (see Figure 8.1) in the northern Cascades and subsequent events provided the first opportunity to study responses to Cascade volcano activity since the Lassen Peak eruptions of 1914–1917. With the support of the National Science Foundation, the authors undertook a reconnaissance study in the late summer of 1976 of the public response to the Mount Baker activity. The investigation was subsequently broadened to include a portion of the island of Hawaii because of a prediction of an eruption of Mauna Loa that might affect portions of the city of Hilo and the renewed activity of Kilauea that threatened a nearby community with a lava flow.[2]

The two cases were very different. Hawaii represented a high-probability–

---

[1]Prepared with the support of National Science Foundation grant No. ENV-76-20735. Opinions, findings, and conclusions expressed are those of the authors and do not necessarily reflect the views of the National Science Foundation. For the full report, see Hodge *et al.* (1978).

[2]Another study, of a hypothetical Mount Rainier event, was also carried out. See Cullen (1978).

*VOLCANIC ACTIVITY AND HUMAN ECOLOGY*

**FIGURE 8.1.** Mount Baker and Baker Lake Reservoir. Boulder Creek and Boulder Glacier are in the center; crater is in the notch immediately to the left of and below the summit dome. (Photo by David Frank, courtesy of U.S. Geological Survey.)

low-risk case; Mount Baker, a low-probability–potentially-high-risk case, though the risk is tempered by the relative isolation of the mountain from permanent settlements. In the actual event, a lava flow from Kilauea stopped short of the community of Kalapana, Mauna Loa has not erupted, and Mount Baker continues to vent unusual amounts of steam from its crater, but there has been no other untoward event.

The primary purpose of the study was to use the rare opportunity afforded by the Mount Baker case to provide empirical feedback on public attitudes for the guidance of officials who may be involved in the future in the management of volcanic events and predictions, especially in the Cascades, where experience is so limited. The attitudes and perceptions of people in the affected areas were recorded and analyzed while the predictions and events were relatively fresh in the public mind. Both studies dealt with reactions both to the physical events and predictions and to the associated management events. The findings, which are discussed in the following sections, are ambiguous, as were the events, but they clearly suggest that much remains to be done before official response and public attitudes can be brought into harmony during postprediction, pre-event periods. The amount of tolerance for official actions in the Mount Baker case surprised the investigators, and the role of ethnicity and inadequacy of communication with some segments of the public became obvious during the interviewing in Hawaii. Perhaps the most disturbing findings, which came as a byproduct, not a direct outcome, of the formal study, were that the Mount Baker experience has not been built into U.S. Forest Service operating procedures, and that the volcano hazard is not considered in land management and development even on the island of Hawaii, where lava flows are recurring phenomena.

## MOUNT BAKER EXPERIENCE

### History and Setting

Recent interest in Mount Baker was sparked by the observation on 10 March, 1975 of an unusually long and dark gray steam plume emitted from Mount Baker. Geologists from the U.S. Geologic Survey (USGS), the University of Washington, and Washington State University quickly converged on Mount Baker and confirmed the unusual nature of the new activity. Air photos documented that substantial changes were taking place in the crater glacier, that new fumaroles had been developed, and that the fumaroles were emitting not only steam but some kind of debris as well. During the next 3 weeks numerous on site measurements were taken and monitoring equipment was put into place to provide continual surveillance of the volcano's activity. These measurements showed that steam discharge was at least 10 times normal levels and that hydrogen sulfide discharge was more than 20 times normal levels.

In addition to a serious but very uncertain concern that the volcano might be building toward an eruption, the results of these investigations led USGS scientists to conclude that the increased thermal activity probably indicated a tenfold increase in the likelihood that a large and destructive mudflow might be generated on the slopes of Mount Baker. During the past several decades small amounts of debris have in fact flowed out of Sherman crater an average of once every 4 years. Twice in the past several hundred years large mudflows have reached the Baker River Valley, including one flow down Boulder Creek which covered an area that is now a Forest Service Campground (Figure 8.2).[3] According to USGS scientists, the worst possible hazard, given the geologic history of Mount Baker and the volcano's new level of activity, would be an airborne avalanche large enough and fast enough to reach Baker Lake,

---

[3]For a more complete geologic description of Mount Baker, see Crandell *et al.*, Chapter 7.

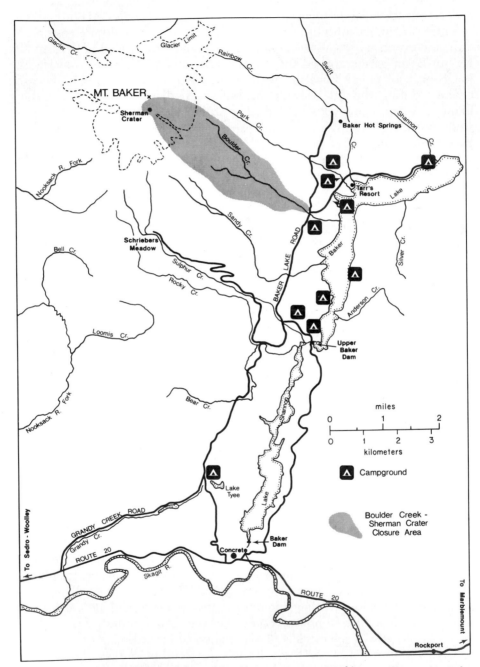

**FIGURE 8.2.** Mount Baker, Baker Lake, and Concrete area, Washington. (Source: Mt. Baker-Snoqualmie Forest Map, Forest Service, 1975.)

generating a huge wave that would inundate those campgrounds surrounding the lake and top the dam, flooding areas downstream.

In response to this perceived hazard an interagency meeting was called on 8 April to consider what actions, if any, might be necessary to protect the public. This meeting marked the beginning of the second stage of reaction in which various officials responded to the warnings issued by geologists. The participants at this first meeting included representatives of USGS, the U.S. Forest Service (USFS), which was responsible for managing most of the territory potentially affected by a large mudflow, and the Puget Sound Power and Light Company, which manages Baker Lake. A consensus was quickly reached that immediate steps should be taken to prohibit entry to the Boulder Creek Campground, Sherman Crater, and the Boulder Creek drainage basin. The USFS officially closed entry to those areas on 24 April.

Interagency meetings continued throughout May and were expanded to include the U.S. Army Corps of Engineers; National Park Service; Washington State Departments of Ecology, Emergency Services, Community Relations, and Natural Resources; as well as the Skagit Regional Council, which included commissioners, mayors, and city managers of the communities downstream from the Baker Lake area. The Department of Ecology, because of its interest in dam safety, requested estimates from USFS regarding the "maximum plausible" mudflow or avalanche that might be expected. Puget Sound Power and Light, owners of Baker Lake Dam and power plant, also contracted a geologic consulting firm to arrive at independent estimates of the worst event. Reaction to the increased thermal activity on Mount Baker entered a third stage when on 23 June the Interagency Task Force reconvened to consider these results and further analysis by USGS scientists monitoring the volcano. As a consequence of these reports the task force recommended that the USFS close public access to the Baker Lake Recreation Area (nearly 4000 ha) and that Puget Power lower the level of the reservoir by 10 m in order to accommodate the "worst possible" debris flow, both of which actions were immediately taken.[4]

Public outcry against the closure was swift and loud. Leading the opponents of the closure were business leaders in Concrete, Washington. Concrete, located about 15 km south of Mount Baker, is itself in little danger from all but the most catastrophic volcano event. Still, many Concrete residents and businessmen were frequent recreational visitors of the area to be closed. In addition, many businessmen thought that the closure would strain an already faltering economy. The evidence appeared to confirm their suspicion as sales tax revenues, indicating retail sales, fell nearly 25% from the previous summer, a decline business leaders attributed to the closure.

---

[4]Actually, Baker Lake was in a seasonal low-level period and simply was not allowed to refill to normal levels. Lost output was reported to the investigators by Mr. William Finnegan of Puget Sound Power and Light as 43 average megawatts for 1 month and 1 to 4 average megawatts for 5 other months. The authors estimate, with liberal assumptions, that this might total 38,750,000 kilowatt-hours. If replacement energy cost 2 cents per KWH, the dollar loss borne by the utility would be in the order of $775,000.

Although the claims of local businessmen are in part true, they exaggerate the economic impact felt by Concrete as a result of the closures. Our investigation revealed that several other factors accounted for at least 20 of the 25% drop in sales. The summer of 1975 was exceptionally rainy and all recreation areas experienced a slump; the novelty of the North Cascades Highway was wearing off, reducing traffic through Concrete; and the lumber industry was in a depression. Moreover, interviews with recreationists showed that most do not purchase any goods or services in Concrete and those who do spend little. After allowing for these factors, the real impact of the closure was probably less than a 5% drop in sales. Concrete businessmen simply had overestimated the importance of Mount Baker area recreationists to the economic vitality of their community. However, it is significant that they *believed* the impact was great, since their behavior reflected that belief rather than facts.

To a large extent, the exaggerated perception of economic hardship on the local economy reflected the dramatic impact of the closure on one local business, Tarr's Resort, a private resort operated on the shore of Baker Lake on National Forest land by permission of the USFS. The resort had been newly purchased in the fall of 1974 by two California businessmen who wanted to live at and operate the resort themselves. The closure of the entire area and the lowering of the lake meant that their entire investment lay idle until the perceived danger had passed. The two men and their families were allowed to remain at the resort to protect the property, but all other activities were shut down. Eventually one partner was forced out but the other has remained and, although not bitter about the USFS decisions themselves, is upset that no provisions were made to mitigate his losses since he was forced to absorb the major impact of the closure decision.

The plight of Tarr's Resort became a significant focal point in local opposition to the closures and to the USFS itself. An editorial in the *Concrete Herald* in October of 1975 reflected the spirit of much of the local opposition to the closures.

> More than three months have passed since the hasty decision to close the Baker Lake Recreation Area was made. We think the time has come, in fact is long past due, to re-examine the original decision and immediately re-open the area to the public since it never should have been closed in the first place. . . . As we have said before, all that is really needed is a warning sign that there is a slight possibility of a mudslide in the area and allow recreationists to enter at their own risk. . . . Maybe it's time to take a lot closer look at the bureaucratic decisions being made by some of our governmental agencies and to start reducing their powers back to where the citizens control instead of being controlled [*Concrete Herald*, October 10, 1975].

The closure had cost the Forest Service a lot of local good will.

The closure of Mount Baker Recreation Area and the lowering of Baker Lake lasted nearly 1 year, during which time no mudflow or other hazard event occurred. On 6 April 1976, the USGS issued a reassessment of the potential danger, concluding

(1) that there was now no clear evidence of forthcoming eruption, whereas an eruption could have been imminent at the time of the earlier statement; (2) that the degree of mudflow or avalanche hazard now probably approached the level of that in an average year within the last 10,000 years; and (3) that the probability of the occurrence of a large fast avalanche had therefore somewhat diminished [Frank *et al.* 1977].

This statement marked the fourth stage of response to the increased thermal activity on Mount Baker as the USFS reacted to the new assessment by reopening all areas but Sherman Crater and the Boulder Creek drainage to recreation use, and Puget Power refilled Baker Lake.

Finally, on 15 February 1977, the Forest Service reopened Sherman Crater and the Boulder Creek drainage, thus completing the lifting of all closures except for one campground, which remained closed because of its deteriorating condition, not because of hazard. Some local residents were critical of the relative lack of publicity concerning reopening as compared with the initial publicity about the threat and the closures. They felt that more forceful statements regarding the reopening were necessary to restimulate tourism and restore the local economy.

## Survey Analysis

The increased thermal activity on Mount Baker provided a rare opportunity to study individual, community, and official response to a very uncertain, but potentially dangerous, volcanic event. In addition to piecing together a history of the experience and documenting the economic impact of the closures, a research team spent nearly 2 months during the summer of 1976 interviewing Concrete residents and Mount Baker recreationists in an attempt to assess the social impact of the volcanic activity and the closures. Three questions were of paramount importance:

1. What did individuals perceive as the likelihood of dangerous volcanic activity?
2. What were their reactions to those actions taken by managers of the Mount Baker environment?
3. What would they do in the event of future warnings from USGS and USFS?

Ideally the interviewing should have taken place in both 1975 and 1976 so that changes in opinions could have been monitored and post hoc reasoning reduced. There can be little doubt that the perceptions of residents and recreationists in 1976 were significantly influenced by the events of 1975. Warnings of volcanic disaster had come and gone with no actual disaster. Yet the closure was a reality that had directly or indirectly affected many of the respondents. It was also anticipated that residents of Concrete would have different perceptions than recreationists from outside the area, since Mount Baker is a part of their immediate environment. Following the interviews extensive analysis was undertaken in an effort to answer these various questions and verify the suspected relationships.

Both Concrete residents and Mount Baker recreationists were asked to predict the likelihood of several potential volcanic events or volcanic-induced events occurring in the next 10 years on a scale of 1 to 5 with 5 indicating that such an event would definitely occur within the next 10 years and 1 indicating that it definitely would not occur. Results of the analysis (Table 8.1) confirm that substantial differences exist between residents and recreationists. Although Concrete residents admit some possibility of volcanic events occurring, most residents attach much lower proabilities to these events than do their recreationist counterparts. Certainly their greater familiarity with the Mount Baker area and longer experience with reports of steaming and heating reduce their expectations. This reasoning is supported by further analysis that revealed numerous statistical differences with respect to the likelihood of volcanic events between those who had seen steam compared with those who had not, and between young and old in both the resident and the recreationist group. In each instance experience and/or age results in greater skepticism of the occurrence of volcanic events.

Perhaps more important than the question of the perceived likelihood of volcanic events is the question of potential danger from such events. If the events of the worst scenario described by USGS had occurred during a period

**TABLE 8.1**
**Perceptions of the Likelihood of Volcanic Events on Mount Baker**

| Event | Mean perceived likelihood[a] | | | Tests of factors affecting perceptions[b] | | | | | |
| | | | | Concrete residents (N = 135) | | | Mount Baker recreationists (N = 227) | | |
| | Concrete Res. | Mount Baker rec. | Sig.[b] dif. | Age | Seen steam[c] | Know actions[d] | Age | Seen steam[c] | Know actions[d] |
|---|---|---|---|---|---|---|---|---|---|
| Mudflow | 2.65 | 3.05 | .01 | .01 | — | — | .01 | .10 | .05 |
| Ash fall | 2.61 | 2.83 | — | — | .01 | .01 | — | — | — |
| Falling rock | 2.33 | 3.16 | .01 | .05 | — | .05 | .05 | .05 | — |
| Forest fire | 2.25 | 2.87 | .01 | — | .01 | .10 | .01 | — | — |
| Earth tremor | 2.87 | 3.27 | .01 | — | — | — | .05 | — | .10 |
| Blowing steam | 4.30 | 4.29 | — | — | .05 | — | .01 | .01 | .05 |
| Lava flow | 2.15 | 2.60 | .01 | — | .01 | .01 | — | — | .05 |
| Avalanche | 2.79 | 3.05 | .05 | — | — | — | .05 | — | .10 |
| Flood | 2.44 | 2.08 | .01 | .05 | .05 | .01 | — | — | .01 |

[a]Answers to the question, "In your opinion, how likely is it that volcanic activity on Mount Baker will cause any of th[e] events to occur sometime during the next 10 years?" Five-point scale, with 1 = Definitely will not occur and 5 = Definitely occur.

[b]t-Tests of differences of means—significance levels indicated.

[c]Tests between those who had and those who had not seen steam emissions from Mount Baker.

[d]Tests between those who were aware and those who were unaware of closures.

of heavy recreation use, the results could have been disastrous. Do recreationists perceive such hazards? Are they aware of what could happen where? Most recreationists (58%) acknowledged that should any of the hypothesized events occur, they would have found escape difficult. Significant differences were observed between the proportion of recreationists located on the southern end of Baker Lake who thought escape would be difficult (58%) and the proportion of recreationists on the northern, more remote end who thought escape would be difficult (64%). Even though all recreationists in campgrounds around Baker Lake would be endangered by a mudflow down Boulder Creek and into the lake, those recreationists located at the more remote end of the lake appeared to recognize their vulnerability and lack of escape alternatives (see Figure 8.2).

As stated earlier, a significant dimension of the human response to volcanic hazards on Mount Baker is the action taken by USFS officials, principally the closure of the Mount Baker Recreation Area. In part those who objected most to the closures expressed little concern for potential danger, a position consistent with their attitudes toward the closures. In addition, many, many persons may have developed skeptical attitudes as a result of the lack of volcanic activity providing credibility to the warnings. On the other hand, some people who previously failed to appreciate any danger from Mount Baker may have developed some concern as a result of those warnings.

As might be suspected, Concrete residents were more critical of the USFS than were recreationists. Nearly two-thirds of the residents (65%) thought that the official actions taken had been an overreaction to the danger, whereas less than half (45%) of the recreationists felt that way. Still, both figures indicate substantial unhappiness with USFS decisions.

The reaction of Concrete residents is related in part to perceptions of the economic impact of the closures. Of those who supported the closures 76% believed that there had been no noticeable economic impact on Concrete, while in contrast a majority (58%) of those opposed to the actions believed that there had been an adverse effect. It was also found that those who opposed the actions were more fatalistic and conservative. On a 7-point Likert statement where 1 is "strongly agree" and 7 is "strongly disagree" with the statement "Natural disasters are acts of God," the average score of those who opposed USFS actions was 2.7 (indicating strong agreement), whereas those supporting USFS actions averaged 4.2 (indicating mild disagreement). Similarly, responses to a Likert statement measuring political conservatism, "Individuals themselves, not government, should determine whether they ought to visit places that might be dangerous," displayed differences between those who opposed the closures (mean score of 2.9) and those who supported the closures (mean score of 4.1). Age also seems to be a factor influencing responses. Young people were significantly more tolerant than older people in their attitude toward the closures.

The most significant variable separating supporters of the closure from nonsupporters among the Mount Baker recreationists was experience with the

**TABLE 8.2**
**Relationship between Number of Visits to Mount Baker Region and Attitudes toward Official Actions**

|  | Number of visits (including present visit)[a] | | |
|---|---|---|---|
|  | 1–5 (N=87) | 6–25 (N=56) | 25+ (N=34) |
| Actions were warranted | 36% | 25% | 21% |
| Actions were an overreaction | 33% | 64% | 65% |
| Not sure | 31% | 11% | 14% |

[a]Missing cases = 50.

area, indicated by the number of visits (Table 8.2). Consistent with other findings regarding perceptions of the likelihood of volcanic events on Mount Baker, those recreationists with substantial experience were overwhelming in their opposition to the closures. Only those persons with five or fewer total visits supported USFS. Unlike Concrete residents, the Likert statement measuring fatalism ("Natural disasters are acts of God") failed to differentiate between supporters and opponents of the closures. However, as with residents, those who opposed the closures were more conservative (mean score 4.3) than those who supported USFS (mean score 2.9).

The analyses above have demonstrated that human response to volcanic hazards on Mount Baker is conditioned by experience with the physical environment and reaction to official actions taken to protect the public from these hazards. What remains, then, is the question of expected future response to warnings of such hazards. Both recreationists and Concrete residents were asked what they would do in the event of warnings issued in the future. Once again the dichotomy between Mount Baker recreationists and Concrete residents is clear (Table 8.3). Concrete residents would mostly ignore any warnings. Their responses to these questions are virtually identical with responses to the questions regarding the appropriateness of USFS actions. This pattern is partially repeated with recreationists with some important exceptions. It ap-

**TABLE 8.3**
**Anticipated Response to Future Warnings of Volcanic Hazards**

|  | Do nothing[a] (No)[b] | Do something[a] (Yes)[b] | Not sure |
|---|---|---|---|
| Concrete residents[a] (N = 135) | 71.3% | 28.7% | — |
| Recreationists[b] (N = 227) | 25.2% | 39.9% | 34.9% |

[a]Answers to the question, "Would you do anything if another warning were given?"
[b]Answers to the question, "If you had heard or read a report before leaving on your trip here that volcanic activity had increased on Mount Baker, would you have come?"

pears that nearly half of those people who thought that the closure had been an overreaction nonetheless were not adamant about ignoring future warnings, most indicating at least some uncertainty with respect to such warnings. It would seem that recreationists are more likely to consider adjustments—and they of course have readily available alternatives—in the face of environmental uncertainty than are residents (or nearby residents) of the affected area. However, it is also possible that the response reflects differences in perception of the environmental uncertainty more than differences in acceptable adjustment strategies.

### Conclusions

In summary, interest in Mount Baker was stimulated as a result of a sudden increase in thermal activity. The scientific community, lacking past experience with such events in the Cascade Range, provided uncertain conclusions regarding possible volcanic activity. Managers of the Mount Baker Recreation Area and Baker Lake Reservoir responded to these conclusions by closing those areas that might reasonably be endangered. Subsequent interviews of residents of the nearest permanent settlement, Concrete, and of recreationists visiting the reopened area showed some belief that serious volcano hazards exist but that the likelihood of such events is not great. Differences in these perceptions were observed, however, between residents and recreationists, between young and old, and between those who had actually seen steam plumes and those who had not. In general, the greater the experience with or exposure to the Mount Baker recreation area, the lower the perceived probability of serious volcanic events occurring. The same relationship holds true for attitudes toward actions taken by USFS and Puget Power. Those persons most familiar with the Mount Baker environment were the most strongly opposed to the closures. Finally, it was observed that expected future responses to possible new warnings differ between residents and recreationists. Concrete residents say that they would respond only to mandatory controls, not to warnings, whereas most recreationists say that they would either not visit the area if warnings were issued or would at least be uncertain as to their reaction. Such differences very much reflect the range of human response to volcanic hazards on Mount Baker.

## HAWAIIAN EXPERIENCE

### Introduction

The Hawaiian island chain was built by hundreds of thousands of lava flows from volcanic eruptions occurring over many millennia. The Big Island of Hawaii has the distinction of containing the only volcanoes that remain active. Three of Hawaii's five volcanoes have erupted since the island was settled and have played major roles in shaping the lives of Hawaii's population.

Hawaii has been inhabited for about 1000 years. Its people have lived successfully with the volcanoes throughout this period. Early Hawaiians quickly learned to adjust their lives to the threat of volcanic activity. They either avoided certain areas or entered them only briefly. Even so, their activities were sometimes abruptly curtailed or otherwise affected by lava flows and other volcanic activity. More recent residents of Hawaii also recognize the susceptibility of certain parts of the Big Island to volcanic disturbance that will affect human activities. Because of the common sense of the settlers and preservation of some vulnerable areas as a national park, many of these areas have never been inhabited.

In recent years, however, population growth on Hawaii, especially in the Hilo and Puna districts, has enticed people to move into previously unsettled areas. This has been true of both permanent residents and off-island purchasers of vacation land. Under the pressures of a growing economy—especially during prolonged lulls in volcanic activity—the potential hazards tend to be forgotten or underestimated (USGS 1976b). Without strict controls on land uses in potentially hazardous areas, Hawaii's residential and vacation developments have begun to encroach on areas previously deemed unfit for human habitation because of the threat of volcanic disruption.

One recent study looked at the way residents of Puna District, especially the Pahoa-Kalapana area (see Figure 8.3), perceive, evaluate, and adjust to the ever-present volcanic activity (Murton and Shimabukuro 1974). The study had three major objectives:

> (1) to assess the importance of volcanic activity to the economy and social organization of eastern Puna; (2) to judge whether volcanic activity should be taken account of in land-use planning; (3) to understand the circumstances in which people in eastern Puna make decisions to cope with hazards [p. 152].

Interviews were conducted with 101 heads of households in the study area. Analysis of the survey results showed that experience—as reflected by age, length of residence in the area, and personal encounters with volcanic hazards—was the most important variable in understanding how people perceive and evaluate the threat of volcanic activity (Murton and Shimabukuro 1974:159). Most of the people interviewed, however, did not perceive volcanic eruptions as hazardous, or at least not to people. In terms of awareness of adjustments to eruptive activity, experience was again found to be the critical factor, though simply bearing the loss and evacuation were the only adjustments volunteered as reasonable. People living in close proximity to Kilauea's rift zone (less than 3 km) appeared to have a better understanding of the unpredictable nature of eruptions than epople living farther away.

Murton and Shimabukuro, however, did not find any other social or economic variables significant in predicting residents' perceptions of and anticipated adjustments to volcanic hazards, though they suggest that interethnic variations should be studied further. They also did not attempt to objectively

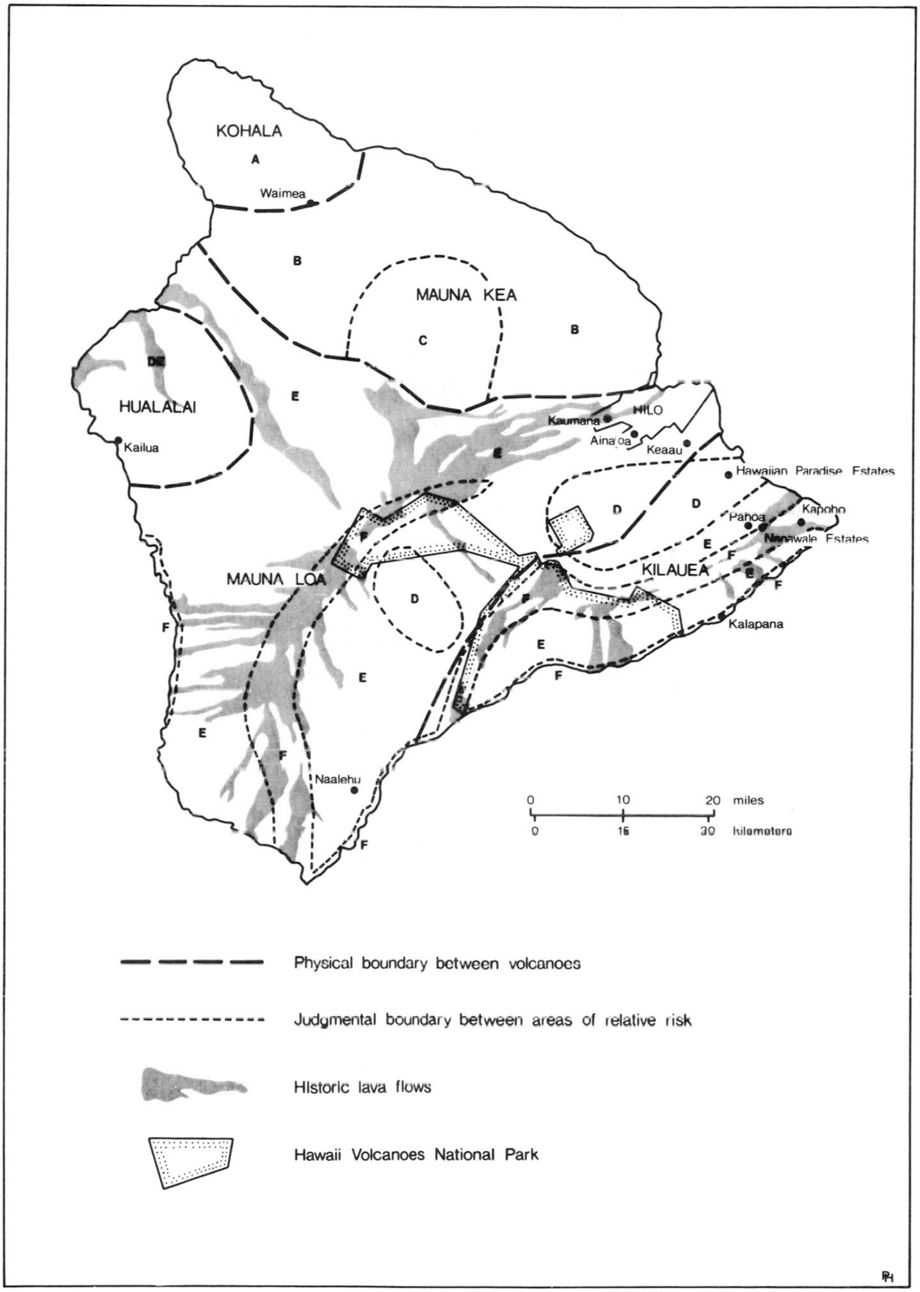
KOHALA
A
Waimea
B
MAUNA KEA
C
B
HUALALAI
Kailua
E
HILO
Kaumana
Aina'oa
Keaau
Hawaiian Paradise Estates
D
D
Pahoa
Kapoho
Nanawale Estates
E
F
KILAUEA
MAUNA LOA
D
F
E
E
F
E
F
Kalapana
E
F
Naalehu
F
0 10 20 miles
0 15 30 kilometers
Physical boundary between volcanoes
Judgmental boundary between areas of relative risk
Historic lava flows
Hawaii Volcanoes National Park

**FIGURE 8.3 Island of Hawaii—volcanic features. Zones of increasing hazard, A through F. (Source: USGS, 1976.)**

measure differences in knowledge about volcanic activity or attitudes toward controlling the effects of natural hazards. The study reported herein attempts to cover these omissions as well as being territorially broader in scope and directed toward a predicted eruption rather than historical experience.

## Physical Setting

The island of Hawaii is made up of five individual volcanoes: Kohala, Mauna Kea, Hualalai, Mauna Loa, and Kilauea (Figure 8.3). All are shield-type volcanoes with characteristic broad gentle slopes made up of thousands of individual lava flows. Most of the flows are long, relatively narrow lobes of dark basaltic lava that slope away from either the central summit area or from rift zones that descend the volcanoes' flanks (Mullineaux and Peterson 1974:11). Hawaii's volcanoes differ markedly in frequency of recent volcanic activity. Kohala has probably not erupted for several tens of thousands of years. Mauna Kea shows some evidence of having erupted a few times within the past 5000 years, though not within historic time. The probability is very low that Mauna Kea will erupt again within the next several decades, but it will probably erupt again at some time in the future. Hualalai, which last erupted during 1800 and 1801, can be expected to erupt again, but probably not in the immediate future (Mullineaux and Peterson 1974:7).

Mauna Loa and Kilauea are both highly active volcanoes. Between 1830 and 1950 Mauna Loa erupted on the average of once every $3\frac{1}{2}$ years. For the next 25 years it was quiet. Kilauea is believed to have been continuously active during most of the period from 1823 to 1924. Twenty-one separate eruptions were recorded between 1924 and 1965, and since that time Kilauea has been active about 80% of the time (Mullineaux and Peterson 1974:11). Continued eruptive activity from Mauna Loa and Kilauea can be expected in the future, though not always affecting areas beyond their summits and rift zones.

More than half of the island of Hawaii is subject to volcanic activity from Mauna Loa and Kilauea. This area includes the city of Hilo—the economic, political, and transportation center of Hawaii, where approximately half the island's 70,000 people live. Since 1880 four lava flows have extended to within 16 km of Hilo's city limits, one reaching into areas now occupied by residences. To the southeast of Hilo, lava flows from the east rift zone of Kilauea caused considerable damage to personal property in 1955 and 1960. The 1960 eruption destroyed more than 80 buildings in the town of Kapoho and covered some 1000 ha of land, much of which had been cultivated (Richter et al. 1970). Frequent eruptions can be expected from both Kilauea and Mauna Loa in the future. Based on past experience, eruptions will vary from brief voluminous outpourings of lava to long-lived eruptions that produce lava at moderate rates. This lava will be emitted chiefly from vents in the summit areas and from along the rift zones, though entire flanks of both volcanoes could potentially be covered by lava flows.

The Hawaiian volcanoes have been watched and monitored by scientists at the Hawaiian Volcano Observatory for over 60 years, but specific predictions of locations, timing, and severity of volcanic activity are still not always possible. Past records suggest that about 64–192 km² of land will be inundated by lava from Kilauea and Mauna Loa during the next 20 years (Mullineaux and Peterson 1974:25). Judging by the historical pattern of eruptive activity, scientists predicted an eruption from Mauna Loa's northeast rift zone to occur between January 1976 and July 1978 as a follow-up to a small summit eruption in July 1975 (USGS 1976a). The predicted eruption did not occur within the time frame specified, nor since.

## Social Setting

In 1970 the U. S. Bureau of the Census reported a total population of 63,468 for the island of Hawaii, marking the first time since 1930 that the island's population increased. Historically, much of Hawaii's 10,000 km² and economic activity have been devoted to the sugar industry. With the mechanization of sugar-cane cultivation and processing during the past 40 years, employment opportunities declined precipitously. When tourism began to emerge as a major industry on Hawaii during the late 1960s, the island's population again began to increase. Between 1960 and 1970 Hawaii's population showed an increase of 3.5%; from 1970 to 1975 a 17.7% increase occurred (County of Hawaii 1976).

Approximately half of the island's population is concentrated in the Hilo area on the east side of the island (1970 city population of 26,353). The city is typical of American small urban centers, with new suburbs sprawling up the lower slopes of Mauna Kea and Mauna Loa, both of which are visible from many parts of the city. One neighborhood of new, upper-middle-income homes, Kaumana, is mostly constructed on an 1881 lava flow.

Hawaii Island, like the rest of the state, has no ethnic majority. According to the 1970 census, the population's ethnic makeup has the following percentage distribution: Japanese, 37.5; Caucasian, 28.8; Filipino, 16.5; Hawaiian, 12.3; Chinese, 2.9; and other, 2.0. Except for the Hawaiians and Caucasians, most of Hawaii's ethnic groups were introduced to the island as contract laborers. Permanent Caucasian residents first came as missionaries in the early 1800s. In the last 10 years, Caucasian immigration to Hawaii from the U.S. mainland has resulted in an increase of over 10% in their share of the island's population—from 17.8% Caucasian in 1960 to 28.8% in 1970 (County of Hawaii 1976).

The ethnic diversity on Hawaii is by no means consistent across the island. The dominant ethnic group in any particular part of the island is primarily a function of that area's economic activity and history. For example, whereas Japanese dominate much of the eastern side of the island, especially around Hilo, Caucasians are in the majority in the tourist areas around Kailua on the west coast; Filipinos make up relatively larger proportions of the population in the large sugar-cane plantation areas along the island's northern and southern coasts; Hawaiians are more concentrated in the northwestern portion of the island.[5]

Within the eastern quarter of Hawaii, where this study was undertaken, considerable local ethnic variation also exists. South Hilo district, dominated by the city of Hilo, is primarily Japanese and Caucasian. The Kaumana area of Hilo is almost evenly divided between Japanese and Caucasians (though in separate settlements), whereas the Ainaloa area includes both a Japanese-dominated subdivision and a densely clustered Portuguese community (see Figure 8.3 for locations). In Puna District, just south of Hilo, Keaau town is dominated by Japanese and Filipino employees of a large sugar mill and cane lands; Hawaiian Paradise Park and other recent speculator-oriented subdivisions are primarily Caucasian; Pahoa's agricultural base is dominated by Japanese, with a cluster of Filipino workers' homes; Nanawale Estates is basically a Caucasian retirement community, but adjacent to it is the resettlement area from the Kapoho eruption (mostly Hawaiians and Filipinos); Kalapana is the home of several older Hawaiian families and newly arrived Caucasian retirees. Each community is distinctive ethnically as well as economically.

## Volcanic Hazards on Hawaii

Volcanoes have always been an important element in the lives of Hawaii's inhabitants. Early Hawaiians feared and revered the volcanoes and believed Pele, the volcano goddess, to be one of the strongest and most malevolent gods in the islands (*National Geographic* 1892). Many Hawaiians would not venture nearer than 30 km from the craters of Kilauea and Mauna Loa, fearful that they might offend Pele, causing her to "either rise out of the crater in volumes of smoke, send up large stones to fall upon them and kill them or cause darkness and rain to overtake them," not to mention sending a stream of lava down upon them (Martin 1970). The missionaries and other Caucasians who arrived on the island in the early 1800s took a different view of the volcanoes; they found them intriguing and worthy of careful study and observation. The reverends Ellis, Lyman, and Coan, missionaries of the American Board, were particularly zealous in their efforts to understand the volcanoes. Each new eruption sent them in search of the source and provided additional subject matter for their journals (Coan c.1882; Ellis 1917).

A 1974 study by the U.S. Geological Survey identified the following products of eruptions to be the primary volcanic hazards on the island of Hawaii: lava flows, falling fragments, gases, and particle-and-gas clouds (Mullineaux and Peterson 1944). Falling fragments and particle-and-gas clouds can be substantial hazards to life but are relatively rare except in the case of debris from fountaining. Although the vast majority of Hawaiian eruptions are mild,

---

[5]Based on statistics in the 1970 U.S. Census and County of Hawaii (1976).

twice within recorded history Kilauea has erupted explosively, propelling solid or molten material and gases into the air. The chief hazard to property, however, comes from lava flows. These occur frequently and often cover broad areas. General geographic zones of differing relative risk from volcanic hazards, primarily lava flows, are illustrated in Figure 8.3; Table 8.4 summarizes each area's volcanic history. The highest-risk areas include the summits and rift zones of Mauna Loa and Kilauea. Relative risk of volcanic hazard then generally declines with distance from these highest-risk areas. The degree of risk also varies within zones; in some the risk decreases gradationally across the entire area, while in others it varies with topography or other physical characteristics.

Scientific predictions based on the 1975 summit eruption of Mauna Loa suggested a future flank eruption from the northeast rift zone, the lava probably flowing in the direction of Hilo (Lockwood et al. 1976). Kilauea's east rift has been consistently active over the last 30 years, resulting in inundation by lava of thousands of acres of land in Puna District. Thus, a band of communities from Hilo to Kalapana, all lying in the South Hilo and Puna districts, was selected for study (see Figure 8.3). These communities vary considerably in relative risk from volcanic hazard, as defined by the U.S. Geological Survey, as well as in the economic and social characteristics of their populations (see Table 8.5). Note that Nanawale Estates and Kalapana lie within the highest risk areas; Pahoa, Keaau, and Hilo all lie in Zone E; and Hawaiian Paradise Park lies in Zone D.

**TABLE 8.4**
**Number of Eruptions Originating within Hazard Zones and Number of Times Lava Flows Have Covered Land within Hazard Zones during Historic and Recent Prehistoric Time**

| | Historic time (since approximately 1800) | | | Recent prehistoric time (5000-year interval prior to 1800) | |
|---|---|---|---|---|---|
| one | Number of times vents have erupted within area | Number of times lava flows have covered land within area | Percentage of land covered within area | Number of times vents have erupted within area (estimated) | Number of times lava flows have covered land within area (estimated) |
| | 0 | 0 | 0 | 0 | 0 |
| | 0 | 0 | 0 | 0 | <5 |
| | 0 | 0 | 0 | <5 | <5 |
| | 0 | 0 | 0 | 0 | >10[a] |
| E | 1 | 2 | 6 | >10 | >10 |
| | 1 | 35[a] | 15 | ≈10 | >100[a] |
| | 80 | >80 | 50 | ≈2000 | >2000 |

Source: U.S. Geological Survey, 1976.
[a]Most lava flows that entered Zones D and E erupted from vents in Zone F.

# Table 8.5
## Community Characteristics

| | No. of interviews | Ethnic distribution | | | | | | Percentage answering yes to whether home threatened by volcanic activity | Volcanic risk zone (see Fig. 7.3) |
|---|---|---|---|---|---|---|---|---|---|
| | | Japanese | Caucasian | Hawaiian | Filipino | Portuguese | Other | | |
| Kaumana | 45 | 33.4 | 33.4 | 13.3 | 6.7 | 8.9 | 4.4 | 63.6 | E |
| Ainaloa | 55 | 34.5 | 18.2 | 9.1 | 12.7 | 18.2 | 7.3 | 83.9 | E |
| Keaau | 25 | 41.7 | 20.8 | 8.3 | 25.0 | 4.2 | 0 | 58.3 | E |
| Hawaiian Paradise Park | 25 | 7.1 | 53.6 | 17.9 | 10.7 | 3.6 | 7.1 | 48.3 | D |
| Pahoa | 32 | 71.0 | 3.2 | 3.2 | 16.1 | 3.2 | 3.3 | 74.2 | E |
| Nanawale | 20 | 10.5 | 26.3 | 31.6 | 21.1 | 5.3 | 5.7 | 45.0 | F |
| Kalapana | 8 | 0 | 62.5 | 25.0 | 12.5 | 0 | 0 | 100.0 | F |

## Hazard Perceptions

Interviews were conducted with 210 residents of the seven communities from Kaumana to Kalapana (see Figure 8.3). Background information, including age, sex, ethnicity, occupation, length of residence in Hawaii, and homeownership, was collected in each interview in addition to the subject's experience with volcanic activity, knowledge of government actions related to volcanic activity, perceptions of different volcanic hazards, and attitudes toward government and self-determination. Overall, approximately 30% of the residents perceived volcanic hazards as a "bad quality" of Hilo and Puna; most had seen the steam from Mauna Loa and Kilauea and had read or heard radio or television accounts of volcanic activity. And even though two-thirds of the respondents stated that their home would be threatened by some form of volcanic activity if it occurred in the next 10 years, over half (56%) felt there was no cause for concern. Less than 5% of those interviewed had changed any of their activities since the announcement of Mauna Loa's swelling; most perceived no real danger.

The following question was used to elicit individuals' perceptions of volcanic hazard potential:

> Experiences elsewhere in the world have shown that several different kinds of hazards might result from volcanic activity. In *your* opinion how likely is it that *volcanic activity at Mauna Loa or Kilauea* will cause any of these events to occur sometime during the *next ten years*? Please circle a number between 1 (definitely will not occur) and 5 (definitely will occur) for each hazard.

The following list of hazards was then presented to each respondent for evaluation: mudflow, ashfall, falling rock, earth tremor, blowing steam, lava flow, avalanche, flood. Notice that the question relates to volcanic activity at Mauna Loa and Kilauea in general, and not as it would affect each community or area. According to the U.S. Geological Survey 1974 report and interviews with scientists at the Hawaiian Volcano Observatory, mudflows are the least likely of the named hazards to occur anywhere on Hawaii (although one was recorded on the southwest flank of Mauna Loa in 1868 [Martin 1970]); falling rock, avalanches, and floods are somewhat more likely to occur, but probably not within the next 10 years; ashfall, forest fire, or blowing steam are still more likely to occur; and earth tremors and lava flows will almost certainly occur within the next 10 years.

For all respondents, the "no opinion" option was chosen most frequently for the low-probability events, especially mudflow, avalanche, and flood (all greater than 20%) (see Table 8.6). In contrast, less than 3% of the respondents answered "no opinion" to earth tremor and lava flow. Information about these latter high-probability events is apparently more available and/or comprehensible, allowing definite perceptions to be formed. However, Portuguese and Filipino residents had fewer strong opinions of event likelihood than did other ethnic groups; females chose "no opinion" almost twice as often as males across all hazards. These distinctions suggest that information about volcanic hazards on Hawaii has not been disseminated to all residents equally.

Table 8.6
Locational Variations in Perception of Probability of Various Volcanic Events during the Next 10 Years in Hawaii (composite scores)[a]

| Event | Kaumana | Ainaloa | Keeau | Hawaiian Paradise Park | Pahoa | Nanawale | Kalapana | Total | % No opinion |
|---|---|---|---|---|---|---|---|---|---|
| Mudflow | 2.49 | 1.90 | 1.69 | 1.78 | 1.96 | 2.07 | 1.33 | 2.03 | 21.8 |
| Ashfall | 3.49 | 3.43 | 3.10 | 3.15 | 3.39 | 3.29 | 2.75 | 3.34 | 11.4 |
| Falling rock | 3.05 | 3.10 | 3.16 | 2.73 | 2.89 | 3.32 | 3.00 | 3.03 | 12.2 |
| Forest fire | 4.07 | 3.71 | 3.77 | 3.96 | 3.84 | 3.58 | 4.14 | 3.85 | 6.1 |
| Earth tremor | 4.55 | 4.44 | 4.52 | 4.79 | 4.28 | 4.15 | 4.75 | 4.48 | 2.8 |
| Blowing steam | 4.02 | 3.69 | 4.10 | 3.79 | 3.79 | 3.35 | 4.71 | 3.84 | 8.0 |
| Lava flow | 4.36 | 4.40 | 4.59 | 4.41 | 4.25 | 4.75 | 4.50 | 4.43 | 2.8 |
| Avalanche | 2.39 | 2.30 | 2.88 | 2.22 | 2.36 | 2.06 | 1.80 | 2.33 | 20.0 |
| Flood | 2.33 | 2.24 | 2.50 | 2.44 | 1.86 | 1.94 | 1.00 | 2.21 | 20.0 |

[a]Scale: 1 = definitely will **not** occur
2 = might not occur
3 = neutral
4 = might occur
5 = definitely **will** occur

As might be expected, long-term residents (more than 15 years) generally had more accurate perceptions of hazard probabilities than more recent arrivals. Caucasian and Japanese perceptions were more closely related to official perceptions than those of other ethnic groups (see Table 8.7). However, no significant difference was found in the accuracy of perceptions by community (Table 8.6). Especially for the lower-probability events, variation within communities was often greater than between communities, with almost even distributions of responses in the five likelihood categories. Such internal variation was similarly great for groups defined by length of residence in Hawaii, but not for ethnic groups.

The misperception of volcanic hazards does not present a simple pattern. In general, all groups overestimated the likelihood of low-probability events such as mudflows, falling rocks, forest fires, and blowing steam, and underestimated the higher-probability hazards. Specifically, females were more likely to overestimate hazards than males, Filipinos and Portuguese more than other ethnic groups (see Table 8.7), and Kaumana and Nanawale residents slightly more than other communities (see Table 8.6). However, the Portuguese also underestimated the high-probability events by much more than other ethnic groups. The pattern of overestimation by females, Filipinos, and Portuguese is consistent with the lack of knowledge among these groups, as shown by the use of the "no opinion" answer. With less information about a hazard, the tendency is to assume that it might occur. This also leads to greater fear; 52% of Filipinos, compared with 24% of all residents, described the general attitude toward Mauna Loa in their neighborhood as being at least somewhat fearful.

Residents were also questioned as to whether they felt that any of these volcanic hazards might threaten them or their home. According to the scientists, the areas most likely to be affected by such activity are Nanawale,

**TABLE 8.7**
Ethnic Variations in Perceptions of Probability of Various Volcanic Events during the Next 10 Years in Hawaii (composite scores)[a]

| Event | Japanese | Hawaiian | Portuguese | Caucasian | Filipino | Total[b] | No opinion[b] |
|---|---|---|---|---|---|---|---|
| Mudflow | 1.88 | 2.14 | 2.09 | 2.07 | 2.32 | 2.02 | 21.8% |
| Ashfall | 3.23 | 3.10 | 3.21 | 3.34 | 3.61 | 3.34 | 11.2% |
| Falling rock | 2.70 | 3.28 | 3.36 | 2.96 | 3.35 | 3.02 | 12.0% |
| Forest fire | 3.85 | 3.73 | 3.78 | 3.81 | 3.92 | 3.85 | 6.3% |
| Earth tremor | 4.45 | 4.44 | 4.36 | 4.51 | 4.52 | 4.48 | 2.9% |
| Blowing steam | 3.67 | 4.19 | 3.50 | 3.89 | 3.83 | 3.83 | 8.2% |
| Lava flow | 4.45 | 4.70 | 3.93 | 4.28 | 4.56 | 4.43 | 2.9% |
| Avalanche | 2.14 | 2.52 | 2.69 | 2.17 | 2.56 | 2.31 | 20.5% |
| Flood | 1.90 | 2.61 | 2.45 | 2.17 | 2.56 | 2.19 | 20.5% |

[a]Scale: 1 = definitely will **not** occur
    2 = might not occur
    3 = neutral
    4 = might occur
    5 = definitely **will** occur
[b]Slight discrepancies from Table 8.6 are due to rounding.

Kalapana, and Pahoa. Keaau and Hawaiian Paradise Park are least likely to be affected. The responses given by residents, however, do not match this pattern (Table 8.5). Nanawale residents were the most convinced that they would not be affected, thus vastly underestimating the risk involved at their location. Keaau and Hawaiian Paradise Park residents somewhat overpredicted the risk in their communities, as did residents of Ainaloa, where the risk is indicated as only low to moderate. The difference in risk perception between Ainaloa and Kaumana, both suburbs of Hilo, is particularly interesting. Kaumana is built on relatively recent (less than 100 years old) lava flows from Mauna Loa and, of the neighborhoods involved in this study, is the one that scientists at the Hawaiian Volcano Observatory predict will most likely be affected if a major eruption of Mauna Loa's northeast rift zone occurs. Ainaloa, though also built on the side of Mauna Loa, does not face similar risk because of the local topography. However, 20% more Ainaloa than Kaumana residents perceived that their homes would be threatened by volcanic activity (Table 8.5). Cultural differences between the two areas account for some of this perceptual variation, but it also appears that Kaumana's residents are ignoring or denying the scientific information that has been presented. One attitude commonly voiced by Kaumana residents at public hearings on emergency planning for volcanic hazards in Hawaii is that Civil Defense and other agencies should be responsible for protecting their homes and possessions in the event of an eruption. The assumption that they will be protected may lead these residents to feel less threatened by volcanic activity and therefore downgrade the hazard in their minds.

### Hazard Responses

Predicting human responses to events that have not yet occurred is extremely difficult. Though we can ask people what they think they will do sometime in the future, we can never be sure that they can foretell accurately what their actions will be. Fully 95% of the residents interviewed stated that they did not change their activities or employment in any way as a result of the reported increased swelling of Mauna Loa in 1975–1976. If a warning of increased volcanic activity and danger of eruption were given, a quarter of the residents predicted that they would do nothing, either because they feel that there would be no danger to them or because there is nothing that could be done. Another quarter said that if evacuated by government officials, they would go; 22% would take measures for their personal protection; 17% would require personal confirmation of the hazard before taking action. Only 12% of the interviewed population answered an unqualified yes to the question: "If a warning of increased volcanic activity and danger of eruption were given, would you do anything about it?"

The September 1977 eruption of Kilauea's east rift provides a good example of actual response to volcanic hazards. The flow of lava began on 23 September, heading toward the village of Kalapana. For several days the flow continued toward Kalapana, and on 29 September the decision was made by Civil Defense officials, with the assistance of Hawaiian Volcano Observatory scientists, to evacuate the town. There was little doubt in their minds that the 150-m-wide river of lava would eventually inundate the town and flow into the ocean. However, many of the local residents perceived the situation differently. Pele remains an honored and respected goddess in the minds of many of Kalapana's people, and the belief that she harms no one who has not harmed her justified their defiance of the initial evacuation order. Although most of the residents were eventually evacuated, their faith in Pele was upheld when the lava flow suddenly stopped about a quarter mile above the nearest residence on 1 October, causing those who believed in Pele to become even more steadfast in their faith. Such attitudes on the part of Kalapana residents also affected strategies to control the lava flow. When the residents voted in favor of letting the flow continue on its way without interruption, plans for building dikes or bombing in an attempt to divert the flow were canceled (Baclig 1977).

These responses to the lava flow were at least in part predictable. In responses to statements designed to measure one's attitudes toward fatalism and the role of government in predicting and controlling natural hazards, Kalapana residents ranked below all other communities in their willingness to depend on government for decisions and protection from natural hazards (Table 8.8). They, along with Nanawale residents, also were consistently more fatalistic with respect to their ability to control their own fate. In contrast, the Kaumana, Pahoa, and Keaau residents were less fatalistic and more prone to put their faith in government for protection from natural hazards. A response similar to that of the Kalapana residents in 1977 to attempting to control the lava flow would certainly not be expected in Kaumana, where the residents expect to be protected from natural hazards by whatever means are currently available.

Attitudes toward fatalism and the role of government show even greater differences between ethnic groups (Table 8.9). On the whole, the Japanese and Caucasians believe that individuals to a large extent can control their own future. They differ, however, in their attitude toward government; the Japanese and Filipinos believed that people should trust their elected officials to make decisions for them, whereas the Caucasians (and to a lesser extent Portuguese) felt the opposite—that individuals should make their own decisions.

Given these general attitudes and an understanding of Hawaiian spiritual beliefs, the response of some of the Kalapana residents—especially the older Hawaiians—to the threat of inundation by lava and subsequent relief efforts by local government officials is not surprising. Similar responses might be expected in other Hawaiian communities with essentially opposite responses in Japanese and Filipino areas. Most Caucasians, though probably obeying evacuation or similar Civil Defense orders, could also be expected to show more individual initiative in protecting their own belongings and dependents. Some residents, especially those living in areas where they do not perceive any

**TABLE 8.8**
Locational Variations in Attitudes toward Controlling Natural Disasters (composite scores)[a]

| | Attitude | Kaumana N=45 | Ainaola N=56 | Keaau N=24 | HPP N=68 | Pahoa N=20 | Nanawale N=20 | Kalapana N=8 | Total N=210 |
|---|---|---|---|---|---|---|---|---|---|
| Role of government | Public is informed enough to take part in most governmental decisions. | 4.14 | 3.98 | 4.13 | 3.97 | 3.28 | 3.45 | 4.00 | 3.88 |
| | Government's decisions should be left to elected officials and their professional advisers. | 5.00 | 4.00 | 4.39 | 4.14 | 3.21 | 3.84 | 5.50 | 4.20 |
| | Government should be responsible for protecting people against natural disasters. | 2.52 | 2.45 | 2.26 | 3.00 | 2.24 | 2.80 | 3.75 | 2.57 |
| | Government has too much say over lives of individuals. | 3.64 | 3.50 | 3.52 | 2.90 | 3.03 | 3.00 | 3.50 | 3.33 |
| | Individuals themselves, not government, should determine whether they ought to visit places that might be dangerous. | 3.95 | 4.54 | 3.61 | 3.48 | 3.86 | 4.20 | 4.00 | 4.04 |
| Fatalism | Natural disasters are acts of God. | 2.98 | 2.76 | 3.57 | 2.76 | 3.10 | 2.60 | 2.88 | 2.93 |
| | There are many things in life that science can never fully explain. | 2.64 | 2.04 | 2.26 | 2.34 | 2.23 | 2.05 | 2.38 | 2.27 |
| | Scientists will someday be able to predict accurately most natural disasters. | 2.93 | 3.11 | 2.77 | 3.14 | 2.36 | 3.40 | 2.88 | 2.96 |
| | There is little we can do to change our fate. | 3.91 | 4.00 | 3.96 | 4.10 | 3.86 | 3.55 | 3.75 | 3.92 |
| | We can rely on science to solve our future problems. | 3.79 | 3.73 | 3.35 | 3.83 | 3.66 | 4.20 | 4.25 | 3.77 |

[a]Where 1 = strongly agree; 7 = strongly disagree.

**Table 8.9**

**Ethnic Variations in Attitudes toward Controlling Natural Disasters (composite scores)[a]**

| | Attitude | Japanese N=68 | Hawaiian N=26 | Portuguese N=18 | Haole N=54 | Filipino N=27 | Total N=203 |
|---|---|---|---|---|---|---|---|
| Role of government | Public is informed enough to take part in most governmental decisions. | 3.53 | 3.08 | 3.94 | 4.81 | 3.59 | 3.89 |
| | Government's decisions should be left to elected officials and their professional advisers. | 4.00 | 4.27 | 4.56 | 4.76 | 3.80 | 4.23 |
| | Government should be responsible for protecting people against natural disasters. | 2.42 | 2.30 | 2.22 | 3.22 | 2.38 | 2.60 |
| | Government has too much say over lives of individuals. | 3.54 | 3.63 | 3.39 | 2.85 | 3.58 | 3.33 |
| | Individuals themselves, not government, should determine whether they ought to visit places that might be dangerous | 3.77 | 4.19 | 3.78 | 4.63 | 4.19 | 4.08 |
| Fatalism | Natural disasters are acts of God. | 3.46 | 2.81 | 2.35 | 2.65 | 2.81 | 2.93 |
| | There are many things in life that science can never fully explain. | 2.39 | 2.19 | 1.89 | 2.20 | 2.41 | 2.25 |
| | Scientists will someday be able to predict accurately most natural disasters. | 2.69 | 2.85 | 3.17 | 3.34 | 2.85 | 2.98 |
| | There is little we can do to change our fate. | 3.88 | 3.27 | 3.83 | 4.48 | 3.62 | 3.91 |
| | We can rely on science to solve our future problems. | 3.41 | 3.92 | 3.94 | 4.19 | 3.77 | 3.79 |

[a] Where 1 = strongly agree; 7 = strongly disagree.

hazard, will have to be convinced that a hazard does in fact exist, rather than responding to all warnings with immediate action. The example of Kalapana should serve as proof that attitudes and perceptions can play an important role in determining responses to natural hazards. Decision makers should be fueled with information on perceptions and attitudes toward various hazards in order to predict more accurately the responses of all communities before actual disaster occurs.

## SUMMARY COMMENT

The lessons of the Mount Baker and the Hawaii experiences are, like living with low-probability hazards, ambiguous. The protective steps taken by officials were clearly indicated and responsible, given the evidence at hand. These steps were unpopular locally and in the Mount Baker case cost the U.S. Forest Service some of its local stock of good will. Community leaders in Concrete overestimated the economic damage suffered as a result of the protective closure, but the sense of grievance is understandable. Clearly the commercial resort on Baker Lake suffered substantial economic damage from the loss of an entire summer's business, and the utility company suffered a loss of hydroelectric output.

Both the Mount Baker and the Hawaii experiences suggest that many prefer individual coping to governmental intervention, at least up to the point of clear and present danger. The residents of Kalapana accepted evacuation, in view of the approaching lava flow, but suggestions for bombing or diking to divert the threatening lava flow were rejected because, at least in part, of religious conviction. That the lava flow stopped short of the village without intervention confirmed the faith of those who believe in religion as a means of coping. The preference of the residents of Concrete, near the foot of Mount Baker, for coping instead of official intervention was not put to as severe a test because neither their lives nor their homes were actually threatened. No objections were expressed, either in the interviews or in informal comments, to the precautionary lowering of Baker Lake, but there was substantial disagreement with the general area closure. Many expressed the opinion that simply placing an appropriate warning sign would have sufficed in lieu of closure of roads and trails. The Forest Service position, however, is that merely posting warnings does not meet the legal burden of protecting the public.

Responding to an uncertain prediction of an uncertain hazard is a difficult managerial problem. Whether dealing with predictions of volcano or other natural hazard activity, the manager must distill out of the scientists' probability realm a definite course of action—or nonaction. In effect, the manager must translate probabilities into absolutes in order to discharge his or her responsibilities to provide or not provide protection. The legal and institutional environment precludes managers from allowing individuals to assume risks that may appear reasonable to the individuals, and pressures managers to take steps that are unpopular at the time.

Mount Baker was such a case, although the interviews revealed an amount of local tolerance for the closures that surprised the investigators. Mauna Loa would be similar, except no restrictions have been imposed on the public. Kilauea was different; it provided visible threat in the form of an approaching lava flow. There are some interesting similarities between volcano prediction and earthquake prediction. Both rest on scientific predictive methodology that is relatively new, with limited opportunity for empirical testing. Both involve uncertain probabilities, magnitudes, and time frames; and there is no obvious all-clear point when the potential danger has passed. In the absence of change in premonitory activity, concern about Mount Baker subsided over time, and after about 1 year the restrictions were removed. The lifting of restrictions is in itself a probabilistic statement because the steaming activity that gave rise to the closures persists as of this writing (summer, 1979).

Given the propensity to cope in the face of an uncertain hazard, and the reservations that were expressed by local residents concerning official actions, it is very likely that greater resistance to area closure, evacuation, or other measures would be encountered the second time around, especially at Mount Baker. Should there be a need for precautionary measures to be repeated, officials should probably involve local citizens directly in a vigorous public information and decision-making process. Direct involvement of informed citizens in decision making would have the dual advantage of achieving broader consensus on the steps to be taken and diffusing the responsibility for steps taken and not taken. The information dissemination program should include local "town meetings" to lay the scientific background and provide regular progress reports. It could also include special programs for school students with the side benefit of motivating an interest in geologic science and public policy considerations. In short, the "management" of threatening volcanoes should be a community endeavor as well as an official responsibility.

As greater amounts of settlement occur in areas of potential volcanic risk and as USGS and other parties devote increasing attention to forecasting volcanic events, the scenarios described in this study are likely to occur with increasing frequency. Continued study of the physical aspects of natural hazards and of prediction technology is important, of course. But given the problems posed by public response to prediction of such events, it is obvious that further attention needs to be given to the management of hazard prediction, as well as to the hazard itself.

## REFERENCES

Baclig, A.
    1977   Kalapana in path—Kilauea flow slows. *Hawaii Tribune-Herald* (September 30):1.

Coan, T.
  c.1882  *Life in Hawaii, an autobiographic sketch of mission life and labors (1835–1881).* New York:
          Anson D. F. Randolph.
County of Hawaii, Department of Research and Development
  1976    *Hawaii County data book 1976.* Hilo, Hawaii: County of Hawaii.
Cullen, J. M.
  1978    *Impact of a major eruption of Mount Rainier on public service delivery systems in the
          Puyallup Valley, Washington.* Unpublished master's thesis, Dept. of Urban Planning,
          University of Washington, Seattle.
Ellis, W.
  1917    *A narrative of a tour through Hawaii, or Owhyee; with remarks on the history, traditions,
          manners, customs and language of the inhabitants on the Sandwich Islands.* Honolulu:
          Hawaiian Gazette Co., Ltd.
Frank, D., M. Meier, and D. Swanson
  1977    *Assessment of increased thermal activity at Mount Baker, Washington, March 1975–March
          1976.* USGS Paper 1022-A. Washington, D.C.: U.S. Government Printing Office.
Hodge, D. C. and V. L. Sharp with M. E. Marts, F. E. Sheridan and J. L. MacGregor
  1978    *Social implications of the volcano hazard: Case studies in the Washington Cascades and
          Hawaii,* Volume I. Final Report to the National Science Foundation–RANN Program,
          July 31.
Lockwood, J. P., R. Y. Koyanagi, R. I. Tilling, R. T. Holcomb, and D. W. Peterson
  1976    Mauna Loa threatening. *Geotimes* 21(6):12–15.
Martin, M. G. (compiler)
  1970    *Sarah Joiner Lyman of Hawaii—Her own story.* Hilo: Lyman House Memorial Museum.
Mullineaux, D., and D. W. Peterson
  1974    *Volcanic hazards on the Island of Hawaii.* Washington, D.C.: U.S. Geological Survey
          Open-file Report 74-239.
Murton, B. J., and S. Shimabukuro
  1974    *Human adjustment to volcanic hazard in Puna District, Hawaii.* In *Natural hazards,
          local, national global,* edited by G. F. White. New York: Oxford University Press.
National Geographic Magazine
  1892    Kilauea. *National Geographic Magazine* (June):563–576.
Richter, D. H., J. P. Eaton, K. J. Murata, W. U. Ault, and H. L. Krivoy
  1970    *Chronological narrative of the 1959–60 eruption of Kilauea Volcano, Hawaii.* Washington,
          D.C.: U.S. Geological Survey Professional Paper 537-E.
U.S. Geological Survey
  1976a   *Mauna Loa Volcano—The past, present, and prospects for the future.* Volcano, Hawaii:
          U.S. Geological Survey, Hawaiian Volcano Observatory.
  1976b   *Natural hazards on the Island of Hawaii.* Washington, D.C.: U.S. Government Printing
          Office.
Warrick, R. A.
  1975    *Volcano hazard in the United States: A research assessment.* Boulder: Institute of Be-
          havioral Science, University of Colorado.

# 9

# Effects of the Eruption of Parícutin Volcano on Landforms, Vegetation, and Human Occupancy

JOHN D. REES

## INTRODUCTION

The environment and rural economy of the Itzícuaro Valley, Michoacán, in west–central Mexico was disrupted on 20 February 1943, by the eruption of a volcano 2 km from the Tarascan Indian village of Parícutin. The onset of the Parícutin eruption was preceded by 2 weeks of mild earthquakes that grew in intensity before the outbreak, after which the seisms and underground noises ceased (Foshag and González 1956:355). The eruption took place at about 4:30 p.m. on a maize field located in a small shallow valley 2 km southeast of the village. The village of Parícutin is located 320 km west of Mexico City in the state of Michoacán in the west–central portion of the Mexican volcanic axis known as the Meseta Tarasca. The Meseta Tarasca is formed of Tertiary and Quaternary volcanic rock materials overlying older sediments and extrusives of the southern escarpment of the Central Mexican Plateau.

The cone of Parícutin Volcano is situated on the northern flank of Tancítaro, in the upper watershed of the westward-trending Itzícuaro Valley (Figure 9.1). Tancítaro is an eroded major Pleistocene volcano that rises to over 3845 m to make it the dominant volcano of the region. To the northeast of the Parícutin cone, another volcano, the twin peaks of Angahuan rise to over 3290 m. The Parícutin cone is the most recent of the more than 150 cinder cones that are identified within approximately 1600 km² around the newest cone (Williams 1950:168). The majority of the cones are composed of semi-consolidated bedded ash, lapilli, and volcanic bombs; lava flows make up a

*VOLCANIC ACTIVITY AND HUMAN ECOLOGY*

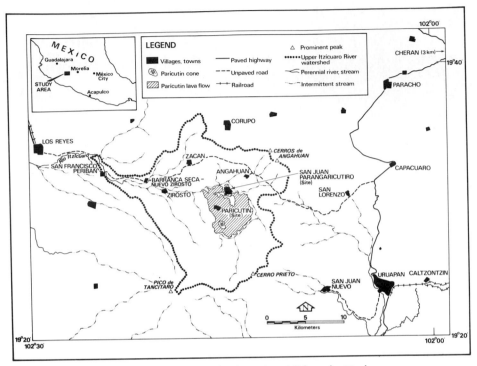

**FIGURE 9.1. Map of Parícutin region, Michoacán, Mexico.**

small area of the surface of most cinder cones. Two types of cone shapes are common: (*a*) those that are conical with a flat-floored crater, and (*b*) those that have been breached by volcanic explosions or lava flows and subsequently eroded, leaving one or more sides removed. Flat-floored craters are sometimes cleared of timber and utilized as sites for long-fallow, short-cropping maize cultivation. Long fallow is a result of the less fertile and more droughty soils of crater floors.

Although volcanoes and cinder cones are the most imposing landforms, lava flows cover large areas of the Itzícuaro Valley. The lower slopes of Parícutin and many older cones are buried by extensive lava fields. In some cases, the lava flows are narrow on the higher slopes and fan out on lower ones. Flows from Tancítaro and from cinder cones have produced a bench-and-cliff topography. The majority of the ancient flows have undergone considerable weathering and possess a true soil surface that is forested or is in long-fallow, short-cropping plow agriculture. The more recent flows, however, are lava badlands with an extremely rough, blocky rubble of vasicular basalt that is difficult to cross on foot and usually prevents entry by cattle and other hoofed animals. Although not used for crops or grazing, these lava badlands produce timber that is exploited by Tarascan Indian woodsmen, who are forced to backpack timber products to sites from which pack animals can operate.

Examination of air photos taken in 1934 for the Departmento Agrario shows that the bottom of the Itzícuaro Valley prior to the eruption of Parícutin was composed of older lava benches and other depositional surfaces. These depositional surfaces were initially produced by airborne ash and lapilli, later augmented by alluvial deposits from nearby slopes. Before the Parícutin eruption, older lava benches and depositional areas made up the cultivable lands, whereas the steep bench cliffs, recent flows, cinder cone slopes, and upper slopes of Tancítaro and Angahuan were heavily wooded.


## SUMMARY HISTORY OF
## THE PARÍCUTIN ERUPTION

On the day of the eruption, a resident of the village of Parícutin planting crops at the site claimed that he witnessed a small explosion on the maize field from a vent that had suddenly opened in a shallow pit where storm runoff had drained during past rainy seasons. The initial explosion was followed by emission of steam and sulfurous gases and a small eruptive column of fine dust and small incandescent stones. The 30-cm vent widened with the slumping of molten materials, and the eruptive column gradually increased in size. By midnight, the fissure activity had become violent. A large eruptive column was accompanied by a roaring sound, lightning flashes, and the violent ejection of great numbers of incandescent rocks (Foshag and González 1956:385). At midnight of the first day the cone measured approximately 6 m high; 6 days later, it became 167 m; by the end of 1 year, the cone had grown to a height of 325 m. Intermittent eruptions continued until in 1952 the cone rose to a height of 410 m above the former maize field; its height, however, is not apparent because of repeated lava flows that have covered the lower slopes of the cone (Foshag and González 1956:355).

The first month of Parícutin's activity was dominated by the explosive ejection of volcanic bombs, blocks, and lapilli that brought about the rapid cone building. Most of the ejecta were blocks (angular chunks of old rock from walls of the magma conduit), and bombs (globs of magma, some of which were still plastic enough to alter their shapes on impact). Most bombs were from 30 cm to 1 m in diameter and fragmented on impact with the ground (Bullard 1976:355). Every few seconds showers of incandescent rock material were ejected from the crater. The explosive violence could be heard as far as Guanajuato, 350 km distant (Bullard 1976:355). By 6 March 1943, about 83 million m³ of bombs, blocks, lapilli, and ash had been ejected from the crater vents of Parícutin; ejection of pyroclastic material averaged 6 million m³ daily in this early period and then decreased to an average of 76,000 m³ thereafter for several months (Fries 1953:606–607).

Lava flowed from the cone vents from the first day and by 24 February issued in a broad sheet 700 m long and advanced at a rate of 5 m/hr. Surges of lava flow repeatedly breached the weak walls of the cone during the first days of eruption. Pauses in the flow allowed the cone wall to build up again by accumulation of volcanic bombs, lapilli, and ash. The shape of the cone thus alternated from a horseshoe to a cone. After the first month of activity the buildup of the cone mass was such that lava no longer broke through the walls from the crater. Instead, surges of lava flowed from vents on the outer flank of the cone, causing slumping and partial collapse of a cone section that was later filled in by the bombs of the next explosive period (Foshag and González 1956:395).

The heavy cineric phase began on 18 March and lasted until 9 June 1943. The change from a volcanic bomb stage (characterized by little ash but great amount of bombs) to the succeeding cineric phase was a result in part of choking of the vents in the cone by debris as volcanic bombs rolled back down the inside slopes of the cone (Foshag and González 1956:430). During the cineric phase a nearly continuous eruptive column of ash rose to heights of 8000 m above the cone. During this period the ashfall was the greatest in Parícutin history. Winds carried the ash far to the north and east. Fine ash fell in Mexico City (320 km away), while in Uruapan (25 km to the east) street lamps and auto headlights were necessary during the daytime. In nearby San Juan Parangaricutiro roofs of houses were cleared every 2 or 3 days to prevent collapse (Foshag and González 1956:399). During the summer of 1943 ash emission decreased, and the lavas rose in the central cone higher than in any subsequent period. Soon after, lava cascaded from a lava fountain 100 m below the crater rim. In July other flows injected under earlier flows carried the crusted earlier flows along for 400 m (Ordóñez 1947:41, 42). In October 1943, a parasitic cone developed at the northeast base of Parícutin and erupted with vigor until January 1944. This second cone, called Sapichu, rose to over 70 m and developed a horseshow shape as lava flows breached the northeast side. In January 1944 the main cone renewed explosive activity after the relative quiet of the Sapichu period. New lava vents opened at the southwest base of the Parícutin cone; vents in this area produced all of the lava flows for the next 3 years.

From January to August 1944 these vents yielded the San Juan flow that moved east, north, and then west, to flow over and past the village of San Juan Parangaricutiro. This flow was the most voluminous of the Parícutin flows and extended over 10 km in length. From September to November 1944 the Parícutin flow advanced at a rate of 6 m/hr northward through Parícutin village and completely covered the village site, and reached the earlier San Juan flow (Bullard 1976:363). During mid-November 1944, additional vents opened on the southwest base of the cone; lavas from these vents flowed mostly over the San Juan and Parícutin lavas. Similar activity from these southwest vents continued through 1945 and 1946.

In January 1947 the cone walls of the main cone slumped on the northeast and southwest sides, and new vents adjacent to the slumping began emitting lava that flowed intermittently during 1947. From December 1947 until the end of activity in 1952, most of the major lava flows came from the northeast vents, the most prominent of which was called new Sapichu. Flows from this vent were nearly continuous from February 1948 to the sudden cessation of lava emissions on 25 February 1952 (Wilcox 1954:287–288). Flows of the later years piled up in depth around the cone and in some cases moved far enough north to cover more uncovered ground enclosed by the San Juan and Parícutin flows of 1944. During the last years of the eruption, ash emissions and explosive ejection of bombs continued to occur at intervals. After lava eruptions ceased, only weak explosions occurred and eruption of pyroclastics ended on 4 March 1952 (Wilcox, 1954:289). The only form of activity since that time has been the gentle emission of warm acidic water vapor from fumarole vents near the cone rim and from 12 vents on the lava flow immediately west of the cone.

In the 9 years and 11 days of eruption history, the Parícutin lava flows covered 24.8 km² of old ground area; the flows have a calculated total volume of 700 million m³ (Fries 1953:611). Lava thickness is at a maximum (over 242 m) near the cone, where lava piled up over previous flows, and a minimum, 3 m, at the outer edge of the lava field (Williams 1950:227). Although sizable, the Parícutin lava field is much smaller than some geologically recent flows of nearby cinder cones. For example, the Capacuaro cones (northeast of Parícutin) produced a lava field of 155 km² with an average lava thickness of over 60 m. This is 6.25 times the area and over 32 times the volume of the Parícutin lavas (Williams 1950:203).

The Parícutin lavas are generally classified as the Aa type. The surface of the Parícutin flows are a jumble of irregularly shaped blocks with jagged and broken surfaces and with a scoria texture; color is from dark brown to black. The lavas of February 1943 are olivine-bearing basaltic andesite containing 55% silica; succeeding ejecta are progressively more salic until in 1952 the ejecta contain over 60% silica (Wilcox 1954:281). This increase in silica, acidity, and viscocity has been often observed in other eruptions. All of the Parícutin lavas and pyroclasts are classified as intermediate between the acidic (having over 66% silica) and basic (having less than 52%) (Bullard 1976:51). The Parícutin basaltic andesites are similar to ejecta from the younger volcanoes in the nearby area (Williams 1950:165). It can be expected that the Parícutin lavas and pyroclasts will eventually weather into soils similar to those found in the region.

For the Parícutin eruptions, Fries calculated the yearly total weights and volumes and average daily weights of lava and pyroclasts (see Table 9.1). The table gauges the persistence of the lava and pyroclast eruptions and highlights the sharp decline in eruption volume after the second year. Fries estimated that these rock materials would have occupied about 1.4 km³ in the magma chamber. Lava was estimated to make up about 27% of the total (by weight), with pyroclasts making up the balance (water not included in the calculations) (Fries 1953:611). Measured in cubic kilometers of ejected rock material, the Parícutin eruption is minor as compared with more major eruptions in recent historic time. For example, in the Mount Katmai, Alaska eruption in 1912, 11 km³ of dense horizontal ash flow erupted in less than 20 hours; an additional vertical ash emission of 16 km³ reached Kodiak Island some 160 km away and dropped 30 cm of ash. In an earlier eruption occurring in April 1815, the

TABLE 9.1
Total Weight and Average Daily Weight of Pyroclastic Material and Lava Erupted at Parícutin, by Years, from 1943 through 1951[a]

| Year | Duration (days) | Ash Volume (million m³) | Ash Weight (million metric tons) | Lava Volume (million m³) | Lava Weight (million metric tons) | Total weight (million metric tons) | Average daily weight (thousand metric tons) |
|------|------|------|------|------|------|------|------|
| 1943 | 314 | 601 | 1022 | 107 | 203 | 1225 | 3900 |
| 1944 | 366 | 280 | 476 | 122 | 232 | 708 | 1930 |
| 1945 | 365 | 148 | 252 | 98 | 186 | 438 | 1200 |
| 1946 | 365 | 103 | 175 | 87 | 165 | 340 | 930 |
| 1947 | 365 | 98 | 167 | 79 | 150 | 317 | 870 |
| 1948 | 366 | 28 | 48 | 76 | 144 | 192 | 520 |
| 1949 | 365 | 21 | 36 | 48 | 91 | 127 | 350 |
| 1950 | 365 | 16 | 27 | 40 | 76 | 103 | 280 |
| 1951 | 365 | 14 | 24 | 43 | 82 | 106 | 290 |
| Total | 3236 | 1309 | 2227 | 700 | 1329 | 3556 | 1100 |

[a]Source: Fries, 1953:611.

Tambora Volcano on the island of Sumbaya, Indonesia, yielded about 100 km³ of pyroclasts. This eruption resulted in the death of 92,000 people (Macdonald 1975:110, 118).

There is no official record of human death caused directly by the Parícutin eruption but the devastation of villages and agricultural and forest lands forced the migration of thousands of inhabitants from the valley. Two villages were engulfed by lava; three others were partially abandoned owing to heavy ashfalls on roofs, crops, pastures, and forests. Incomplete success and refugee discontent at government-sponsored resettlement sites caused many refugees to return to their former lands where they attempted crop growing and/or forestry.

## REVIEW OF LITERATURE DEALING WITH PARÍCUTIN

The eruption of a new volcano in a populated farming area attracted considerable scientific attention; literature dealing with the eruption is voluminous. Hatt (1948) listed over 150 items in his bibliography of Parícutin Volcano, most of which, however, are descriptions of eruption behavior based on brief field visits. Foshag and González (1956) presented a detailed description of the first 2½ years of the volcano. Bullard (1976) described the entire eruption history in a chapter on birth of new volcanoes. Williams (1950) studied the other volcanoes of the region, and Wilcox (1954) analyzed the petrology and speculated on the nature of the magma body. Segerstrom (1950, 1960, 1961, 1966) dealt with the erosion of the ash mantle and modified drainage, revegetation, and other effects. Work done on the effects of volcanic activity on plant life was published by Eggler (1948, 1959, 1963), and colonization of the cone and lava was recorded by Beaman (1960, 1961). A folk narrative of responses to events associated with the volcano is presented by Gutiérrez (1972). Impressions of individual and group reaction to conditions brought on by the eruption are presented by Nolan (1972). Studies by Rees (1961, 1970) describe the impact of the eruption on the ecology and economy of the area. Estimates on the damage to forests and crops during 1943 and 1944 were made by Arias Portillo (1945).

The present chapter will review the eruption history and pre-eruption conditions, and discuss the subsequent changes in landforms, vegetation, rural economy, and human settlement resulting from the eruption. Field reconnaissance of the affected area was conducted from October to December 1957, and brief visits paid in July 1965, December 1967, October 1972, and September 1978. Portions of the chapter have been adapted from Rees (1970), with additional information on vegetation, soils, agriculture, and forestry based on research conducted in the region in 1967–1968 (Rees 1971), and in the devastated area in 1972.

## PRE-ERUPTION SETTLEMENT AND ECONOMY

The villages of the Itzícuaro Valley are located on valley flatlands at sites not too distant from seeps or springs to serve for domestic use and stock watering. Most villages were founded in the sixteenth century by Spanish monks who resettled the Tarascan Indians from their mountain hamlets to sites on valley flatlands better suited for plow agriculture and close supervision.

The modern Tarascan culture is a mixture of pre-Columbian, colonial, and modern Mexican traits. Pre-Columbian characteristics predominate in the Tarascan language, in major food crops, in the diet and food preparation, and in ceremonial use of certain wild plants. Colonial Spanish influence survives in a conservative form of Roman Catholicism, and in the social structure with its dual political and religious hierarchy and obligatory rituals and feasts that form the prime basis for social prestige. Many of the economic and material aspects of Tarascan lifestyle also date to colonial times. Examples include the village grid plan, the house types, the land tenure system, sixteenth-century Mediterranean agricultural technology, the presence of some European orchard and field crops, and the cottage production of handicrafts (West 1948:33–71).

Villages whose residents consisted of Tarascan Indians or their Spanish-speaking descendants and who were later disrupted by the Parícutin eruption had the following population data in their 1940 census: San Juan Parangaricutiro (1895), Zirosto (1314), Angahuan (1098), Zacán (876), and Parícutin (733). At the time of the eruption, the villages were in communication with other areas of Mexico. Their main contacts were through Uruapan and Los Reyes, the regional market towns where the villagers sold their small-crop surpluses and forestry and crafts products, and where they purchased factory-made goods in return. Contacts beyond the immediate region were made by traders and those who did harvest labor beyond the Tarascan-language region. These outside contacts, and the government primary school present in each village, helped to expand the use of the Spanish language and accelerated the partial integration of the Tarascans into the national Mexican culture. Use of the Tarascan language had noticeably declined by the time of the 1940 census in two of the settlements affected by Parícutin: Zirosto and San Juan Parangaricutiro (West 1948:19).

Until the late nineteenth century all lands in the Upper Itzícuaro Valley were communal, with title deeded to individual villages. The village community held title to all the land, and village residents inherited or purchased use rights to village house lots, to parcels on the open fields, and to parcels in the forests. After the beginning of the twentieth century, individuals were allowed to register their individual usufruct lands; thus, conversion of village land to deeded private property took place. This was called to a halt when the Agrarian Code was enforced in the 1940s. In the 1940s, however, the bulk of the cropland was not registered, and few of the registered parcels were sold to outsiders.

At the time of the eruption, two types of traditional land ownership, communal and private, existed in the villages. A third type, the *ejido*, had been introduced by the federal government at Zacán prior to the eruption. (The large Mexican private estate, the *hacienda*, existed only in non-Tarascan areas beyond the upper Itzícuaro Valley.) The *ejido* is a type of communal land tenure in which the federal government holds the title, and the use rights are held by individuals belonging to the village *ejido* organization. This new type of land tenure was introduced at the resettlement village of Zirosto Nuevo, 2 km northwest of Zirosto, and at resettlement villages beyond the Itzícuaro Valley.

At the time of the eruption, as is still the case today, most village families practiced subsistence agriculture supplemented by forestry and grazing. Crop production methods were traditional, involving manual labor and use of the wooden plow pulled by two oxen. Improved hybrid maize varieties and chemical fertilizers were not accepted by most families. Maize was produced, often intercropped with beans and squash, on unfenced plots on the open fields, on fenced plots on nearby slopes, and on shifting cultivation sites within the forest. Most open field plots and fenced plots on adjoining slopes were fallowed every other year. The use of the digging stick and the hoe was limited to house-lot gardens and steep shifting cultivation sites. The shifting cultivation sites usually required longer fallow periods, and the less fertile plots often were allowed to return to forest. On the open fields the ground was prepared by plowing under the weed growth of the previous fallow during late summer or early fall, and then cross-plowed before seeding took place in late April or early May. The maize was plowed at least once to ensure good drainage for the seedlings and to suppress weeds; a second weeding cultivation was sometimes carried out. The maize and bean harvest, generally in late December, was done by the entire family and some hired labor from a nearby village. Upon the agreed date of the end of harvest, the outside fences of the open fields were let down and grazing animals soon removed all signs of cultivated plant life, except the furrows.

The presence of human and animal refuse on house-lot gardens permitted intensive cultivation of horticulture. The women of the family attended to several vegetables, which included roasting ear and ceremonial maize, European cabbage, squash, broad beans, chayote (*Sechium edule*), and husk tomato (*Physalis angulata*). Medicinal and culinary herbs and ornamental flowers were grown. Deciduous fruit trees such as peach, pear, apple, *tejocote* (Mexican crab apple—*Crataegus mexicana*), and *capulín*, a native cherry (*Prunus capuli*), grew unattended. Fruit was the only house-lot garden crop sold outside the village but was usually undersized, and the larger varieties were commonly hail-damaged.

Subsistence agriculture provided the principal, though not the only, source of income for the villagers. Forestry was important as a supplementary source for most agriculturalists, and the primary source of income for the landless. The production of handicrafts in the villages occupied a few full-time specialists but was the part-time occupation of many who were engaged in other economic activities; the total output for sale outside the village was of little significance. Hunting and gathering existed but were of minor importance.

In the forest, women of the village gathered edible greens, mushrooms, and various other edible and medicinal plants for family use and for sale in Uruapan. At Christmas time, decorative plants and foliage were gathered for sale to Uruapan residents for making Nativity scenes. In the forest, too, most of the cattle grazed, as did poor-quality sheep kept for their meat and wool. Almost all the families in the village owned a burro, a dog, some pigs, and several chickens. Horses and mules were not common because of their high initial cost and the need for higher-quality pasturage. Most farming families owned or had access to a pair of oxen; cattle were highly valued as animals for plowing and as capital investment.

Dead time in the agricultural cycle had always been taken up by forestry activities. Pine beams were hand-hewed, and pine and fir shingles were hand-split. Lumber was hand-sawed or hand-hewed into planks or boards. Pine resin was collected most commonly from *Pinus oocarpa, Pinus leiophylla, Pinus michoacana*, and *Pinus montezumae*, and was shipped to turpentine distilleries in Uruapan. Oak charcoal was produced exclusively for non-Tarascan market towns. All these originated from timber growing on pure communal lands or on forest parcels controlled by individual use rights. Villagers lacking use rights to croplands often depended entirely on forestry for subsistence.

At the time of the eruption, government policy limited the scale of timber production. Existing forestry regulations and new laws enforced in March 1943 required written governmental approval for the cutting of timber to be sold, taxes to be paid on timber sold, and documents accompanying each shipment.

The individual Tarascan cutting only an occasional pine found the regulations difficult to comply with, and lacking documentation, his burro-loads of timber products became contraband. The federal government, not wanting to provoke violence, permitted small-scale illegal production and sale, but prohibited larger shipment of timber to other areas of Mexico.

## IMPACT OF AIRBORNE AND WATER-TRANSPORTED ASH ON THE RURAL ECONOMY FROM 1943 TO 1945

Ashfall was greatest during the period 18 March–9 June 1943. During this period, known as the heavy cineric phase, great damage was done to buildings, croplands, and forests. As far as 5 km to the north of the Parícutin cone, large tree branches were broken, and throughout the area saplings were bent over under the burden of ash. In the fields within 8 km of the cone, crops were killed and agriculture rendered impossible. But famine did not follow, largely because the eruption began 60 days after the end of the annual maize harvest, at a time when most families had already stored several months' supply of grain in their houses.

The eruption began in February and the sowing of maize usually takes place in March and April. On sites where maize seeds were planted after the eruption, germinating seedlings met with a harsh environment. In areas near the cone, ash was falling faster than maize could grow through it, whereas throughout the ashfall zone, undernourished seedlings were subjected to abrasion by both falling ash and ash blown across the surface by wind. Maize was killed where the plants were buried or where fungus diseases entered ash-bruised tissues. Attempts to plant seedlings in shallow pure ash where roots might penetrate down to the old soil failed as a result of fungus attack during the heavy ashfall. Parcels on the open fields with more than a 14-cm ash depth could not be cultivated because the wooden plow in general use at the time did not reach the old soil surface. Small-scale plantings of maize, however, were possible on house lots where ash was removed by hand and where abrasion by windborne ash was less. In sum, heavy ashfall during the first 2 years of the eruption dissuaded most of the Tarascan farmers from planting maize or other crops. On the other hand, Arias Portillo (1945:28) reported excellent crops of wheat and barley beyond the Itzícuaro Valley, where only 3 cm of ash fell.

Fruit crops in areas as far away as 48 km were adversely affected by the heavy ashfalls of 1943 and 1944. Within the area where ash depth averages at least 25 cm, fruit trees were defoliated by falling ash, whereas beyond the Itzícuaro Valley (around Uruapan), fine ash entered avocado flowers, preventing pollination. The avocado crop loss around Uruapan was more than U.S. $300,000 by the end of 1943 (Ordóñez 1947:33).

Ashfalls also upset the ecological balance between insects and crops. Near Los Reyes (25 km west of Parícutin), subtropical and tropical fruits were able to grow more successfully for several years following the heavy cineric phase because of the temporary elimination of a destructive fruit fly by the falling ash. On the other hand, ashfall killed a beneficial insect that preyed upon a sugar-cane boring insect in the Los Reyes region. From January 1944 to May 1945, an estimated loss of 80–90% of the sugar-cane crop resulting from a plague by the cane-boring insects was reported; a total area of 1263 ha was affected (Segerstrom 1950:22).

Damage to agriculture was compounded by water-transported ash. With the onset of the summer rains of 1943, the unconsolidated ash was moved by landsliding, sheetwash, and channel erosion into ash-choked streams with increased cutting power provided by the ash. Large volumes of ash and soil were removed from the upper Itzícuaro Valley westward and deposited on the flood plains near Los Reyes. These floods destroyed the irrigation system by breaking up dams and silting up canals, and laying down beds of ash over 887 ha of sugar-cane land. A total of 2500 ha of sugar cane and rice was devastated (Arias Portillo 1945:27). With the incidental blocking off of half the Rio Itzícuaro watershed by the San Juan lava flow of 1944, no additional sugar-cane land had since been silted over. Though more than half the ash-silted cane land was in cultivation by 1944, cane fields were still being reclaimed by 1946 (Segerstrom 1950:22). Where water-transported ash had filled the basins and valleys in the upper Itzícuaro watershed, ash playas were formed on many of the open fields, having an ash cover of considerable depth that prevented crop plant seedlings from tapping resources from the deeply buried soil.

Ashfall during the heavy cineric phase caused death to many livestock. A loss of over 4500 cattle and 500 horses was reported (Arias Portillo 1945:30), and an undetermined number of sheep and goats died as a result of breathing volcanic ash. Animals that were initially exposed to, and later removed from, the deep-ash zone often died months later; animals thus weakened were quickly disposed of by their Tarascan owners to outside buyers at greatly reduced prices. Upon postmortem examination, it was revealed that the lungs of animals that breathed volcanic ash were congested with an ash–mucous coating that interfered with respiration. The teeth of many dead animals also showed excessive wear from grazing on ash-coated forage. The loss of cattle amounted to a disaster to the Tarascan farmer: First, it represented an inability to plow even if conditions were permissible; second, it deprived a young farmer of the means to accumulate enough capital to purchase use rights on the open fields and to set up separate housekeeping for his family.

After the end of the heavy cineric phase (June 1943), an intermittent ashfall continued. During this period an increasing number of villagers turned to the poorest-paying of traditional forestry activities: the collecting and selling of firewood. The huge volumes of broken branches during the heavy cineric phase were made available to local residents. However, there was the problem of transportation because of the decreased numbers of pack animals and the scarcity of forage near the volcano. Firewood collection and the sale of firewood increased in the local region until prices were driven down to near uneconomic levels. The low prices for firewood rendered shipping the surplus to more distant towns unfeasible. Similarly, there was a drop in local price of charcoal. Sale beyond the immediate market towns was curtailed by existing federal forestry regulations.

The collection of pine resin, traditionally a poorly paying part-time occupation, came to a complete halt during the heavy cineric phase. Ash fell into the collecting cups of the pine, and resin that was thus contaminated was found unacceptable by the distilleries in Uruapan.

Exploitation of the standing timber proved to be the only form of resource utilization that sustained a sizable population in areas of heavy ashfall. As the months of eruption wore on, increasing quantities of hand-hewn lumber products were made and sold illegally, both locally and to other areas of Mexico.

## CHANGES IN HUMAN SETTLEMENT CAUSED BY ASHFALL AND LAVA INVASION

During the nearly 3 months of the heavy cineric phase, the people of the upper Itzícuaro Valley chose to remain in spite of the fact that most land within 14 km of the volcano was made useless for agriculture and grazing. The officials of the affected villagers, however, took steps to seek federal and state aid. Surveys of possible refugee resettlement areas were made by the Secretaría de Recursos Hidráulicos, Secretaría de Agricultura y Ganadería, and other agencies. Government officials considered several privately owned haciendas for expropriation or purchase. Early in May 1943 the federal government made a purchase of a hacienda at Caltzontzin (5.5 km east of Uruapan) with 100 ha of arable land, 160 ha of woodland, and 5 ha of avocado orchards, together with several buildings. Existing nearby *ejidos* were also reviewed for parcels with enough surplus arable land to sustain a relocated village. A transfer of over 200 ha from the adjacent San Francisco Uruapan *ejido* was made to provide additional cropland, timberland, and pasture.

On 15 June 1943, the village of Parícutin was threatened by lavas that moved to within 20 m of the outskirts of the village (Foshag and González 1956:408). The threat of inundation by lava forced the residents into taking action. The government helped to resettle the entire village on the hacienda at Caltzontzin, and by 10 July, the entire population of Parícutin and its possessions were successfully relocated on the new site. Former wooden houses were dismantled and moved to the new site, where heavy slab timbers were milled into lighter-weight lumber to be utilized for later construction; temporary stablelike sheds were constructed for use as shelter during the initial stage. By September, a classic Mexican village with a rectangular grid plan was laid out, and the construction of *mestizo*, or non-Indian, houses was completed. The communal *ejido* organization was formed by March 1944, individual parcels were assigned to male heads of families, and the resettlement was complete.

In August 1943, a second relocation took place when 363 families, mostly from Zirosto and San Juan Parangaricutiro, were moved to Miguel Silva, located 45 km southeast of Uruapan. Miguel Silva was developed on 2616 ha transferred from another *ejido* for refugee settlement; only 400 ha were arable land. All the relocated families received a village house lot, but many were without the additional allotment of cropland. At Miguel Silva, the government supervised the construction of adobe and shingle-roofed houses. The climate and soils of Miguel Silva, however, were completely different from that of the Itzícuaro Valley. The clay soils and the warm subtropical climate did not accommodate the Sierran crop plants. In addition, potable water was not available until 1946, and over a hundred died from typhoid and other diseases between 1943 and 1946. There was also resentment of the landless against those with allotted cropland; distrust developed between factions from Zirosto and San Juan Parangaricutiro. By 1945, many of the former villagers from Zirosto returned to their homeland; eventually, almost all of the refugees from San Juan Parangaricutiro left for the new village of Nuevo San Juan.

In June 1944 San Juan Parangaricutiro (the county seat of the *municipio* of the same name) became the second village to be threatened by lava flows. Prior to the lava, there had been a steady movement of residents to other areas, not only to Miguel Silva, but also to ranchos located on Parangaricutiro's ample communal lands to the southeast of the volcano, to Uruapan, and to other settlements. In June the government moved the remaining inhabitants of the destroyed village to the new site of Nuevo San Juan established on the former hacienda Los Conejos (6 km west of Uruapan). The villagers were given house lots of the same size and at the same relative location as they had in the former village. Not all the timber used at Nuevo San Juan was remilled, as it had been at Cultzontzin and Miguel Silva. Some dismantled wooden houses were reconstructed when the timbers were small enough to be trucked to the new site.

The town of Zirosto, located 7.7 km northwest of the volcano, suffered through the heavy ashfall of 1943 that devastated its agricultural and forest lands. Many of the residents went elsewhere temporarily until it became possible for them to return to cultivate crops again. Six hundred went to Miguel Silva, where the outbreak of diseases, internal dissension, and the lack of cropland caused them to return to Itzícuaro Valley. Many of the inhabitants, however, remained in Zirosto and in nearby Barranca Seca, where an outside contractor established a lumber mill in 1944 to exploit timber killed by the eruption. By 1945, an estimated 20% of the inhabitants had moved from Zirosto and from the resettlements at Miguel Silva to Uruapan, Charapan, Tijuana, Mexicali, and other distant localities. Most of those who migrated at that time have not returned. Barranca Seca later became known as Zirosto Nuevo when in 1952 the federal government transferred 371 ha from another *ejido*, and purchased 184 additional ha from private owners.

Nearby Zacán, located 7.8 km northwest of the volcano, was similarly affected by ashfall but not by lava. Zacán was more fortunate than other villages of the valley in that it received less ash, and that the ash that was stripped from its slopes during the rains of 1943 and 1944 was deposited on lands of the less fortunate. The majority of its residents, however, left during the ashfall of 1943 for other nearby villages and towns of the Meseta Tarasca. The government's offer to relocate them on a former hacienda near Morelia was turned down by the village officials on grounds that the location was beyond the Tarascan cultural region. The villagers were fearful that the move might be a repetition of the Miguel Silva disaster. Instead, they went into temporary residence among the Meseta Tarascans and managed to maintain their separate identity. A majority of them moved back to Zacán at the first opportunity in 1944 when remittances from men doing bracero labor in the United States permitted their families to return; still more returned in 1945 when cropping conditions improved.

Angahuan, located 5.8 km north of the volcano, received almost the same amount of ashfall as Zacán. As a result of the heavy inwash of ash from surrounding hill slopes, most of Angahuan's open fields were laid to waste. Very few families, however, left the village during the years 1943–1950. Village elders estimated that only 20 persons had permanently left the village for residence in San Lorenzo, San Felipe, and Uruapan. During this period, few men did temporary bracero labor; crop growing on distant slopes northwest of the village and forestry activities continued to provide a supplementary livelihood. A new resource, namely tourism, was created on account of the eruption and provided Angahuan with a vital source of income from horse rental and guide services. Wilcox reports that the tourism business during 1946–1948 was shared by alternating groups of guides from Angahuan, Caltzontzin, and San Juan Nuevo; each village group worked a week or two alone, alternating in succession (Wilcox, personal communication). Since the mid-1950s the much diminished tourist trade has been served almost entirely by men from Angahuan.

## CHANGES IN DRAINAGE AND GROUNDWATER CAUSED BY THE LAVA FIELD

The scarcity of water flow in the form of permanent streams and lakes is a distinctive feature of the Parícutin region. Despite the average annual precipitation of over 1500 mm, most surface moisture is quickly absorbed into the porous volcanic soil. Repeated drilling for subsurface water to depths of over 100 m has failed to locate exploitable water supplies. Only at a few canyon sites, where intermittent stream erosion has cut through ancient ashfalls and reached impervious clay beds, have seeps been formed that provide the only source of water supply for many villages in the Parícutin region. Altered drainage began with the initial cone building and lava flows, but these did not block the massive floods of 1943 from transporting ash of the heavy cineric phase from the upper Itzícuaro watershed. The massive San Juan and Parícutin flows of 1944 filled the channels of three intermittent tributaries of the

Itzícuaro River and covered over the lower watershed. The border of the lava field then became a barrier to drainage of the upper watershed. The height of the lava border, local slope conditions, and permeability of the lava determined whether a closed basin was formed or gullying took place.

During the first 3 years of eruption, the edges of the lava field were very permeable and streams 10 cm deep and 1 m wide entered the edge of the flow without forming ponds. Much larger streams entered the lava with only slight pond formation. By 1946, sediment was filling in spaces in the lava flow and blocking openings in the lava edge; increasingly, the lava succeeded in damming up flood waters, and many closed basins were formed. On gentle terrain adjacent to the north and northeast border of the lava field, extensive shallow lake basins developed, the largest lake being 1500 m long with a width varying from 50 to 500 m wide (Segerstrom 1950:32).

The steep slopes south of the lava field favored the development of V-shaped troughlike basins between the edge of the lava and the hill slope. During the eruption these troughs, filled up with ash washed down from upslope, would then be covered by the next flow, and the cycle repeated.

By 1953, over 60 m of ash, soil, dead timber, and debris had been redeposited behind a formidable dam of lava from later flows to the southwest of the cone (Segerstrom 1960:141). Decreasing permeability still caused temporary ponds and sink conditions on the margins of the flows in 1978. Despite the amount of runoff flowing at the border of the lava field and a yearly rainfall of 1.5 m, the water does not reappear as springs at the lava margin because of the extreme permeability of the lava. However, groundwater flow has increased downstream from the lava field.

Ash-choked floodwaters sought new exit routes around the edges of the lava field. By 1957 large areas north and west of the lava field that had been closed basins in 1946 were reconnected, through basin filling and downcutting, to the Itzícuaro River during flash floods. Most of the remaining blocked area south and east of the lava field became a single connected system by 1957 (Segerstrom 1960:5). Sediment-laden floodwaters from this connected system were channeled onto a low-lying section of the lava field by farmers, who constructed a 1.5-m-high rock dam that directs floodwaters onto the lava and away from the next closed basin. This small section of the lava field (about 9 ha in 1965) is covered by a storm debris fill of about 3 m in depth (Segerstrom 1966:C96). In 1978 a long brush wall protected maize growing on this fill and guided runoff onto exposed lava areas. The long-term trend is for the closed basin levels to rise with increments of fill and for storm waters to flow over or cut through low divides and thus reestablish normal open drainage.

Other factors also affected the flow of groundwater in seeps and springs. Some old springs and seeps dried up, possibly from earthquakes associated with the onset of the eruption; others became silted up with ash; still others dried up because of a lowering of the local water table caused by the deepening of barranca channels by ash-charged floodwaters. Surface water flow was reduced in the principal barranca channels because of ash redeposited on the channel floor. On the other hand, subsurface flow in the barrancas probably increased because of the greater water flow observed in many springs during the eruption. It is believed that the permeable ash mantle increased infiltration and reduced evaporation (Segerstrom 1950:25–27).

## ASH DEPOSITION AND REMOVAL UP TO 1965

Massive quantities of airborne ash were deposited on the terrain of the Parícutin region. About 90% of the airborne ash was deposited during the first 2½-years of eruption (Fries 1953:605). The ash blanketed all the landscape but the distribution was not even. Ash depth decreased rapidly with increasing distance from the cone vents (e.g., from a maximum of 6 m near the cone to 25 cm at Angahuan, 7 km distant). Wind direction, intensity, frequency, ash particle size, and slope gradient all influenced how much ash fell at a given site. Depth of air-deposited ash on level ground is shown in Figure 9.2 as isopach

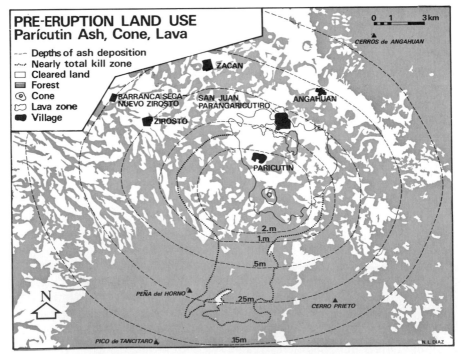

**FIGURE 9.2. Map of pre-eruption land use, Parícutin ash depths, cone, lava. Data sources: base map and former land use from air photos taken in 1934 by Cía. Mexicana Aerofoto S.A. for the Banco de Crédito Agrícola; cone and border of lava field from Segerstrom (1960:Plate 1); isopachs of airborne ash to October 1946 from Segerstrom (1950:plate 1).**

lines of equal ash thickness; these lines form a concentric ellipse centered west of the cone vents. The concentric ellipses point out the decrease in ash depth with increasing distance from the cone, that east and west winds predominated during ash fall, and that east winds during the rainy season caused a greater ash fall to the west of the cone.

The volcanic ash deposited on the terrain was very susceptable to erosion, transportation, and redeposition. The bare ash surface lacked a protective cover of vegetation or leaf litter. The ash was unconsolidated—that is, uncemented—and the beds had only slight cohesiveness. The greatest erosion and transportation (stripping) of ash took place during the rainy season of the first 2 years of eruption. A much reduced rate of stripping is still taking place through several processes: (a) soil creep, (b) landslides, (c) splash erosion and sheetflow, (d) rill and channel erosion, and (e) wind deflation.

Slow landsliding, called soil creep, was common on slopes steeper than 32°. Soil creep is apparent in deeper ash because standing trees lean downhill. Landslides, the sudden descent of material from a steep slope, were also very common during the first years in deeper ash, especially where stream channel erosion undercut a cone slope whose upper beds of ash were rainwater-logged. Scars may be observed on many of the steep slopes where ash and underlying soil, ancient bedded ash, and rocks were removed by landslides.

Stripping by water action has removed much of the ash mantle on slopes of the region and eroded underlying soil, ancient ash, and rock materials, and redeposited them at nearby and distant downslope sites of lesser gradient. Accelerated erosion occurs when an intense rain falls on a bare ash surface with a gradient. Falling raindrops cause splash erosion that slowly shifts particles downhill. As overland flow increases and coalesces and flows as a sheet, grains of ash are removed in rather uniform thin layers and deposited at the base of a slope or into rills and channels downslope. Segerstrom states that a large proportion of the ash eroded from the region has been stripped off by sheet flow. Interestingly, sheet and rill erosion did not take place on Parícutin cone as late as 1965; the cone surfaces were too permeable with coarse material for surface runoff to occur. Segerstrom believes that the coarse inner slopes of Parícutin will not show evident water erosion for at least 200 years (1966:C94).

Sheet flow often changes to rill erosion when runoff is heavy and the slope steepens. Rills are frequently initiated by irregularities in the pre-eruption terrain such as plowing furrows, or by runoff from tree trunks and volcanic bombs that formed small pools that overflowed and cut a rill downslope. Rill and channel erosion was spectacular in deep ash in 1943 and 1944. Massive volumes of ash eroded as rill channels enlarged and were broken up by stream coalescing and complex branching that altered with each heavy rain. Accelerated erosion produced high narrow interfluves of ash that were removed by lateral cutting and development of large gulleys (barrancas) (Segerstrom 1950:91).

During the first 3 years of eruption, stripping of ash from unprotected slopes yielded a high sediment load in floodwaters. The upstream Rio Itzícuaro channel in flood measured 79% solids with a specific gravity of about 2.0 (in contrast to streams transporting only pre-eruption sediments with a load of 16% and a specific gravity of 1.25 [Segerstrom 1950:107]). High-density flood-waters moved large volumes of ash as well as logs, debris, and many boulders as large as 2 m in diameter. Boulders gave the illusion of floating as they bobbed to the surface of floodwaters.

Channels eroded in ash are usually box-shaped with vertical walls rarely more than 2–3 m high because of lateral undercutting of the weak ash. Where gradient increases sharply, the posteruption drainage channels cut large bar-rancas. On formerly closed basins south of the Parícutin cone, barrancas 35 m deep and 50 m wide were cut into prehistoric ash and lava by the year 1957. Barranca cutting took place elsewhere when floodwaters dammed by Parícutin lava found a new exit by cutting through ash beds.

Deposition of water-transported sediments produced alluvial fans at the bases of all the hill slopes blanketed by ash. Sediments stripped from slopes into channels and into a principal gully were transported out if channel gradient and water volume permitted. Large gully watersheds produced large fans up to hundreds of meters long.

Lake deposits resulted when floodwaters formed ponds behind lava dams. Because all the sediments are trapped and finer sediments do not escape downstream, lake sediments contain more fine particles than those of alluvial fans. Lake-deposited ash is often covered by a thin film of soil as the pond water carrying suspended silt dries up.

Massive volumes of eroded ash were transported from the channels of the upper Itzícuaro Valley into the Itzícuaro River. Downstream, where the gra-dient lessens, the floodwaters dropped the ashload on the flood plain near Los Reyes. During the destructive flood of 1943, an estimated 800,000 m$^3$ was deposited in a 4-km$^2$ area (Segerstrom 1950:109). Flood deposition was less in later years.

Deceleration of the high rate of stripping began after the 1944 rainy season and was more apparent after ashfalls ended in 1952. Erosion and redeposition slowed because (a) the most vulnerable ash had been largely stripped from the steepest sites and main stream channels, and (b) vegetation colonized soil exposed by erosion and ash and soil-ash sites (Segerstrom 1961:D225). In 1978 gully erosion was observed in deep ash on a steep slope adjacent to the Parícutin cone, and sheetwash and rill erosion on plowed cropping sites south-west of Angahuan. However, on sites with a forest cover, ground litter, and herb layer, erosion appears to be no greater than other areas beyond the affected cone.

Wind erosion and deposition occur during the December to April dry months but on an insignificant scale. Wind-drifted ash has not built dunes, but wind-moved ash did pile up behind field walls on exposed open fields. The reduced visibility noticed by Segerstrom in 1946 did not seem to be a problem in 1972.

## PRE-ERUPTION VEGETATION

The three types of forest vegetation present in the Itzícuaro Valley were adversely affected by the Parícutin volcanism: (*a*) the fir (*Abies*) forest located on Tancítaro and the Cerros of Angahuan; (*b*) the mixed pine–oak forest that covered ash-derived soils and old lava surfaces on slopes and valley-flats below the fir zone; and (*c*) the broadleaf subtropical forest restricted to humid barrancas.

### The Fir Forest

The fir forest in the affected zone was largely restricted to the upper slopes of Tancítaro and the Cerros of Angahuan between 2700 and 3040 m in elevation. Prior to the eruption the fir forest suffered only a limited interference by man; killing frosts discouraged maize cultivation, and distance from the villages inhibited timber exploitation. Fir forest is common at these elevations on deep, well-drained humid volcanic soils in many areas of central Mexico. High atmospheric humidities and more than adequate rainfall encouraged a dense forest dominated by *Abies religiosa* that attained a height of 45 m.

Shrub, herb, and moss–lichen strata were present in the fir forest with the maximum heights at 5.0 m, 1.5 m, and 5.0 cm, respectively. Fewer species were present in the flora of the fir forest than in the pine–oak community found at lower elevations because of winters with extreme minimums from −5.0 to −11°C (Madrigal Sanchez 1964:17). Oak, pine, and other tree species had a minor role except in the transition to pine–oak forest, where two oak species, *Pinus montezumae*, and other tree species were intermixed with fir.

### The Mixed Pine–Oak Forest

The subtropical mixed pine–oak forest covered most of the pre-eruption landscape below 2700 m elevation that was not in crops of pasture. A less humid atmosphere than in the fir forest, the concentration of rainfall in the months between June and November, and droughts during the less rainy season favor pine–oak forest over fir or broadleaf subtropical forest.

The pine–oak forest adjacent to pre-eruption settlements was heavily exploited and altered. A complex patch-quilt of differing timber stands resulted from heavy and usually premature forest exploitation for lumber, charcoal, firewood, pine resin, shifting cultivation, and grazing. Most pine–oak stands on ash-derived soils were former cropping sites usually under individual usufruct; species composition and age and size of trees usually reflected the forest occupation of the man controlling the woodlot. For example, shingle makers removed large-diameter pines, and charcoal makers removed medium-sized oaks.

Some forested sites were cleared for shifting cultivation. This altered the soils and accelerated the erosion process, resulting in impoverished and de-layed forest succession. Repeated fires on many abandoned cropping sites suppressed broadleaf trees and shrubs and favored the development of even-aged pine woodlots with an underlying dense sward of grass.

Within the pine–oak forest different soil (edaphic) conditions produce four different subtypes of vegetation: Subtype 1 is found on well-drained ash-derived soils of the hill slopes; Subtype 2 grows on humid ash-derived soils of the canyons and draws of cinder cones and hill slopes; Subtype 3 is found on humid moss-covered blocky lava *malpais*; and Subtype 4 occurs on weathered lava flows with dry (xeric) rocky shallow soils. The description of these four subtypes is based on a study of the pine–oak forest in the Itzícuaro Valley some 8 km east of the Parícutin cone (Rees 1971:23–33).

Subtype 1 is the most extensive and is found on slopes with deep well-drained ash-derived soils where *Pinus montezumae* and *P. pseudostrobus* with two *Quercus* spp. form a canopy to 30 m in height in mature stands subjected to infrequent fires. Underneath, a short-tree stratum of *Alnus jorullensis*, *Clethra mexicana*, *Arbutus xalapensis*, *Crataegus mexicana*, and others grows to a height of 10–15 m. A tall-shrub stratum to 6 m in height is present with *Arctostaphylos rupestris*, *Ceanothus coeruleus*, and others. Also present is a shrub stratum to 3 m with *Lupinus elegans*, *Monnina xalapensis*, *Cestrum thyrsoideum*, *Arracacia atropurpurea*, *Baccharis* spp., *Solanum* spp., *Crotalaria* sp., *Salvia* spp., *Fuchsia* spp., *Rubus adenotrichos*, and others not identified. The short-tree and shrub layers increase their diversity with increase in the oak crown cover.

The herb stratum in Subtype 1 is less varied under the dense forest, where only about six grass species predominate, but becomes more varied in forest openings where can be found *Lopezia recemosa*, *Oxalis divergens*, *Geranium mexicanum*, *Cuphea* spp., *Viola* spp. *Portulaca oleracea*, *Sigesbeckia jorullen-sis*, and others. On old forest clearings, fields, and field edges, the herb flora becomes even richer with many Compositae such as *Eryngium carlinae*, *Stevia* spp., *Tagetes* spp., *Heterotheca inuliodes*, *Bidens* spp., *Cosmos* spp., *Dahlia* spp., *Galinsoga* spp., *Heterotheca* spp., *Tithonia* sp., and *Vigueria* sp. Also present are the crucifers *Eruca sativa* and *Lepidium virginicum*, several legumes including *Dalea* spp., Solanaceae such as *Saracha procumbens*, and also an *Amaranthus* hybrid. Among the most prominent weeds on fallow fields are *Argemone platiceras* and *Mirabilis jalapa*, and on infertile clearings *Cestrum thyrsoideum* and a spiny *Solanum* sp. Many other herbaceous plants are also present.

Subtype 2 is found on humid, ash-derived soils in the canyons and draws of cinder cones and hill slopes that favor a greater development of the short tree and shrub strata. More frequently encountered than in the nearby mesic sites are the trees *Prunus capuli*, *Garrya laurifolia*, *Arbutus xalapensis*, *Clethra mexicana*, and other broadleaf trees. Shrubs likely to be present are *Satureja laevigata*, *Monnina xalapensis*, *Arctostaphylos rupestris*, *Fuchsia* spp., *Ceanothus coeruleus*, and *Salvia* spp., among others.

Subtype 3 vegetation grows on the humid blocky lava *malpais*; trees and shrubs grow in crevices between blocks of lava that are often entirely covered by low vegetation. The shaded lava block surfaces are commonly covered by algae, mosses, and ferns of several genera including *Pellaea, Pityrogramma, Pteridium, Adiantum, Asplenium, Dryopteris, Cystopteris*, and others. Protected sites between rocks having a thicker mat of soil and organic matter support succulents such as *Agave* sp., *Echeveria obtusifolia, Sedum tortuosum*, and an orchid, *Bletia* sp. Small herbs include five grass species, and herbs *Lopezia racemosa, Stevia rhombifolia, Geranium mexicanum*, and several others not identified.

The dominant trees growing on the blocky *malpais* are *Pinus pseudostrobus* and two *Quercus* spp., with the former being heavily exploited for timber. The short-tree and shrub stratum is well developed and includes two very distinctive tree species found only on the blocky *malpais, Clusia salvinii* and *Balmea stormae*. Other tree and shrub species include those mentioned for Subtype 2.

Noteworthy is the mass of epiphytes growing on broadleaf trees in the blocky *malpais*. Mosses and ferns make up the bulk, but angiosperms present are two Crassulaceae, *Echeveria obtusifolia* and *Sedum tortuosum*, a single large strap-leafed bromelaid, *Tillandsia prodigiosa*, one *Peperomia* sp., two unidentified cacti, and a rich selection of orchids, including the very common *Laelia autumnalis, Cattleya citrina, Odontoglossum cervantesii, O. insleayii, Oncidium cavendishianum, O. tigrinum, Stanhopea* sp., *Erycina* sp., and others.

Subtype 4 occurs on older xeric lava flows that have weathered to a generally flat surface with thin, droughty soil and rock occasionally jutting upward. Cut over for timber, used for grazing, and sometimes used for crop production if soil depth permits, most of these sites have been subjected to severe interference by man, so that a description of "normal" vegetation is difficult. *Pinus leiophylla* and a *Quercus peduncularis* predominate, with *Pinus pseudostrobus* increasing on less rocky and less logged-out sites.

Although *Pinus leiophylla* is dominant on such xeric sites, this may be only a successional stage, since other trees are well represented as seedlings (e.g., *Crataegus mexicana, Arbutus xalapensis, Quercus crassipes*). Most sites are characterized by the openness of the canopy—actually a parkland—with much space between middle-sized 30 cm diameter at breast height (D.B.H.) trees. Mature shrubs are infrequent and the herb stratum is rich in species as compared with densely forested areas. Epiphytes are poorly developed.

## The Broadleaf Subtropical Forest

The third major vegetation type is the broadleaf subtropical forest. This forest type is restricted to a few deep barrancas with long, steep slopes where increased soil moisture, lower light levels, and good air drainage result in a highly equable environment favoring mesophytic trees over adjacent pines.

Tropical and humid–subtropical trees, woody lianas, and shrubs (usually associated with subtropical cloud forest at much lower elevations) are present in deep canyons and especially prominent at infrequent seeps.

The dominant trees that create a shaded canopy from 15 to 25 m in height in the broadleaf subtropical forest include *Carpinus carolineana*, *Fraxinus uhdei*, *Prunus capuli*, *Meliosma dentata*, *Quercus* spp., and *Pinus pseudostrobus*. A short-tree and tall-shrub layer of 7–15 m is composed of *Tilia houghii*, *Clethra mexicana*, *Cornus disciflora*, *Garrya laurifolia*, *Symplocos pringlii*, *Styrax ramirezii*, *Ilex brandegeana*, *Crataegus mexicana*, the araliads *Dendropanax arboreus* and *Oreopanax echinops*, and others not identified. The low-shrub and herb strata include unusual species. Among them are a tall, coarse fern (*Woodwardia* sp.?) found near the seeps, and subtropical elements such as a large flowering solanacous shrub and a deciduous massive woody liana. Larger trees are covered with epiphytes similar to those found on broadleaf trees in the humid *malpais* environment (Rees 1971:22–32).

The subtropical broadleaf forest is usually well cut over, as the wood of these broadleaf trees sells for high prices. At seep sites utilized for village water supply the forest may be found in good condition since village officials have prohibited timber cutting lest the felling disturb the water flow.


## VEGETATION DESTRUCTION AND RECOVERY

The Parícutin eruption brought about drastic changes to the vegetation of the Upper Itzícuaro Valley. The effects of the volcanism upon vegetation are recognizable by the following four zones:

1. *Total-kill Zone*: Coincident with the cinder cone and lava flows.
2. *Nearly Total-kill Zone*: Where most individuals of all size classes of all species were eliminated, including an area of fir kill on Tancítaro where other species survived in relatively shallow ash. Sites in this zone average 1.5 m of remaining ash depth.
3. *First Zone of Partial Survival*: Characterized by tree damage and heavy kill of shrubs and herbs; sites with 0.5–1.5 m of ash deposition.
4. *Second Zone of Partial Survival*: Characterized by slight tree damage and partial survival of shrubs and herbs; sites with 0.15–0.5 m of ash deposition.

The extent of these zones can be seen in Figure 9.2. The zone of total kill and that of nearly total kill are indicated. The zones of partial survival can be correlated with isopachs of ash deposition. Table 9.2 indicates the total amount of pre-eruption forest land, cleared land, and village sites within these zones.

Within the total-kill zone, the cinder cone obliterated everything within 600 m of the original vent. Adjacent to the cone, moving lava flows crushed and buried vegetation with a wall of cooled lava rubble. Burning of vegetation was not common; only in a few instances did incandescent lava engulf living vegetation.

**TABLE 9.2**
**Zones of Vegetation Devastation and Recovery**

| | | Area (in hectares) | | | | | | |
|---|---|---|---|---|---|---|---|---|
| | | Forest (pre-eruption) | | Cleared[a] (pre-eruption) | | Villages (pre-eruption) | | |
| Zones | Total area | Area | Percentage | Area | Percentage | Area | Percentage |
| Total-kill Zone | 2480 | 975 | 39.3 | 1374 | 55.4 | 131 | 5.3 |
| Nearly Total-kill Zone | 4375 | 3435 | 78.5 | 940 | 21.5 | 0 | 0 |
| First Zone of Partial Survival | 5245 | 3866 | 73.7 | 1379 | 26.3 | 0 | 0 |
| Second Zone of Partial Survival | 26,110 | 17,363 | 66.5 | 8695 | 33.3 | 52 | 0.2 |

[a]Cleared land = lands under cultivation, in annual fallow, in long fallow with shifting cultivation without timber.
    Sources: Data computed from map of pre-eruption land use, Parícutin ash, cone, lava prepared by author. Data sources: base map and former land use from air photos taken in 1934 by Cía Mexicana Aereofoto S.A. for the Banco de Crédito Agrícola; cone and border of lava field from Segerstrom (1960:plate 1); isopachs of airborne ash to October 1946 from Segerstrom (1950:plate 1).

Volcanic ash caused the death or decline of most natural and cultivated vegetation. Four obvious causes are suggested:

1. Complete burial of the plant
2. Partial burial restricting root access to oxygen
3. Defoliation and prolonged absence of leaves
4. Ash covering foliage surfaces, clogging stomata and also blocking out sunlight.

Death or decline of plants may also have resulted from unrecorded acid-bearing rains, volcanic gases, or the buildup of toxic acids in ash and underlying soil.

Beyond the area of total kill, the major environmental factor determining the survival of vegetation was the depth of airborne ash (Table 9.3). Other factors were general response of a species to ash deposition, size class of individual plant, and site characteristics influencing stripping of ash.

Many species adapted poorly to ash deposition—for example, the moss-lichen stratum and the stratum of herbs became buried. Epiphytes on broadleaf trees perished under conditions of ashfall, partial burial, and strong sunlight as trees defoliated.

With few exceptions, most shrubs died in deeper ash. In 1945 Eggler (1948:431) noted the survival of several shrub species growing through 40 cm to 1.3 m of ash, some growing continuously, others following injury. Included were *Baccharis pteronoides* and *Fuchsia pringlei* in ash to 50 cm, and *Cestrum terminale* in 1.3 m of ash. By the late 1950s, there was little evidence of pre-eruption shrubs' surviving where over 50 cm of ash remained; however, they did survive on sites partially or entirely stripped of ash.

Two species of herbs, *Argemone platyceras* and *Mirabilis jalapa*, and three grasses, *Cynodon dactylon*, *Epicampes* sp., and *Digitaria velutina*, are com-

**TABLE 9.3**
**Zones of Thickness of Ash Deposition to October 1946**

| | | Area (in hectares) | | | | | |
|---|---|---|---|---|---|---|---|
| | | Forest (pre-eruption) | | Cleared[a] (pre-eruption) | | Villages (pre-eruption) | |
| Zones | Total area | Area | Percentage | Area | Percentage | Area | Percentage |
| Ash over 2.0 m | 1430[b] | 1035 | 72.4 | 395 | 27.6 | 0 | 0 |
| Ash 2.0–1.0 m | 2290 | 1685 | 73.6 | 605 | 26.4 | 0 | 0 |
| Ash 1.0–0.5 m | 4830 | 3009 | 62.3 | 1792 | 37.1 | 29 | 0.6 |
| Ash 0.5–0.25 m | 9810 | 6429 | 67.6 | 3015 | 31.7 | 67 | 0.7 |
| Ash 0.25–0.15 m | 16600 | 10890 | 65.6 | 5710 | 34.4 | 0 | 0 |

[a]Cleared land = lands under cultivation, in annual fallow, in long fallow with shifting cultivation without timber.
[b]Lava field excluded.
Sources: Data computed from map of pre-eruption land use, Parícotin cone and lava prepared by author. Data sources: base map and former land use from air photos taken in 1934 by Cía Mexicana Aereofoto S.A. for the Banco de Crédito Agrícola; cone and border of lava field from Segerstrom (1960:plate 1); isopachs of airborne ash to October 1946 from Segerstrom (1950:plate 1)

monly known to have survived 50-cm ash because of their ability to grow up through the deepening ash. Some of these draw strength from tubers; others develop adventitious roots in the ash.

It might be expected that mature trees would be little affected by ash since the foliage mass would be above the ash deposits; this proved not to have been the case. Survival rates varied for each species. *Alnus* was found to have survived in 1945 (Eggler 1948:431), but no individuals were seen in the ash areas in 1957. *Arbutus* seemed to have nearly died out, although Eggler (1963:45) noted some regrowth, probably from old roots, at a site on Tancítaro. *Abies* on the upper slopes of Tancítaro was killed, apparently by light but frequent ash-falls upon foliage; other species were scarcely affected under similar conditions. Located downwind and at a higher elevation, *Abies* possibly was intolerant of volcanic gases and acidic water vapor rising from Parícutin. This is only conjecture, however. Evaluation of the extent of *Abies* survival is difficult to assess owing to salvage lumbering of living, along with dead or dying, timber during the late 1940s.

Few trees of the broadleaf subtropical forest seem to have survived, even up to 1945. *Clethra* and *Symplocos* survived in 10 cm of ash in 1945 but were not seen later (Eggler 1948:429). Only *Tilia* has survived on uncovered soil at sites stripped of ash (Eggler 1963:40). Nothing is known of *Carpinus, Meliosma, Cornus, Garrya, Styrax, Illex, Dendropanax,* and *Oreopanax*; defoliation during the heavy cineric phase may have prevented the identification of ash-killed individuals.

The death or survival of pine and oak has been correlated to the basal-diameter size class of the individual tree, as well as to the depth of the ash mantle (Eggler 1948:430; 1963:45, 49). Among the three common species of pine, seedlings and small trees were killed owing to excessive bending and burial, whereas large, mature timber suffered from branch breakage under loads of ash. The 10–30-cm basal-diameter size class of pine was the last to succumb because of stems strong enough to resist excessive bending yet sufficiently flexible to dump part of the ash load and avoid breakage. However, almost all pine died when ash deposition exceeded 2 m, regardless of rates of removal of ash from foliage or stripping of surface ash by accelerated erosion.

The several species of oak were uniformly slower to die than the pines; basal-diameter size class survival, interestingly enough, was similar to that of pine, except that fewer oak survived on sites with ash depths over 1.5 m. Other tree species, including *Alnus*, *Arbutus*, and *Abies*, have not survived as well. An exception is *Crataegus*: Old gnarled trees survived in abandoned gardens in San Juan in 75 cm of ashfall.

Uncovering of the soil through removal of ash stimulated the growth of vegetation. Soil creep and particularly landslide often destroyed vegetation on a steep site, but removal of ash then created a favorable environment for seedling development.

Most stripping away of ash took place through channel erosion. By 1967 most steeper sites had been severely gullied and many even cleared of ash. Exposure of roots, stumps, or standing tree trunks at the old soil level during the 1940s encouraged sprouting of many shrubs and trees that were thought to have been killed but later regrew. On the steep slopes of Cuaxandaran, 4 km northwest of Parícutin, complete stripping of 1.5–2.0 m permitted the survival of medium-sized pine and oak. On the slopes of Tiripan, 5 km northwest of Parícutin, isolated individuals of *Tilia* and *Alnus* are growing back after 1.0 m of ash was removed by gullying. Near the summit of Tancítaro, growth of plants continued all through the eruption as ash was removed almost as fast as it fell (Eggler 1963:45). In all cases where old soil has been exposed, plants have established themselves. By 1959 many areas that experienced deep ashfall (from 1.5 to 3.0 m) had plants growing on uncovered soil in gullies. All of the herbs and most of the shrubs growing in the gullies were posteruption seedlings (Figure 9.3). The more common shrubs growing in 1959 were *Senecio* and *Piqueria*, followed by *Eupatorium*, *Baccharis* (all four are composites), *Ceanothus*, and *Salix*. Herbs present were *Dalea*, *Crotolaria*, *Desmodium*, *Hosackia* (all papilianaceous legumes), *Penstemon*, *Castilleja*, and *Solanum* (Eggler 1963:45). Two species of *Buddleia* were also becoming common by 1963 (Eggler, written communication to Segerstrom, October 1965). In 1965 Segerstrom (1966:C97) observed six species of *Eupatorium*, two *Baccharis* species, and three species of *Senecio* growing on exposed soil in gullies.

Protection by trees or other living or dead vegetation has permitted the survival of occasional shrubs, herbs, and grasses. In 1957 most live shrubs observed in deeper ash grew at the base of living or dead trees (Rees 1961:50). The trees offered protection against windblown ash and provided sources for moisture and nutrients.

In areas with less than 50-cm ash deposits, the rate of survival and seedling

FIGURE 9.3. Landscape typical of Parícutin region. Older forested cinder cones in background. Parícutin lava and ash playa colonized by seedlings in middle. Eroded remnants of ash in foreground with surviving pine at left, posteruption pine, oak, shrubs, and herbs on soil exposed by erosion, 1968.

development improved. Sufficient individuals of medium sizes of pine and oak survived to provide a leaf litter that protected the sites from erosion as well as supported seedling growth. Medium-sized pine, some oaks, and *tejocote* (*Crataegus mexicana*) survived at Cuezeño, a site with rolling terrain 5 km north of Parícutin. At the same location, seedlings of pine and other plants were observed in 1950; by 1959 over 6 species of trees, 7 shrubs and 17 species of herbs, 4 grasses, 3 ferns, 1 sedge, and various mosses were present (Eggler 1963:51). Pine was well represented in all diameter-size classes. Eggler considered that as far as pine is concerned, the forest at Cuezeño had practically recovered from the effects of volcanism 14 years after the maximum ashfall had ceased.

Other forested sites with less than 50 cm of ash similarly provided a leaf mulch supporting seedlings. Vegetation succession is not uniform and the seedling composition varies considerably within short distances; in part, this reflects the survival of nearby propagule sources.

The great open fields are poor sites for plant survival and colonization. Herbs and grasses that grew on annual fallow land were in most cases buried and killed in 1943. Planting sites of that year had been twice plowed in preparation for seeding and were largely devoid of vegetation. Most of the open fields became ash playas during the heavy cineric phase and remain so today. Less erosion has taken place on the open fields because of slight gradients, and many sites have received insufficient eroded soil or organic material to sustain plant life. Scattered plants found in deep pure ash are often survivors that have grown up through the ash, such as the grasses *Cynodon dactylon* and *Digitaria velutina* and the herbs *Argemone platyceras* and *Mirabilis jalapa*.

Few seedlings have established themselves on open fields retaining a deeper ash mantle. Sterility of the ash is the main limitation, but other factors make the ash fields a difficult environment for growth of seedlings: The crust formed on ash surface after rainfalls makes both seed burial and seedling emergence difficult; also windblown surface ash is abrasive to seedling tissue. Despite these difficulties, a ground cover of seedlings was able to colonize open fields where the ash is shallow (under 15 cm depth), or where ash is enriched with water-transported soil or organic materials such as leaf litter, dead timber, or animal dung (Figure 9.4).

In summary, the vegetation in the region mantled by more than 15 cm of

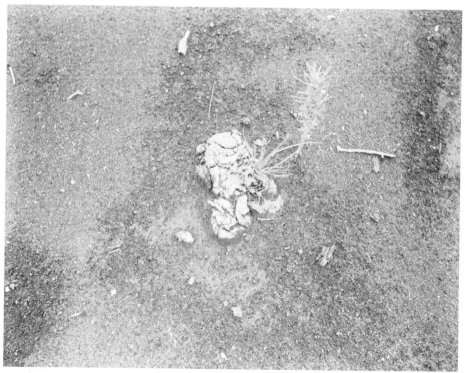

FIGURE 9.4. Sterile ash playa enriched with animal manure supporting seedlings of *Geranium, Conyza,* and a *Gramineae,* 1968.

airborne ash has not recovered from the eruption. Only a small fraction of the tree and shrub species native to the Itzícuaro Valley have successfully survived or colonized the ash or the ash-cleared sites. An even smaller proportion of the small herb species common in fallow fields and forest openings 8 km east of the cone are present in the devastated areas. Most of the herbs and many of the shrubs present in the devastated area are plants usually found elsewhere in the region on disturbed or less fertile soils. By 1972 the vegetation of the devastated area was undergoing delayed successional stages leading to the normal species composition. The absence of seed source, inadequate plant nutrition, and excessive grazing contributes to the delay of the succession.

Vegetation rapidly colonized the lava flows and the cone, encouraged by favorable moisture conditions. In large part, the rapid colonization was made possible because of the humid climatic condition for most of the year and a lava surface that retains moisture in vesicles, hollows, and narrow crevices. By 1950 blue green alga, crustose lichen, moss, and ferns were reported growing on the 1944 flows (Eggler 1963:54). (This early colonization parallels the 1883 Krakatoa eruption, where blue green algae colonized pumice 3 years after the eruption [Poli and Giacomini 1970:140].) By 1960 the Parícutin lavas were supporting 33 species: 2 pines, 15 angiosperms, 12 ferns, and 4 mosses (Eggler 1963:57). Few individuals of any species were present except for the mosses and ferns that were more abundant (Figure 9.5).

Eighteen years later vegetation cover remained sparse on the lava, with most trees and shrubs growing in crevices (Figure 9.6). However, by 1978 three trees were frequently seen on the lava, and all three grew to 4 m: *Buddleia cordata* had become the most common and visually prominent, *Clethra mexicana* was also vigorous, and *Pinus leiophylla* was present (some individuals appeared chlorotic, possibly from nitrogen deficiency). Eggler observed that the flows of 1944 supported more plants than those of 1945 and 1950; he concluded that it was not age but the character of the crevices to collect more ash and moisture that made the 1944 flows a more favorable environment for plants. Segerstrom (1966:C95) appropriately noted that the early flows received considerable ashfall and the later flows less.

As the lava weathers and vegetation cover expands, an increasingly organic and mesic soil environment will develop. In several hundred years vegetation succession should result in a humid, blocky lava *malpais* subtype of pine–oak forest similar to that found on the Capacuaro flow.

By 1957, 5 years after the end of the eruptions, the cone rim was colonized. Lichen and two angiosperm species were found growing on coarse pyroclastic materials moistened by warm acidic fumarole water vapor issuing from fissures (Segerstrom 1960:16). Fourteen species of higher plants grew on the cone rim by 1958; included were a pine and several shrubs (Beaman 1960:179–182). Many of the plants were single individuals; most grew near fissures or in depressions with additional moisture from fumarole water vapor and rain channeling (Eggler 1963:64).

The limiting of plant colonization to the most favorable environments on

FIGURE 9.5. 1944 lava flow supporting moss, lichen, ferns, and two Angiosperms in 1972.

the cone and lava makes it likely that it will be centuries before the Parícutin cone and lava surfaces weather sufficiently to support a dense plant cover similar to pre-1943 vegetation.

## RECOVERY OF GRAZING FOLLOWING RETURN OF VEGETATION

A partial recovery of forage following the heavy cineric phase brought about a return of grazing animals to reclaimed and new pasturelands. Although the pastures did not initially produce as much forage as in pre-eruption times, forage conditions soon improved and, at the time, were more than adequate for the reduced number of animals grazing.

Grazing for a limited number of livestock developed on five types of sites

**FIGURE 9.6.** *Buddleia cordata, Clethra mexicana* thriving on 1944 lava flow. Height of trees 4 m in 1978.

within the ashland: (*a*) forested sites where leaf litter supports seedling growth; (*b*) eroded slopes where soil is exposed; (*c*) deposition areas with soil–ash mixtures; (*d*) croplands after harvest or in fallow; and (*e*) open-field croplands subjected to deposition and erosion.

Grazing is poor and unevenly distributed on the open fields because of the patchy, thin colonization of seedlings on the varying depths of the ash mantle. The open fields near Zirosto, for example, provide poor grazing on hummocks of surviving *Cynodon* grass and a sparse cover of seedlings of other species that have invaded the eroding ash beds.

Soil–ash mixtures on alluvial fans and playas support a good growth of herb and shrub seedlings, but these areas are heavily grazed and often fenced and converted to croplands open to grazing after harvest or during fallow. In general, palatable seedlings growing near settlements are heavily grazed, whereas the nonpalatable species, such as *Argemone*, are left by the grazing animals to grow to full size.

Beginning with the rains of 1943–1944, continual erosion of ash from the slopelands permitted a plant cover to develop on areas of bare soil, and on sites where ash has been mixed with soil or plant litter. Following the rains of

1943–1944, pastureland has been developing on sites near the volcano. However, forage production and animal capacity near the volcano are limited because large areas still retain an ash mantle. In addition, the 2480 ha of lava and cone represent a permanent loss of usable land.

The most heavily grazed sites are forested areas near villages where leaf litter supports seedling growth. On the more distant forested slopes of Tancítaro, the light grazing after 1943 and the greater retention of soil moisture resulting from the thin mantle of remaining ash allowed for a more luxuriant herb and grass cover. Grazing capacity was further augmented by timber felling on the slopes south of the volcano between Tancítaro and Cerro Prieto. Over 800 ha was cleared through the removal of timber on the communal lands of Parícutin and Parangaricutiro during the late 1940s and early 1950s. These clearings support a dense cover of annuals and grasses that is utilized for long-term pasturage in the fir forest zone; at lower elevations they serve for pasturage before, and after maize cultivation.

In the late 1940s, sizable numbers of sheep were herded on the lands of Angahuan and Parícutin. Sheep were purchased in nearby Corupo. Lack of funds encouraged the purchase of sheep rather than cattle. Also, sheep could get along on poorer forage than cattle. However, by 1965 the higher economic and prestige value of cattle, combined with the continuing improvement of forage as a result of increased plant cover, favored cattle grazing. People with little capital were now able to exploit forage resources on village communal lands by livestock production. In Caltzontzin, for example, several individuals each owned more than 100 cattle—many more than in 1945. In Caltzontzin, most of the cattle were pastured on Parícutin communal lands on the slopes of Tancítaro; the cattle population in 1965 was 2500 as against 450 in 1950.

In the mid-1950s, the raising of cattle yielded income not only from the sale of cattle in villages and market towns, but also from dairy products. Traditionally, milk is consumed only by children among the Tarascans. Since the mid-1950s, milk has been converted into cheese and sold in the market towns.

Cattle owned by the villagers on communal lands of Parícutin and San Juan Parangaricutiro are generally low-grade animals and have low milk production. These cows produce on the average of 4–5 liters of milk per day for a period of 2–3 months, whereas those owned by Tarascans in Cherán yield 8–12 liters per day for more than 7 months. Low milk production is caused by inadequate watering and nutrition. No supplementary grain or forage or water is supplied by the villagers to their cattle, even during droughts that occur in March and April. On top of all these setbacks, the distances and difficulties of transportation make the sale of fresh milk impractical even today.

Despite their low milk production, the Tarascan cattle are still highly prized as an easily salable, high-return speculative investment. Nonetheless, the primary importance of cattle continues to be in their value as work animals in agriculture.

## RECOVERY OF AGRICULTURE

Successful reestablishment of agriculture is closely related to conditions permitting survival, recovery, and reestablishment of native vegetation. Ash characteristics that inhibited native vegetation were detrimental to economic crops. As mentioned before, during the heavy cineric phase, maize seedlings near the cone were buried; those not buried died as a result of the infertility and droughtiness of the ash beds, and from fungus attacks through bruised tissues from windblown ash. Of all the negative factors, the lack of nutrients was by far the most important.

Pure ash devoid of vegetation had only 55 ppm of nitrogen at 2 cm of depth, as compared with 350 ppm in the underlying old soil. Ash with a pine-needle litter or supporting grass had higher nitrogen values (105 ppm) at surface levels, and lower values with increasing depth until true soil was reached (Eggler 1963:66). Farmers found that the nutrients in a thin layer of pine-needle litter plowed into pure ash were not sufficient to mature a maize crop.

Experiments carried out in the 1950s by agronomist Eduardo Limón (personal communication) showed that the ash must be heavily fertilized and continually supplied with nitrogenous materials to produce a maize crop. Enriching ash with commercial fertilizers seemed economically feasible only if the farmer were in a position to sell the maize grain, plow under all the remaining plants, and reinvest the sale proceeds in purchasing fertilizer for the next year. By the end of the third year, the enriched ash should be sufficient to sustain a maize crop without further fertilizing, although stubble should still be plowed under for ash enrichment. The entire process would require a capital outlay before the first crop was in; it presumed, therefore, that the farmer had sufficient resources to sustain his family for almost 3 years without having to rely on income from the farmland. In reality, the average Tarascan farmer did not have the money even to pay for the first year's fertilizers.

Fertilizing pure ash allows the farmer to produce a mature maize crop, but the yield is not sufficient to both feed the farmer and pay for fertilizer. On the other hand, fertilizing soil–ash mixture results in a dramatic increase in productivity that could provide subsistence for the farmer, plus crop surpluses to pay for fertilizer costs. Crop experiments in 1948 by Miller showed that traditional maize grown in fertilized soil–ash mixture on open field cropland at Angahuan yielded 2.025 metric tons of shelled grain per hectare when fertilized with a treatment of 40 kg N and 40 kg $P_2O_5$ per hectare. This is in sharp contrast with the unfertilized plots that produced only 0.633 metric tons (896 liters) per hectare (Miller et al. 1949:271). Miller's data indicate that the Angahuan plots were well suited for heavy fertilizing owing to the more than adequate rainfall during the critical months of maize growth and ear development (Miller et al. 1949:272).

Crop experiments by Tarascan farmers 20 years later in nearby San

Lorenzo showed that fertilizing soil–ash mixture yielded 2800 kg of maize per hectare. The 1700-kg increase over the usual 1080-kg yields sold for 1360 pesos, more than double the cost of fertilizing (Rees 1971:97). Unfortunately, effective use of commercial fertilizer, insecticides, and close planting did not begin to take place in the Itzícuaro Valley until as late as 1973.

In the absence of government technical support and funding for a commercial fertilization program, Limón suggested enriching the ash with the planting of two native herbs: one, *Lupinus leucophyllus*, a native legume, which failed to germinate properly in ash, and the other, *Mirabilis jalapa*, an herb with a large tuberous root, which succeeded. But *Mirabilis* was rejected by the Tarascan farmer as being a troublesome weed on maize lands.

The Tarascan farmer then attempted several approaches of his own: (*a*) the wholesale removal of ash by sluicing; (*b*) removal by bulldozing; (*c*) manual removal by digging of pits to reach old soil; (*d*) deep plowing by tractor; (*e*) furrow plowing on the ash mantle using animal traction, a wooden plow, and later, a steel plow. The first four approaches met with various difficulties; the last, furrow plowing, was successful.

The sluicing away of ash by directing floodwaters across the fields, used in experiments in Angahuan during the heavy cineric phase, resulted in deep gullying of the old soil, with high ridges of ash remaining. This technique has been little used since the mid-1940s. However, in 1978, at least one Angahuan farmer removed ash by canalizing storm runoff across a maize plot (Figure 9.7).

The bulldozing of the Angahuan fields by the government in 1953 initially cleared 25 ha of land but was discontinued because there were no satisfactory dumping sites for the ash spoil; ash dumped into ravines and water courses created problems downstream. The method of digging pits down to old soil and growing crops in them proved successful only on house lots. As early as 1945, maize, beans, and squash were produced on Angahuan house plots, but pit growing of crops was unsuited to open fields for three reasons: (*a*) labor and time required to dig and maintain the pits; (*b*) large dumping areas required to contain the ash spoil; and (*c*) filling of pits by wind- and rain-moved ash. The deep plowing of ash to bring soil to the surface was briefly tried near Zirosto in 1947 (Wilcox, personal communication). The effort was discontinued for reasons not known.

Furrow plowing, the sowing of maize in the old soil and soil–ash mixture, accounted for most of the productive acreage in the Itzícuaro Valley after 1945. Furrow plowing enables old soil, weeds, and crop stubble to be mixed into the ash mantle, thus providing some nitrogenous enrichment for each successive crop. The plowing of old soil into ash, however, does not guarantee the growth of maize evenly throughout a single field. Maize growth was uneven because of the differing depths of water-moved ash within a field. In many instances, plots with shallow ash produced harvested crops soon after the heavy cineric phase, whereas adjacent plots with deeper water-moved ash did not produce their first successful crop until 1960–1962, after a series of earlier

FIGURE 9.7. Accelerated gully erosion created by deliberate canalization of runoff to remove ash. Angahuan, rainy season 1978.

failures. In general, most fields under furrow plowing with an ash depth of less than 15 cm were cropped as early as 1945; however, they did not come up to pre-eruption productivity until the mid-1960s.

Plowing using the traditional ox-drawn wooden plow limited successful maize growing to sites where true soil existed within 14 cm below the ash surface. Plowing with burros further limited cropping to where soil was found within 8 cm beneath the ash. The scarcity of cattle, compounded with the high prices for horses and mules and the lack of pasturage following the heavy cineric phase, all contributed to limiting the reclamation of land by the traditional wooden plow. The steel plow, which permits plowing to 22 cm of ash, was not in general use until the late 1940s, when forestry activities and earnings sent home by bracero workers made its purchase possible.

In areas of deep ash where the plow did not reach the old soil, two solutions were applied: (a) successive cropping, and (b) waiting until natural vegetation developed before applying the first plowing. With the first method, maize was seeded into pure ash where the lack of nutrients caused the seedlings to yellow and die; the field was then laid fallow for a year, and the procedure repeated until humus and weed growth provided sufficient nutrients to mature the maize plants. This, however, proved to be time-consuming, painstaking, and costly in maize seed, and was therefore avoided by most farmers. The second method, waiting for the return of a natural pasture before applying the first plowing, required no active participation on the part of the farmer and was therefore preferred.

Some former cropping sites buried under ash cannot be easily reclaimed on account of the presence of a hummocky landscape that prevents plowing. On a site near Zirosto, for example, mounds were created by windblown ash caught and retained by the presence of *Cynadon* grass; ground between mounds was then lowered by channel and wind erosion, and the irregular surface thus formed made the reclamation of the cropping site difficult.

Most former cropping sites buried under 15–25 cm of initial ash deposition are presently under cultivation, with the exception of sties buried under water-transported ash. Many fields with an initial ash deposit of 25 cm were, by 1958–1962, eroded down to a remaining ash depth of 15 cm and were support- ing a pasture reclaimed for crop growing during the same period. Fertility was often low during the initial years and many maize crops did not mature. Fallow periods of 2 or 3 years were common, and grain production was usually less than half of that in a contemporaneous forest clearing devoid of ash.

On sites where the initial ash deposition was between 25 and 50 cm, the recovery of crop growing vaires conspicuously from field to field, reflecting differences in either erosion, or enrichment of the ash with deposition of soil and organic material. By 1965, some plots were in the first years of cultivation, whereas others were still in the pasture stage. Over the years few farmers have attempted to enrich the ash by direct application of organic debris or animal fertilizer. Direct enrichment of the ash with a small handful of animal manure placed with the maize seed at planting time was practiced by some farmers at Angahuan in 1978. Figure 9.8 shows a plot with 30 cm of remaining ash in the third year of maize cultivation; the first crop was planted in 1973. The fertilized plants are undernourished, resulting in undersized maize ears not satisfactory for dry grain but picked early to prepare *atole* (a corn soup). Adjacent unfer- tilized maize was stunted and failed to produce maize ears.

In areas where the former croplands received more than 50 cm of ash, the recovery of crop growing was limited to plots where the ash mantle had been removed, or where crops were sustained by soil and organic debris eroded from adjacent slopes. By 1978, maize sustained by eroded soil and organic debris grew on seven debris basin sites at the margins of the lava field. The first site cultivated was Choritiro (at the northeast edge of the lava field) in 1946; most other sites came into cultivation in the 1950s (Segerstrom 1961:D226). Gener- ally, these sites are fertile enough to sustain cropping every other year without animal or chemical fertilizer. The maize growing at Choritiro in September 1978 looked healthy and grew to 2 m. Unfortunately, the Choritiro site receives more runoff, sediment, and debris than desirable from the arroyo that drains a large area east of Angahuan. The runoff cut destructive swaths 4 to 15 m wide through the growing maize. Past construction of timber, rock, and brush walls have not controlled runoff at Choritiro as successfully as at the other sites.

FIGURE 9.8. Maize on an ash playa 0.5 km southwest of Angahuan, 1978. Unfertilized plot at left, animal-manure enriched soil on right in third year of planting.

However, the proportion of crop lost to flooding is small. Productivity at Choritiro is described as a seed-to-harvest ratio of 1:45 (approximately equivalent to 0.630 metric tons per hectare).

Slopelands that were partially cleared of deeper ash by sheet, rill, and channel erosion saw a return to crop productivity as early as the spring of 1945. For example, the upper portion of the inclined open fields adjacent to Zacán was reclaimed in the mid-1940s because the initial ash deposition of 25 cm had already been eroded to a depth that permitted the plowing of old soil into the ash.

As a result of early crop failures on most of the open fields, crop growing has greatly expanded on the slopelands since the spring of 1945. Planting of maize was intensified on existing permanent fields, shifting cultivation sites, and pasturelands located on the slopes of Tancítaro and the Cerros of Angahuan (Figure 9.9). Most existing clearings were enlarged and new clearings created where sites could remain undetected by the federal foestry police. Most of these sites are located within communal forests (San Juan Parangaricutiro, Parícutin, Zirosto, Angahuan) where the initial ash deposition was shallow, and where erosion of ash was accelerated. Salvage lumbering and deforestation on the slopes of Tancítaro following the heavy cineric phase increased the

FIGURE 9.9. Partially deforested slopes near Angahuan supporting shifting cultivation and grazing. Alluvial fan partially reclaimed for maize cultivation. Eroded trail flanked by ash playa with *Lupinus, Argemone, Epicampes*, 1972.

availability of lands cleared for cropping. Almost continuous cropping of slope-land clearings has occurred since 1945, resulting, in many cases, in poor yields, shorter fallow periods, and severe erosion.

None of the pre-eruption cleared land covered by exposed lava (a total of 1370 ha) will ever be reclaimed. Unlike most of the ash mantle, the lava surface will not be cultivable for several thousand years. Most of the ash-covered cropland could have returned to pre-eruption productivity shortly after the heavy cineric phase (1943–1945) if subsoil plows drawn by tractors had been used. The subsoil plow is used in older irrigation districts in other areas of Mexico, and is capable of plowing to depths exceeding 1.8 m.

Reclamation has been speeded up recently because of changes in government policy and increased profitability in avocado production. Since 1973, government financing and technical support for reclamation of cropland has rapidly improved agriculture in Angahuan and Zirosto. The federal government has provided credit for chemical fertilizer to a group of 70 Angahuan farmers who are producing successful maize crops on soil–ash mixtures. The Michoacan State Forestry Commission has provided budded *membrillo* (*Cydonia oblongata*) and other fruit trees and technical aid to farmers willing to plant, protect, and care for trees planted on unreclaimed ash-covered fields where the ash is not deep. Some of the deeper ash sites are being reclaimed as the Forestry Commission encourages planting of well-developed pine seedlings provided without cost. By 1978 most pine planted on Angahuan lands appeared to be healthy and fast-growing.

A portion of the ash-covered fields of Zirosto now support avocado orchards. The federal government supervises contracts between village officials and outside private capital that leases the land and develops and operates the orchards using local labor. The villagers receive land rental income, wage income, and eventual ownership of the trees. The high profits from avocado production encouraged growers to pay the cost of creating soil–ash mixture for each tree, a heavy fertilization program, and other development costs. Winter frosts will probably prevent the planting of avocados on the deeper ash areas of the higher elevations nearer to Parícutin. Without outside subsidy, the deeper ash areas will continue to remain uncultivated until plant succession enriches the ash with sufficient nutrients to mature a maize crop.

## POSTERUPTION FORESTRY ACTIVITIES

Prior to the eruption, timber exploitation in the upper Itzícuaro Valley had always been on a small scale, with sales aimed at the village and local market, and usually not included in official statistics. From the beginning of the eruption, during the time when normal agriculture and grazing were arrested, villagers earned cash by greatly expanding the illegal cutting of dead and dying timber. Large quantities of firewood, charcoal, shingles, and hand-hewn and hand-sawn lumber and timbers were produced up to 1944, when large-scale lumbering and commercial milling were introduced to exploit timber devastated by ash. By 1954, the excessive cutting of timber brought commercial timber production to an end.

During and following the heavy cineric phase, firewood collection and oak charcoal production experienced an initial boom. The low prices of firewood limited its sale to local regions, but the government allowed large volumes of charcoal to be shipped to urban markets in other parts of Mexico. Today, firewood collection for sale to market towns is a minor activity. The ample supply of milling scrapwood from lumber mills in Uruapan has led to low firewood prices, and there is a reduction in market demand for firewood because of the shift to use of petroleum fuel by the major consumers in Uruapan—the bakeries and the public baths. The long-distance hauling of firewood by burros from areas near the volcano has become the exclusive occupation of old men, since it pays less than the legal minimum wage. The production of oak charcoal, aimed exclusively at market towns, has continued on a much reduced scale since the 1940s. Market demand, however, remains constant in Uruapan despite the shift to bottled gas as cooking fuel, because charcoal is still being used by older residents and impoverished immigrants.

Presently, oak charcoal continues to be supplied from the forests of Angahuan and San Juan Parangaricutiro, where there is ample oak timber beyond the devastated zone.

During 1943 and 1944, an unusually large number of pine and fir shingles were produced for sale outside the devastated zone. Traditional hand-tooled lumber products such as beams and other squared timber were sold to outside buyers, usually to be milled later into small lumber. Today, a limited number of lumber products are made. Continued efforts are being made by the federal forestry officials to restrict timber exploitation to quantities consumed in each village. Timber cutting at Zacán, Zirosto, and Barranca Seca is now limited to production of lumber for village house construction. Beams, squared timber, and pine shingles, produced at Angahuan, are sold locally or to buyers from more distant markets, with or without official permission.

Carpentry that uses unseasoned pine and nails to make chairs, cabinets, and panel doors was a part-time activity of some individuals during the early years of the eruption. In 1965, full-time carpenters could be found in Caltzontzin, Nuevo San Juan, and Angahuan, and part-time carpenters in Zirosto, Barranca Seca, and Zacán. Cabinetry and furniture are the only products of wood that can be sold legally beyond the village without prior government permission. Carpentry items are sold locally and to market venders in Uruapan; however, lack of knowledge of marketing and of easy access to large trucks prevents furniture from being sold to Guadalajara or Mexico City, where prices are twice those of Uruapan.

The illegal cutting and selling of timber by villagers in 1943 and 1944 was allowed by federal forestry officials for fear of public criticism if laws against it were enforced. However, by 1944, villagers were required to apply for permits to cut timber. In this way, the first legal small-scale timber exploitation began to take place on the forest lands of Zirosto, San Juan Parangaricutiro, and Parícutin.

From 1944 to 1954, commercial lumber mills operated on the forest lands of Zirosto, San Juan Parangaricutiro, Parícutin, and Angahuan. Commercial lumbering was introduced when forestry officials decided to maximize utilization of salvaged timber. The wastefulness of using high-quality large-diameter pine for traditional shingle making encouraged the government to introduce commercial milling. Contracts were awarded to the Tarascan communal mill at San Felipe, and to non-Tarascan contractors to establish mills at Barranca Seca and Pantzingo (a hamlet located on the communal lands of San Juan Parangaricutiro) and to existing mills in Uruapan. Power sources at these mills varied from electricity to steam power to gasoline motors. Commercial lumbering increased the rate of cutting and salvaged huge volumes of damaged timber that would otherwise have decomposed and been wasted. It also permitted closer federal control over timber cutting and tax payments, and provided for payment for timber from communal forests to village communal funds.

In the initial phase of lumbering for the mills, hand-hewn squared timber was produced in the forest and converted into lumber at the mill. This procedure kept the villagers employed and diverted maximum income to the Tarascan laborers rather than to mill operators. However, timber squared with axes wastes too much wood in chips. Motor-driven saws are usually less wasteful. Therefore, government forestry officials decided to encourage the hand-tool production of logs and cants (short logs for boxwood) for later milling into lumber at the mills. The mills made contracts with individual villages to purchase felled timber. Contracts specified payment of the felled timber to the communal fund of each village and payment to villagers working in the logging and milling operations. Wages paid to the Tarascan woodcutter were on a piecework basis; those paid to men working in road building, timber hauling, and milling were on a fixed daily-wage basis. Income to the village communal funds, believed to have amounted to over U.S. $200,000, was held in trust by the federal government and helped to pay for resettlement—as in the case of Barranca Seca—and for general improvements, such as electricity and piped water in all the villages.

From 1944 to 1954, men from Zirosto, Miguel Silva, Neuvo San Juan, Caltzontzin, and Angahuan alternated from crop growing on their own fields to lumber operations to contract farm labor in the United States. The men of Zacán, not involved in commercial lumbering, farmed on their own reclaimed fields or migrated as harvest laborers to the United States or to other parts of Mexico.

By 1954, most of the devastated timber had been removed. Federal forestry officials were aware that much live timber was being illegally cut on the communal lands of Zirosto; permission to cut timber was withdrawn and commercial lumbering in the Itzícuaro Valley came to an end.

Resin production, traditionally a secondary occupation for most farmers, came to a complete halt during the heavy cineric phase because falling ash polluted the bleeding faces of the pine, as well as the collection cups. Resin production was resumed when the ashfall subsided. Ash-polluted pine was bled for resin as early as 1946 on private holdings of residents of San Juan Parangaricutiro. Production on communal lands was delayed because wages paid to resin collectors were low as compared with lumbering wages and wages paid for harvest outside the region. By the mid-1950s, the government renewed or approved new contracts with existing turpentine distilleries in Uruapan, and the acreage in pine resin production steadily increased. By the mid-1960s, more men were employed in resin collection than in any other forest activity. Most men in Zirosto, Zacán, Barranca Seca, and Angahuan supplemented their income by working approximately 6 days per month in resin collection. The majority of men possessed collection rights averaging between 200 and 500 trees; a few had rights to 1000 or over. Each tree yielded between 3 and 5 kg of resin annually. Government further stimulated production by establishing a government-sponsored cooperative turpentine distillery in the Tarascan village of Cherán. In 1968, the Cherán distillery paid members of its resin collection cooperative one peso per kilo, whereas privately owned distilleries were paying only 45–67 centavos per kilo.

The recovery of the pine forests and the higher price for resin account for the present success of resin activities. Unfortunately, the quest for resin income has resulted in poor management of young pine stands. Crowding and poor nutrition produce narrow crowns and small-diameter trunks—the poorest characteristics for resin production. Many of these trees are under 20 cm in diameter at breast height and show three bleeding faces. The exploitation of premature young pines will ultimately result in reduced resin yields and lowered quality of wood as a whole for use as lumber.

## SUMMARY AND CONCLUSION

The Parícutin eruption created a cinder cone that rises 200 m above the surrounding lava field. Beyond the lava field, the landscape was covered by volcanic ash that varied in depth from 11 m near the cone to 25 cm at Angahuan 7 km distant. Accelerated erosion has transported much of the ash from slopes to basins and alluvial fans, or onto the floodplains of the lower Itzícuaro Valley. The lava blocked prior drainage, and in areas of gentle terrain created large lake basins that filled with eroded ash and debris. Upslope from the lava field, isolated pre-eruption basins also became filled with ash and debris. Floodwaters that flowed out of these basins cut barrancas and created a new drainage system that connects with the intermittent streams on the border of the lava field. Vegetation was totally killed under the cinder and lava and in areas of deep ash. Grazing and agriculture were devastated during the heavy ashfalls of 1943.

The heavy ashfalls of 1943 forced the majority of residents of Zirosto to migrate to Miguel Silva, a distant resettlement site, or to other Tarascan villages. Soon after, many of the former residents from Zirosto returned to work in salvage lumbering at Barranca Seca. When additional lands were provided by the government, Barranca Seca was transformed into Zirosto Nuevo. The former residents of Zacán returned to reclaim their ash-eroded land as early as 1945. Angahuan alone remained intact during the ashfall. Sustained by its more distant forests and croplands, Angahuan's economy was further augmented by tourist income following the eruption. The invasion of lava over the villages of Parícutin and San Juan Parangaricutiro in 1944 forced the inhabitants to relocate permanently at Caltzontzin and San Juan Nuevo outside the Itzícuaro Valley.

Although the croplands at Nuevo San Juan are limited, the relocated residents remained there permanently on account of the ample undevastated communal and private lands stretching to within 2 km of the new village. The income that Nuevo San Juan derives from crops, grazing, and forestry is further augmented by tourist income from the annual pilgrimage to the Miraculous Image of Christ.

Most of the former residents of Parícutin remained permanently at Caltzontzin. The croplands at Caltzontzin produce sufficient maize to keep the population from leaving, but supplementary income is necessary for most families. Most men are forced to do harvest work outside the village or work as unskilled laborers in nearby Uruapan. Some families return for days to months to hamlets on Parícutin communal lands to work in forestry, cheese production, and maize cultivation on forest clearings. As a result, the resident population fluctuates seasonally at the hamlets of La Escondida, San Salvador Teruto, and La Capilla, located south of the Parícutin cone.

With the direct intervention of the outside world, the resettlement villages were modernized 10–20 years ahead of other Tarascan villages in the Meseta Terasca. The government introduced water storage and piping installations, electricity, medical clinics, and poultry-raising projects into the resettlement villages. Traditional village communal funds, enriched by salvage lumbering, were channeled into financing these improvements and encouraged cottage industry that produced hand-woven cotton textiles, cabinetry, leather sandals, fired bricks, etc. With the exception of cotton textiles at Caltzontzin, most of these enterprises failed through insufficient technical training and generally low quality of the products.

As conditions stand now, the upper Itzícuaro Valley has not yet recovered from the Parícutin eruption. A delayed vegetation succession is slowly taking place on the ash playas and the lava and cone surfaces. With time, the lava field should develop into a humid blocky lava *malpais* subtype of pine–oak forest, such as that found on the geologically recent Capacuaro flows. The ash playas and cone surfaces can be expected to weather to productive soils more rapidly than the lava, but excessive drainage on the cone, insufficient nutrients, and overgrazing will inhibit a rapid colonization and succession to a pine–oak forest with a balanced species composition similar to pre-eruption times. Distance to seed sources, and the stripping away of humus-rich soil, will retard the return of the broadleaf subtropical forest to more humid barranca environments.

Resin production has become the most important type of forest exploitation since the eruption ended; it will continue to be important if resin prices remain subsidized by the government. Timber cutting by individual villagers will continue on a small scale, as in pre-eruption times. However, because of excessive posteruption cutting and smaller-diameter class size of pines grown, it is unlikely that large-scale privately owned lumber mills will return to the valley in the near future. The smaller-diameter pines, however, are being used for boxwood production in Capacuaro. Pine timber will continue to be used by villagers for carpentry and for clandestine production of common lumber for village construction and outside sale.

Grazing conditions remain poor on the barren ash playas, but forested hill slopes and more distant sites support cattle grazing. Forage conditions will improve as ash and soil–ash mixture continue to be colonized by plants. Overgrazing, however, will delay recovery of vegetation on sites near settlements.

Changes in agricultural productivity are just beginning. In the early 1970s some of the wealthier villagers began experimenting on their best lands with inexpensive chemical fertilizers on traditional maize varieties, resulting in the triple yields that more than cover the cost of fertilizing. In the 1960s budded fruit trees—peach, pear, and apple—obtained from the Forestry Commission of the state of Michoacán, were planted on a few house-lot gardens; but most were neglected and few survive for lack of pruning, fertilizer, and of insect, gopher, and fungus abatement.

Agricultural production could have been in operation earlier if technical assistance and credits for fertilizers and fruit trees had been available and insect and gopher pests put under control by chemicals. The still prevalent resistance to fertilizing and hybrid maize, augmented by lack of experience in efficient production and sale of market crops, hinders the full use of existing farmlands. It may be years before the farmers will shift from subsistence production of maize to market production of cabbage, potatoes, and carrots—crops that do well in the region and have a ready market in the cities. Presently, for instance, potatoes are earning substantial profits for Tarascans at Cherán outside the valley, where technical assistance is available and where potatoes and cabbage have found a market in Guadalajara.

Since 1973, government financing and technical support has accelerated reclamation of cropland in the Itzícuaro Valley. At Angahuan, ash playas are being transformed into orchards and small pine plantations. Credits for chemical fertilizer have raised maize yields. Outside capital is transforming Zirosto ash playas into avocado orchards.

Most pre-eruption cleared lands are lost to cropping where buried under 50 cm or more of ash (2790 ha); exceptions are those sites with soil–ash mixture or where sites are eroded down to original soil. Long-term loss of pre eruption cleared lands—those under cultivation, in annual fallow, in long fallow with shifting cultivation—amounts to approximately 4160 ha. Of these, over 1374 are under lava and will remain totally uncultivable for thousands of years.

## REFERENCES

Arias Portillo, P.
  1945  La region devastada por el Volcán de Parícutin. Thesis presented for the title of Agronomist, specialist in Forestry, Escuela Nacional de Agricultura, Chapingo, Mexico.
Beaman, J. H.
  1960  Vascular plants on the cinder cone of Parícutin Volcano in 1958. *Rhodora* 62:175–186.
  1961  Vascular plants on the cinder cone of Parícutin Volcano in 1960. *Rhodora* 63:340–344.
Bullard, F. M.
  1976  *Volcanoes of the earth.* Austin: University of Texas Press.
Eggler, W. A.
  1948  Plant communities in the vicinity of the Volcano El Parícutin, Mexico, after two and a half years of eruption. *Ecology* 29:415–436.
  1959  Manner of invasion of volcanic deposits by plants with further evidence from Parícutin and Jorullo. *Ecological Monographs* 29:268–284.
  1963  Plant life of Parícutin Volcano, Mexico, eight years after activity ceased. *American Midland Naturalist* 69:38–68.

Foshag, W. F. and Jenaro González R.
   1956   Birth and development of Parícutin Volcano, Mexico. Geological Survey Bulletin 965-D.
          Washington, D.C.: U.S. Government Printing Office.
Fries, C. Jr.
   1953   Volumes and weights of pyroclastic material, lava, and water erupted by Parícutin
          Volcano, Michoacán, Mexico. *Transactions, American Geophysical Union* 34:603–616.
Gutiérrez, C.
   1972   A narrative of human response to natural disaster; The eruption of Parícutin. In *San Juan
          Nuevo Parangaricutiro: Memories of past years*, edited by Mary Lee Nolan. Environmen-
          tal Quality Note No. 6. College Station: Texas A & M University.
Hatt, R. T.
   1948   A bibliography of Parícutin Volcano. *Papers of the Michigan Academy of Science, Arts,
          and Letters* 34:227–237.
Macdonald, G. A.
   1975   Hazards from volcanoes. In *Geological hazards*, by B. A. Bolt, W. L. Horn, G. A.
          Macdonald, and R. F. Scott. New York. Springer-Verlag.
Madrigal Sanchez, X.
   1964   Contribución al conocimiento de la ecología de los bosques de Oyamel (Abies Religiosa H.
          B. K.) en el Valle de Mexico. Thesis presented for the title of Biologist, Escuela Nacional
          de Ciencias Biologicas, Instituto Polytechnico Nacional, Mexico D. C.
Miller, E. B., J. B. Pitner, Ricardo Villa J., Carlos Romo G.
   1949   Population density of unirrigated maize and its influence upon fertilizer efficiency in
          central Mexico. *Soil Science Society of America, Proceedings* 14:270–275.
Nolan, M. L.
   1972   The towns of the Volcano: A study of the human consequences of the eruption of
          Parícutin Volcano. Doctoral dissertation, Dept. of Geography, Texas A & M University.
Ordóñez, E.
   1947   *El Volcán de Parícutin*. Mexico: Editorial Fantasiá.
Poli, E., and Valerio Giacomini
   1970   Vulkane and pflanzenleben. In *Vulkane*, edited by Christoph Kruger. Vienna: Anton
          Schroll & Co.
Rees, J. D.
   1961   Changes in Tarascan settlement and economy related to the eruption of Parícutin.
          Master's thesis, Dept. of Geography, University of California, Los Angeles.
   1970   Parícutin revisisted: A review of man's attempts to adapt to ecological changes resulting
          from volcanic catastrophe. *Geoforum* 4:7–25.
   1971   Forest utilization by Tarascan agriculturists in Michoacán, Mexico. Doctoral disserta-
          tion, Dept. of Geography, University of California, Los Angeles.
Segerstrom, K.
   1950   Erosion studies at Parícutin. Geological Survey Bulletin 965-A. Washington D.C.: Gov-
          erment Printing Office.
   1960   Erosion and related phenomena at Parícutin in 1957. Geological Survey Bulletin 1104-A.
          Washington D.C.: U.S. Government Printing Office.
   1961   Deceleration of erosion at Parícutin, Mexico. Article 370 in U.S. Geological Survey
          Professional Paper 424D. Washington D.C.: U.S. Government Printing Office.
   1966   Parícutin, 1965—Aftermath of eruption. U.S. Geological Survey Professional Paper
          550-C. Washington D.C.: U.S. Government Printing Office.
West, R. C.
   1948   Cultural geography of the modern Tarascan area. Smithsonian Institution. Institute of
          Social Anthropology No. 7.
Wilcox, R. E.
   1954   Petrology of Parícutin Volcano, Mexico. Geological Survey Bulletin 965-C. Washington
          D.C.: U.S. Government Printing Office.
Williams, H.
   1950   Volcanoes of the Parícutin Region, Mexico. Geological Survey Bulletin 965-B. Washing-
          ton D.C.: U.S. Government Printing Office.

# 10

# Impact of Parícutin on Five Communities

MARY LEE NOLAN

*Now one sees an admirable flow of fire covering the traces of our last footsteps and the works of man that he made during the life that God permitted him.*

—CELEDONIO GUTIÉRREZ

The tower of a ruined village church rises above a mass of lava in the Sierra Tarasca of Michoacán, Mexico (Figure 10.1). Once the site of regional religious pilgrimages, the church is now visited by tourists on excursions from Uruapan, a rapidly growing provincial city about 25 km away. They travel the last 2 km on foot or horseback through a forest of young pines that spring from ground blackened by pyroclastics. When the tourists reach the edge of the lava flow, they must clamber over the sharp, black rock to the church, which marks the site of San Juan Parangaricutiro.

San Juan was a thriving, although somewhat isolated, local political center when, on the afternoon of 20 February 1943, a dust-belching hole in a cornfield 3.5 km to the south began to grow into the lava- and ash-spurting Parícutin volcano (Figure 10.2). When the eruptions ended on 4 March 1952, the cinder cone rose 410 m above the former cornfield surface and a 233-km² area had been devastated by lava and ash (Fries and Gutiérrez 1954:490). Lava had claimed San Juan and nearby Parícutin village. Pyroclastics, called "the sands of the volcano" by the local people, had covered the lands of three other peasant farm communities, Angahuan, Zacán, and Zirosto (see Figures 10.3 and 10.4).

The inhabitants of San Juan and Parícutin had relocated in refugee settlements near Uruapan. The communities of Zacán and Angahuan remained in place. Zirosto had fragmented into three parts, with some residents remaining in the original settlement while others moved to the refugee settlements of Miguel Silva and New Zirosto. Thus, the five communities that lay

*VOLCANIC ACTIVITY AND HUMAN ECOLOGY*

FIGURE 10.1. Ruins of the pilgrimage church at San Juan Viejo Parangaricutiro. The façade survived because it was encircled by the lava flow. [From M. L. Nolan, "The Mexican Pilgrimage Tradition," *Pioneer America*, Vol. V, No. 2, 1973.]

FIGURE 10.2. Parícutin Volcano, 1972.

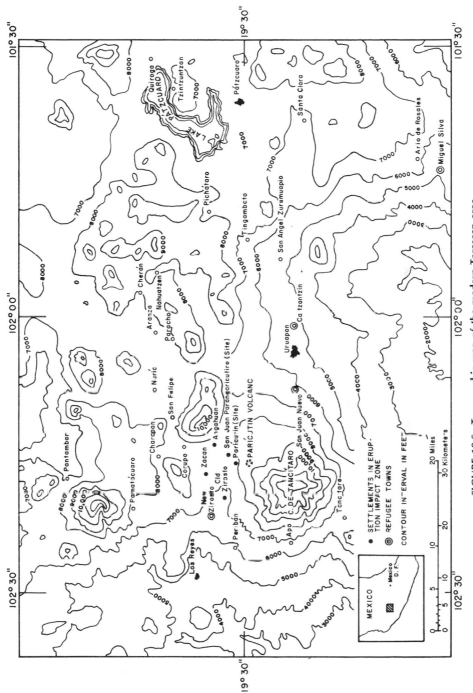

FIGURE 10.3. Topographic map of the modern Tarascan area.

295

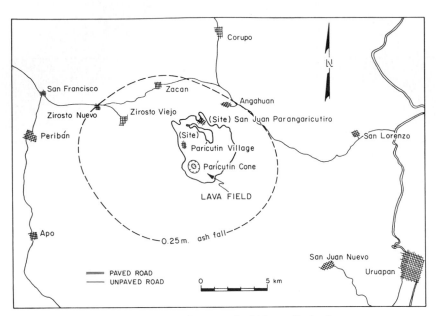

FIGURE 10.4. Region near the Volcano Parícutin.

within Parícutin's zone of devastation (Williams 1950: Plate 9), are now seven quite diverse and geographically scattered settlements known in the region as "the towns of the volcano" (Nolan 1972:2).

The new volcano's birth initiated drastic changes in the environment to which the people of these communities were accustomed. The forces of social, political, economic, and technical change that were already at work in Mexico were focused on these villages because of destruction or severe damage to the resource base and traditional places of habitation. As individuals and communities, the people were faced with the challenge of adapting to rapidly changing physical and social situations. Ostensibly similar in lifestyle and cultural background, the people revealed underlying differences in the variety of adaptive choices they made and the extent to which they allowed forces for change to determine their futures. Had the volcanic zone population been truly homogeneous in cultural orientation, and had the physical force of the volcano been the only change agent at work, then the story might be simple to recount. However, a volcanic eruption in a rapidly changing society creates a situation of great complexity.

This case study of the communities most affected by Parícutin volcano seeks to provide insights by examining the situation before the eruptions, the sequence of events and human responses during the 9-year period of volcanic activity, and the characteristics of the resulting seven communities nearly 30 years after the eruptions began. Attention is then focused on implications of the Parícutin eruption in the context of current research on human response to natural hazards and disasters and followed by a brief speculation on the significance this study holds for archaeological investigations in zones of volcanic devastation.

## COMMUNITY LIFE BEFORE THE VOLCANO

In folk mythology the time before the volcano was a golden age when the region was beautiful and prosperous. Folk songs describe flowers and singing birds, the rains of summer and the green fields—a time when all was pleasure and happiness. The region lay along the northern foot of the Cerros de Tancítaro, an old shield volcano that loomed above a landscape dotted with hundreds of younger and smaller volcanic forms. These were interlaced with a network of valleys that lay at an elevation of about 2100 m. The valleys and gentle slopes around the villages were cultivated, or allowed to grow in pastures that supported locally owned livestock. The ancient cinder cones and the rocky badlands of old lava flows were covered with forests of pine and oak (Dorf 1945:258). These forests provided cover for the region's wild fauna, which primarily consisted of deer, rabbits, and many species of birds, and wild bees valued for their honey.

The life of the region's people was not so idyllic as the folk songs suggest, but it was not a bad life. The very lack of abundant resources and transportation networks had left it a place where village farmers, many of whom spoke the Tarascan language, still followed traditional patterns as subsistence tillers of the soil. Most of the valley lands were in private ownership but, according to custom in most communities, were not supposed to be sold to outsiders without the consent of community elders who had paid their dues to society by sponsoring religious fiestas. Forest lands were generally in community ownership. They were utilized for timber, with which village homes were built, and for resin sold to nearby turpentine distilleries. Each village had its own craft traditions, as was the custom in the Sierra Tarasca, and goods were exchanged in nearby market towns and during the fairs held in association with local fiestas. One hacienda had been established in the region, but its sphere of influence affected only the Zacán community. In many ways, life before the volcano followed the socioeconomic model laid down by the Spanish bishop Vasco de Quiroga some four centuries earlier in the wake of the Spanish conquest.

In a general way, the communities to be affected by the volcano were similar. All were designated as *pueblos*, a term that can mean either village or town. Populations were small, ranging from 733 in Parícutin village to 1895 in San Juan Parangaricutiro (hereafter referred to as San Juan) (Secretaria de la Economía Nacional 1943). All had been founded as Tarascan Indian settlements. In 1940 some people in each of the communities spoke the Tarascan language, although the number varied from 100% in Angahuan to only 20% in Zirosto, indicating that the communities lay along the margin of a shrinking Tarascan culture area (West 1948:19). The settlements looked much alike. All were laid out in grid patterns centered on churches. Styles of architecture were much the same, although some communities had a higher proportion of stone and adobe houses than did others (West 1948: Map 19). Most dwelling lots contained a wooden *troje*, or family ceremonial structure, and a *cocina*, or kitchen structure where the family cooked, ate and slept (Beals *et al.* 1944:21). The primary resource base for these small communities was composed of the village fields, pastures, and forests. They were relatively isolated physically. The bus trip from San Juan to Uruapan took 3 hours over an extremely bad road (Secretaria de Hacienda 1940:57), and the region's only telephone had been installed in San Juan only a year or so before the eruptions. San Juan was also the only town with electricity, produced by a small generator that powered lighting for the church and government offices.

Despite their similarities, these communities were distinctive in numerous ways, including orientations toward the larger Mexican society. Some of the communities were, in a sense, preadapted to take advantage of new opportunities, whereas others were either ill prepared for, or particularly resistant to, massive socioeconomic changes. The differences rather than the similarities best explain the unique adaptation made by each community in the wake of an abrupt convergence of physical and social forces for change.

### San Juan

San Juan, the largest, most important town of the region, was *cabecera* of the *municipio* that contained all but one of the communities. Mestizo or Spanish-speaking families were present in the town by the late eighteenth century. Gradually, the percentage of Spanish speakers increased, so that by 1940 they constituted a majority of the population of 1895. However, the Tarascan-speaking 32% numbered 725 persons and constituted a community within a community. San Juan was the only town described as having two distinct ethnic groups (Gutiérrez 1972:8; Nolan 1972:74). This ethnic division between Indians and mestizos was, however, mitigated by San Juan's possession of an extremely powerful icon in the form of a crucifix called the Lord of the Miracles. The image was revered by Indian and mestizo alike and both groups valued the profits and status that resulted from being a major regional pilgrimage shrine. Meeting religious obligations was still important to the local achievement of secular power. In 1945, San Juan was a mestizo town with a significant Tarascan minority and a traditional local symbol capable of serving to rally both groups in an effort to preserve the community.

### Zirosto

The 1314 people of Zirosto had been oriented toward mestizo culture for several generations. The process began with the rise of mule driving, mule-string ownership, and moneylending in the late eighteenth century. By the early twentieth century there was a marked difference between rich and poor families in terms of access to resources and styles of living, although Spanish did not replace Tarascan as the local language until the second decade of the twentieth century. The 20% of the population that claimed to speak Tarascan in 1940 included men from the mule-driving families who used the language for trade in the Sierra, as well as older people and some women. All who discussed Zirosto claimed that there had been no ethnic difference between those who could speak Tarascan in 1940 and those who could not, although genealogical research indicates that there were a few poor and conservative families with Tarascan surnames who were more Indian in orientation than the majority of Zirosto's citizens.

Meanwhile, Zirosto had suffered economic decline as the importance of mule trade was undercut by railroads and highways. The traditional system for sponsoring fiestas was given up in the 1930s when members of the wealthy families refused to accept the burdens, and there had been an exodus of the younger members of these families. When the volcano erupted, Zirosto faced catastrophe with a nineteenth-century mestizo orientation much like that described by Wolf (1959:236–246). Compared with the other communities, Zirosto seems to have lacked community solidarity prior to the eruptions.

## Angahuan

Angahuan was the most Indian of the villages. All of its 1098 people spoke Tarascan, and many were monolinguals. Their version of Tarascan was notably different from that spoken in other places (Beals et al. 1944:9; Foster 1948:22). The people of Angahuan generally kept to themselves, and few men had ventured far from the boundaries of the community lands prior to 1943. The rule of endogamy was strong, although the village contained a few women who had married in from other communities. Two of these women, who had come to Angahuan prior to 1910, were still known locally as "the old lady from San Juan" and the "old lady from Paracho." Community lands were not sold to outsiders and people from other places were not allowed to live permanently in Angahuan. It is said that school teachers sent by the government during the 1930s usually did not remain long, largely because of local hostility.

Before the volcano, Angahuan seems to have closely fit Eric Wolf's (1955) model of the closed corporate community. It restricted its social and economic interactions with other communities to an extent that seems to have been unusual for Sierra Tarasca towns in the early 1940s.

## Parícutin

The people of Parícutin village were also Tarascan speakers but were considerably more open to the outside world than the people of Angahuan. This greater receptivity to new people and ideas was indicated by fairly frequent exogamous marriages, a desire for education, willingness to accept school teachers in the community, and an active agrarian movement. Fifteen percent of the population claimed not to speak Tarascan in 1940, although they probably did, because the Indian tongue was the language of street and household. In spite of these portents of change, the *mayordomía* was still in effect. Relationships with San Juan were poor because of a longstanding land feud with the neighboring community. As a result of actions by Parícutin's "agrarianists" that were seen as hostile and sacrilegious, much of the blame for the volcanic eruption was eventually placed on this small community.

## Zacán

Zacán's 876 people were transitional. The half who claimed to speak Tarascan were mostly older people. Men of the village had worked in other parts of the nation since the eighteenth century and some had worked in the United States in the early twentieth century. Apparently, these travelers had been welcomed back into the community, and the tales of their adventures were a part of the community life. The community decision to acculturate toward the norms of the larger society was made before the volcano. The transition from old to new orientations was symbolized in 1941 when the *mayordomía* system of sponsoring local fiestas was abandoned on the advice of a priest and the locally born head of the school, who had traveled widely and had also achieved traditional status in the community by serving as *mayordomo* for the Saint's Day fiesta.

'The traditional system demanded hard work and self-sacrifice because the *mayordomo* of a village fiesta had to serve in expensive lesser positions of ceremonial responsibility before he could hope to expend his life savings to pay most of the costs of a major fiesta. The reward was local status and a strong voice in decisions affecting the community (Wolf 1959:212; Foster 1967:206). Thus, in Zacán the focus of community service changed from the sponsoring of fiestas toward the goal of producing educated children "for Mexico." This transition had just begun when Parícutin devastated the village lands and forced adjustments, which included a period of life in Mexican towns and cities for many families. There men and women had firsthand experience with the benefits of education, and they were to bring this message back to the village.

## THE TIME OF THE ERUPTIONS

The birth of Parícutin was witnessed by an Indian farmer named Dionisio Pulido, who, in the company of his wife and a hired hand, was preparing his land for spring planting. As the ground began to heave and smoke poured out of a small hole and along a fissure that opened across the newly plowed earth, Pulido and his companions fled in terror (Foshag and González 1956:375–377). The initial eruption column was soon sighted by residents of the surrounding villages as "a simple little smoke which grew little by little into strange grey vapor silently making its course toward the southeast [Gutiérrez 1972:14]."

For the preceding 15 days, the people of the region had experienced nearly continuous earth tremors, including a few strong shocks that cracked the walls of homes and public buildings. The worry and fear with which people responded to these earthquakes were intensified by prophecies that sprang from recent sacrilegious acts. These violations of religion and custom were the front wave of a coming social change, and, in retrospect, were considered serious enough to have caused the volcano as an act of the wrath of God on a sinful people (Foshag and González 1956:372–373; Gutiérrez 1972:10).

When the eruption column appeared, some men thought it a forest fire and others suggested it was a new sawmill that would bring jobs at good wages. However, it had an unusual appearance and men were sent out to investigate from the larger towns of San Juan and Zirosto. As the eruption column grew in size, people clustered around the community leaders in the town plazas and waited for reports. The San Juan investigators, who had been blessed by the priest before setting out on their mission, returned as dusk dimmed the symmetrical forms of the old cinder cones beyond the town. The priest listened to their description of events in Pulido's cornfield, consulted a book on Vesuvius in the church library, then solemnly announced the birth of a new volcano (Foshag and González 1956:380). Word was sent from San Juan by telegraph and telephone to Uruapan and to the state governor in Morelia, the National Secretary of Defense, former president Lázaro Cárdenas, and Ávila Camacho, President of Mexico.

## Convergence

Even as many of the local people fled, scientists converged on the scene. The Mexican geologist don Ezequiel Ordóñez arrived on the evening of February 22, and was soon joined by other geologists and natural scientists representing a variety of disciplines. Although North American social scientists were in Michoacán for the purpose of studying Tarascan culture, they continued their studies in other villages. The least studied aspect of the Parícutin eruptions was the impact on the region's people. Nevertheless, the people affected by the volcano were the subject of considerable attention and were exposed to an unprecedented number of people from beyond the region. By 25 February, not only scientists, but tourists, reporters, curiosity seekers, and even peasants from other villages were arriving to witness the spectacle of Parícutin's eruption. Thus, from the very beginning of the physical catastrophe that destroyed their lands, the people affected by Parícutin volcano were also exposed to a radically changed social environment. As refugees, tourist guides, and even field research assistants, they came into contact with a great variety of people from beyond their region, especially North Americans and urban Mexicans. They became subjects of concern and planned social change for officials representing the rapidly modernizing Mexican nation. Within the first week governmental agencies began a search for vacant lands where people from the eruption zone could be relocated (Departamento de Asuntos Agrarios y Colonización No. 846).

Behavior patterns during the period of earthquakes and the early days of the eruptions were similar to those observed in connection with other catastrophic events that provide warning before direct impact. Curiosity leading to investigation and to remaining in the area to watch (Moore 1964:56; Drabek 1969:377), exaggerated fantasies in an unfamiliar situation (Janis 1962:74), and individual differences in choice of a time to leave the region or even to leave at all (Perret 1924:54; Moore 1964:36; Drabek 1969:341) are apparently common reactions. As is often found in disaster studies (Young 1954:383–391; Hill and Hansen 1962:203; Trainer et al. 1977:147–206), the family was an important unit of action and members of the extended family residing outside the impact zone provided a source of aid.

### Events of the Quitzocho Period

The period of volcanic activity that lasted from February until October 1943 began with explosive activity followed, after March 18, by a heavy fall of cineric materials. When the seasonal rains began in May, the "rain of sand" was replaced by a terrible "rain of mud." Both ruined attempts at farming over a wide area, and forests in the eruption zone began to die. The pyroclastic fall also created highly uncomfortable conditions for humans within a radius of about 25 km from the cinder cone, but particularly within the five villages in the immediate area (Foshag and González, 1956:399; Gutiérrez 1972:20).

Many of the people who first fled the eruption zone in fear soon returned home; however, within the first 3 months after the volcanic outbreak, a more permanent exodus began. It was composed mostly of those who had kinsmen in other communities or skills allowing reasonable employment in other places and/or resources that could be liquidated to pay for relocation. In other words, it was generally the richer, better-educated, more exogamous, Spanish-speaking mestizos who first left their home communities. However, some of the richer families stayed because their wealth was in land and homes that could not be disposed of without sacrifice, and they felt they had too much to lose by leaving. A few of the early migrants were among the poor but ambitious Spanish speakers who had already been thinking of leaving their villages to seek better opportunities elsewhere.

Hardly anyone left the Indian-oriented communities of Parícutin and Angahuan during this period, but by mid-June 1943, the situation in Parícutin village was desperate. Water sources had disappeared. Fields were buried to great depths under ash and, in some cases, lava. The heavy pyroclastic fall had destroyed the roof of the church and many of the homes. When lava began flowing down an arroyo in the direction of the village, government officials and geologists agreed that the town should be evacuated, even though the geologists were reasonably certain that the settlement was not in immediate danger of inundation by lava.

Accounts of the evacuation of Parícutin village by the Mexican army range from stories of people forced into government trucks at gunpoint to dramatic accounts of a peasant population gladly following General Lázaro Cárdenas to a new land, a new destiny, and a better future. All the accounts apparently contain some truth, because many of Parícutin's younger citizens did welcome new opportunities, whereas many others, especially among the older people, "did not wish to leave and . . . preferred to die covered with lava rather than abandon their homes [Gutiérrez 1972:26]." Dying under the lava was not permitted either by nature or other men. Lava did not cover Parícutin village until more than a year after its evacuation, and then only slowly.

The new settlement of Caltzonzin was 5 km from Uruapan and lay at an altitude of about 1525 m. There the formerly isolated highland Tarascans of Parícutin were given lands in *ejido*, new houses that they did not like, and new shoes that most were not accustomed to wearing. Many well-meant efforts were made to help them adjust to their new situation, but little attention was paid to some of their deepest, noneconomic needs. It would have been a simple matter to change the name of the refugee settlement to Parícutin Nuevo, but this was not done and the loss of the community name was symbolic of many other losses in the continuity of life. Culture shock stemming from too much change too fast plagued the Caltzontzin refugee settlement for more than a generation (Rees 1961:14ff; Nolan 1972:218–309).

## The Volcanic Zone

Meanwhile, in the volcanic zone conditions worsened as the reserves from the 1942 harvest were consumed and it became obvious there would be no harvest in 1943. Resin collection was no longer possible and more trees were dying. The livestock, their bellies full of pyroclastics from consumption of ash-covered vegetation, also began to die (Gutiérrez, 1972:52). Wild fruits and berries, bees, and wild game disappeared from the landscape.

Families continued to leave the sierra. Some, especially from Zacán and San Juan, went to other Mexican towns and cities. Others, primarily from Zirosto, headed for lands in the *municipio* of Ario de Rosales, where the government was reported to be planning a major refugee settlement. Salvaging the dead and dying forests, road building, and guiding tourists and scientists provided a meager subsistence for those who remained in the volcanic zone through the summer of 1943. In San Juan boys sold rocks from the volcano to visitors, and men and women established refreshment stands dangerously near the eruptions. A Red Cross station was established in San Juan in May 1943, and famine was averted through the donation of nearly 800,000 pesos worth of food and other goods by various relief agencies (Segerstrom 1950:25).

The heaviest and most destructive period of pyroclastic fall in the volcano's short history ceased on 9 June 1943, but the sands did not remain where they had fallen. They were shifted and reshifted by wind and water, deeply eroding slopes and burying some former fields to great depths.

Numerous town meetings were held in San Juan to consider relocation, but the remaining people "were not able to make the effort to uproot their hearts and take themselves to another place far away [Gutiérrez 1972:32]." General Cárdenas came in person to urge resettlement. Finally, the men of San Juan were given governmental permission to locate an acceptable site for a refugee settlement, but resistance to relocation continued.

In spite of 6 months of "hell under the shadows of the vapors, black clouds, cold rains of sticky sand, cinders and ashes [Gutiérrez 1972:33]" the people of San Juan prepared a final September fiesta for the Lord of Miracles. Pilgrims came by the thousands, from Michoacán and elsewhere in Mexico, and even from other nations. Because they suspected that the great pilgrimage shrine was doomed and that 14 September 1943 marked the last year of fiesta in San Juan, "Men and women with tears in their eyes kissed the divine feet of Our Lord of the Miracles. With loud sobs they kissed the altars and the sacred face [Gutiérrez 1972:36]."

## The Sapichu Period

October 17, 1943 to 8 January 1944 has been termed the Sapichu period by geologists, after the name given to a small subsidiary cone on the northeast base of the main cone. Eruptions from Sapichu provided a spectacle for tourists, but the main cone became so quiet that it was climbed several times during November and December 1943 (Foshag and González 1956:422).

The principal human event of the period was the founding of Miguel Silva (Figure 10.5). The search for a resettlement location in the *municipio* of Ario de Rosales, about 80 km southeast of the volcanic zone, was initiated during the first week of the eruptions (Departamento de Asuntos Agrarios y Colonización No. 846). Several Zirosto families moved into the area on their own initiative during the summer of 1943. They settled near a rancho and attempted to farm vacant lands that had belonged to the Hacienda de las Animas prior to its expropriation in 1938. On 6 October 1943, the ex-hacienda was formally selected as a refugee location and the lands were listed for settlement on 17 October. Refugees from Zirosto and San Juan were offered 2616 ha, and from late 1943 through the spring of 1944, 1000–1200 refugees poured into the settlement. Government trucks provided transportation and the Banco Nacional de Credito Ejital issued loans, although no *ejido* grants had yet been made (Departamento de Asuntos Agrarios y Colonización No. 846).

Secular leadership was composed of men from both communities. Probably 80% of Zirosto's population relocated, but the number was considerably smaller from San Juan. It is usually explained that the many Tarascan speakers of the latter community were the most resistant to resettlement, although Zirosto's fewer Tarascan speakers were represented.

Unfortunately, the hacienda lands had not been adequately surveyed, and when the survey was completed, it was found that only 350 ha of the vast area were suitable for agriculture. By this time 310 men, most with families, were

FIGURE 10.5. Miguel Silva, founded primarily by refugees from Zirosto, lies at an elevation of about 1500 m. A tradition of pioneering in a strange environment is part of the community iconography.

already there (Departamento de Asuntos Agrarios y Colonización No. 846). Obviously, the major problem with the Miguel Silva settlement was that more refugees arrived than the land could support. In early 1944, the local rancheros killed the secular leaders of the refugee settlement and threatened the priest, who was recalled to another post. They also killed the surviving livestock and otherwise waged war on the huge number of refugees who had descended upon them.

With local leadership gone, refugees from San Juan and Zirosto began to fight among themselves. The climate, although not excessively hot at an altitude of about 1500 m, was unfamiliar to these highland people. Many of the seeds they brought failed to grow. In addition, the water was bad and most of the refugees were sick from intestinal disorders and malaria. As many as one-tenth of the refugees died, especially the older people, who are said to have lacked the will to live under such conditions.

Within a year, many of the survivors left Miguel Silva, and by 1946 the population had decreased to about 300 (West 1948:23). Most of the San Juan refugees went to New San Juan after its founding in May 1944. Zirosto families either went to the United States or returned to the volcanic zone. The few from Zirosto who stayed cut themselves off emotionally from the mother community and came to view themselves as the pioneer founders of a new settlement on a difficult frontier. Many looked with disdain on those who returned to the Sierra "to eat the sands of the volcano."

## The Taqui Period

In January 1944, two new lava vents opened on the southwest base of the original cone (Foshag and González, 1956:444). The flows from these Taqui vents began on 7 January and continued until the end of July 1945, when scientific observations were temporarily discontinued. The lava flows from these vents ultimately covered San Juan.

As the wall of molten lava slowly approached in March 1944, the people remaining in San Juan reached an agreement. Although the site of a new settlement had already been chosen at Rancho Los Conejos by the men of the community, it was decided that no one would leave the town until the lava flow reached the cemetery (Gutiérrez, 1972:50–51). In April, the San Juan flow moved under a previous lava flow, and men hoped that the town would be spared. Faith that the Lord of the Miracles would save San Juan was reinforced by the belief that science would bring salvation. The latter belief stemmed from the establishment of a seismographic station by Mexican geologists in February 1944 (Gutiérrez, 1972:49). However, both the clergy and the scientists were active in the attempt to reconcile the townsmen to the eventual necessity of relocation.

On 14 April 1944, the lava flow sprang from the tunnels that had hidden its progress. By 24 April, a burst of molten lava from the 9-m-high summit of the flow reached the San Juan–Uruapan road, and the water lines to San Juan were endangered and moved. The road, which passed through a narrow valley, was as yet the only good access to San Juan for motorized traffic, and in a spurt of activity the road linking San Juan to Angahuan and thence to the Uruapan highway was completed (Gutiérrez 1972:53).

The 9-m-high wall of the main lava front continued to move at a rate of 4–5 m/hr as tongues of lava surged forth more rapidly through the pass along the old road and the steep-walled arroyo that ran along the town's eastern edge. In early May, the lava flow reached the cemetery, where it slowed in its progression to a few centimeters per hour. Families came to the foot of the lava front, and there, kneeling on the graves of their ancestors, they prayed. In respect for the March agreement, few asked that the town be spared, but only for time to dismantle their church and their homes and take with them all they could of their beloved town. They "cried out to the Lord of the Miracles offering to go with Him, even to the place in which He would make us think we should remain [Gutiérrez, 1972:54]."

On 7 May, the bishop arrived from Zamora accompanied by several priests and other church officials. The next day, the bishop celebrated a solemn mass and confirmed the local children, and an old priest arose to address the people of San Juan. He was Father Luís Gómez, who had served in the community between 1895 and 1917, and it was under his guidance that the colonial church had been dismantled and a new church begun. He spoke of his sadness in seeing the still unfinished church threatened with destruction, but he added that "the Lord God who had allowed the temple to be built was now allowing it to vanish and we should not regret it so much because that loss was nothing compared to the loss of a human being, or worse of the inhabitants of an entire town [Gutiérrez 1972:55]." When Father Gómez finished speaking, "everyone wept, women and men, and perhaps these were the last sounds of voices to be heard in that place. . . . The sound of the echo filled the space and then vibrated for a long time inside the church which soon would be destroyed by the lava [Gutiérrez 1972:55]."

Even in this emotional moment, the San Juan community was divided because there were still a few who wanted to keep the image of the Lord of the Miracles in the church in hope of a miraculous salvation from the advancing lava. To counteract this belief and avoid possible disaster, the bishop lifted the image of Christ from its place above the altar on 9 May. Accompanied by church officials and supportive townsmen, he held the image high in the beginning of a small procession. A few people threw themselves in the path of the procession, but their fellow townsmen pulled them aside and joined the march behind the most sacred symbol of their community. Soon all were following, and the people spent the night in Angahuan, where the Lord of the Miracles rested in the company of the image of Santiago, patron of that town.

On 10 May 1944, the procession continued toward Uruapan. All along the route groups of pilgrims met the people of San Juan. "They were weeping, seeing that the Holy Christ had left the town of San Juan and was being carried away without their knowledge of where He would be transferred [Gutiérrez 1972:56]." The viewers who lined the roads came out to kiss the image and give water to those who walked behind. When the refugees reached the Uruapan highway, they found great crowds of people who offered them food and water and walked beside the procession shooting off fireworks.

At the edge of Uruapan, the multitude was so great that people could hardly move as the priests of the city came out to meet the procession. "On seeing that people could no longer walk on the streets of the city, some men were asked to pass the word that no one was to move from the place he was occupying. Hundreds of men formed an arm chain in order to take Our Lord in between [Gutiérrez 1972:57–58]." The city was decorated as for fiesta and the people were shouting, "Long live Christ the King." Thus, along the route of procession, a feeling of hope built among the people of San Juan and they "felt a kind of comfort and were sure that we were traveling on a road along which Our Lord was guiding us [Gutiérrez 1972:58]."

On the third day, the people of San Juan reached the valley they had chosen as a new home. Because of their insistence on choosing a new home, they received only a town site and no agricultural lands. They had lost their position as head town of a *municipio*. Many of the richer and more prominent citizens had left the community early in the eruption period, and in the early days in the new settlement people had to do without things to which they had long been accustomed, such as schools, piped water, and electricity in the main plaza. For a time, many lived in tents, although a small chapel for the Lord of the Miracles was built immediately. The new place was called Rancho

Los Conejos, "the little hamlet of the rabbits." People in Uruapan still make jokes about San Juan of the Rabbits, but government documents indicate that the men of San Juan had the name of the settlement formally changed to San Juan Nuevo Parangaricutiro on 9 July 1944 (Departamento de Asuntos Agrarios y Colonización No. 1973). By that time, old San Juan and its church had been buried under lava. People from San Juan who had previously left for Mexican cities or had attempted pioneering at Miguel Silva soon joined the New San Juan community.

## The Bracero Program

By early 1944, a new economic opportunity emerged as men from communities affected by the volcano were urged to enlist as contract laborers in the United States. The importance of the bracero program as a stimulus for change is unquestioned. The braceros experienced life in a very different cultural setting, and the opportunities for capital accumulation were enormous. It has been estimated that during the World War II years, bracero workers could make 15 to 20 times as much as they could have earned from the same labor in Mexico (Simpson 1952:311). In contrast with the situation in most other Mexican communities, where quotas were imposed and this opportunity was open only to a few men with good connections or a lucky lottery ticket, all able-bodied men from the towns in the volcanic zone and the refugee settlements could enlist as braceros.

The communities varied in the extent to which opportunities for bracero work were accepted. Although exact figures are not available, it is probable that a large majority of the men of San Juan and Caltzontzin worked at least once as braceros. Many also went from Zacán, but there was a somewhat greater tendency for men of that community to seek wage labor in nearby Mexican cities and towns. A large number of those who had attempted to settle at Miguel Silva also enlisted as braceros, but few of those from Zirosto who had not gone first to Miguel Silva. Hardly any of Angahuan's men accepted the opportunity. This was partly due to a general reluctance to leave the community, and perhaps to some extent to the fact that Angahuan had inherited the traffic in scientists and tourists after the evacuation of San Juan. It is said locally that men could not become braceros because they could not speak much, if any, Spanish, but this is not a completely adequate reason, because numerous Tarascan monolinguals went as braceros from San Juan and Caltzontzin.

## The Later Years of Eruptions

By the volcano's second anniversary, February 1945, the period of high drama was over. Never again were the Parícutin eruptions so catastrophic or, for very long, so spectacular. The lava field was outlined by December 1944, and later eruptions added depth and buried exposed areas within the field.

People remaining in the volcanic zone tried various strategies of subsistence. They felled the forests, including areas that had not been damaged by the eruptions, and in Angahuan they guided tourists. Seed was planted in pure ash but did not survive. Ash shoveled off small plots by hand produced some crops in sheltered house lots but was not effective in the open fields because the sand drifted back over the clearings. Slopes blown or washed free from ash were put into cultivation, including high mountain lands where crops were subject to frost hazard (Segerstrom 1950:20; Eggler 1959:273; Rees 1961:202).

By 1946, the wild bees were returning to the sierra and the wild berries and crab apples near the grave of San Juan bore abundant fruits (Segerstrom 1950:22–23). Crops ripened in painstakingly uncovered soil on lands near Zacán and Zirosto, and some wheat ripened on the lands of Angahuan (Segerstrom, 1950:20). Fighting broke out between the men of Zacán and the rancho of Las Palmas over lands that Zacán had claimed for generations but that had become particularly desirable because they lay away from the zone of devastation (Departamento de Asuntos Agrarios y Colonización No. 154). In Zirosto, men began taking the treasures of the ancient monastic church to sell in the cities.

Meanwhile, adjustments were underway in the refugee settlements. Caltzontzin's citizens began rebuilding their government-designed homes to fit their tastes (Rees 1961:175). Miguel Silva got a pure-water system and favorable government intervention in the feud with the local rancheros (Departamento de Asuntos Agrarios y Colonización No. 846). The population stabilized as the survivors began to think of themselves as pioneers rather than refugees. In San Juan (Nuevo Paragaricutiro or New San Juan), the ground was blessed for the consecration of a new church for the Lord of the Miracles, shown in Figure 10.6 as it is today (Parroquia de San Juan Nuevo ca. 1970).

Life was resuming a more normal rhythm, which included the volcano's occasional periods of major activity. In 1947, it was noted that the cone had shrunk in apparent height because of the lava piled up at its base, and toward the end of 1948, the responsibility for continued scientific observation was placed in the hands of Celedonio Gutiérrez of San Juan (Fries and Gutiérrez 1950:406–418). Dionisio Pulido died in Caltzontzin in 1949, and the *ejido* grant to Miguel Silva was confirmed. Zacán found a new technique for waging war with Las Palmas by claiming that old boundaries had been obscured by the ash. In the same year a new species of grass not previously known in the region began invading areas of deep ash.

The year 1950 began with strong, fairly frequent explosions from Parícutin and a few earth tremors. Dust storms were severe during the dry spring and damaged vegetation that was moving back into the zone of devastation. Mosses, lichens, and ferns were found in moist places on the thinner lavas (Fries and Gutiérrez 1951a:212–221). The people of Angahuan were putting corn lands back into production, and as pasturage revived, the region's livestock population increased. Around Zacán, lands had been washed sufficiently free of ash to permit a return to nearly normal agriculture (Fries and Gutiérrez

FIGURE 10.6. The pilgrimage church at San Juan Nuevo Parangaricutiro. Thousands of pilgrims from central Mexico and even the United States visit the new church of the Lord of the Miracles each year. [From M. L. Nolan, "The Mexican Pilgrimage Tradition," *Pioneer America*, Vol. V, No. 2, 1973.]

1951b:572–581), although the land war with Las Palmas continued (Departamento de Asuntos Agrarios y Colonización No. 154).

In 1951 it was dry in the spring as usual. In the summer it rained. The volcano erupted occasionally. Life had, indeed, returned to normal. For some geologists, as judged from their writings, the volcano became boring because nothing new happened. For others it became more interesting because it was repetitious and therefore predictable. For Celedonio Gutiérrez, "Every day for nine years the volcano was different. The earthquakes were recorded and the march of the lavas. All this, for me, was very interesting [Nolan 1972:204]." Then, on 4 March 1952, Parícutin was dead. Spring brought damaging dust storms, but the following summer was unusually wet and many areas were washed clean of ash. The best harvests since the birth of the volcano were gathered in August 1952 (Fries and Gutiérrez, 1954:486–494). Figure 10.7 shows the progression of the lava field from December 1943 through March 1952.

### The Immediate Aftermath

For most people in the refugee towns, the end of the eruptions did not matter very much. They had long before come to terms with new destinies. Nor was it of much importance in Zacán, where agriculture was already nearly

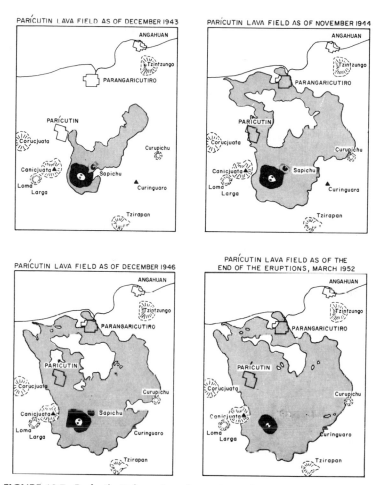

**FIGURE 10.7. Parícutin Volcano lava flows, December 1943 to March 1952.**

normal. Angahuan, however, had become dependent economically on tourists who came to see the eruptions. As word spread that the show was over, the tourist traffic virtually disappeared. Some men were finally forced to leave the town as temporary laborers because the lands were still too unproductive to support a population that had increased slightly during the eruption years. Although a new generation of tourists eventually returned to see the volcano they had read about in school textbooks, Angahuan tourist guides still look wistfully at the conical black form of Parícutin. Most probably feel what one expressed: "It would be nice if the volcano would erupt again—just a little bit."

In Zirosto, the last act of a social tragedy was played out, when the community fragmented into the spatially discrete settlements of Old and New Zirosto. Things had been going badly for a long time, and men had fought among themselves for the few available resources. In about 1949, the church of the once important colonial monastery burned (Figure 10.8), and the bapistry

**FIGURE 10.8. Ruins of the monastic church at Old Zirosto. This colonial structure burned in about 1949. The bells, long a symbol of the Zirosto community, were retained by those who stayed in the old settlement in 1953.**

**FIGURE 10.9. The main road from Uruapan to Los Reyes bisects New Zirosto, a town established on the edge of the devastation zone in 1953 by residents of Zirosto.**

was fragmented into three pieces by a lightning bolt from the eruption column. When the eruptions ended, the Mexican government encouraged the Zirosto population to relocate a few kilometers away at the site of Rancho Barranco Seca which was on the Uruapan–Los Reyes road (Figure 10.9). In return for relocation, the people were promised an *ejido* grant, a 6-year school, bus service along the road, electricity, piped water, and street lights.

A majority went to New Zirosto in 1953, but some adamantly refused to leave the old town. Brothers made different choices, placing strain on family relations and the solidarity of kinship networks. It is said in New Zirosto that the move would have been truly successful if all had gone together, taking with them the old bells that symbolized the community. In Old Zirosto, those who left are blamed. It is argued that if all had stayed, the Mexican government would eventually have rerouted the road, provided electricity, water, 6 years of school, and all the other benefits. In terms of social disruption Zirosto became, as is said locally, the town "most destroyed by the volcano [Nolan 1972:209–213]."

## COMMUNITY LIFE AFTER THE ERUPTIONS

As the years passed, plant and animal life continued its regeneration in the zone of early devastation. Slowly, plants invaded the lava fields and appeared in scattered, humble forms on the dead cinder cone. Although the lava field will remain agriculturally useless for centuries, areas of light ashfall were back in production before the eruptions ceased. Areas of heavy ashfall and places where ash had washed down on former fields, covering them to depths of several meters, remained useless for agriculture in the early 1970s (Nolan 1972:449; Rees 1970:25). However, urban entrepreneurs had discovered an economic value for pyroclastic deposits and some deep sand deposits were being mined for use in concrete manufacture.

As fertility returned to areas of medium ashfall, both naturally and through human efforts, the conditions generated by the eruptions continued to affect human life. Incidents of aggression and land wars with loss of human life became more intense in the region as formerly useless lands regained agricultural value. These conflicts were particularly difficult to settle because the eruptions had destroyed landmarks traditionally used as boundaries between one village or individual and another.

The longstanding struggle between Zacán and Rancho La Palma was resolved in 1957 when Zacán was allowed to annex the rancho and thus became politically responsible for it. Sporadic sniping continued for several more years (Departamento de Asuntos Agrarios y Colonización No. 154).

The most serious later hostilities involved New San Juan, Caltzontzin, and the people of ranchos founded by both refugee settlements on community lands in the volcanic zone. Court battles were in process by the early 1950s and incidents of aggression were reported in 1959 and 1960. In 1965, the feud intensified and reports of hired gunmen, killings, and substantial property destruction were filed with government agencies by San Juan and Caltzontzin. In 1969, representatives of the two communities agreed to let trained surveyors set the disputed boundaries. The exact toll in lives and property is impossible to estimate because the record consists largely of unsupported claim and counterclaim from the feuding communities. In April 1967, however, a government investigator documented destruction of harvest, killing of livestock, and the burning of 128 houses (casas) in small volcanic zone settlements (Departamento de Asuntos Agrarios y Colonización No. 154).

The San Juan–Caltzontzin land war clearly retarded the course of permanent resettlement in the volcanic zone. In 1971, both communities contained a number of families who had left the volcanic zone ranchos during the height of the conflict.

## The Towns of the Volcano in the Early 1970s

Although the eruptions ended in 1952, social pressures for change continued to affect the three eruption-zone settlements and the four refugee communities. In 1971, these seven communities were quite different in spite of their original similarities and their shared history of experience with a catastrophic natural event. To a considerable extent they represented a cross section of rural Mexico under the pressures of a rapidly changing socioeconomic order.

New San Juan symbolized the success story of the modernizing rural *cabecera*, or *municipío* head town, capable of retaining meaningful traditions in the course of change. It also provided a case study of one of Mexico's numerous thriving pilgrimage centers. In contrast, Old Zirosto exemplified the community that declines even as others progress. New Zirosto was the community with insufficient roots and communal purpose. Miguel Silva could be compared with many pioneer settlements that have developed on newly opened lands in recent years. Caltzontzin had its counterparts in the numerous small communities that are being transformed into urban barrios as cities expand across the countryside. Zacán was the village that gave the best of its young to the growing urban middle class and declined in the process. Angahuan was the adamantly Indian town, greatly changed, yet clinging to outward expression of tradition, a condition complicated by the interests of tourists and folk-craft developers.

The general trend of change was toward greater interaction with the larger Mexican society accompanied by loss of Tarascan traditions. This generalization, however, obscures a multitude of complex variations in the nature of the new societal adjustments.

Except in Angahuan, where everyone still spoke the Indian language, use of Tarascan was declining. No exact figures are available, but only a few of the elderly spoke Tarascan in Zacán, Miguel Silva, and the Zirostos, although a few poets and songwriters in Zacán and Miguel Silva cultivated the language for artistic purposes. Tarascan was sometimes heard on the streets of New San Juan, but one of the local priests estimated that only about 20% of the population, mostly elderly, could use the language fluently. Tarascan was more frequently heard in Caltzontzin, and was deliberately retained as a household language by several relatively affluent families, but for the most part, Spanish was the first language of the younger generation.

In 1971, fiestas were celebrated in all the communities, but the traditional *mayordomía* system survived only in New San Juan. There, the great fiesta for the Lord of the Miracles had long been sponsored by the community as a whole, but individual sponsorship of the fiesta of the town's patron saint remained a route to local power and prestige in the early 1970s. It was only one such route, however.

Traditional Tarascan dress, particularly for women, was common in Angahuan, not particularly unusual in San Juan and Caltzontzin (Figure 10.10), and either rare or nonexistent in Zacán, the Zirostos, and Miguel Silva. As might be expected, traditional housing was most notable in Angahuan, Zacán, and old Zirosto, the three settlements that remained in place in the volcanic zone.

By 1971 all settlements but Old Zirosto had federally sponsored electric power, a piped supply of potable water, and a 6-year elementary school. Only Old Zirosto lacked a post office and bus service to the center of town. Comparison between the study towns and three other Michoacán communities along

**FIGURE 10.10.** Elderly Tarascan woman in Caltzontzin, the refugee community founded by people from Parícutin village. The large wooden crosses were a distinctive culture trait in the prevolcanic community, and are also found in Caltzontzin.

the lines of "institutional differentiation" (Graves *et al.* 1969) is presented in Table 10.1.

The list of facilities and services was derived by means of Gutman scaling techniques applied by Young and Fujimoto (1965) to information contained in early 1940s ethnographic accounts of several small Mexican communities including the Michoacán towns of Quiroga (Brand 1951), Cherán (Beals 1946), and Tzintzuntzan (Foster 1948). The second point of reference for these communities comes from 1967 data collected by Graves *et al.* (1969). Data for the existence of these traits in the communities affected by the volcano in 1943 and 1971 were obtained through interviews with knowledgeable older citizens during the 1971 field season. The results of these interviews were checked by administering the same set of questions to at least two other people in each town. Probably because the volcano provided such an excellent time marker, there were few inconsistencies in response about what was present in 1943.

Comparison of communities in the 1940s indicates that Zacán, Parícutin, and Angahuan were relatively undeveloped. They also had the smallest populations. San Juan and Zirosto, in contrast, were well-developed towns for the region. Both compared favorably with the much larger community of Cherán, which had recently won the advantage of location on a paved highway.

By 1971, the item list no longer "scaled" but still served as a useful measure of comparative change. San Juan was already close to the top of this scale in 1943, as were Quiroga, Cherán, and Zirosto. San Juan, which lost many of these facilities and services during the eruption years, had regained its high position by 1971. None of the fragments of Zirosto had reached the level of the mother community in the prevolcanic period. Caltzontzin, Zacán, and Angahuan had acquired appreciably more new facilities and services during the time period involved than had Tzintzuntzan, an intensively studied mestizo community with a Tarascan heritage, located on the shores of Lake Patzcuaro.

Another comparative measure based on 24 facilities and services (Graves *et al.* 1969) allows comparison between the communities affected by Parícutin and a selection of other small Mexican towns and villages (Table 10.2). As in the first comparison, the most obvious features are the failure of any of the Zirosto-derived communities to reach the 1943 level of the mother community, and the major increases in services and facilities achieved by Caltzontzin, Zacán, and Angahuan. Prevolcanic San Juan was too close to the top of the scale to show much change along this dimension. If such items as dry-goods stores, banking facilities, and farm-supply outlets were added, the amount of change in San Juan could be better evaluated.

In 1971, the towns were still predominantly agricultural communities (Figure 10.11). According to the 1970 Mexican census (Secretaria de Industria y Comercio 1971), more than three-fourths of the work force in Angahuan, the Zirostos, and Miguel Silva was engaged in primary activities. Well over 60% of the workers in San Juan and Caltzontzin labored in field and forest. In those towns, as in Angahuan, there was a substantial amount of small industry, mostly backyard enterprises, and in Caltzontzin's case, a plant in which hand-

**TABLE 10.1**

**Comparison of the Study Towns with Other Michoacán Communities**

| Community | Population | Items[a] | | | | | | | | | | | | | |
|---|---|---|---|---|---|---|---|---|---|---|---|---|---|---|---|
| | | Autonomous and named | Elementary school | Plaza or square | Government organization | Bar or cantina | Bakery | Barber shop | Butcher shop | Resident priest | Hotel or inn | Pool hall | Resident doctor | Movie theater | Gas station |
| Quiroga | 1940:3009 | x | x | x | x | x | x | x | x | x | x | x | x | x | x |
| Quiroga | 1970:7129 | x | x | x | x | x | x | x | x | x | x | x | x | x | x |
| Old San Juan | 1940:1895 | x | x | x | x | x | x | x | x | x | x | x | x | — | — |
| New San Juan | 1970:4689 | x | x | x | x | x | x | x | x | x | x | x | x | x | — |
| Cherán | 1940:3358 | x | x | x | x | x | x | x | x | x | x | x | — | — | — |
| Cherán | 1970:7793 | x | x | x | x | x | x | x | x | x | x | x | x | x | x |
| Zirosto | 1940:1314 | x | x | x | x | ? | x | x | x | x | x | x | f | — | — |
| Old Zirosto | 1970: 434 | x | x | x | — | — | — | — | — | — | — | — | — | — | — |
| New Zirosto | 1970:1085 | x | x | — | x | — | x | x | x | — | — | f | — | — | — |
| Miguel Silva | 1970: 648 | x | x | x | x | — | — | — | x | x | — | — | — | — | — |
| Tzintzuntzan | 1940:1077 | x | x | x | x | x | x | — | — | x | — | — | — | — | — |
| Tzintzuntzan | 1970:2174 | x | x | x | x | x | x | — | — | x | — | x | — | — | — |
| Zacán | 1940: 876 | x | x | x | x | — | — | — | — | — | — | — | — | — | — |
| Zacán | 1970: 926 | x | x | x | x | — | x | x | x | — | — | x | — | — | — |
| Parícutin | 1940: 733 | x | x | x | — | — | ? | — | — | — | — | — | — | — | — |
| Caltzontzin | 1970:1295 | x | x | x | x | — | x | — | x | x | — | x | — | — | — |
| Angahuan | 1940:1098 | x | * | x | — | — | — | — | — | — | — | — | — | — | — |
| Angahuan | 1970:1762 | x | x | x | x | x | x | — | x | x | — | — | — | — | — |

[a]The 1940s item list for Quiroga, Cherán, and Tzintzuntzan was compiled by Young and Fujimoto (1965) from ethnographic accounts. The second item count for these towns comes from the 1967 data collected by Graves et al. (1969) Items for the towns of the volcano reflect condition: in 1943 and 1971 (Nolan 1972).

x = item present

? = conflicting information

f = item had recently been present

* = item present off and on

**TABLE 10.2**
**Comparison of the Study Towns with Other Mexican Communities**

| Number[a] of items | Selected Mexican towns as of 1967 (1960 population figures) | Study towns, 1943 (1940 population) | Study towns, 1971 (1970 population) |
|---|---|---|---|
| 24 | Quiroga, Michoacán (5336) | | |
| 23 | Mitla, Oaxaca (3651) | | San Juan (4698) |
| 21 | Cheran, Michoacán (5651) | | |
| 20 | | San Juan (1895) | |
| 19 | Cajititlan, Jalisco (1880) | | |
| 18 | Aldama, Guanajuato (1919) | | |
| 17 | Lagunillas, Michoacán (1981) | | |
| 16 | Yalalag, Oaxaca (3117) | Zorosto (1314) | Caltzontzin (1295) |
| 15 | San Pedro, Tobasco (1500) | | Zacan (926) and Angahuan (1762) |
| 14 | Tzintzuntzan, Michoacán (1840) | | |
| 13 | Teotitlan, Oaxaca (2849) | | |
| 12 | Amatenango, Chiapas (1832) | | New Zirosto (1085) and Miguel Silva (648) |
| 11 | Santa Rosa, Guanajuato (632) | | |
| 9 | Zangarro, Michoacán (327) | | |
| 8 | Atzompa, Oaxaca (1726) | | |
| 7 | Santa Cruz Etla, Oaxaca (613) | | Old Zirosto (434) |
| 6 | San Feli Rigo, Puebla (701) | Zacan (876) and Parícutin (733) | |
| 5 | Chachalacas, Veracruz (208) | | |
| 4 | Kikeil, Yucatan (230) | Angahuan (1098) | |
| 2 | Tatacuatitla, Hidalgo (153) | | |
| 1 | Carmen, Sinaloa (221) | | |

[a]Items drawn from Graves *et al.* (1969) are elementary school, grocery store, mass once a year, church, square or plaza, government organization, government official, public transportation, bakery, butcher shop, shipping service, newspaper delivery, barber shop, resident priest, bar or cantina, telephone, billiard parlor, movie theater, resident doctor, filling station, restaurant, hotel or inn, secular organization, and secondary school. Only the number of traits present is considered in this table.

loomed textiles were produced. The largest percentage of the work force engaged in commerce and services was in Zacán, where slightly less than half of the population was employed in the primary sector of the economy. Since there was relatively little local development of commerce and services, it seems likely that these figures describe people who maintained the community as a home base but were employed in Uruapan and elsewhere. In keeping with an old Zacán tradition, some were musicians who primarily made a living by playing in urban nightspots during the weekends. Periods of work in the United States, often without permits, were common, especially among younger men from San Juan and Zacán. Migrant labor of various kinds in other parts of Mexico also resulted in contributions to the local economies.

There had been a general trend toward population growth, as was occurring all over Mexico during the same time period. However, only the San Juan community had more than doubled in size, and none of the offshoots of

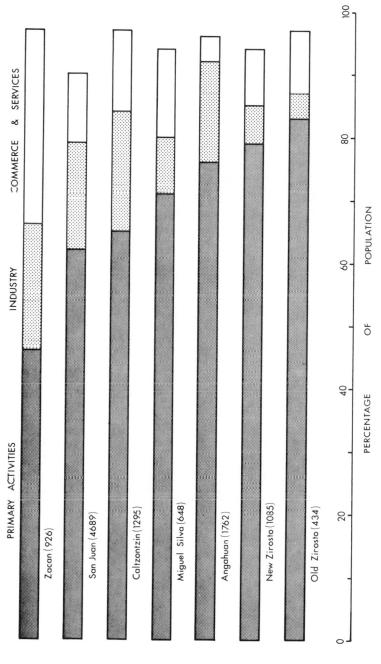

PRIMARY ACTIVITIES    INDUSTRY    COMMERCE & SERVICES

Zacan (926)

San Juan (4689)

Caltzontzin (1295)

Miguel Silva (648)

Angahuan (1762)

New Zirosto (1085)

Old Zirosto (434)

0          20          40          60          80          100

PERCENTAGE    OF    POPULATION

FIGURE 10.11. Occupational categories in the communities as of 1970.

Zirosto had reached the size of the prevolcanic town. Comparative population figures are included in Table 10.1.

In a nation like Mexico, where migration of the young from rural areas to the cities is a major means of adaptation to changing conditions and rural population pressures, the viability of the rural community is, to some degree, indicated by the out-migrants' ability to achieve positions of economic and social reward in the larger society. The custom of keeping track of professionals, common in many small Mexican communities, provides an indicator of achievement along these lines. Examination of Figure 10.12 shows that Zacán and Caltzontzin were significantly more productive of offspring with careers in teaching, medicine, law, engineering, and other professions requiring educational certification than any of the other communities. For more than a generation, Caltzontzin young people had been located a short bus ride from Uruapan schools. In 1971, Zacán children still had to leave their home community for larger towns to continue their studies beyond the sixth grade, a fact that makes the Zacán achievement more remarkable than that of Caltzontzin.

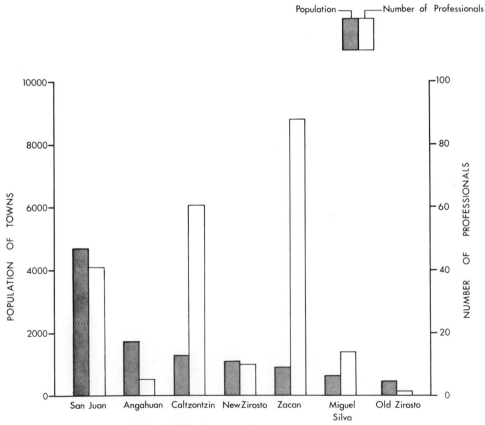

FIGURE 10.12. Number of offspring with professional positions, 1971. (Figures for Angahuan may be an underestimate.)

Educational levels of local residents also showed differences that may well influence continued viability of adjustments. Three-fourths of the adults in San Juan, Zacán, and Miguel Silva were literate, according to 1970 census figures, whereas fewer than half the adults in old Zirosto could read and write. The other communities were intermediate. Miguel Silva had the highest percentage of children aged 6–14 in elementary school, followed closely by Zacán. The lowest percentage was represented by Old Zirosto, which only had three grades of school available locally. Twenty-nine percent of the Zacán adults held degrees from primary school, whereas only 4% of the adults in the Zirostos had graduated from a 6-year educational course.

Objective attempts to measure change and current circumstances convey only part of the complex reality that geographers sometimes refer to as the "personality" of places. The uniqueness of each town was reflected in some degree by the way people spoke of their community when first questioned. Comments collected from a diverse, although not systematically sampled, variety of people carried much the same message within a town, but differed appreciably from community to community. The consistency of initial statements about past community experience and future possibilities showed no particular variation with the individual's socioeconomic status within the community, and was affected to only a slight degree by age and experience in the prevolcanic community.

In-depth interviews with selected people resulted in more complex interpretations of community but rarely were completely inconsistent with initial statements. Examination of first-reaction statements recorded in field notes suggested that there were certain things that people in each community more or less automatically expressed to outsiders on initial acquaintance. Although the extent to which these statements reflected internalized perceptions of the community, its history, and its prospects is unknown, their very consistency within a town was certain to affect the perceptions of outsiders, some of whom might make decisions that affected the community. Further research along these lines should be undertaken, because if the differences in local perception of community success in dealing with conditions created by the eruptions and present problems were as great as they seemed, the implications are important.

In general, people in San Juan and Miguel Silva expressed the most positive views of community. In both cases, these views related to a perception of past success in meeting the challenges of the eruption years. In Zacán there was a strong expression of community educational achievement, thought to have resulted from an opening of horizons during the volcanic years, but there was also concern about community future because few of the educated young could work in this small, still relatively remote community. The basic note was positive, perhaps best expressed in a local woman's statement that "we are giving our children to Mexico." In both Old and New Zirosto, emphasis was placed on the "death" of the old community, sometimes followed by a casting of blame on the other highland settlement. There was much ambivalence in Caltzontzin, but in spite of individual achievements and what seemed to be a new sense of hope and community spirit, there was a downbeat cast to many comments. Half the people spoken with almost immediately volunteered information that the town was poor, and, in contrast to New San Juan and Miguel Silva, there was a tendency to state that the old town had been better than the new. The notion that "we had to come here" was often reflected, and the positive note, if there was one, might best be paraphrased as "we endured in spite of all the things that happened to us." In Angahuan, the constant refrain was, "We are indigenous. We are Tarascans." It prefaced and/or ended many other statements about community.

## Comparisons

The two largest, richest, and most mestizo of the original communities were San Juan and Zirosto. In both cases, agricultural lands near the communities were destroyed, although the devastation of many San Juan fields by lava represents a more permanent loss of land. Both communities also had vast forest reserves, including stands of timber not badly damaged during the eruptions. The most significant difference in the experience of the two was that whereas San Juan had to be evacuated in the face of the lava flow, old Zirosto, though virtually buried in sand, was never threatened by lava; thus the choice of remaining in the home community stayed open. Accounts of growing social stress in San Juan prior to the evacuation indicate that the community may well have been saved by the lava that destroyed the town site (Gutiérrez 1972:45–46).

In any event, the San Juan community survived and eventually prospered while Zirosto fragmented into small social units. None of these pieces of Zirosto had regained either the population or the importance of the prevolcanic community. Old and New Zirosto, which exist a few kilometers apart in the highlands, were not highly viable as communities, and there was bitterness between the two settlements. In contrast, those of Zirosto who endured the hardships of pioneering in Miguel Silva were proud of their new community. They had reinterpreted the old traditions and established new ones around their common frontier experience. They were also better educated and more open to new ideas than the people of either Old or New Zirosto. However, their town was small, their land base poor, and the opportunities for growth severely limited.

The building of New San Juan was an epic story. This community of nearly 5000 people was again the head town of *municipio*. The battle for lands in *ejido* was won in 1968, and the new lands connected the town site with community lands in the volcanic zone, thus creating a geographical unit. San Juan graduated its first class from secondary school in 1971. In 1973 the road to Uruapan was paved, reducing an hour's journey by bus or car to about 10 min. Possibly because the growing community offered local opportunity for ambitious young people, educational orientations stressed literacy but not advanced degrees. Perhaps most important symbolically, the people of San

Juan had built a great brick church for the Lord of the Miracles. The community was again a major shrine visited by more pilgrims than ever before and from greater distances. Many came as before to ask for health or pay respects to the Lord of the Miracles, but increasing numbers came to witness the "Miracle of San Juan," interpreted as a miracle of achievement through faith and hard work in the face of heavy odds.

The former difference between Indian and mestizo had greatly diminished, and San Juan's traditions, old and new, were thriving and profitable. In New San Juan it was said, "We are the same town and the same people. Only the place is different." The people were exceptionally proud, resourceful, and confident of their ability to cope with new changes.

New San Juan can also be compared with Parícutin–Caltzontzin, the other group to be resettled near Uruapan. The Caltzontzin refugee settlement seemed like a perfect opportunity to make Mexicans out of Indians overnight, and little effort was spared in furthering this dream of Lázaro Cárdenas and other change agents. The people received much more government assistance than those of San Juan, and the *ejido* grant was the most generous received by any refugee settlement in terms of good farmlands relative to population. The new location on the outskirts of Uruapan provided proximity to jobs and schools. New San Juan was several kilometers more distant from the city, and the road, which ran over mountainous terrain, was extremely bad until 1973; thus San Juan's access to urban opportunities was more limited than Caltzontzin's. Nevertheless, Caltzontzin was not an especially happy community in 1971, and its people often referred to local poverty. Because the majority of Caltzontzin's citizens seemed reasonably well off for Mexican villagers, poverty seems to have meant being poorer than their urban neighbors. According to a 17-year-old girl whose father owned two brickyards, a truck, and a brick house, people in Caltzontzin were so impoverished that they could not afford lavish fifteenth-birthday parties for their daughters.

However, the true poverty that existed in Caltzontzin was more obvious than in the other communities. The obviously poor were ragged in dress, ill-housed, generally illiterate, and often unemployed, and they were particularly notable because they composed a relatively small minority of the population. The mark of poverty had little to do with retention of certain aspects of Tarascan traditionalism. Although some of the poor were among the most conservative inhabitants of the town, others had lost all roots in the Indian heritage, including the language. In contrast, some of the more prosperous families with the best-educated children spoke Tarascan at home and preferred houses modeled on the traditional style, although there was a widespread tendency to regard symbols of Tarascanness as obsolete.

Several lines of evidence suggested that this originally homogeneous community had developed a definite class structure within 28 years. It was marked by differences in dress, quality of housing, educational level, and patterns of mate selection. The symbols of class position were much like those evident in the nearby city. Uruapan people sometimes spoke of Caltzontzin as just another *barrio*, or district, of the city. The edge of Uruapan was growing ever nearer to Caltzontzin, and eventually the larger community would absorb the smaller physically as well as in social patterns.

In spite of numerous somewhat fatalistic comments about past and future community destiny, there were some strong positive notes in 1971. A new generation that had never known life in Parícutin village identified with the new town, and the period of massive culture shock that followed relocation had largely ended. A young priest who arrived in 1967 had worked diligently toward the creation of community pride and solidarity. He reinstituted an annual fiesta for the patron saint sponsored by the community at large. The celebrations stressed sports events rather than traditional dances (Figure 10.13). He raised money for a new church, but the men of the town told him they would rather have an enclosed basketball court, and he agreed. As he explained this decision, it was what the people needed. The church could be built later. Local men finished the basketball court in the summer of 1971 in time for the fiesta games. The name of the team, emblazoned on the shirts of the players, was "Parícutin."

As was the case in all the towns, Caltzontzin was plagued with too little land for its population and shortages of outside work. However, many families had encouraged their children to seek education. By 1971, 55 young people had taken advantage of scholarships and/or proximity to urban schools to win professional degrees. Most of these were schoolteachers, but among those still in school there was an increasing trend toward university educations in engineering and other practical fields. The proximity of Uruapan meant that young people with good educations did not necessarily have to leave the home region to find employment, although many were established in more distant towns and cities.

Angahuan and Zacán also offered an interesting comparison. Both were small, isolated villages before the volcano. Zacán's agricultural lands were not as badly damaged as those of Angahuan, but the latter town was richer in forest resources. In addition, Angahuan had income from tourism during the volcanic period and afterwards, whereas Zacán did not. Both communities remained in their original locations, and they looked much alike because of the traditional nature of the housing. Angahuan was an hour by road from Uruapan, whereas it took an additional 15–20 min by car to reach Zacán. Each highland community had six grades of school available locally, and in 1971 there were no secondary schools within commuting distance. Both Angahuan and Zacán had faced the same geophysical ordeal and had much the same options for response. Therefore, the differences are astonishing unless one considers the very different social orientations of these two villages in the early 1940s.

During the most difficult years of the eruptions, many Zacán families moved to nearby towns and cities, and some of the men worked as braceros. Strong ties with the home community were, however, retained. The realization of a need to educate the children grew stronger. Within 20 years, this

FIGURE 10.13. Instead of having traditional dances, the people of Caltzontzin celebrate the annual patron saint's fiesta with footraces and crowning of a fiesta queen.

FIGURE 10.14. View of Zacán, with Parícutin in the background. Although the house types remain traditional, this community produced 88 offspring with professional degrees between the early 1940s and 1971.

small, economically devastated community with six grades of school available locally, boasted 88 children with professional degrees (Figure 10.14). Many other offspring held white-collar and technical positions in Mexican cities. An organization of the professionals of Zacán, based in Uruapan, raised scholarship money for local children, and the kinsmen and friends who had established themselves in urban areas provided additional aid. Because of lack of local opportunity, most of Zacán's educated children did not return home permanently. The approximate 25% of the population that did not send children to local elementary school was inheriting the village. For others, the decline of Zacán was considered a tragedy offset by the belief that their educated children could contribute to the larger society and find good lives in the process. Zacán in 1971 appeared traditional in terms of material culture because the parents and grandparents of the educated offspring were using available funds to educate more children, rather than build new homes. It seemed likely that any family who attempted to display prosperity but refused to provide educational opportunities for offspring would meet the kinds of local sanctions that in the past induced the more affluent to sponsor fiestas.

Angahuan was considered different, closed and apart in 1943, and it retained this reputation in 1971 (Figure 10.15). The Tarascans of Angahuan met the volcanic crisis with a staunch refusal to move, a decision made possible because the town was not reached by the lava field. Very few of its men accepted the opportunity to work as braceros and almost none of its people took up temporary residence elsewhere in Mexico. When competition was eliminated by the evacuation of San Juan, Angahuan found economic salvation through catering to tourists.

An institution of tourist guiding had emerged by 1971. The guides were considered marginal by several local elders because their work did not involve traditions of obtaining sustenance from field and forest. Their primary function seemed to be a combination of allowing a flow of income into the community and ensuring the least possible contact between locals and outsiders. Tourist income primarily benefited hotelkeepers and restaurateurs in Uruapan because Angahuan had no such facilities. Even the direct sale of craft items was limited. It was mediated by the tourist guides, who carefully screened prospective buyers before taking them into homes where wares were displayed. Because crafts were not displayed on the streets, considerable revenue was probably lost to the community. Many tourists who passed through to see the volcano did not know that crafts were available locally. In 1971 one shop, owned by a local power figure, displayed the local goods, but it was tucked away on a corner off the main square. Craft entrepreneurs who sold to shop owners in Uruapan and other cities did not advertise the goods available within their house lots or display such goods in the little shops that some of them owned along the main tourist access into town. It seemed likely that powerful social sanctions were operating against an economic activity that would lead toward obvious differences in wealth, although it seemed equally likely that subjugation of craft sales to maintenance of "an image of limited good [Foster 1967:

FIGURE 10.15. School boys en route to the lava field walk through the streets of conservative Angahuan.

123–125]" would not last much longer. Mechanized carpentry had already become important and was resulting in economic differentiation, although both the activity and the differences in prosperity deriving from it could be hidden behind the high walls of the house lots.

Perhaps most critical for the future of Angahuan was the increasing commercialization of visible aspects of the Indian heritage. The town's importance as a tourist attraction near Uruapan was based on its proximity to the scene of volcanic devastation and its importance as a staging point for guides and horses. However, by the late 1960s it was being viewed by urban entrepreneurs as a quaint Tarascan village and thus a tourist attraction in its own right. The national thrust to make Mexicans out of Indians, which had contributed to social–psychological traumas in Caltzontzin, was waning. A new theme emphasized respect for indigenous traditions. Undoubtedly good as a basic idea, this emphasis was potentially damaging for the community which found its new place in the larger society as an enclave of "professional Indians" needed to meet tourist expectations. There was reason to believe that Angahuan was in danger of falling into this category.

In spite of its outward appearance of traditionalism, Angahuan had changed drastically. As Celedonio Gutiérrez of New San Juan stated it, "They have changed far more than we have." The community was linked economically to the larger society to a far greater extent than before the volcano. The underlying roots of traditional life were greatly diminished, although the exterior manifestations of house type, dress, and fiesta dances remained. The population had grown substantially and was placing a strain on the resource base. Some community lands had been sold to outsiders, who had begun capital-intensive development of these lands with the aid of employees from the village. Such people were viewed by many as patrons in the traditional sense. Thus, the first line of the "corporate community's" defense against the world beyond had been broken (Wolf 1955).

Although literacy rates and percentages of young children in schools were about average for the communities affected by the volcano, there seemed to exist relatively little insight into the value of education except as a defense mechanism against the encroaching Spanish-speaking society. As one elderly local leader put it, there was no point in much education because schools did not teach the young how to work on the land. Yet, in another conversation, the same man said there was a need in Mexico for more engineers and agricultural technicians. In contrast to those in Zacán, few parents in Angahuan seem to have grasped the idea that their children could be the engineers, teachers, and agricultural technicians of the future given a certain native ability, hard work, parental sacrifice, and good luck (Figure 10.16).

## THE PARÍCUTIN ERUPTION AS A HAZARD EVENT

Most geographical literature on human response to environmental hazards focuses on specific hazard conditions. Works such as that by Hewitt and Burton (1971) that deal with all the hazards faced by a particular people in a given place are rare. Yet people in hazard zones do not respond exclusively to events labeled "floods," "earthquakes," or "volcanic eruptions," but to the perceived totality of their constantly changing life situations. Even during the impact period of a particular hazard event, the environment may be charged with other hazardous conditions, including social hazards. Convergence of hazard events is often a factor in major disasters, and volcanic eruptions are particularly complex because they generate a complex set of environmental problems.

As is the case during any volcanic eruption, the people affected by Parícutin dealt with a large variety of hazards during the eruption years. These included hazards such as tornadolike storms, which ordinarily affect the region; hazards such as lightning and earthquakes, which were intensified during the eruption period; and hazards associated uniquely with volcanic activity, such as pyroclastic fall and lava flows. Decisions to remain in the volcanic zone or relocate were made in the context of real and perceived social and biotic hazards outside the zone, not just in response to volcanic conditions. For some people, particularly those of Angahuan, the world beyond the familiar region was seen as more threatening than the volcano, and this perception affected human action (Nolan 1972:33). The hazards of relocation in Miguel Silva were greater than those in the volcanic zone. Only three people, all struck by lightning thought to be associated with volcanic activity, were killed as a direct

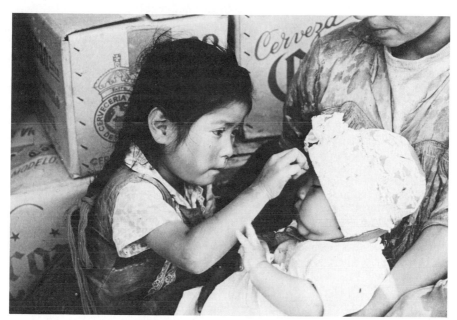

FIGURE 10.16. Angahuan children.

FIGURE 10.17. An Angahuan youth herds cattle in the volcanic zone. The rocks in the foreground are volcanic bombs. In the middle ground are cornstalks—an area that is in agricultural production. Parícutin can be seen in the distance, and beyond lies the Cerro de Tancitaro.

**TABLE 10.3**
**Hazards Faced by the Affected People during the Eruption Period**

I. Volcanic eruption

II. Hazards directly or indirectly connected with the eruptions or the products of the eruptions plus other variables
   1. *Earthquakes* in addition to those to which the region is ordinarily subject
   2. *Air pollution* from pyroclastic fall
   3. *Shifting sands*, after pyroclastic fall
   4. *Sandstorms*, especially in windy spring months
   5. *Mineral deficiencies and excesses* in the pyroclastic materials covering formerly fertile fields
   6. *Landslides* and mudslides relating to the shifting of pyroclastics
   7. *Lightning* associated with the eruption column, in addition to that associated with thunderstorms in the region
   8. *Floods* due to topographic changes resulting from the eruptions
   9. *Drought*, including failure of the regular water supply because of covering of springs with pyroclastics and changes in groundwater levels and extremely droughty soils resulting from high concentrations of volcanic sands
   10. *Agricultural frost*, always a hazard at high elevations, but greatly intensified during the eruption period by more extensive plantings in high, relatively ash-free lands

III. Other physical environmental hazards common to the region but not associated with the eruptions
   1. *Tornadolike storms*, strong enough to fell trees and deroof homes
   2. *Hail*
   3. *Fog*, which may have been more intense during the eruptions, but which did not constitute a major hazard

IV. Biological hazards recorded in the region
   1. *Insect plague*, including locust plague, which damaged crops before the eruptions and was described as the "first punishment" for the sins that produced the volcano (Gutiérrez 1972:13)
   2. *Epidemic disease*, apparently not too serious since the major cholera years of the nineteenth century
   3. *Death of game and livestock* from ingestion of vegetation covered with pyroclastics directly related to the eruptions

V. Biological hazards in Miguel Silva
   1. *Polluted water*
   2. *Malaria*
   3. *Nonviability* of highland seed corn and other plants

VI. Sociocultural hazards related to the eruptions and relocations
   1. *Culture shock* and/or serious psychological breakdown including loss of will to live, especially among the elderly (Nolan 1972:179)
   2. *Perception of the world* beyond the region as alien, perhaps hostile
   3. *Low economic viability* outside the region due in some cases to lack of literacy often combined with little or no knowledge of Spanish, and lack of training in skills other than those of subsistence agriculture
   4. *Hostility* of local rancheros at Miguel Silva, leading to at least two murders
   5. *Land wars* in the volcanic zone during and after the eruptions, resulting in loss of life and property

consequence of the eruptions (Nolan 1973:29). In contrast, the colonization of Miguel Silva took an estimated 100 lives (Rees 1970:15) from a combination of malarial conditions, polluted water supplies, loss of will to live—especially among the elderly sick—and the actions of hostile natives. The multihazard nature of the eruption period is summarized in Table 10.3, which is based on hazard lists presented by UNESCO (1970:6–7), Burton and Kates (1964:415) and Burton *et al.* (1978:21).

Early work in environmental hazard research that focused on the United States and other urban–industrial countries suggested that people in a hazard zone seldom recognize the full range of theoretically possible adjustments (Burton *et al.* 1968:11). However, the people affected by Parícutin responded with an exceptionally wide range of adjustments. Results of a major collaborative research program focused on human response to a variety of hazards in many countries (White 1974) have since indicated that peasant agriculturalists tend to show more adaptive ingenuity than the urbanites of more advanced societies (Figure 10.17). Burton *et al.* (1978) have concluded that "the pattern of folk response is a large number of adjustments and a high rate of adoptions among individuals and communities" as compared with the "favored adjustments" of modern industrial society, which tend to be "uniform in application, inflexible, and difficult to change [p. 216]." Thus, results of research in the Parícutin area that ran counter to accepted theoretical trends in the early 1970s (Nolan 1973) now fall into the category of what should be expected.

The difference in response between ethnically and economically similar communities affected by Parícutin has its parallel in a study of five Eskimo villages affected by the Alaska earthquake (Davis 1970). This study also supports the finding that an essentially religious response to catastrophe, combined with other actions, can be highly adaptive, especially during the recovery period. Old notions of peasant fatalism in the face of hazard and disaster such as those presented by Kendrick (1957:113), Kingdon-Ward (1951:130–131), and Sjoberg (1962:363) probably need revision. One must always be careful, however, to avoid confusion of past and present. Evidence that modern peasants generally are not highly fatalistic does not prove that this orientation was not widespread in the past. It is only suggestive.

There may have been prehistoric episodes of volcanic activity that pitted man against nature in situations little influenced by outside social conditions, but this was obviously not the case during the Parícutin eruptions. The convergence of social pressures for change with a physically destructive environmental event make it impossible meaningfully to distinguish sociocultural changes induced by the eruptions from changes resulting from response to the modernizing larger society. As has been found in other studies of disaster-impacted communities, the rate of change was accelerated, but the essential direction of change showed little variance from expected patterns (Kates 1977:263).

For most of the affected communities, the changes were in the direction of greater integration with the modern Mexican socioeconomic system and many would have occurred even had there been no volcano. New roads, the building of dams, penetration of a region by change agents such as school-teachers, discovery of a place by promoters of tourism, or decisions concerning location of industrial plants can also be catalysts for social change in traditional communities. There was nothing unique about Parícutin as a volcano in its impact on the eruption-zone communities except for the fact that it abruptly destroyed their traditional means of livelihood at a time when the larger society could encourage both relocation and alternative ways to make a living.

It should be emphasized that not all change was in the direction of general societal trends. The remnant population of Old Zirosto was probably more isolated from Mexican life in 1971 than was the town's population in 1943, and neither New Zirosto nor Miguel Silva had regained the mother town's prevol-canic position in terms of local facilities and services. This is not so unusual, however. Recent North American history is full of examples of communities that declined even as others progressed. Near ghost towns now stand where once-prosperous communities were bypassed by railroads and later highways. The abandoned buildings of crossroads farm-market towns attest to loss of function after the spread of the automobile carried the rural population to larger trade centers. Zirosto, with its prosperity tied to mule driving, was in a state of economic decline before the eruptions (Nolan 1972:100–104). Had there been no volcano, the community would not have fragmented, and the decline might have been reversed. But there is no way to know exactly what would have happened in any of these communities if Parícutin had not ap-peared when and where it did.

In examining change that coincides with a catastrophic event, a distinc-tion should be made between the concepts of eventual result and direct cause. Many things happened as a result of the volcano, yet hardly anything other than the destruction of farmlands and two town sites was actually *caused* by the eruptions—that is, was an inevitable consequence. The distinction between result and causation can be illustrated with the story of Manuel, the Zirosto-born principal of a school in a medium-sized Michoacán town. According to Manuel, he would not be a schoolteacher today if Parícutin had not erupted. His family was poor and lived some distance from the center of Zirosto. He had no schooling until his family moved to Miguel Silva when he was 11 years old. There he began elementary school and because of societal attention focused on the children of refugees, he received scholarships that allowed him to complete normal school. This man and many others who were children in the volcanic zone during the eruption years believe with good reason that they are school-teachers because of the volcano.

Thus, the volcano is an important part of the explanation for the course of individual lives, just as it is a critical episode in the history of seven Mexican communities. Many lives would have been different had there been no vol-cano, and the affected communities would have had a different history. Yet, as a man of New San Juan pointed out after listing the advantages of the new town site, "We would have had all those things anyway by now, in the old town. It really didn't make any difference."

## DISCUSSION AND CONCLUSIONS

### Adaptation in the Communities

The Parícutin eruptions severely damaged the traditional resource base of five farm communities. The changed physical environment forced new adaptations, which included migration and the acceptance of new ways of making a living in the volcanic zone and elsewhere. Those who continued as volcanic-zone agriculturalists developed strategies for dealing with ash-covered lands, and farmers in the refugee settlements of Caltzontzin—and particularly Miguel Silva—were forced to adapt their traditional agricultural technology to lands of lower elevation. In two communities, one in the volcanic zone and the other in a relocation site near a city, advanced education for the younger generation was an important means of dealing with new conditions. However, even in these towns only a minority of the young achieved advanced degrees, and most of the highly educated left the community.

The fact that individuals and communities responded differently to similar physical and social forces for change suggests that neither the volcano nor the larger society determined the exact nature of the new adaptations. Change occurred because people adapted to a changed environment, but the variety of adaptation was too great to be discussed in terms of simple cause-and-effect relationships. However, the idea of preadaptation is useful. Some individuals and populations are better suited to certain kinds of environmental change than are others. When change occurs, they may prosper even as those best adapted to the former environmental conditions flounder in search of appropriate new responses. Within the framework of this case study, some communities were more receptive to the new opportunities offered by a modernizing Mexico than were others (Nolan 1974:47–49).

Perhaps more important for the future is the community tradition of success or failure in controlling destiny. In this situation of forced response, some individuals and community units made deliberate choices. Others simply reacted, thus becoming pawns in the hands of nature and the larger society. At the individual level, choice reflected personality and family influence as well as the kaleidoscope of physical and social factors (Nolan 1975). Communities chose or failed to choose in accordance with their traditions, the quality of local leadership, and agreed-upon perceptions of the total situation.

The people of communities with a tradition of choice and a shared belief in the desirability of the outcome seemed to display a greater sense of potential control over future events than those of communities that had a tradition of simply coping with outside events perceived as uncontrollable.

San Juan and Caltzontzin exemplified the difference between choice and making the decision work, and lack of choice combined with continual coping. Although community differences in perception of control over destiny probably go back much earlier, the early eruption period proved a crucial testing point of local will. The lava flows represented a natural event that could not be controlled and that permitted no adaptation in place. Yet, although the flows covered the sites of two settlements, only the people of San Juan were truly forced to evacuate because of the lava. Some people of that community watched as the lava covered their house lots, then turned away to join their fellow citizens in a refugee settlement that they had insisted on choosing. When slow-moving lava eventually covered the site of Parícutin village more than a year after its evacuation, only a few scientists were present to record the event. The people of Parícutin had long before obeyed government orders to move to a strange place with an alien name, which they had not chosen. As a reward for their cooperativeness, the people of Parícutin–Caltzontzin received lands in *ejido* and much special attention. As a result of their stubbornness, the people of San Juan were initially given no *ejido* lands and much less special governmental attention.

Nearly 28 years later, people in Caltzontzin were explaining that they had to come to the new place, and some were complaining about past governmental unfairness and insufficiency of current aid. Even educating children was occasionally described as something that had to be done because the community had few resources. The ideal of educating offspring for a better way of life and as a contribution to the nation, frequently expressed in Zacán, was not very evident in Caltzontzin. As recognized by the priest and some local secular leaders, the community needed to build a more positive self-image.

In contrast, people in San Juan pointed out the good features of the new place and often emphasized their communal wisdom in choosing it. Rather than complaining about lack of government assistance, they boasted about their achievement in building the new town by themselves, although they were quick to explain how they had won concessions from larger, more powerful political entities.

It seemed that there was an important difference between the people of communities which had a tradition of control over destiny and those which had a tradition of coping, or, as in the case of the Zirostos, a history of failure to meet the challenge of the past. The greatest consideration of future community choices in the face of new problems was found in San Juan, Zacán, and Miguel Silva, the communities with the greatest apparent sense of control over past events. The significance of these microcultural differences is supported by recent experimental studies with human subjects and animals that indicate that even the illusion of control improves performance in several kinds of situations (Perlmuter and Monty 1977:759).

### Archaeological Implications

The archaeologist must often deal with only remnants of material culture and evidence of past physical environmental conditions. Unless historical records are also available, extrapolations must be made from that base. Assuming that no historical documentation existed, future archaeological explorations in the Parícutin volcanic zone would show that two settlements had been overwhelmed by lava. With adequate funding for surveys, archaeologists could probably establish much the same zone of devastation defined by geologists in the 1940s. Because no human remains would be found in the volcanic layer, it could be assumed that people from the lava-covered settlements had relocated. Perhaps the sparsity of material remains around the stone churches, house foundations, and walls of those former settlements would suggest that the migrants took most of their possessions with them, which was the case.

Presumably, however, the few things left behind would indicate the general level of technology at the time of the eruptions. New settlements on layers of heavy ash, such as the ranchos established by people originally from the region as the lands became more productive, would show in their remains an increased level of technological sophistication. An observer viewing the remains of a rancho burned during the 1960s land wars commented on sadness over the loss of "hand-operated tortilla presses, transistor radios, sewing machines . . . national flags and sports equipment [Departamento de Asuntos Agarios y Colonización No. 1973]." It would be evident that major changes in material culture had occurred between the onset of the eruptions and the period of recolonization; however, more extensive investigations would show the same radios, sewing machines, and sports equipment appearing at about the same time in other sierra communities far beyond the eruption zone. Examination of only the volcanic zone could lead to speculations about migrations of new peoples, but a broader survey would suggest that the eruptions occurred during a period of rapid and extensive culture change.

This case study, based on written and oral accounts collected largely from individuals who witnessed the eruptions of Parícutin, may provide insights that can help in the interpretation of material cultural evidence. As a general rule, the Parícutin case suggests that volcanic eruptions do cause change in patterns of human life and habitation, but that the eruption events do not determine what the changes will be. It may also be taken as a generalization that similar groups of people affected by the same physical event may adjust to changed conditions in quite different ways and that the perceived success or failure of choices made during times of crisis will affect future decisions and the ongoing course of community history.

## REFERENCES

Beals, R.
  1946  Cherán: A Sierra Tarascan village. Washington: Smithsonian Institution. Institute of Social Anthropology, No. 2.
Beals, Ralph R., Pedro Carrasco, and T. McCorkle
  1944  Houses and house use of the Sierra Tarascans. Washington: Smithsonian Institution. Institute of Social Anthropology, No. 1.
Brand, D.
  1951  Quiroga: A Mexican municipio. Washington: Smithsonian Institution. Institute of Social Anthropology, No. 11.

Bullard, F. M.
  1962  *Volcanoes in history, in theory, in eruption*. Austin: University of Texas Press.
Burton, I., and R. W. Kates
  1964  The perception of natural hazards in resource management. *Natural Resource Journal*
        3:412–441.
Burton, I., R. W. Kates, and G. F. White
  1968  The human ecology of extreme geophysical events. Working Paper No. 1. Toronto:
        University of Toronto, Natural Hazard Research.
  1978  *The environment as hazard*. New York: Oxford University Press.
Davis, N. Y.
  1970  The role of the Russian Orthodox Church in five Pacific Eskimo villages as revealed by
        the earthquake. In National Research Council *The great Alaska earthquake of 1964:*
        *Human Ecology*. Washington: National Academy of Sciences.
Departamento de Asuntos Agarios y Colonización
  1973  Morelia: Archives Nos. 154,846, and 1973.
Dorf, E.
  1945  Observations on the preservation of plants in the Parícutin area. *American Geophysical*
        *Union Transactions* 26:257–260.
Drabek, T. E.
  1969  Social processes in disaster: Family evacuation. *Social Problems* 16:336–349.
Eggler, W. A.
  1959  Manner of invasion of volcanic deposits by plants with further evidence from Parícutin
        and Jorullo. *Ecological Monographs* 29(3):268–284.
Foster, G. M.
  1948  Empire's children: The people of Tzintzuntzan. Washington: Smithsonian Institution,
        Institute of Social Anthropology, No. 13.
  1967  *Tzintzuntzan: Mexican peasants in a changing world*. Boston: Little, Brown.
Foshag, W. F., and Jenaro González R.
  1956  Birth and development of Parícutin Volcano, Mexico. Geological Survey Bulletin 965-D.
        Washington, D.C.: U.S. Government Printing Office.
Fries, C. Jr., and C. Gutiérrez
  1950  Activity of Parícutin Volcano from August 1, 1948 to June 30, 1949. *American Geophysi-*
        *cal Union Transactions* 3:406–418.
  1951a Activity of Parícutin Volcano from January 1 to June 30, 1950. *American Geophysical*
        *Union Transactions* 32:212–221.
  1951b Activity of Parícutin Volcano from July 1 to December 31, 1950. *American Geophysical*
        *Union Transactions* 32:572–81.
  1954  Activity of Parícutin Volcano in 1952. *American Geophysical Union Transactions*
        35:486–494.
Graves, T. D., N. B. Graves, and M. J. Kobrin
  1969  Historical inferences from Guttman scales: The return of age-area magic. *Current An-*
        *thropology* 10:317–338.
Gutiérrez, C.
  1972  A narrative of human response to natural disaster: The eruption of Parícutin. In *San Juan*
        *Nuevo Parangaricutiro: Memories of past years*, edited by Mary Lee Nolan, trans. by
        Carlos Monsanto. Environmental Quality Note No. 06. College Station: Texas A & M
        University.
Hewitt, K., and I. Burton
  1971  *The hazardousness of a place: A regional ecology of damaging events*. University of
        Toronto, Geography Department Research Publication No. 6. Toronto: University of
        Toronto Press.
Hill, R., and D. A. Hansen
  1962  Families in disaster. In *Man and society in disaster*, edited by George W. Baker and
        Dwight W. Chapman. New York: Basic Books.

Janis, I. L.
  1962   Psychological effects of warnings. In *Man and society in disaster*, edited by George W. Baker and Dwight W. Chapman. New York: Basic Books.
Kates, R. W.
  1977   Major insights: A summary and recommendations. In *Reconstruction following disaster*, edited by J. Eugene Haas, Robert W. Kates, and Martyn J. Bowden. Cambridge, Mass.: The MIT Press.
Kendrick, T. D.
  1957   *The Lisbon earthquake.* Philadelphia: Lippincott.
Kingdon-Ward, Capt. F.
  1951   Notes on the Assam earthquake. *Nature* 167:130–131.
Moore, H. E.
  1964   *And the winds blew.* Hogg Foundation for Mental Health. Austin: University of Texas Press.
Nolan, M. L.
  1972   The towns of the volcano: A study of the human consequences of the eruption of Parícutin Volcano. Doctoral dissertation, Dept. of Geography, Texas A&M University.
  1973   *Research on disaster and environmental hazard as viewed from the perspective of response to the eruption of the Volcano Parícutin in Michoacán, Mexico.* Environmental Quality Note No. 14. College Station: Texas A&M University Environmental Quality Program.
  1974   The reality of difference between small communities in Michoacán, Mexico. *American Anthropologist* 76:47–49.
  1975   Familes originating in the Sierra Tarasca: Variations in modernization. Paper presented at the Second Tarascan Symposium: The Tarascan/Non-Tarascan Interface, Seventy-fourth Annual Meeting of the American Anthropological Association, San Francisco.
Parroquia de San Juan Nuevo
ca.1970   Datos recopilados por la Parroquia de San Juan Nuevo con el fin de publicar proximamente la historia del Señor de los Milagros y su pueblo. Notice distributed by the Parroquia de San Juan Nuevo.
Perlmuter, L. C., and R. A. Monty
  1977   The importance of perceived control: Fact or fantasy. *American Scientist* 65:759–765.
Perret, F. A.
  1924   *The Vesuvius eruption of 1906.* Washington, D.C.: Carnegie Institution.
Rees, J. D.
  1961   Changes in Tarascan settlement and economy related to the eruption of Parícutin: Master's thesis, Dept. of Geography, University of California, L.A.
  1970   Parícutin revisited: A review of man's attempts to adapt to ecological changes resulting from volcanic catastrophe. *Geographical Forum* 4:7–25.
Secretaría de la Economía Nacional
  1943   Estados Unidos Mexicanos sexto censo de población, 1940. Michoacán, Mexico D. F.: Secretaria de la Economía Nacional, Dirección General de Estadistica.
Secretaria de Hacienda y Credito Publico
  1940   Estudios Historico–economico–fiscales sobre los estados de la republica: Vol. III. Michoacán, Mexico, D. F.: Secretaria de Hacienda y Credito Publico.
Secretaria de Industria y Comercio
  1971   IX censo general de población, 1970. Estado de Michoacán, Mexico D. F.: Secretaria de Industria y Comercio, Dirección General de Estadistica.
Segerstrom, K.
  1950   Erosion studies at Parícutin, state of Michoacán, Mexico. Geological Survey Bulletin 965-A. Washington, D.C.: U.S. Government Printing Office.
Simpson, L. B.
  1952   *Many Mexicos.* Berkeley: University of California Press.
Sjoberg, G.
  1962   Disasters and social change. In *Man and society in disaster*, edited by George W. Baker and Dwight W. Chapman. New York: Basic Books.

Trainer, P. B., R. Bolin, and R. Ramos
   1977   Reestablishing homes and jobs: Families. In *Reconstruction following disaster*, edited by
          J. Eugene Haas, Robert W. Kates, and Martyn J. Bowden. Cambridge, Mass.: The MIT
          Press.
UNESCO Seminar of Natural Hazards
   1970   Summary report. Godollo, Hungary: UNESCO.
West, R. C.
   1948   Cultural geography of the modern Tarascan area. Smithsonian Institution, Institute of
          Social Anthropology No. 7.
White, G. F. (Ed.)
   1974   *Natural hazards: Local, national, global*. New York: Oxford University Press.
Williams, H.
   1950   Volcanoes of the Parícutin region, Mexico. U.S. Geological Survey Bulletin, 965-B.
          Washington, D.C.: U.S. Government Printing Office.
Wolf, E. R.
   1955   Types of Latin American peasantry: Preliminary discussion. *American Anthropologist*
          57:452–469.
   1959   *Sons of the shaking earth: The people of Mexico and Guatemala—Their land, history and
          culture*. Chicago: Phoenix Books, The University of Chicago Press.
Young, M.
   1954   The role of the extended family in a disaster. *Human Relations* 7:383–391.
Young, F. W. and I. Fujimoto
   1965   Social differentiation in Latin American communities. *Economic Development and Cul-
          tural Change* 13:344–352.

**11**

# The Significance of Volcanism in the Prehistory of Subarctic Northwest North America

WILLIAM B. WORKMAN

## INTRODUCTION

Frequently about the North Pacific coast of Alaska one encounters in archaeological sites and stratigraphic sections bands of volcanic ash, mute testimony to recurring episodes of Holocene volcanism. Similar evidence is encountered, albeit more rarely, in the vast forested interior of Alaska and northwest Canada. Consideration of available archaeological evidence probably leads us to underrate the frequency of past volcanic eruptions, since thin ash deposits may have been dispersed by past human activity on archaeological sites and may thus go unrecognized. Older ashes weather quickly to deposits superficially resembling clay in wet coastal sites (Wilcox 1959:461).

This chapter will evaluate the probable ecological and cultural–historical significance of past volcanism in several contrasting environmental zones. After considering the history of research on this topic we will proceed to a discussion of general characteristics of Alaskan volcanism, which may serve as an introduction to the regional studies that follow. In these regional sketches we will characterize the general nature of Holocene cultural adaptations, the uses made of volcanic products, and ethnohistoric traditions (if any) relating to past and contemporary volcanism in the forested interior of Northwest Canada, the western Alaska Peninsula, and the eastern Aleutians. We will then discuss the distribution of volcanoes and geological and historical evidence for their activity, combining this evidence with what archaeological and environmental evidence we have suggestive of the probable environmental impact of

339

these past volcanic events (often drastic but almost invariably short-lived), and their impact on regional culture history. We will conclude by comparing and contrasting the impact of volcanism on these environmental zones and diverse indigenous food economies and will offer suggestions about profitable lines of future research. It should be noted here that studies of this nature are in their infancy in our area and that logic and historic precedent are often given more weight than hard field evidence in reaching our tentative conclusions. This chapter is, and is intended to be, suggestive rather than conclusive.

## HISTORY OF RESEARCH

Much of the face of southern Alaska has been shaped over long ages by volcanic activity. Early travelers and sojourners were much impressed by the many active volcanoes in the Aleutian Islands and on the Alaska Peninsula (Veniaminov 1840a:29ff). There is a voluminous geological literature on Alaskan volcanism considered primarily from a stratigraphic and lithological point of view (Péwé 1975:11, 78–80), but relatively little attention has been paid to the impact of volcanism on the living world of which man is a part. Thus Nowak (1968) has treated volcanic ashfalls on the eastern Alaska Peninsula almost entirely from a stratigraphic point of view, whereas Lerbekmo and Campbell (1969) and Lerbekmo et al. (1975) have treated the prehistoric White River Ash of northwest Canada in useful detail but without consideration of its possible ecological significance. Souther (1970) has provided a summary of volcanism in interior British Columbia with some discussion of its probable impact on the aboriginal inhabitants. Further afield, stratigraphic considerations appear primary in reviews of volcanic ash studies in North America by Wilcox (1965) and Westgate et al. (1970), and in Japan by Kotani (1969).

Although studies of historic volcanic activity have focused on volcanoes in areas possessing agricultural economies, we do have a detailed pioneering study of the devastating Katmai eruption of 1912 by Griggs and his colleagues (Griggs 1922). To my knowledge, the literature of volcanism contains no study of the effects of volcanic eruptions in a subarctic terrestrial boreal environment.

Although the major emphasis to date has been on stratigraphic correlations of volcanic ashes and their chronological potential, the geologist Ray Wilcox published a paper in 1959 that contains much useful information about the impact of historic volcanic ashfalls, and the geologist Harold Malde included a brief but very useful discussion of the cultural impact of volcanism in a paper published in 1964. Robert Black, also a geologist, has dealt in a series of papers with the role of geological processes (including volcanism) in the population history of the Aleutian Islands (Black 1974, 1975, 1976).

With the partial exception of the Aleutian Islands, only passing attention has been paid by northern archaeologists to the possible significance of volcanism in the environmental or cultural history of their research areas. In

1974, I published a paper in which I considered the probable effect of the emplacement of the eastern lobe of the White River Ash on the lives of the prehistoric Indian inhabitants of northwest Canada and suggested the possibility that this event might be linked to the southern dispersion of ancestral speakers of Southern and Pacific Athapaskan languages (Workman 1974). Working independently, the late David Derry reached somewhat similar conclusions in a paper presented at the same 1972 symposium but not published until 1975. I undertook the present chapter in part because I was encouraged by results of similar, if more sophisticated and field-data-oriented work in other areas (e.g., Sheets 1976), in part to evaluate rather nebulous comparative conclusions drawn in the original paper (1974:246), and to determine if this particular line of inquiry is worth following further in the north. The activities of Black and others in the Aleutians provide modest encouragement for the view that an ecological consideration of past volcanism and the attempt to relate it to northern culture history is worthwhile, if fraught with difficulties.

## GENERAL CHARACTERISTICS AND CONSEQUENCES OF ALASKAN VOLCANISM

Although our consideration of volcanism in the boreal interior takes us into northwest Canada, all volcanoes considered in this report are located within the boundary of the modern state of Alaska. Most of Alaska's young volcanoes occur along a 2500-km arc between Mount Spurr, located about 150 km west of Anchorage, and Buldir Island in the western Aleutians. The majority are found on the Alaska Peninsula and in its island continuation, the Aleutian Archipelago. At least 60 volcanic mountains in this arc have been active in geologically recent times with at least 40 showing signs of activity in historic times (see Figure 11.1). About 10% of the world's identified volcanoes are located in Alaska (Alaska Geographic Society 1976:8). Geologically young volcanoes also occur in the Wrangell Mountains of south central Alaska; Mount Edgecombe, near Sitka in southeastern Alaska, though currently inactive, erupted with great violence in the early Holocene. Only Mount Wrangell appears to be active in the Wrangell Mountains (Miller 1976:28), but the source of the White River Ash was located nearby in the St. Elias Mountains. Certain lava flows on the Seward Peninsula are also geologically recent, but since they represent another type of volcanism than the Alaska norm and are geographically isolated, they will not be considered here.

Alaskan volcanoes characteristically erupt with great violence, distributing pumice, ash, and other pyroclastic debris over large areas (Wilcox 1959:415; Miller 1976:26). They appear to be of the type characterized by Tazieff (1971:59–60) as acid explosive, typified by relatively cool, highly viscous magma that virtually guarantees violent and spectacular explosions. Toon and Pollack (1977:14) note that Pacific volcanoes usually have viscous magmas resulting in explosive displays such as those exemplified by Krakatoa in Indonesia, which

1. Afognak Island, Kodiak Group
2. Akun Island, Eastern Aleutians
3. Akutan Island, Eastern Aleutians
4. Alsek River, southwestern Yukon Territory, southeastern Alaska
5. Anchorage, Upper Cook Inlet
6. Aniakchak Volcano, Alaska Peninsula
7. Bogoslof Island, Eastern Aleutians
8. Buldir Island, Western Aleutians
9. Chernabura Island, western Alaska Peninsula
10. Deer Island, western Alaska Peninsula
11. Dezadeash Valley, southwestern Yukon Territory
12. English Bay, Kenai Peninsula
13. False Pass, westernmost Alaska Peninsula
14. Fisher Caldera, Unimak Island, Eastern Aleutians
15. Izembek Lagoon, western Alaska Peninsula
16. Kaflia Bay, eastern Alaska Peninsula
17. Kamishak Bay, eastern Alaska Peninsula
18. Katmai Village, eastern Alaska Peninsula
19. Katmai Volcano, eastern Alaska Peninsula
20. Kodiak (town), Kodiak Island
21. Lake Iliamna, eastern Alaska Peninsula
22. Mount Augustine (opposite tip of Kenai Peninsula)
23. Mount Edgecombe, southeastern Alaska
24. Mount Spurr, Upper Cook Inlet
25. Mount Wrangell, Wrangell Mountains
26. Makushin Volcano, Unalaska Island, Eastern Aleutians
27. Port Moller, western Alaska Peninsula
28. Samalga Pass, west of Umnak Island, Eastern Aleutians
29. Sanak Island, southeast of Unimak Island, Eastern Aleutians
30. Savanoski village, eastern Alaska Peninsula
31. Shishaldin Volcano, Unimak Island, Eastern Aleutians
32. Unga Island, western Alaska Peninsula
33. Veniaminof Volcano, eastern Alaska Peninsula

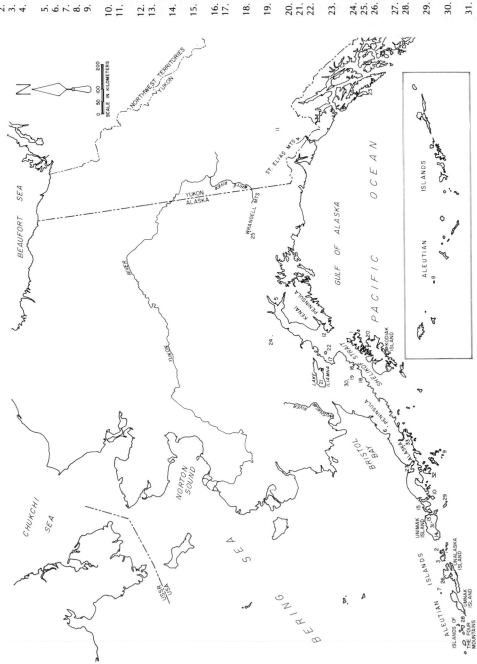

FIGURE 11.1. Index map of selected localities mentioned in this chapter.

threw ash 50 km into the air on its eruption in 1883. Major Alaskan eruptions often appear to be of the Plinian eruptive type (Macdonald 1972:236), characterized by the explosive emission of at least tens of cubic kilometers of magmas resulting in ash and pumice rains and eventual collapse into calderas. Glowing avalanches often occur as well, being associated with 17 of 49 major Quaternary eruptions on the Alaska Peninsula. Major glowing avalanches may radiate as far as 50 km from the source, annihilating all life even in seemingly protected areas far from the volcano (Miller and Smith 1977:174; Kienle and Forbes 1977:46). Lava flows may or may not accompany such explosive eruptions.

These violent pyroclastic eruptions have certain features and impacts in common, regardless of the specific environments in which they occur. Thus the relatively minor eruption of Mount Spurr in July of 1953 presented a truly appalling spectacle as viewed by military pilots. Convulsed by explosions, the mountain generated a mushroom cloud 18,000 to 21,000 m high and 50 km wide at its widest point in 40 min after initial observation. Lightning flashes occurred about every 30 sec on first observation, increasing to about every 3 sec as the enormous cloud of ash and gases began to disperse downwind (Wilcox 1959:421).

Volcanic ash, a hallmark of Alaskan volcanism, is really stone dust created in highly viscous magmas by the shattering of trapped gas bubbles, leaving glass splinters suspended in gas (Tazieff 1971:82). Clouds of ash disperse very quickly with the prevailing winds, with speeds between 24 and 97 km/hr on record (Macdonald 1972:137). Thunder, lightning, and rain commonly accompany the cloud on its dispersion downwind and torrential rains and sometimes hailstorms are to be expected (Wilcox 1959:427, 441, 451). It should be noted that thunder and lightning are almost never experienced under normal conditions along the Alaska coast and occur relatively rarely in the interior, adding a dimension of novelty to the terror of being immersed in an ash cloud. Total darkness, which may endure for days, also accompanies the ash cloud. The town of Kodiak experienced two days and three nights of almost unbroken darkness as the ash plume from Katmai Volcano deposited a foot of ash there in 1912 (Griggs 1922:10). The nature of this experience is indicated by Griggs in the following terms. "None of those who went through those days of terror fail to mention . . . the aweful darkness which is universally described as something so far beyond the darkness of the blackest night that it cannot be comprehended by those who have not experienced it [p. 10]." As an example it was indicated that it was almost impossible to see a lantern held at arm's length.

Various noxious substances fall out of ash clouds in the form of "acid rains" that may cause damage to plants and irritation to animals and men several hundred kilometers from the source. Analysis suggests that a major irritant is sulfuric acid, although different eruptions from the same vent vary significantly in acid content (Wilcox 1959:427, 467; Cadle and Mroz 1978:456; Hobbs et al. 1978). Griggs reports the experiences of a man who had difficulty in traveling on the Alaska Peninsula 6 months after the main Katmai eruption, suffering acid damage to skin and clothing from minor emissions of smoke and fumes. Dogs in the same area were reported to have been blinded by the acid rains (Griggs 1922:21). Ash-induced damage to the eyes, sometimes resulting in blindness for animals, is mentioned in other reports as well (Jaggar 1945:87–88; Freeman 1977:22). Once fallen to earth, the ash would be repeatedly redistributed by the winds or stirred up by animals and humans walking across ash-laden terrain. Although assuming that short-term exposure to volcanic dust would not be dangerous (presumably to the lungs), Wilcox suggests that long-term exposure might have a more harmful effect. There are informal reports of head, throat, and lung pains associated with the fumes and dust of the Katmai eruption (Jaggar 1945:88), and at least one consumptive died on Kodiak during the eruption (Freeman 1977:22). Surely constant exposure to glassy shards of ash in one's clothing, one's dwelling, and one's food would be a fact of life for several years following the deposition of any significant quantity of ash. It seems reasonable to suppose that lung problems, eye problems, and skin problems might well increase in an area for several years after a significant ashfall.

Significant ashfalls would contaminate water sources for a time, both through physical contamination with ash particles and through a temporary rise in acidity due to the adherence of noxious substances to the ash grains. Increase in acidity appears to be short-lived, although contamination can be reinforced by posteruptive runoff from ash-laden terrain (Whetstone 1955). Contamination of drinking water for even a few days would raise serious problems for man and beast alike (Wilcox 1959:445; Malde 1964:9; Macdonald 1972:426). Griggs quotes an eyewitness account from a temporary resident of Kaflia Bay about 50 km from the source of the Katmai eruption: "It is terrible, and we are expecting death at any moment, and we have no water. All the rivers are covered with ashes. Just ashes mixed with water [p. 19]."

One of the most significant short-term impacts of volcanic ashfalls is the intensification of erosion, landslides, and mudslides. In principle, intensified erosion is the consequence of any significant ashfall, with the danger persisting for some time after the event. Heavy rains are almost inevitable during or after the ashfall and are said almost to guarantee mudflows and floods. These dire consequences are rooted in the fact that ground covered with porous ash near the source can hold only a fraction of the runoff it could contain with its normal vegetational mat. Water saturates the interstices in coarse ash, providing lubrication on slopes and triggering landslides and mudflows. Finer deposits farther downwind are less porous and permeable, accelerating runoff and erosion. Floods following an ashfall carry a great suspended load, enhancing their ability to accelerate gully cutting and slope erosion, move boulders that unsaturated waters had left in place, etc. (Wilcox 1959:430, 449; Malde 1964:9–10; Workman 1974:249).

Ash falling on snow during the winter months may cause disastrous "early springs" with intensified flooding. Significant ashfalls on snow also interfere greatly with travel by sled or snowshoes (Suslov 1961:392–394). Under some conditions deep ashfalls retard rather than accelerate the melting of underlying snow, however, and sometimes, by insulating them, preserve snowbanks for several years at low elevations (Griggs 1922:149). When considering ashfalls on terrestrial environments in particular, it would be most useful to determine the season of the year during which the ash fell and whether or not there was snow cover.

It is recognized that in the long run volcanic ashfalls enhance the growth of plants in the affected area (see Veniaminov 1840a:216 for an early statement of this view; Griggs 1922:45ff for a detailed description of the process of recovery after Katmai). With equal fairness it can be stated that the short-term impact is disastrous.

Volcanic eruptions emit great quantities of noxious fumes such as sulfur dioxide, fluorine, and chlorine, which may do great damage to the vegetation as far downwind as 32 km from the source. This effect is greater in warmer climes, but evergreen trees apparently are more susceptible to damage from fumes than deciduous tress (Malde 1964:7). By adhering to the ash particles, these gases travel farther and can damage vegetation at great distances (Wilcox 1959:415), searing vegetation locally even farther downwind than where significant deposits of ash are formed (Griggs 1922:25). Wilcox (1959:451ff) has catalogued the various factors damaging to vegetation. These include mechanical overloading and breakage of broad-leaf plants and trees, smothering of vegetation by falls of ash 5 cm or greater in thickness, destruction by ash-induced hailstorms, mudslides, floods, etc., and weakening of overloaded trees, making them more susceptible to later attacks by insects and disease. The cutting effects of windblown ash are recognized as a strong deterrent to growth of young plants as well (Wilcox 1959:463). Suslov notes that changes in the drainage patterns of subarctic soils induced by ashfalls may change the local floral composition for some years (1961:395). Streams may be filled with ash, creating dangerous quicksand deposits (Ball 1914:62; Evermann 1914:65; Freeman 1977:23). Thus, in the short-term extensive volcanism is disastrous to terrestrial plants over vast areas, creating biological deserts or impoverished environments and causing significant difficulties for herbivorous mammals and the ultimate predator, man.

Ashfalls also appear to have significant if short-term impact upon freshwater ecosystems. Humans would feel this impact insofar as it affected the supply of lake, stream, and anadromous fish. Refined paleoenvironmental studies at Lost Trail Bog in Montana have detected a significant short-term reduction in two genera of lacustrine algae that is attributed to the emplacement of the Mazama ashfall (Mehringer et al. 1977:260). Although there were no base-line studies for comparison, streams and lakes on the western side of Afognak Island were found to be virtually devoid of invertebrate fish food in the summer following the Katmai eruption. Sticklebacks and salmon fingerlings were found to be starving in the lakes (Ball 1914:62–63).

In areas burdened by heavy ashfalls, streams, rivers and small ponds would be choked for years with ash, which might be hard on the delicate gill structures of fish (Ball 1914:62; Eicher and Rounsefell 1957:72; Malde 1964:11; Souther 1970:63). On Kodiak after the Katmai eruption lakes up to 150 cm deep disappeared entirely (Freeman 1977:22). One also wonders whether the increase in water acidity noted above might not be damaging in small ponds and streams, even if these abnormal conditions lasted only a few hours. Large lakes would probably feel less effect, but there was little aboriginal exploitation of deep-water lacustrine resources, and many economically significant fish spend at least part of their life cycle in streams (Workman 1974:250).

Red salmon were beginning their annual run when Katmai erupted in June 1912. Salmon already in the streams of Afognak Island stayed there until they suffocated with their gills filled with liquid mud. About 4000 perished in the Litnik Stream Hatchery. Rains kept the lakes and streams muddy, preventing or delaying salmon from reaching their spawning grounds. A heavy rain in mid-August put so much ash in the water that salmon suffocated as the June run had. In this later run salmon were observed ascending the polluted streams a short way, going back to sea, then trying to ascend again, repeating this erratic movement a number of times (Ball 1914:61, 62, 64).

Clearly the salmon run of 1912 was drastically affected by the Katmai eruption. By 1913 the streams were largely scoured clear of ash, and an unusually large number of young silver (cohoe) salmon were noted in all streams. However, drastic reduction of the previous year's red salmon run was confirmed by the very small number of year-old fingerlings taken in tests in the Litnik area (Evermann 1914:65). Eicher and Rounsefell have pointed out that red salmon, whose fry spend a year or more in freshwater lakes, would be particularly vulnerable to volcanic impact on their freshwater habitat. Their study indicated that the Katmai disaster of 1912 was clearly reflected in small runs (roughly half normal size) of sockeyes between 1916 and 1920 in three Afognak Island streams (1957:70, 73). Rapid full recovery thereafter led them to the conclusion that in the long term significant ashfalls enhance red salmon productivity by injecting new nutrients into the aquatic ecosystem, a finding closely paralleling the conclusions about the long-term impact of volcanism on terrestrial vegetation. Some confirmation for this suggestion comes from the study of an ash-impacted lake in Kamchatka (Kurenkov 1966).

Not all fish life would necessarily be eliminated even in heavily impacted areas. Griggs noted the somewhat anomalous occurrence of Dolly Varden char in 1917 in a stream near Katmai that had experienced a 46-cm ashfall. Further study indicated that these fish had been spawned in 1915, so Griggs surmises that their ancestors must have escaped destruction in deep pools. Neighboring brooks were devoid of fish, and one wonders what the survivors ate unless they retreated to sea. The same author noted an anomalous run of red salmon in a stream without a lake outlet in 1917, but in 1919 no salmon were observed there (Griggs 1922:61–63).

It has long been suspected that cataclysmic volcanism, by injecting vast amounts of volcanic dust and sulfuric acid into the stratosphere, might be implicated in at least short-term climatic cooling. Detailed consideration of this possibility lies beyond the scope of this chapter, but it should be noted that recent studies indicate that large volcanic eruptions may cool the surface of the earth by 1–2°F for several years, although such phenomena must be short-lived and cannot be held responsible in isolation for significant long-term climatic cooling (Macdonald 1972:136; Kennett and Thunell 1975:501; Bray 1977; Toon and Pollack 1977; Hein et al. 1978:140–141).

When considering the ecological impact of volcanic ashfalls one needs to determine if possible the minimum thickness of ash sufficient to impact the environment. Malde noted that after the Hekla eruption of 1947 in Iceland pastureland covered by as little as 1.9 cm of ash was noted to harm livestock (Malde 1964:9). Wilcox has indicated that grasses and low-lying plants may be smothered by as little as 5 cm of ash where falls of much greater thickness can weight down and smother low bushes (1959:451). After the Katmai eruption small birds, waterfowl, and small mammals died and caribou sickened in an area near Lake Iliamna that had received a fall of 2.5–10 cm of ash (Jagger 1945:88). In my earlier study I concluded that the "some effect" zone could be extended out as far as the 2.5-cm isopach (Workman 1974:248–249).

Studies of historic volcanism indicate that in many cases the major damage and suffering may well be more psychological than physical, caused by panic and fear of the unknown (Wilcox 1959:441–442; Workman 1974:248). The Katmai eruption of 1912, in which not a single life was lost, is a case in point (Griggs 1922:7ff). If this is the case among people fortified in part by a scientific understanding of the phenomena in question, one might well suppose that the effects would have been even more drastic in prehistoric times, especially where volcanic outbursts were relatively rare. With the exception of Griggs's thorough discussion of the fear and suffering accompanying the Katmai eruption, the only ethnographic reference I have seen pertains to a very minor fall of volcanic ash upon the territory of a plateau fishing and gathering people, the Sanpoil-Nespelem. An informant told Ray (1954):

> When my grandmother was a small girl a heavy rain of white ashes fell. The people called it snow. . . . The ashes fell several inches deep all along the Columbia and far along both sides. Everybody was so badly scared that the whole summer was spent in praying. The people danced—something they never did except in winter. They didn't gather any food but what they had to live on. That winter many people starved to death [p. 108].

While this isolated instance cannot be pressed too far, it is suggestive of a possible impact that is only indirectly related to environmental considerations.

A further issue regarding native traditions deserves some consideration. Though it has been demonstrated that the oral traditions of certain northern peoples can transmit accurate information for some centuries (Laguna 1958), one wonders how far into the past such accurate cultural memory might extend. Evidence reviewed below suggests that the emplacement of the East Lobe of the White River Ash some 12 centuries ago has been totally forgotten.

## VOLCANISM IN THE BOREAL INTERIOR:
## THE WHITE RIVER ASH

The area affected by the emplacement of the East Lobe of the White River Ash (Figure 11.2) includes the historic territory of the Han (Osgood 1971), the poorly known Northern Tutchone, some Southern Tutchone (McClellan 1975), and some Upper Tanana Athapaskans (McKennan 1959). The basic ethnographic pattern, which can be extrapolated into the prehistoric past, is that of thinly scattered small groups of highly mobile hunters and fishermen lacking access to significant salmon runs. Big game animals such as moose, caribou, and mountain sheep supplemented by various smaller mammals, limited vegetal resources, birds, and freshwater fish provided the subsistence base. Various rough estimates of population density range between one individual per 250 km² (99 mi²) and one individual per 100 km² (40 mi²) (Workman 1974:260).

Volcanic activity is an unusual occurrence in this area, with two substantial pyroclastic eruptions in the first millennium A.D. providing the only record of Holocene volcanism. Use of volcanic products was confined to obsidian and other volcanic stones. Probably no Indians were close enough to the source area in the rugged and inaccessible eastern St. Elias Mountains to be affected directly by possible short-range effects (clouds of glowing ash, etc.) of these two catastrophic explosions; hence all discussion of ecological impact hinges on the

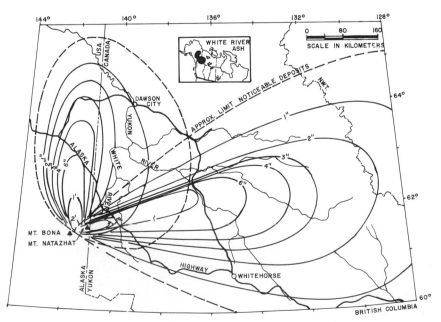

**FIGURE 11.2. Isopach map of the distribution of the White River Ash. (After Lerbekmo et al. 1975.)**

consequence of associated ashfalls. Discussion is limited here to the emplacement of the larger East Lobe of the White River Ash.

With one dubious exception, no local traditions seem to convey even a dim memory of this cataclysmic event of 12 centuries ago. Indeed, southwest Yukon Indians appear not to have speculated on the significance of the striking band of white volcanic ash that is encountered everywhere in roadcuts and natural exposures. A fragmentary tradition of the Han recorded by Osgood (1971:39) refers to a supernatural being called "fire man who lives in a house built of white clay and stone. In the house is a small table with a big hole behind it." Though the atypical architecture (white clay = white ash?) and the association with a fire being are suggestive, this fragment was recorded totally out of context. It is as likely to refer to experiences with the thick ash and pumice deposits of the upper White River country that lies in the territory Osgood attributes to the Han as it is to reflect some dim memory of 1200-year-old volcanism.

Recent field and laboratory studies by Lerbekmo and his colleagues have refined our knowledge of the White River volcanic events first brought to the attention of the scientific world by Capps over 60 years ago (Capps 1915; Lerbekmo and Campbell 1969; Lerbekmo et al. 1975). The White River Ash is a bilobate formation covering well over 250,000 km², mainly in the Yukon Territory of Canada. The East Lobe has an axis of about 1000 km. A conservative estimate indicates that the volume of this ejecta is at least 25 km³ (6 mi³) (Lerbekmo and Campbell 1969:109–110, 115; Lerbekmo et al. 1975:204). Study of the properties of the ash confirms other evidence suggesting that the East Lobe eruption was a very violent short-lived event (Lerbekmo et al. 1975:208). The source has been localized to within a few kilometers of a pumice mound on the northeast slope of Mount Bona about 24 km west of the International Boundary in the St. Elias Mountains. About 90% of this area is under permanent ice fields today, and Lerbekmo and his colleagues feel that the vent of origin for the White River Ash is almost certainly buried beneath the ice (1975:204). Although I used a 1400 B.P. date for the emplacement of the East Lobe of the White River Ash in my 1974 paper, new radiocarbon evidence suggests a span between 1175 and 1390 B.P., with a mean of around 1250 years B.P. Nine new radiocarbon dates indicate that the North Lobe was placed around 1890 years ago (range 1750–2005 B.P.), thus confirming a separation of several centuries between these events (Lerbekmo et al. 1975).

Review of the available paleoenvironmental evidence (Workman 1974: 243–245; 1978:26ff) indicates that the world on which the ash fell differed in some significant details from that of the present day. An ice dam across the Alsek River had impounded a large neoglacial lake that covered much of what is now the floor of Dezadeash Valley in southwest Yukon. The lake would have removed a significant area of pastureland, and the ice dam would have cut off access of salmon from the Pacific Ocean to the far interior. Other large lakes still in existence today may have been somewhat enlarged at the time the ash fell. The white spruce forest was spreading outward from the streams and water courses in southwest Yukon, but field evidence (Johnson and Raup 1964:111–116) indicates that spruce forest had not come to dominate as it does today and that extensive areas were still grassland. Revision of the dating of the White River Ash (see above) would place this event in the relatively mild Scandic climatic episode rather than, as previously suggested, at the end of the possibly more severe sub-Atlantic (Bryson and Wendland 1967:280, 294; Workman 1974:246).

Professor Lerbekmo has recently shared with me information based on work done by his former student Lawrence Hanson that suggests that the East Lobe eruption occurred in winter, perhaps in early winter (personal communication, April 1977). High-atmosphere winds generally blow eastward in January in this area today, swinging around to the north by July. Assuming that the same pattern prevailed in the past, one might suggest that the probabilities favored a winter emplacement of the East Lobe and a summer emplacement of the earlier North Lobe. In their 1969 publication Lerbekmo and Campbell noted the anomalous occurrence of thick deposits of ash on slopes up to 40° (p. 110). This observation coupled with a strongly bimodal distribution of ash particle size (Hanson 1965) led Lerbekmo and Hanson to the conclusion that the ash may have served as nucleation points for condensing water vapor. The observed stability of ash deposits on slopes leads to the interpretation that it may well have been snowing during the eruption, with the ash compacting under a snow load rather than being washed off steep slopes in a torrential rain. Preservation of ash on steep slopes might even raise the possibility that the ash accompanied an early winter snowfall, ensuring maximum preservation during the spring thaw.

Winter emplacement, strongly suggested by this new evidence, would have wrought maximum hardship upon the Indian hunters upon whom the ash fell, effectively immobilizing them in the ash-laden dark at a time of year when food was always scarce. Several days of immersion in the inky darkness of an ash cloud at a time of year when daylight was already at a premium might have added to the psychological impact of the event as well (Workman 1974:247, 249).

Little field evidence bears on the ecological impact of this ash. Stumps of trees killed by 60–150 cm of ash have been reported by Rampton near Klutlan Glacier (1971:976). The few pollen sections that incorporate the White River Ash reflect no significant vegetal changes. The area needs to be restudied using the refined techniques recently developed by Mehringer et al. to detect short-term vegetation changes (1977). The distribution of the ash clearly indicates that the prevailing wind at the time of the eruption blew steadily to the east away from the inhospitable ice-clad peaks of the St. Elias Mountains and across territory inhabited by prehistoric man. We can legitimately infer that the huge plume of ash was accompanied by torrential rains or snow, noxious fumes, etc. Hypothesized intensified erosion after the ash fell should be detectable by geological studies, but to my knowledge none have been made with this particular orientation. The lightning-riven unnatural darkness caused by the ash cloud would have been an appalling spectacle, especially in this area where volcanic outbursts were far from common.

Consideration of the impact of the ash on the living world, especially the fauna, is inferential, based largely on studies of the impact of historic volcanism on domestic livestock. It should be remembered that most domestic livestock are grazers (as is the caribou) while the moose, an important food resource in this area, is a browser. Ashfalls have a serious impact on domestic herbivores. Cattle grazing on ash-laden pastures often sicken, bloat, and die (Wilcox 1959:455), and fluorine adhering to ash damages the teeth and joints of sheep so that eventually they are unable to chew or walk. Fluorine-contaminated ash killed about 15,000 ewes and over 6000 lambs (of about 95,000 exposed sheep) after the 1970 eruption of Hekla Volcano in Iceland (Bauer 1971). Bad effects have been noted from an ash cover as thin as 1.9 cm (Malde 1964:9). Griggs suggests without totally satisfactory documentation that the teeth of both moose and caribou were so worn by ash adhering to their plant food after the Katmai eruption that many are said to have perished through inability to feed properly (1922:314). A similar statement was made by a witness to the May 1931 eruption of Aniakchak Volcano on the Alaska Peninsula with reference to the caribou of the Nushagak River area (Trowbridge 1976:73). Trowbridge also notes that this spring eruption caused the Nushagak herd to abandon their ash-laden calving grounds, leaving the newly born calves to die. The food supply itself is also adversely affected by ashfall. Reindeer moss is notoriously slow growing and probably would be affected by minimal ashfalls and regenerate only very slowly, although windswept upland pastures might be freed in a timely fashion by wind scouring. Berries, a seasonally significant food resource, appear to have been greatly affected by the Katmai eruption (Ball 1914:63; Freeman 1977:22). Moose, as browsers, would probably be less affected by light ashfalls than caribou but might be most vulnerable to the effects of ash contamination on the vegetation of small ponds, which appear to provide a significant portion of their summer diet.

Since waterfowl were a significant, if minor, resource in our area, we should note in closing Trowbridge's (1976:73) report that after the Aniakchak eruption the rivers of the Alaska Peninsula contained the bodies of geese, ducks, swans, and other birds that were said to have been killed by ingesting ash. This assertion, though plausible, is not documented. Ptarmigan, a year-round food resource that lives, nests, and feeds on the ground, appears to have been particularly hard hit by the Katmai eruption (Ball 1914:62–63; Freeman 1977:22).

Only the southern portion of the area affected by the East Lobe of the White River Ash has been studied in enough detail to permit the formulation of preliminary regional archaeological sequences (Johnson and Raup 1964; Mac-Neish 1964; Workman 1978, 1977). MacNeish postulated a cultural break and lack of cultural continuity between his final pre-ash Taye Lake phase of culture and the post-ash Aishihik phase (1964:322). Elsewhere he devoted a single sentence to the possible environmental significance of the volcanic ash

(1964:304), but he did not attempt to link these two observations. A restudy of some of MacNeish's material and additional information has led me to the interpretation of strong cultural continuity between the Taye Lake and Aishihik phases, despite the considerable ecological impact that I am prepared to attribute to the emplacement of the White River Ash (Workman 1974:250).

In this study I concluded that territory covered by as little as 2.5 cm of ash would have undergone considerable reduction in carrying capacity and that the affected area was large enough that a significant number of human beings (between 60 and 1000; see Workman 1974:260) would have been displaced for a time. Emigration would have been either in a northerly or southerly direction. It is doubtful that out-migration on this scale could have been accommodated peacefully by neighboring peoples, so I suggested the likelihood that there was strife and discord following the ashfall as various small groups jockeyed for position, with some groups peripheral to the area affected by the ashfall perhaps being displaced, to continue the process of disruption of equilibrium as they sought new homes elsewhere.

I view the emplacement of the White River Ash as a likely triggering mechanism that may ultimately have detached certain small groups of ancestral Athapaskans to the south to become the ancestors of the Navajos and Apaches, and the Pacific Athapaskans (see Workman 1974:254–255 for details). I was encouraged in this view by tentative (and inherently controversial) glottochronological estimates suggesting that ancestral speakers of Southern Athapaskan left their northern homeland about 1200 years ago, whereas ancestral speakers of Pacific Athapaskan departed slightly later from a different staging area or areas (Krauss 1972:159 and personal communication, January 1973). Recent revision of the date for the East Lobe from about 1400 to about 1250 B.P. would seem to bring these events into even closer alignment.

I consider this a reasonable idea but one that cannot be regarded as confirmed on archaeological grounds (see Workman 1978:430), although data from the southern Brooks Range as interpreted by Derry (1975) and from the Chilcotin area in British Columbia as interpreted by Wilmeth (1977) are compatible with the general framework. I suggest that, if this idea is correct, further archaeological research on areas peripheral to the ashfall and perhaps far beyond should reveal considerable evidence for culture change and intensified cultural contacts in the second half of the first millennium A.D. as archaeological indicators of a time of flux and small-scale population movements. My interpretation of cultural continuity across the time of the great ashfall suggests that the area was not depopulated for any lengthy span of years and that we must search for subtler archaeological evidence, which I have had only limited success in finding in the sparse record from southwest Yukon. I persist both in my interpretation of the East Lobe of the White River Ash as an event of potential cultural historical significance and in my desire to see studies specifically designed to test the viability of this hypothesis as a significant problem in paleoenvironmental interpretation.

## VOLCANISM IN A MIXED TERRESTRIAL AND MARINE ECONOMIC SETTING: THE WESTERN ALASKA PENINSULA

Ideally the western Alaska Peninsula, an active volcanic area, should afford an opportunity to study indigenous subsistence economies transitional between the purely terrestrial and the fully maritime under volcanic stress. Unfortunately there are only minimal data to work with, and I must confine myself to a few remarks based on a hitherto untranslated Russian report (Veniaminov 1840a).

The Alaska Peninsula east to Port Moller was occupied ethnographically by the Aleut (Dumond 1974), whose way of life has never been described in any detail. Veniaminov notes that even in later prehistoric times the western peninsula was sparsely populated by comparison with the Aleutian Islands, with at best only 10 villages, most on the southern or Pacific shore. In 1836 there were only three settlements harboring a total of 210 persons (Veniaminov 1840a:236). Ecologically Unimak Islanders should be included in this transitional category, since they had access to terrestrial resources such as caribou absent in the islands farther to the west, but we know little specific about either Unimak Aleut ethnography or archaeology.

The archaeology of the Alaska Peninsula west of Port Moller has been but little studied, the only substantial work being that of McCartney (1974b) at Izembek Lagoon located about 40 km east of False Pass on the Bering Sea shore of the peninsula. No large sites such as those of the Aleutians proper were found in the Izembek survey area. Only one volcanic ash 7–10 cm thick was noted in the stratigraphic sections (1974b:63).

In Veniaminov's account of historic Aleutian volcanism he indicates that volcanoes on Unimak spread volcanic ash as far as the Alaska Peninsula. There also were sporadically active volcanoes on the peninsula itself (1840a:29ff). Interestingly, for our purposes, he attributes the rapid decline of caribou on the peninsula to ash originating in volcanic eruptions on Unimak in 1825 and 1826. In particular, he notes that the eruption of 1825 caused the caribou to withdraw into the interior of the peninsula followed by wolves and bears (1840a:70) and that caribou, formerly abundant on Deer Island, had by 1828 been virtually exterminated by volcanic eruptions and the crossing over on ice of wolves from the mainland (1840a:247–250). He also notes that in 1823 a pair of swine were introduced on Chernabura Island. By 1825 more than 20 swine were present as a result of natural increase, but over 100 swine (sic) perished in 1826 because of ashfalls and the cold, exterminating the population (1840a:74ff). These statements cannot be taken entirely at face value since other factors such as increased hunting pressures or (in the case of the swine) unsuitability to the environment may have played a role, but Veniaminov was very positive in correlating the timing of the events in question with various ashfalls, and, in discussing other species, he includes variables such as over-hunting and pollution in his explanations of population decline.

It appears that the western Alaska Peninsula was underutilized in ethno-graphic times. Further studies of this area might well focus on an evaluation of the real, as opposed to the apparent, wealth of the ecosystem. Endemic volcanism (including periodic "fallout" from active volcanoes on Unimak Is-land to the west) should be considered as a factor in this evaluation.

## VOLCANISM IN A MARITIME ZONE: THE EASTERN ALEUTIANS

The Aleuts were a populous island folk who subsisted almost entirely on products of the sea. Marine mammals (including large whales), halibut and cod, a variety of birds, salmon, and intertidal invertebrates were mainstays in the traditional economy, which has recently been described in considerable detail by McCartney (1975). Facets of Aleut traditional culture have been described by the great naturalist Ivan Veniaminov from firsthand observation over a period of years (1840a, 1840b) and in secondary compilations by Hrdlička (1945) and Lantis (1970). In the Aleutian Islands a rich environment and a sophisticated extractive technology meshed to permit growth of large popula-tions and a stable village life. Here we will consider primarily the eastern islands of the Aleutian Archipelago, with special reference to Umnak Island, for which a rich archaeological and geological record exists.

The Aleuts made fairly extensive use of volcanic products, including a variety of volcanic stones for flaked stone implements and pumice and scoria for abraders. Perhaps the most significant stone material, because it was rare and highly prized, was obsidian or volcanic glass. A major Aleutian obsidian source is found on the northwest slope of Okmok Caldera on northeastern Umnak (Denniston 1966:90; Laughlin 1972:2; Black 1976:11). Veniaminov indi-cates that Okmok Caldera was the only source in the Aleutians (1840b:94–95), but he indicates in his geographic observations (1840a:43, 193) that obsidian also occurs in huge chunks tumbled down a cliff near three springs on south-eastern Akutan. Conflict over obsidian was a cause of war and discord in protohistoric times in the eastern Aleutians (Veniaminov 1840b:94–95).

The Aleuts used sulfur as a fire starter, apparently collecting it at or near the vents of active volcanoes (Veniaminov 1840a:135, 160–162, 193, 195; Jochelson 1933:7, 11). They also used hot springs, which occur near volcanoes, to cook fish, sea mammals, and edible roots in plaited bags. They thought the hot springs were healthful and many were fond of food cooked in them. Ac-cording to Veniaminov, in aboriginal times they bathed only in cold water, but Jochelson quotes a 1750 account indicating that bathing as well as cooking was done in hot springs in the western Aleutians, indicating either regional varia-tion (Veniaminov's data apply almost entirely to the eastern Aleutians) or early evidence for acculturation to Russian ways (Veniaminov 1840a:135–137, 149;

1840b:114; Jochelson 1933:7, 10). Aleut traditions pertaining to volcanoes are said to deal mainly with specific volcanoes (Veniaminov 1840a:29ff). Veniaminov characterized the traditions as "interesting but not very satisfactory." In another context he describes a tradition dealing with a war between the volcanoes of Unalaska and Umnak during which "they erupted fire, rock and ashes in such quantity that all animals living near them perished [1840b:276–277]." From this it appears that volcanoes were seen as sensate beings and that the Aleuts were familiar with their destructive power.

The eastern Aleutians have been volcanically active in historic times, with volcanoes on all the major and some of the smaller islands. The volcanoes of Unimak were most active in early historic times (Veniaminov 1840a:29ff), but, as we shall see, Umnak has had a colorful and eventful past. The area is also seismically active. Veniaminov records no less than three earthquakes a year on Unalaska during the decade he spent there (1840a:29). Because the Aleutians are volcanic islands with a large population and limited shoreline area suitable for villages, the ancient Aleuts had little choice but to live in sometimes uncomfortable proximity to active or dormant volcanoes. For this reason they were likely to suffer from a variety of eruptive phenomena (meltwater floods, glowing avalanches, lava flows, etc.) of limited geographic extent. Since they always lived near the sea, we must also consider the impact of tsunamis.

In his consideration of Aleut settlements, McCartney has noted (1974a:118) that "Sites are relatively rare on islands made up largely of volcanoes and evidencing extreme past volcanism." He suggests that volcanism limits human settlement in two ways: by the steep topography it creates and by the destructive impact of lava flows, ashfalls, etc. He estimates that only 5–10% (240–480 km) of the total coastline of the Aleutians was available for primary human settlement.

Because of the proximity of volcanoes, the Aleut upon occasion had early warning of forthcoming eruptions. Thus Veniaminov notes that earthquakes were more common and stronger before a volcanic eruption on Unimak in 1825 (1840a:29ff). That the Aleut heeded such warnings is indicated by the case of Shishaldin Village on Unimak (Veniaminov 1840a:219). Aboriginally located on a small cove, with the coming of the Russians it was moved 6–8 km east to a location on Shishaldin River. In Veniaminov's time it was moved back to its aboriginal location, which had a less convenient harbor and a poorer subsistence base, for fear of Shishaldin Volcano, which was smoking mightily at the time. On the Alaska Peninsula strong earthquakes were felt at Katmai Village 32 km from Katmai Volcano 5 days before its cataclysmic eruption in 1912. People became so frightened that they moved 32 km down the coast. Villagers at Savonski 32 km to the northeast also became frightened but did not act on their fears until after the eruption began (Wilcox 1959:417). It is difficult to evaluate the effectiveness of the Aleut early warning system.

The maritime Aleut would have been extremely vulnerable to tsunamis generated by volcanism, although as Macdonald notes (1972:414), energy of volcanically induced waves is not usually enough for them to travel great distances as tectonically induced waves do. Veniaminov notes a terrible "flood" that struck the island of Unga in July 1788 (1840a:29ff) and killed many Aleuts, although he appears justifiably skeptical of a reported wave height of about 106.5 m (50 *sazhen* or 300 ft). He reports without providing details another flood on Sanak Island in 1790 (1840a:25ff). Black notes that tsunamis have carried driftwood up to 30 m above sea level in places along the Pacific shore of Umnak Island (1974:136). In 1957 many low-lying sites on the Pacific shores of this island were inundated by tsunamis as well (Black 1975:167). A tsunami generated by the 1883 eruption of Mount Augustine struck the community of English Bay 80 km east on the Kenai Peninsula, causing great damage to houses and boats. This was a wave 8–10 m high, but fortunately it came at low tide and there was no loss of life (Miller 1976:18). Turner and his colleagues have described evidence suggesting that a tsunami could have accompanied deposition of a volcanic ash at the Chaluka site on Umnak Island about 3000 years ago, inundating the settlement (1974:139; see pages 363–364 of this chapter).

A further danger to those living in proximity to erupting volcanoes is catastrophic floods caused by the melting of mountain snow and ice cover and the drainage of caldera lakes. A 1795 eruption on southwest Unimak caused the perpetual ice cap to slide down both sides of the mountain with much water and lava. Revegetation of the affected area had just started in the 1820s. An eruption on northeastern Unimak in March 1825 created a "horrifying" river of meltwater from ice and snow 4.8–11.3 km (5–10 *versts* or 3–7 mi) wide. Although the eruption was in spring, the sea remained muddy until late fall (Veniaminov 1840a:29ff). The affected area would obviously undergo a change in its productivity from such an inundation, which would probably annihilate all intertidal life.

Turning to geological evidence, we note that a large lake formed in Okmok Caldera drained catastrophically sometime after caldera formation in the early Holocene and before the present glaciers formed. This catastrophic drainage cut a gorge 150 m deep in the side of the mountain (Black 1974:133; 1975:163) and presumably was accompanied by a huge flood. Rechesnoi Volcano on Umnak extruded lava about 3000 years ago, a time when neoglacial ice was in existence (Black 1976:13).

A further danger to people living in proximity to exploding volcanoes would have been glowing avalanches. Okmok Volcano deposited up to 30 m of pyroclastic debris at the coast about 9 km from the caldera rim in the early stages of its cataclysmic outburst in the early Holocene. Lithological studies indicate that this glowing avalanche was accompanied by melting snow and ice cover on the mountain. This event would have literally sterilized northern Unmak Island (Black 1975:163; see pp. 360–361 of this chapter). Fisher Caldera on Unimak erupted in similar fashion at about the same time (Miller and Smith 1977:74).

We must now consider the probable consequences of ashfalls that are well documented in the archaeological record from Umnak Island. At the blade site on Anangula Island near Nikolski Village on southwestern Umnak Island, a shallow pond, one of two freshwater sources on the island, was largely filled by volcanic ash (Ash III; see pp. 360–361), which overlies the occupation (Black 1976:23). Presumably this water source would have been unavailable to the occupants of the site during the eruption and for some time thereafter. It would seem that the acid rains that routinely accompany ashfalls might have been especially hard on the covers of skin boats on which the people depended for transportation, including flight. Construction of these boats represented a substantial investment of effort, and they could not easily be replaced, especially under crisis conditions. In the case of major eruptions, floating pumice might also be thick enough upon the sea to damage skin boats or impede their progress. Macdonald notes that after the eruption of Krakatoa, pumice piled in rafts up to 3 m thick so that modern ships with engines experienced great difficulties forcing their way through it (1972:239). Jaggar (1945:85) reports that pumice rafts thick enough to support a man formed in Shelikof Strait after the Katmai eruption of 1912. A year later patches of floating pumice persisted. Pumice windrows along the beaches reached 60–90 cm in depth (Ball 1914:61). These hindrances, coupled with the total darkness that accompanies many ashfalls, might sometimes have rendered flight impossible until the eruption was over.

Little information exists on specific effects of volcanic explosions on marine ecosystems, which is of course the fundamental question in considering the significance of past volcanism in the Aleutians (see Brongersma-Sanders 1957:974–977 for a brief catalog of volcanic and related impacts on marine environments). The Katmai eruption of 1912 provided an opportunity for such study, but Griggs and his co-workers seem to have concentrated almost exclusively on the terrestrial impact (1922). Several impacts on sea mammals can, however, be predicted with some confidence. Since sea mammals need to surface periodically to breathe and are subject to shocks transmitted through water, it is reasonable to suppose that they might withdraw for a time from the vicinity of active eruptions and ash- and pumice-laden seas. The seals and sea lions would be sensitive to the behavior of the fish on which they prey and would abandon those areas left by them. Sea otters, with their special dependency on sea urchins and kelp beds, would suffer greatly from factors adversely affecting these resources, and their preference for fairly shallow waters might render them especially vulnerable to tsunamis. Since most sea mammals haul up on land to breed and have their young, it seems reasonable to suppose that significant deposits of ash in their rookeries during the breeding season would be an annoyance to them and might cause abandonment of areas or of young. Their sensitivity in this regard should not be overrated, however, since a colony of sea lions was noted hauled up on Bogoslof just a year after the eruption of 1883 had enlarged the island. They apparently avoided the "new land," however (Alaska Geographic Society 1976:49).

Ashfalls appear to be very damaging to birds who attempt to find escape through flight in the ash-laden dark. Mariners enveloped in ash clouds during the Katmai eruption noted birds of all species falling helpless on deck, with many of their companions presumably perishing in the sea (Griggs 1922:11; Alaska Geographic Society 1976:40). Significant ashfalls during the nesting season might affect nesting behavior adversely (Ball 1914:61–62, 64). Sea birds are likely to frequent cliffy shores safe from human or other predators where they might be overwhelmed by floods, glowing avalanches, etc. emanating from nearby volcanoes. Myriad puffins were noted perching in rocks and clefts on Bogoslof a year after its eruption, although it is not specified that they were nesting there (Alaska Geographic Society 1976:49). Birds are highly mobile and could easily abandon an affected area, as they apparently did after Katmai. Migratory waterfowl may bypass an impacted area or terminate their stay abruptly (Ball 1914:61, 64).

Volcanism and associated submarine explosions appear to have an adverse impact on certain economically significant marine fishes. Veniaminov notes that no cod were taken in Unalaska in 1825 and 1826, although previously they had been available in large numbers. An Unimak volcano had erupted in the winter of 1825 and great quantities of cod and sculpins floated to the surface in stunned and half-dead condition, presumably victims of a submarine explosion. None were caught thereafter until 1827 (Veniaminov 1840a:69–70). Ball reports a similar situation near Afognak Island after the Katmai eruption (1914:62). As noted above, significant withdrawal of fish would have a multiple effect, since sea mammal predators would be obliged to move with them. We have also previously discussed the impact of ashfalls on salmon streams. Veniaminov suggested the reduction of salmon he witnessed in some places might be due in part to the change in the configuration of streams due to volcanic activity in addition to overfishing and pollution by man (1840a:69–70). Abortion of a major salmon run or destruction of eggs and fry would have unfortunate consequences for local human populations for years following the outburst that caused it.

Intertidal invertebrates were an important part of the Aleut diet. These are of course among the least mobile of creatures and would not be able to save themselves from altered conditions through flight. On the other hand, they occur in economically useful quantities in reef and strand-flat communities that are often located at a considerable distance from volcanic mountains, so they may not often have been subject to the deleterious effects of glowing avalanches, landslides and mudflows of volcanic origin, and dilution of salt water by freshwater floods. Large numbers of sea urchins died near Afognak Island after the Katmai eruption, and clams and cockles also seemed to suffer (Ball 1914:62). Barnacles and mussels were killed down to low tide in Katmai Bay (Jaggar 1945:88).

Very limited study of the effect of ashfalls on water plants suggests it to be very temporary, with major damage caused by the grinding effect of floating pumice, poison gases, and actual burial in ash (Wilcox 1959:455). Kelp attached to rocks and reefs near Afognak blackened and withered after the Katmai eruption (Ball 1914:64). It has been suggested that submarine volcanism may add many nutrients (compounds of phosphorus, nitrogen, silicon, etc.) to the water in areas of recurring volcanism and may thus in the long run enhance the productivity of the area (Wilcox 1959:470), a minor consolation to hungry people in a devastated post-ash world.

In summary, though few detailed studies of the effect of volcanism on a rich marine ecosystem are available, it seems reasonable to suppose that these effects may have been drastic on a localized basis for a period and that certain consequences (i.e., disruption of salmon runs and harm to intertidal inverte-brates) might have reduced carrying capacity of an area for a period of years. Psychological trauma attendant on being a close witness of some of nature's more appalling spectacles might have prompted a willingness to move as well. Since the Aleutians for at least the last few millennia have had a large popula-tion combined with a limited amount of usable shore area, major volcanic events were almost certain to affect some human settlements. I suspect with-out quantified data that the process of population readjustment would have been easier in this rich area than it would have been in the boreal interior discussed above with its very limited carrying capacity. On the other hand, the marine Aleutian ecosystem was probably much more resilient than the boreal interior ecosystem previously discussed. The flushing action of the vast sea would cleanse all but the greatest damage from ejecta, etc., in a short period of time. No equivalent mechanism exists on land. In a complex paper, Hett and O'Neil (1974:36–37) have constructed an admittedly very crude model of the Aleutian ecosystem, and they find it to be quite resistant to minor environmen-tal fluctuations, perhaps as much as 300 times more resistant than terrestrial ecosystems.

When considering the geographic extent of the Aleutian Islands and the vagaries of the wind, it becomes clear that volcanism is and was a localized phenomenon, unlikely to affect many areas at once. Relatively stable dense populations are probably harder to disrupt in a cultural–historically significant way than are sparsely scattered ones. We tentatively conclude that endemic volcanism in the eastern Aleutians may have had drastic effects on a relatively local scale but that its total cultural historical impact, at least in the last 4000 years, was probably less than the much rarer volcanism in the boreal interior.

We turn now to a brief discussion of the Holocene human and volcanic history of Umnak Island as elucidated over the last few years by the geologist Robert Black and by William Laughlin, Jean Aigner, and their colleagues with special attention to suggestions concerning the significance of volcanism in prehistoric Aleut life there. Umnak Island is divided into three topographic segments (Figure 11.3). The northeast end of the island is dominated by Okmok Caldera, 1000 m high and 12 km in width. The central portion of the island contains the beautifully symmetric Vsevidov Volcano, towering to 2100 m, and the dissected elongate ridge of Recheshnoi, 2000 m high. Southwestern Umnak, the focus of Aleut settlement and archaeological research, is a rolling plain of low relief. All three volcanoes have been active since man first came to Umnak at least 8500 years ago. Okmok is currently active, Vsevidov has been

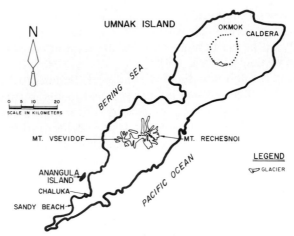

**FIGURE 11.3.** Volcanoes and archaeological sites on Umnak Island, Eastern Aleutians. (After Black 1975.)

active since 1750, and Recheshnoi generated a lava flow that reached the Pacific about 3000 years ago (Black 1976:9–10).

Black has developed the following basic chronology of Holocene events on Umnak (1976:9). The lowlands of southwestern Umnak were deglaciated by about 11,000 years ago, a major ash (Ash I) fell about 10,000 years ago, another major ash fell about 8500 years ago (Ash II), and Ash III was emplaced about 8250 years ago. There followed a long interval of Hypsithermal soil formation, with evidence for neoglacial activity in the mountains, a lava flow from Recheshnoi, and imposition of the last major ashfall (Ash IV) about 3000 years ago. Other minor Holocene eruptions are also reflected in minor ashfalls. Laughlin prefers a somewhat different chronology, with Ash II emplaced 9000 rather than 10,000 years ago and Ash III falling about 7000 rather than 8250 years ago (1975:509; see discussion of the Anangula site on pp. 361–362). Black notes that Ash I thickens to the west and thus presumably has its source west of Samalga Pass, probably in the Islands of the Four Mountains, whereas Ashes II and III presumably came from Okmok Volcano to the northeast (1976:16, 20). Ash IV also came from the northeast and it might be tempting to associate it with the documented activity of Recheshnoi about 3000 years ago, although Black does not attempt this correlation. No sources have been precisely determined lithologically as yet. Although it has not been proved, Black favors the interpretation that Ash III represents the cataclysmic eruption of Okmok that resulted in the formation of the caldera, whereas Ash II represents a lesser eruption from a similar source (1974:137).

Although noting cautiously that some of the geological correlations are not proven (1976:20–21), Black advances the following plausible suggestions about the cultural significance of volcanism on Umnak. Umnak was a peninsula forming the southern terminus of the Bering Land Bridge when late-Pleistocene lowered sea levels brought that land connection between Asia and

America into being. Down to at least 11,000 years ago heavy glaciaton of the mountains of the western Alaska Peninsula and on Umnak would have barred human access (1974:127). If boat-using people had filtered into the Umnak area, they would have come from the north and east following the shore of the land bridge. Virtually continual early Holocene eruptions of Okmok would have made it a significant barrier to pass and likely would have barred in-migration of any large terrestrial herbivores (1974:126). Fisher Caldera on Unimak Island was active at the same time (Miller and Smith 1977:174). Catastrophic volcanism resulting in the creation of Okmok Caldera about 8250 B.P. would have sterilized northeastern Umnak and probably would have severely impacted the settlement at Anangula Island about 70 km from the source (Denniston 1972:21; Black 1975:163; 1976:11; see also discussion of Anangula on pp. 361–362). Littoral obsidian sources on northeast Umnak may have been available to the inhabitants of Anangula Island. These would have been buried by up to 30 m of pyroclastic deposits during this cataclysm.

Mount Vsevidov on central Umnak is a very young volcano with evidence of a long Holocene eruptive history in the form of multiple lava flows, the youngest of which is certainly post-Hypsithermal and possibly post-1760. Black notes that, although none of Vsevidov's eruptions were as large or destructive as those of Okmok, ancient Aleuts living near it would have had to move often, and that ancient settlements might well be covered with lava flows (1975:163). Recheshnoi erupted about 3000 years ago, sending lava flows to the Pacific. It bore glaciers at the time, and Black wonders whether the eruption, in combination with other factors, might not have stimulated neoglacial ice advance here (1976:13). In closing he notes that volcanism and tectonic activity cannot be ignored in considering the extremely complex problem of Holocene sea levels in the Aleutians (Black 1974:16), a topic we need not pursue here.

The first occupation of the blade site on Anangula Island, located in Nikolski Bay about 70 km from Okmok Volcano, is dated to about 8500 years ago (Black 1974:133; S. Laughlin et al. 1975:39). The rich cultural level is sandwiched between Ashes II and III, with a number of lesser ashfalls, including the basal "key ash" noted as well (S. Laughlin et al. 1975:43; McCartney and Turner 1966:37; Black 1975:164; W. Laughlin 1975:510). The occupation layer is overlain by 10–20 cm of Ash III, which Black correlates with the explosion that created Okmok Caldera. Obsidian tools make up approximately 20% of the tool inventory (W. Laughlin 1972:2). This suggests either that the ancient inhabitants braved the dangers of Okmok Caldera or that they had access to a source since buried or lost. Three samples of this obsidian that have been analyzed resemble certain samples from northwest Alaska (Black 1974:137; 1976:11), but this source seems improbable on geographic grounds and I suspect that this interpretation is in error.

At present there is strong disagreement as to the duration of the Anangula occupation. Because of lack of evidence for soil development within the cultural layer, Black believes that Anangula was occupied for a very short time, between one Aleut generation and at most 100 years (1976:25). In 1966

McCartney and Turner suggested a short occupation, several centuries at most (p. 37). Aigner (1974:15; 1976a; 1976b:34) suggests an occupation of 500 years or less. W. Laughlin, on the other hand, interprets 33 radiocarbon dates as suggesting a span of occupation of 1500 years (1972:3; 1975:508). The difference of interpretation (which includes the dating of Ashes II and III) hinges on conflicts over the reliability of radiocarbon samples taken on soil humates and driftwood from the cultural level at Anangula and possibly contaminated by sea water, as opposed to a much smaller series taken from peat deposits undisturbed by man, and the weight to be attached to relative degree of soil development between the various ash levels (Black 1976:17). We need not attempt to resolve this issue here. Far more important for our purposes is the possible significance of the imposition of Ash III and related collapse of Okmok Caldera in the abandonment of the Anangula site.

W. Laughlin has suggested that the blade site was abandoned because of rising sea levels and that Ash III fell after the people left (1972:3). More than a decade ago McCartney and Turner suggested that absence of a sterile layer between the cultural horizon and Ash III at the site indicated that the ashfall may have led to abandonment of the site (1966:37). Black agrees, suggesting that the appalling spectacle of the collapse of Okmok Caldera and the deleterious impact of the associated ashfall and related events would have been sufficient reasons to abandon the site (1976:25; 1975:164). He cites as a personal communication from Aigner that the abandonment of the site appears to have been fairly precipitous, with a number of tools left behind that one might ordinarily expect to have been taken along (1974:139). Another possibility that has to be mentioned is that the occupants perished, either at the site or in the area. I find this intuitively unlikely, but possible. Evidence is lacking since the acid soils of the site preserve no bones. Could the well-known artifactual richness of the site reflect such abandonment? If so, there would perhaps be a greater concentration of tools at the top of the deposit.

Black thinks it likely that the Anangula people were driven either to seek refuge elsewhere on southwestern Umnak or perhaps to disperse to the west across dangerous Samalga Pass in their skin boats to the Islands of the Four Mountains and beyond (Black 1974:139; 1975:164). No archaeological evidence buttresses this view, with the problematical exception of a radiocarbon date of about 8000 B.P. of unspecified association from the Sandy Beach Bay site in the next bay west of Nikolski Bay (Black 1975:164). The prehistoric record on Umnak is blank for several thousand years after the Anangula occupation, but geological complexities (especially changing sea levels) preclude the drawing of conclusions about areal abandonment. Most scholars would probably concur with W. Laughlin (1975) and Aigner et al. (1976) that the Anangula technology is implicated, through one or several possible "transitional industries," in the technological ancestry of later Aleut culture. This indirect reasoning seems sufficient to indicate that it is unlikely that the Anangula people were a unique isolate that perished in the holocaust associated with the formation of Okmok Caldera, but no direct archaeological evidence bears on their post-Anangula history, and even the suggestion that they were driven from the area by volcanism, though in my opinion responsible and quite likely, is certainly not proven.

At the Chaluka site that underlies the contemporary Umnak Aleut community of Nikolski, archaeological excavations have revealed a 4000-year occupational history pertaining to the cultural and biological ancestry of the modern Aleut. Deep in the deposit about 15 cm of ash, pumice, and clay are found that correlate with Ash IV and represent an episode of volcanism dated between about 3200 and 3000 years ago. There is some evidence that the emplacement of this ash was accompanied by a tsunami that inundated part of the site (Turner et al. 1974:133, 139). As noted previously, this ashfall correlates temporally with dated lava flows from Recheshnoi Volcano, and one wonders whether these events might be related. In an earlier interpretation that placed the ashfall (incorrectly, it now appears) some centuries earlier, Lippold noted that fish remains decreased significantly in proportion to sea lion and sea otter bones in the strata that immediately overlay the ash. Although noting continuity in bone tools between sub-ash and super-ash levels, she suggested that there was a hiatus in occupation marked by the ash and that volcanic activity might have been involved (1966:130). The economic shift still requires explanation but present evidence indicates that there was no significant hiatus in occupation. Indeed the most intensive occupation of the eastern portion of the site immediately overlies the volcanic ash with no apparent cultural or chronological break (Turner et al. 1974:139). Although one must remember that midden site stratigraphy and the vagaries of radiocarbon dating do not permit identification of the passage of decades, much less years, this evidence suggests that Chaluka, if it was abandoned temporarily after the ashfall and possible associated tsunami, was reoccupied almost immediately by bearers of the same cultural tradition. Thus available evidence suggests that the historical significance of this disaster was minimal, however inconvenient it may have been. By 3000 years ago cutting of the modern strand-flats and reef system was completed and intertidal invertebrates were available in quantity to the ancient Aleut (Black 1976:9). I suggest that by then, if not before, ancestral Aleut population and settlement density in the eastern Aleutians had reached a level where a localized disaster affecting (or even annihilating) one or several settlements would have little far-reaching cultural or historical impact. A fitting analogy would perhaps be the internecine warfare of late prehistoric times when whole communities might be destroyed with little long-range effect. By this time, from the coarse-grained perspective of the archaeological record, there were simply too many Aleut and Aleut communities for localized extinctions to be recognized in significant cultural losses or modifications. Warfare indeed was probably more significant than natural disaster in this regard.

## SUMMARY

The data reviewed in this chapter suggest that volcanism would have had significantly different effects in the interior of northwestern Canada and in the eastern Aleutians. The boreal interior had a limited carrying capacity in contrast to the highly productive marine ecosystem of the Aleutians. This difference is reflected in the ethnographic population density figures and in archaeological settlement history. The interior was sparsely populated by small, highly mobile human groups, whereas large, sometimes closely spaced permanent villages characterize the Aleutians in places. Certain Aleuts had little choice but to live near active or dormant volcanoes, where they became subject to localized but truly catastrophic volcanic phenomena that could and doubtless sometimes did lead to instant death for numbers of individuals. Interior ancestral Indians lived and hunted far from the rare and inaccessible volcanoes and were subject only to ashfalls and related phenomena. Volcanism was endemic in the eastern Aleutians with major eruptions occurring every few years, whereas only two volcanic outbursts of note are recorded for the entire Holocene in the Yukon. The Aleut would therefore have been much more familiar with volcanic phenomena and possibly more hardened to their impact than the interior peoples, but without question more Aleut peoples must have suffered from their effects throughout Holocene time.

The ecological impact of interior ashfalls would have persisted until the terrestrial vegetation was restored to pre-ash conditions, a process taking minimally several years; the sea protected some marine resources from the initial impact and rapidly cleansed portions of an affected area, mitigating if not undoing the damage done. Accelerated erosion and other severe terrestrial impacts would not have the effect on a sea-oriented people that they would have had on a land-bound one. Limited data appear to support the subjective feeling that subarctic marine ecosystems are far more resilient than terrestrial ones; certainly the variety and quantity of food resources available to the Aleut differ by several orders of magnitude from the resources available to Indians of the interior. Though not wishing to minimize the localized consequences of volcanism in the Aleutian setting, I think it can be fairly stated that a much larger area would have been impacted more drastically and for a longer period of time as a consequence of the emplacement of the East Lobe of the White River Ash than was the case with even the major volcanic explosions in the Aleutians.

Seemingly against the foregoing generalizations, I advance the conclusion that sporadic volcanism in the interior had more historic significance than endemic volcanism in the Aleutians. Because populations were sparse in the interior, relatively few people would have faced death or removal as a result of this event. But because of this population sparsity, the loss of one or several bands would have been more likely to have significant cultural impact than the death of hundreds in the Aleutians. In the best of times, refugees may well have been harder to accommodate, given the meager carrying capacity of the interior, and the wanderings of a few small bands in search of new homes might have caused more strife than the displacement of several Aleutian villages. What happened to these few refugees and the people they may have displaced in turn is in the final analysis far more important than what happened to a few hundred Aleut, quite simply because the interior refugees represented a greater percentage of the bearers of the cultural tradition. The very rareness of catastrophic volcanism in the interior may have enhanced its effect on out-migration, although it is also possible that its novelty might have led to apathy, shock and nonadaptive behavior (see the Sanpoil–Nespelem example cited on page 347).

These generalizations can be only indirectly tested at present. A suggestion has been made that the emplacement of the White River Ash 12 centuries ago may have been significant in the dispersion of certain Athapaskan speakers. Arguments of continuity between the technology represented by the Anangula Blade site and universally accepted ancestral eastern Aleut of 4000 years ago appear to indicate either that the Anangula people escaped extinction in the holocaust accompanying the formation of Okmok Caldera or that there were related populations in the area who escaped annihilation and carried on the cultural tradition. Although not fully documented archaeologically, by 3000 years ago there were probably too many people in too many villages in the eastern Aleutians for localized catastrophes to have had significant long-term impact on the subsequent development of eastern Aleut culture.

## SUGGESTIONS FOR FURTHER WORK IN THE SUBARCTIC

Several profitable lines of further inquiry into the significance of volcanism in northwest North America suggest themselves. As earlier suggested (Workman 1974:252), more useful field data might be generated by incorporating detection of volcanic impact into specific research designs rather than having to deal, as now is usually the case, post facto with data gathered for other purposes. Because of the short-term if spectacular nature of volcanological impact on the living world, finer chronological resolution than is usual at present is needed if paleoenvironmental research is to help us confront some of our basic problems (see Mehringer et al. 1977 for an example). I believe that in the foreseeable future a profitable approach will continue to be the study of the impact of historic volcanic eruptions. A considerable literature, including anecdotal as well as scientific accounts, must be culled for the relatively few observations that have bearing. Then, armed with appropriate geological evidence bearing on the nature and severity of particular volcanic eruptions, environmental information indicating in detail what resources were being exploited by affected populations, and a sophisticated understanding of ecological relationships based on contemporary data, it should be possible to construct satisfactory models indicating, within reasonable bounds, what the ecological effect of a given volcanic eruption must have been. From a series of such modeling exercises dealing with particular volcanic events, we can proceed to make useful statements about the ecological significance of volcanism in a specified region through time.

Establishing the historical, as opposed to the ecological, significance of past volcanism is another matter. Basic to such understanding is a refined and chronologically controlled knowledge of areal prehistory immediately before and immediately after the volcanic episode in question. Hypotheses must then be generated regarding the cultural effects one might anticipate as a consequence of a volcanic eruption of a certain magnitude with a given probable ecological effect. Some of the more obvious effects to be anticipated are abandonment of the area, breaks in the local archaeological sequence, intrusive appearance of cultural complexes or traits derived from the affected area in the peripheries, and evidence of intensified interregional contacts synchronous with the volcanic event in question (see Sheets 1976 for an example). Evidence for immediate reoccupation of affected sites without recognizable cultural breaks is, given a sufficient data base, evidence against profound cultural impact, although the problem must be considered in regional rather than site-specific terms.

On a less ambitious scale, certain other approaches to the data appear to deserve testing. As we have seen, the significant problem of seasonality of eruption can be approached, if one is willing to make some simplifying assumptions, by comparing seasonal high-atmosphere wind patterns of the present with mapped ash distributions. We could advance our knowledge of the impact of ashfalls on Alaskan salmon runs by detailed studies of the appropriate fisheries records. The comparative approach that I have attempted in this chapter could be broadened by considering other subarctic volcanic areas with somewhat different indigenous subsistence economies—for example, Kamchatka with its aboriginal population heavily oriented toward riverine fishing. In considering the social impact of natural catastrophes, the growing disaster study literature could usefully be consulted (see Sheets 1976 for a brief review), although at this late date one can scarcely expect to obtain data on the reaction of hunting and collecting peoples to natural disasters. Acknowledging that geologists have taken the lead in addressing some of the problems considered in this chapter, we should also indicate that the prehistorian may well be in a position to aid the geologist and the volcanologist. A cooperative program of dating ashfalls, in which the archaeologist has at least a stratigraphic interest, and carefully tracing them to their probable source through geological studies and modern lithological techniques, may ultimately provide researchers with a much more thorough and useful record of Holocene volcanic activity in northwest North America than is presently available from limited historic records and geological studies. An accurate history of past volcanism would be of practical value for, among other things, long-range settlement planning (Crandell et al. 1975). The practical value of further study of the effects of volcanism on North Pacific salmon runs should also be obvious (Eicher and Rounsefell 1957:76).

In conclusion, I wish to state that in suggesting that volcanism should be taken seriously in considering the environmental and cultural history of several subarctic areas I do not propose an all-purpose tephra-ex-machina explanation for complex problems, nor do I advocate cultural or environmental catastrophism as a substitute for painstaking multifaceted studies or disciplined scientific thought. I do believe, however, that in highly volcanic areas the impact of past volcanism on human life must be considered.

## ACKNOWLEDGMENTS

Professor Lydia Black of Providence College went far beyond the call of duty in providing lengthy taped excerpts from Veniaminov's untranslated geographic description of the Unalaska District. She also read and commented upon an earlier draft of this paper. Her aid and enthusiasm for the project are gratefully acknowledged. Professors Robert Black (University of Connecticut) and John Lerbekmo (University of Alberta) kindly preserved me from gross geological errors. Professors Steven Langdon (University of Alaska, Anchorage), Allen McCartney (University of Arkansas), and Thomas Lux (Providence College) also read the preliminary draft and made valuable suggestions. Professor Ronald Shimek (University of Alaska, Anchorage) made useful comments on the impact of ashfalls on marine invertebrates. Colleagues and discussants at the Volcanism Symposium in New Orleans provided reinforcement, additional references, and useful comments.

## REFERENCES

Aigner, J. S.
   1974   Studies in the early prehistory of Nikolski Bay: 1937–1971. *Anthropological Papers of the University of Alaska* 16(1):9–25.
   1976a  Dating the early Holocene maritime village of Anangula. *Anthropological Papers of the University of Alaska* 18(1):51–62.
   1976b  Early Holocene evidence for the Aleut maritime adaptation. *Arctic Anthropology* 13(2):32–45.
Aigner, J. S., B. Fullem, D. Veltre, and M. Veltre
   1976   Preliminary report on remains from Sandy Beach Bay, a 4300–5600 B.P. Aleut village. *Arctic Anthropology* 13(2):83–90.
Alaska Geographic Society
   1976   Alaska's volcanoes: Northern link in the ring of fire. *Alaska Geographic* 4(1).
Ball, E. M.
   1914   Investigations of the effect of the eruption of Katmai Volcano upon the fisheries, fur animals and plant life in the Afognak Island reservation. In B. W. Evermann, *Alaska Fishery and Fur Seal Investigations in 1913*. Washington, D.C.: Government Printing Office.
Bauer, P. S.
   1971   Review of Hekla. A notorious volcano by Sigurdur Thorarinsson. *Science* 172:692–693.
Black, R. F.
   1974   Geology and ancient Aleuts, Amchitka and Umnak Islands, Aleutians. *Arctic Anthropology* 11(2):126–140.
   1975   Late-Quaternary geomorphic processes: Effects on the ancient Aleuts of Umnak Island in the Aleutians. *Arctic* 28(3):159–169.
   1976   Geology of Umnak Island eastern Aleutians as related to the Aleuts. *Arctic and Alpine Research* 8(1):7–35.
Bray, J. R.
   1977   Pleistocene volcanism and glacial initiation. *Science* 197:251–254.

Brongersma-Sanders, Margaretha
  1957   Mass mortality in the sea. *Geological Society of America Memoir 67*, Volume 1:941–1010.
Bryson, R. A., and W. M. Wendland
  1967   Tentative climatic patterns for some late glacial and post-glacial episodes in central North
          America. In *Life, land and water. Proceedings of the 1966 Conference on the Glacial Lake
          Agassiz Region*, edited by William J. Mayer-Oakes. Winnipeg: University of Manitoba
          Press.
Cadle, R. D., and E. J. Mroz
  1978   Particles in the eruption cloud from St. Augustine Volcano. *Science* 199:455–457.
Capps, S. H.
  1915   An ancient volcanic eruption in the Upper Yukon Basin. United States Geological Survey
          Professional Paper 95-D. Washington: Government Printing Office.
Crandell, D. R., D. R. Mullineaux, and M. Rubin
  1975   Mount St. Helens Volcano: Recent and future behavior. *Science* 187:438–441.
Denniston, G. B.
  1966   Cultural change at Chaluka, Umnak Island: Stone artifacts and features. *Arctic An-
          thropology* 3(2):84–124.
  1972   Aishishik Point: An economic analysis of a prehistoric Aleutian economy. Doctoral
          dissertation, Dept. of Anthropology, University of Wisconsin.
Derry, D. E.
  1975   Later Athapaskan prehistory: A migration hypothesis. *The Western Canadian Journal of
          Anthropology* 5(3–4):134–147.
Dumond, D. E.
  1974   Prehistoric ethnic boundaries on the Alaska Peninsula. *Anthropological Papers of the
          University of Alaska* 16(1):1–7.
Eicher, G. J. Jr., and G. A. Rounsefell
  1957   Effects of lake fertilization by volcanic activity on abundance of salmon. *Limnology and
          Oceanography* 2(2):70–76.
Evermann, B. W.
  1914   Effects of Katmai eruption evident in 1913. In B. W. Evermann, *Alaska fishery and fur seal
          investigations in 1913*. Washington, D.C.: Government Printing Office.
Freeman, N.
  1977   Eyewitness to disaster. *Alaska Geographic* 4(3):20–23.
Griggs, R. F.
  1922   *The Valley of the Ten Thousand Smokes*. Washington, D.C.: The National Geographic
          Society.
Hanson, L.
  1965   Size distribution of the White River Ash, Yukon Territory. Master's thesis, Dept. of
          Geology, University of Alberta.
Hein, J. R., D. W. Scholl, and J. Miller
  1978   Episodes of Aleutian Ridge explosive volcanism. *Science* 199:137–141.
Hett, J. M., and R. V. O'Neill
  1974   Systems analysis of the Aleut ecosystem. *Arctic Anthropology* 11(1):31–40.
Hobbs, P., L. V. Radke, and J. L. Stith
  1978   Particles in the eruption cloud from St. Augustine Volcano. *Science* 199:457.
Hrdlička, Aleš
  1945   *The Aleutian and Commander Islands and their inhabitants*. Philadelphia: Wistar Insti-
          tute.
Jaggar, T. A.
  1945   *Volcanoes declare war. Logistics and strategy of Pacific volcano science*. Honolulu:
          Paradise of the Pacific Limited.
Jochelson, W.
  1933   History, ethnology and anthropology of the Aleut. Carnegie Institution of Washington
          Publication No. 432.

Johnson, F., and H. M. Raup
    1964   Investigations in southwest Yukon: Geobotanical and archaeological reconnaissance.
           Papers of the Robert S. Peabody Foundation for Archaeology Vol. 6(1). Andover: Phillips
           Academy.
Kennett, J. P., and R. C. Thunell
    1975   Global increase in Quaternary explosive volcanism. Science 187:497–503.
Kienle, J., and R. B. Forbes
    1977   Augustine: Evolution of a volcano. Geophysical Institute, University of Alaska, Annual
           Report 1975–76, pp. 26–48.
Kotani, Y.
    1969   Upper Pleistocene and Holocene environmental conditions in Japan. Arctic Anthropology
           5(2):133–158.
Krauss, M. E.
    1972   Na-Dene. Current Trends in Linguistics 10:146–206.
Kurenkov, I. I.
    1966   The influence of volcanic ashfall on biological processes in a lake. Limnology and
           Oceanography 11:426–429.
Laguna, Frederica de
    1958   Geological confirmation of native traditions, Yakutat, Alaska. American Antiquity
           23(4):434.
Lantis, M.
    1970   The Aleut social system, 1750–1810, from early historical sources. In Ethnohistory in
           southwestern Alaska and the southern Yukon. Method and content, edited by Margaret
           Lantis. Studies in Anthropology No. 7. Lexington: University of Kentucky Press.
Laughlin, S. B., W. S. Laughlin, and M. E. McDowell
    1975   Anangula blade site excavations, 1972 and 1973. Anthropological Papers of the Univer-
           sity of Alaska 17(2): 39–48.
Laughlin, W. S.
    1972   Holocene history of Nikolski Bay, Alaska and Aleut evolution. Abstract (6 pp.) of a paper
           prepared for a Conference on the Bering Land Bridge and Its Role for the History of
           Holarctic Flora and Fauna in the Late Cenozoic. Khabarovsk, U.S.S.R.
    1975   Aleuts: Ecosystem, Holocene history, and Siberian origins. Science 189:507–515.
Lerbekmo, J. F., and F. A. Campbell
    1969   Distribution, composition, and source of the White River Ash, Yukon Territory. Cana-
           dian Journal of Earth Sciences 6:109–116.
Lerbekmo, J. F., J. A. Westgate, D. G. W. Smith, and G. H. Denton
    1975   New data on the character and history of the White River volcanic eruption, Alaska. In
           Quaternary Studies, edited by R. P. Suggate and M. M. Cresswell, Wellington: The Royal
           Society of New Zealand.
Lippold, L. K.
    1966   Chaluka: The economic base. Arctic Anthropology 3(2):125–131.
McClellan, C.
    1975   My old people say: An ethnographic survey of southern Yukon Territory. National
           Museums of Canada Publications in Ethnology, No. 6(1 and 2). Ottawa.
Macdonald, G. A.
    1972   Volcanoes. Engelwood Cliffs, N.J.: Prentice-Hall.
McKennan, R. A.
    1959   The Upper Tanana Indians. Yale University Publications in Anthropology, No. 55. New
           Haven.
MacNeish, R. S.
    1964   Investigations in southwest Yukon: Archaeological excavations, comparisons and specu-
           lations. Papers of the Robert S. Peabody Foundation for Archaeology Vol. 6(2). Andover:
           Phillips Academy.
McCartney, A. P.
    1974a 1972   Archaeological site survey in the Aleutian Islands, Alaska. In International confer-

*ence on the prehistory and paleoecology of western North American Arctic and Subarctic,* edited by Scott Raymond and Peter Schledermann. Calgary: Department of Archaeology.

1974b  Prehistoric cultural integration along the Alaska Peninsula. *Anthropological Papers of the University of Alaska* 16(1):59–84.

1975  Maritime adaptations in cold archipelagoes: An analysis of environment and culture in the Aleutian and other island chains. In *Prehistoric maritime adaptations of the circumpolar zone,* edited by William Fitzhugh. The Hague: Mouton.

McCartney, A. P., and C. G. Turner

1966  Stratigraphy of the Anangula unifacial core and blade site. *Arctic Anthropology* 3(2):28–40.

Malde, H. E.

1964  The ecological significance of some unfamiliar geologic processes. In *The reconstruction of past environments. Proceedings of the Fort Burgwin Research Center on Paleoecology 1962,* assembled by James A. Hester and James Schoenwetter. Fort Burgwin, N.M.: Fort Burgwin Research Center.

Mehringer, P. J., E. Blinman, and K. L. Petersen

1977  Pollen influx and volcanic ash. *Science* 198:257–261.

Miller, T. P.

1976  Augustine volcano. *Alaska Geographic* 4(1):17–30.

Miller, T. P., and R. L. Smith

1977  Spectacular mobility of ash flows around Aniakchak and Fisher Calderas, Alaska. *Geology* 5(3):173–176.

Nowak, M.

1968  Archeological dating by means of volcanic ash strata. Doctoral dissertation, Dept. of Anthropology, University of Oregon.

Osgood, C.

1971  The Han Indians. *Yale University Publications in Anthropology,* No. 74. New Haven.

Péwé, T.

1975  Quaternary geology of Alaska. United States Geological Survey Professional Paper 835. Washington, D.C.: Government Printing Office.

Rampton, V.

1971  Late-Quaternary vegetation and climatic history in the Snag-Klutlan area, southwestern Yukon Territory, Canada. *Geological Society of America Bulletin* 82(4):959–978.

Ray, V. F.

1954  *The Sanpoil and Nespelem.* Reprinted by the Human Relations Area Files. New Haven.

Sheets, P. D.

1976  *Ilopango Volcano and the Maya Protoclassic.* University Museum Studies No. 9. Carbondale: Southern Illinois University.

Souther, J. G.

1970  Recent volcanism and its influence on early native cultures of northwest British Columbia. In *Early man and environments in northwest North America,* edited by R. A. Smith and J. W. Smith. University of Calgary, Dept. of Archaeology.

Suslov, S. P.

1961  *Physical geography of Asiatic Russia.* Translated from the Russian by Noah D. Gershevsky and edited by Joseph E. William. San Francisco: W. H. Freeman.

Tazieff, H.

1971  The nature of volcanic activity. In *Volcanoes,* edited by Christopher Kruger. Toronto: Longmans Canada Limited.

Toon, O. B., and J. P. Pollack

1977  Volcanoes and climate. *Natural History* (January):8–26.

Trowbridge, T.

1976  Aniakchak Crater. *Alaska Geographic* 4(1):71–73.

Turner, C. G. II, J. S. Aigner, and L. R. Richards

1974  Chaluka stratigraphy, Umnak Island, Alaska. *Arctic Anthropology* 11 (supplement):125–142.

Veniaminov, I.
  1840a *Notes on the islands of the Unalaska District. Vol. 1: Physical description of the islands.*
     St. Petersburg. Oral translation of excerpts by Lydia Black.
  1840b *Notes on the islands of the Unalaska District. Vol. 2.* Human relations Area Files Transla-
     tion, New Haven.
Westgate, J. A., D. G. W. Smith, and M. Tomlinson
  1970 Late Quaternary tephra layers in southwestern Canada. In *Early man and environments
     in northwest North America,* edited by R. A. Smith and J. W. Smith. Dept. of Archaeol-
     ogy, University of Calgary.
Whetstone, G. W.
  1955 Effects of volcanic ash from Mt. Spurr on the chemical character of surface waters near
     Anchorage, Alaska (Abstract). *Geological Society of America Bulletin* 66:1709.
Wilcox, R. E.
  1959 Some effects of recent volcanic ash falls with especial reference to Alaska. *United States
     Geological Survey Bulletin* 1028-N.
  1965 Volcanic ash chronology In *The Quaternary of the United States,* edited by H. E. Wright,
     Jr. and D. G. Frey. Princeton: Princeton University Press.
Wilmeth, R.
  1977 Chilcotin archaeology: The direct historic approach. In *Problems in the prehistory of the
     North American Subarctic: The Athapaskan question,* edited by J. W. Helmer, S. Van
     Dyke, F. J. Kense. Dept. of Archaeology, University of Calgary.
Workman, W. B.
  1974 The cultural significance of a volcanic ash which fell in the upper Yukon Basin about 1400
     years ago. In *International Conference on the Prehistory and Paleoecology of the Western
     North American Arctic and Subarctic,* edited by Scott Raymond and Peter Schledermann.
     Dept. of Archaeology, University of Calgary.
  1977 The prehistory of the southern Tutchone area. *Problems in the Prehistory of the North
     American Sub-Arctic: The Athapaskan question,* edited by J. W. Helmer, S. Van Dyke,
     F. J. Kense. Dept. of Archaeology, University of Calgary.
  1978 Prehistory of the Aishihik-Kluane area, southwest Yukon Territory. *Mercury Series: Ar-
     chaeological Survey of Canada Paper,* No. 74. National Museum of Man. Ottawa.

# 12

# People and Pumice on the Alaska Peninsula

DON E. DUMOND

On 6 June 1912, one of the world's greatest volcanic eruptions occurred among a string of volcanoes in the Aleutian Range on the northern Alaska Peninsula, at or in the immediate vicinity of a mountain known as Katmai (Figure 12.1). The Katmai eruption released pumiceous ejecta in the form of ash and larger particles, in a quantity estimated to be as great as 28 km³ (Griggs 1922: xvi, 29; Curtis 1955). Piled in some valleys near the volcano to a depth of 200 m, this ejecta covered an area of 60,000 km² to a depth of 3 cm or more, and brought about the permanent abandonment of at least four separate settlements pertaining to three distinct native communities.

Archaeological investigations in the Naknek River drainage system of the Bering Sea slope of the peninsula, however, have indicated that although the Katmai eruption may have been the most severe ever experienced there by man, it was by no means the first significant eruption to occur; rather, evidence of at least nine earlier explosive eruptions is present in the upper portions of the Naknek drainage in recognizable subsurface strata of pumiceous volcanic ash that date within the past 7400 years. Inasmuch as the same area has also yielded archaeological evidence of human occupations through nearly 5 millennia, and the general region of the northern Alaska Peninsula is known to have been occupied for 9, one must conclude that many of these eruptions were directly witnessed by human beings.

Two questions must be asked. The first, of purely archaeological significance: How useful were these tephra layers in establishing the control that permits the description of cultural chronology and development? The second,

VOLCANIC ACTIVITY AND HUMAN ECOLOGY

Copyright © 1979 by Academic Press, Inc.
All rights of reproduction in any form reserved.
ISBN 0-12-639120-3

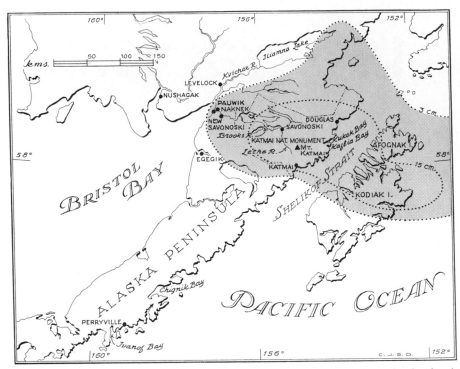

**FIGURE 12.1. The region of the Katmai eruption on the northern Alaska Peninsula, showing approximate area covered by volcanic ash to depths of more than 3 cm and more than 15 cm.**

of major prehistoric interest: How significant were the impacts of these eruptions on the long-term course of development of aboriginal culture? This chapter is divided into two sections, in each of which one of these questions is addressed. The first and shorter part is devoted to a summary of some characteristics of volcanic ash deposits in the upper Naknek drainage, and of attempts to discriminate between them. The second and major section considers the apparent temporal relationships of the volcanic occurrences to human occupation of the Naknek drainage system, turns next to accounts of the Katmai eruption of 1912 and its effects, attempts then to assess the probable impact of the prehistoric eruptions on the local population, and tries finally to generalize concerning the probable effects of significant volcanic eruptions on humans everywhere in comparable circumstances.

## THE ASHES AND THEIR RECOGNITION

A sustained program of archaeological research on the Alaska Peninsula was begun by the University of Oregon in 1960, and by 1975 a total of 10 field seasons of research had been devoted, in whole or part, specifically to the region drained by the Naknek River system. As may be expected in northern areas, the formation of soil in the Katmai region is normally slow, and most excavation profiles reveal the ten or more volcanic ashes of the upper Naknek drainage simply as a single multicolored layer of loesslike texture that is usually no more than 50 cm in thickness. Ash deposits appear visibly separated only in those particular locations where periods between eruptions were times either of human occupation with its inevitable debris, or of deposition of nonvolcanic sediments by wind or water.

For this reason, by the end of the first season at Brooks River in the upper drainage (1960) only three ashes appeared separable, but by the close of the following field season (1961) a total of five separate deposits had been distinguished in the field, and active work was begun to try to discriminate them petrographically with the naive expectation that perhaps by the index of refraction of the glass particles alone they would provide an easily definitive dating framework. After the succeeding season (1963), however, a total of 10 ashes were recognized tentatively, and the next two summers served to make clear the fact that though there were somewhat more than 10 separate deposits being dealt with, the ephemeral nature of some of them allowed only 10 to be recognized consistently in the various sites in the locale. By this time field identification (based on color and stratigraphic position) and dating of at least the most significant of them (by means of interleaved occupation remains yielding radiocarbon dates) were adequate to provide an excellent and quick control in various testing and excavation projects. Attempts to provide a positive identification and discrimination of these 10 proceeded apace, making use now of the identification of heavy mineral suites as well as of the index of refraction of the glass particles. Table 12.1 lists these deposits as they were identified in the Brooks River area and includes those radiocarbon dates from the entire drainage that were collected between 1960 and 1975 and that can confidently be assigned a stratigraphic position in relationship to one or more ash deposits. It also cites a few descriptive aspects, including the modal (although by no means the only) index of refraction of the glass of each ash. Additional information is provided by Nowak (1968).

The thickest of the ash deposits are A, C, F, G, and J, and these also were consistently the most helpful in understanding the stratigraphy of cultural deposits in the field. Although at Brooks River there is sufficient color distinction between certain ash layers to be extremely helpful to recognition, there is some overlap; in particular the black Ashes D and E look much alike, and often merge physically with the very similar Ash F. Even more awkward, there is a significant overlap in many cases between the modal indexes of refraction; if additional and recurrent (but submodal) indexes of all ashes were added, the overlap shown would be much greater. And, unfortunately, the information provided by the identification of the suites of heavy minerals (see Nowak 1968), is no less ambiguous.

This imprecise situation was compounded when concerted attempts were made to relate the Brooks River ash sequence to ash deposits encountered during the seasons of 1964 and 1965 in excavations of two archaeologically

**TABLE 12.1**
**Dates and Characteristics of Recognized Volcanic Ash Deposits at Brooks River**

| Date | Ash designation | Relevant C-14 determinations[a] (B.P.) | Thickness (cm) | Color | Modal index of refraction |
|---|---|---|---|---|---|
| 1912 | A | ------[Katmai eruption]------ | 20 | White | 1.493 |
| 1750 ± 50 | B | ----------------------------------- | 1–2 | Blackish gray | 1.518 |
|  |  | 230 ± 80 (I-209) |  |  |  |
|  |  | 335 ± 85 (SI-1853) |  |  |  |
|  |  | 450 ± 60 (Y-932) |  |  |  |
|  |  | 480 ± 90 (I-532) |  |  |  |
| 1400 ± 100 | C | ----------------------------------- | 10 | Olive gray | 1.523 |
|  |  | 300 ± 75 (I-524) |  |  |  |
|  |  | 670 ± 105 (I-1632) |  |  |  |
|  |  | 680 ± 90 (I-525) |  |  |  |
|  |  | 845 ± 100 (I-1635) |  |  |  |
|  |  | 880 ± 65 (SI-2075) |  |  |  |
|  |  | 975 ± 120 (I-520) |  |  |  |
| 900 ± 100 | D | ----------------------------------- | 1–2 | Black | 1.505 |
|  |  | 1175 ± 125 (I-522) |  |  |  |
|  |  | 1200 ± 170 (I-519) |  |  |  |
|  |  | 1225 ± 130 (I-521) |  |  |  |
| 450 ± 250 | E | ----------------------------------- | 1–2 | Black | 1.503 |
| A.D. |  | 1790 ± 130 (I-1633) |  |  |  |
| B.C. |  | 1895 ± 140 (I-1631) |  |  |  |
|  |  | 2110 ± 350 (I-1158) |  |  |  |
|  |  | 2140 ± 105 (I-1948) |  |  |  |
| 650 ± 450 | F | ----------------------------------- | 3 | Black, reddish | 1.518 |
|  |  | 3052 ± 250 (I-1159) |  |  | 1.558 |
|  |  | 3088 ± 200 (I-1157) |  |  |  |
|  |  | 3100 ± 105 (SI-1857) |  |  |  |
|  |  | 3250 ± 200 (I-518) |  |  |  |
|  |  | 3280 ± 60 (SI-1860) |  |  |  |
|  |  | 3390 ± 110 (I-518) |  |  |  |
|  |  | 3450 ± 110 (I-1947) |  |  |  |
|  |  | 3470 ± 65 (SI-1859) |  |  |  |
|  |  | 3610 ± 85 (SI-1856) |  |  |  |
|  |  | 3900 ± 130 (I-1629) |  |  |  |
|  |  | ------------------------------------ |  |  |  |
| 1900 ± 100 | G | ---[3860 ± 90 (Y-931)]----------- | 3 | Yellow | 1.508 |
|  |  | 3840 ± 130 (I-1630) |  |  |  |
|  |  | 3900 ± 120 (I-3114) |  |  |  |
|  |  | 3972 ± 440 (Y-930) |  |  |  |
| 2100 ± 100 | H | ----------------------------------- | 1–2 | Black, orange, yellow | 1.505 |
|  |  | 4240 ± 250 (I-1634) |  |  | 1.513 |
|  |  | 4430 ± 110 (I-1946) |  |  | 1.523 |
| 3500 ± 1000 | I | ----------------------------------- | 1–2 | Gray | 1.505 |
| 4500 ± 1000 | J | ----------------------------------- | 3 | Brown (clay) | 1.548 |
|  |  | 7360 ± 250 (I-1163) |  |  | 1.525 |

[a]The bracketed date for Ash G was derived from peat in a bog, on material within the layer of Ash G itself. The date below Ash J is believed to be from noncultural charcoal.

important locations on the Pacific coast of Katmai National Monument, and in an extensive archaeological survey of the same coast. Although Brooks River Ashes A, B, D, and E could with at least moderate conviction be equated with four ashes from the Pacific coast, the additional six ashes from Brooks River and the additional three ashes from the Pacific sequence of seven could not be correlated as convincingly, although in no cases were deposits totally dissimilar to others in the sequences. This ambiguity was again confirmed in 1972, when it became possible to attempt a neutron activation analysis for 19 different trace elements in the ashes, from which, in brief, it was concluded that because the ash samples are very similar to one another in the abundances of the elements analyzed, no consistent discrimination of samples was possible.

The conclusion to be drawn is that most or all of the separate ash deposits of the area were from the same or immediately adjacent sources that differed in composition of their ejecta very little. Given this circumstance, the usual procedures for identifying volcanic ash deposits—which amount to assignment of unknown specimens to one of a universe of known sets of ejecta, the characteristics of which are relatively unique—appeared not to be practically operable in the Katmai region. For although it may be theoretically possible to discriminate on the basis of physical characteristics and chemical components all of the separate ash deposits known in the area, it was certainly not economically possible with the resources then—or commonly—available. It was concluded that for definitive dating of actual excavation sites, the radiocarbon method was cheaper, less exasperating, and therefore far preferable.

## THE HUMAN IMPACT OF THE ERUPTIONS

Historians and some prehistorians have long been tempted to invoke the effects of specific, isolated events in adducing the causes—at least the proximate causes—of certain turns in the course of human existence. Thus Sir Arthur Evans suggested that an earthquake and attendant fires on Crete were responsible for the downfall of Minoan civilization (Evans 1922–1935:IV, 942–944), a view that continues to be developed by some of his successors. Or more recently an attempt has been made to relate certain "dark ages" in the early history of southwest Asia to periods of climatic drying (Bell 1971). Indeed, there can be no doubt that certain specific events—happenings arriving, as it were, from offstage—have had decisive effects upon humans and their cultural development.

On the other hand, social scientists are often loath to seek the significant impulses in human affairs among those events in nature that are commonly referred to as acts of God, or among those events unpredictably resulting from relatively idiosyncratic actions of humans—such as, for example, the revolutionary insight of genius, or the unforeseen results of exploration and conquest. Instead, the emphasis in the social sciences is apt to be upon "process"—that is, upon the more gradual adaptation of the human social organism to overreaching, environing circumstances, with the use of conceptions that emphasize a relative stability of environment, rather than the sudden, the shocking, or the idiosyncratic. So there is a tendency to conceive of a vanquished people as having been defeated because their inner fiber was weakened through some evolutionary, "processual," malaise, and not because their adversary was inordinately strong, whereas a "genius" becomes one who fulfills with only slight precocity the inevitable destiny of his culture. "Event" gets short shrift in this conceptual milieu.

Here both of these explanatory devices will be considered in attempting to answer the following questions: Do the volcanic events in the upper Naknek drainage of the Katmai region have apparent repercussions in the archaeological sequence? Are there cultural changes attributable to those events, or can the manifest changes more easily be related to longer-term evolutionary developments—to "process"?

## Prehistoric Ashes and Human Occupation

The Brooks River archaeological sequence has been defined by means of a number of sequential cultural phases, the contents of which have been described fairly extensively elsewhere (e.g., Dumond 1971; Dumond *et al.* 1976). With the phase conceived (after Willey and Phillips 1958:22) as a stable assemblage of cultural traits regularly reflected in artifact collections from related sites of the same time in the same area, it follows that the junctures between phases are the loci of major changes in material culture. Figure 12.2 illustrates the temporal relationship of recognizable ash deposits to phases of occupation at Brooks River.

Of the ashes, those designated B, D, and H at times occur stratigraphically within deposits of single phases, apparently without any serious disruption of the occupation. The position of Ash E is ambiguous but may very well be similar. All of these (see Table 12.1) are rather thin deposits.

Ashes C, F, and G, on the other hand, occur at the points of juncture between phases, a circumstance that suggests that the deposition of ash itself may have evoked responses in which significant cultural changes were embodied, or that it may have caused an interruption of occupation that endured long enough to be associated with evolutionary changes permitting the recognition of separate phases. These three are relatively thick deposits (Table 12.1). Closer consideration, however, lessens the apparent cogency of the following coincidences.

Volcanic Ash G was deposited immediately after the contemporary but desultory occupation of Brooks River by two separate ethnic groups—one (the Beachridge complex) of people largely of the interior, of the so-called Northern Archaic tradition (Anderson 1968; Dumond 1977); the second (the Strand phase) of people pertaining to the Pacific coast of the peninsula, of what has been termed variously a late stage of the Ocean Bay tradition (Clark 1975) or an early stage of the Kodiak tradition (Dumond *et al.* 1976; Dumond 1977). To

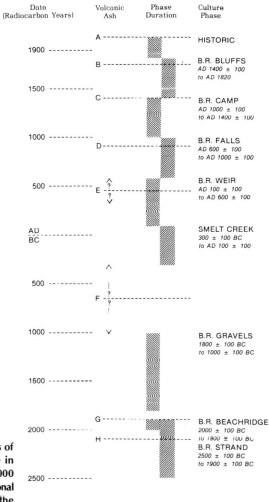

FIGURE 12.2. Diagram of relationships of volcanic ash deposits and phases of culture in the upper Naknek drainage. (From about 2000 to 1900 B.C. the locality was in desultory seasonal use by two distinct groups of people, those of the Beachridge and Strand phases.)

judge by the extensive bone meal associated with the relatively ephemeral campsites in the restricted areas of known settlement, both groups appear to have been in the region primarily for the purpose of hunting caribou.

Immediately above Ash G—that is, at or shortly after 1900 B.C.—there appear the remains of people of the Arctic Small Tool tradition (Brooks River Gravels phase). Besides hunting caribou (the evidence for which is almost entirely inferential), these people also apparently took fish from salmon runs like those that have regularly traversed Brooks River in later time. They instituted a markedly different pattern of settlement from that of their predecessors at Brooks River, for they constructed semipermanent houses at numerous locations along its course, and their plentiful occupational remains indicate a substantial presence over a number of centuries. Furthermore, at about this time or only shortly before (i.e., 2200 to 2000 B.C.) people of this same culture spread across northern Alaska and through northern Canada to Greenland (Dumond 1977, with references).

Thus, although it is possible that the deposit of Volcanic Ash G in the Brooks River area was itself sufficient to discourage further use of the place by people of the Beachridge complex and of the Strand phase, reoccupation after the ash fall was by an entirely different people whose contemporary expansion throughout much of arctic America indicates clearly that their vitality was quite independent of any volcanism on the Alaska Peninsula. The major ethnic break in the cultural sequence that occurs with Ash G, then, cannot reasonably be attributed solely to the volcanic event.

With Volcanic Ash F, on the other hand, the situation is at first glance different, for the ash (Figure 12.2) coincides with a major hiatus in occupation. After disappearance of the Arctic Small Tool people sometime around the end of the second millennium B.C., the drainage was occupied no later than 300 B.C. by people of the ensuing, and apparently related, Norton tradition. Although the dates of this abandonment and reappearance are not entirely fixed, it now appears that there must have been a break in occupation at Brooks River of at least 5 centuries, and perhaps of nearly double this period—roughly time for an additional cultural phase to appear in the sequence. Yet continuity in a number of artifact classes is strikingly clear despite the apparent length of this interruption, and at least some measure of enduring cultural relationship across the hiatus can be postulated until strong evidence to the contrary appears.

This, then, looks like a distinct candidate for a case in which a major volcanic eruption caused a traumatic and lengthy depopulation. There is a good reason, however, why this is probably not the best interpretation. The occupation of Small Tool people seems to have ended almost everywhere in Alaska between about 1500 and 1000 B.C., apparently rather suddenly, suggesting that systematic and pervasive factors were involved—factors much more widespread than the ejecta of a volcano on the Alaska Peninsula. Commonly the disappearance of the Small Tool people is succeeded by a hiatus in occupation comparable to that at Brooks River, even though artifact typology continues to imply that the Norton people were relatively direct descendants of their Arctic Small Tool predecessors (Dumond 1977).

The case of Volcanic Ash C is again different in several ways. In the first place, although the two phases separated so conveniently in the field by this ash are very closely related, each assemblage is distinctive. Thus a collection representing the later aspect of the Brooks River Camp phase is closer to the initial aspect of the same phase, some three centuries earlier, than it is to early collections of the succeeding Brooks River Bluffs phase from which it is separated by less than half as much time. In the second place, the period of transition between these phases is so short that no occupational hiatus can be discovered by dating techniques as imprecise as the radiocarbon method. All together these combine to suggest that if this ash fall caused an abandonment of the Brooks River region at all, it was for a short time only, even though it was a time of evolutionary change in artifact form such as to make the discrimination of separate phases possible. Furthermore, because this change does not seem (at least on the basis of present evidence) to be related to any disruption of occupation of such geographic extent as those mentioned earlier, it reinforces the suspicion that in this case, at least, the volcanic event may be somehow related to the alterations in artifact form.

In short, of the three cases in which volcanic ash deposits certainly coincide with breaks in the regular course of cultural development of such magnitude as to permit the definition of separate phases, only one such break appears to involve relatively localized cultural manifestations and hence to provide a case in which a volcanic eruption may have been integrally related in some fashion to significant changes in material culture.

Additional light may be thrown on the matter through a consideration of possibly analogous events that surrounded the Katmai eruption of June 1912.

## The Native Population in 1912

Upon the establishment of Russian control on the northern Alaska Peninsula—before 1800 on the Pacific coast, and by 1820 on Bristol Bay—the bulk of that region was inhabited by natives termed by the Russians "Aleuts," even though they were recognized to speak a language distinct from that of natives of the Aleutian Islands (e.g., Wrangell 1839:122). These have since been set apart as Peninsular Eskimos (Oswalt 1967b), and placed with natives of Kodiak Island and Prince William Sound among speakers of a form of Eskimo speech referred to as Sugpiaq (Krauss 1974).

In the Bristol Bay settlements at the mouths of the Egegik and Naknek Rivers, however, at the moment of the Russian arrival these local Sugpiaq speakers were being displaced by speakers of the Central Yupik form of Eskimo (Krauss 1974) who were referred to locally as Aglegmiut, a people reputedly deriving from the Kuskokwim River vicinity, or more specifically from Nunivak Island (Oswalt 1967b:4–5, with references; Wrangell 1839:121ff). After the Russian establishment, these peoples and their settlements were mentioned regularly in documents of the Russian Church, particularly in reports of vital statistics (Alaska Russian Church 1816–1936). With few exceptions, the same settlements appear in the early U.S. census reports, the first of which is dated 1880 and was actually derived in large part from Russian Church records (Petroff 1884:v).

On the Pacific coast of the peninsula, the settlements within the area of immediate interest here were Katmai and Douglas; a settlement at Kukak Bay was mentioned much more sporadically and received no entry in U.S. census reports after 1880. On Bristol Bay, all settlements at the mouth of the Naknek River were consistently subsumed under the designation Paugvik or Pauwik— despite the fact that there were at least two settlements (separated by the river), only one of which was properly designated Pauwik according to traditions that survive today; when the village name Naknek appears in the U.S. census of 1900, it apparently designates a settlement on the same side of the river as the village properly called Pauwik. Similarly, the interior community in the upper Naknek drainage was inclusively labeled Severnovsk, later Savonoski; it was composed, however, of at least two separate settlements, and possibly three, on the lower course of the Savonoski River, and may have also included another at Brooks River (Petroff 1884: maps in pocket; Davis 1961).

Nineteenth-century U.S. census figures for these settlements are shown in Table 12.2. Although the reports of 1880 and 1890 contain breakdowns by community, the published census of 1900 contains only totals by somewhat larger civil divisions; the original enumeration sheets of that census (U.S. Census 1900), were drawn upon for the table. Unhappily, the published 1910 census—which would be of special interest here—gives only figures for even larger districts, and the original enumeration sheets have not yet been released.

Traditionally, the native people of the area were relatively sedentary hunters and fishermen who depended upon a wide range of both coastal and interior resources. Though the fact that ocean products were heavily used causes the numerical relationship of population to land area per se to be largely meaningless in a subsistence sense, it nevertheless gives some idea of just how low the overall density was. The most heavily populated portion of the upper Alaska Peninsula was precisely that bounded by the three communities of Pauwik, Douglas, and Katmai, consisting of somewhat more than 8000 km², but certainly not containing all of even the terrestrial sustaining areas of those settlements. With populations as shown on Table 12.2, this yields a maximum possible density of about one person per 12 km² in 1880, and of only one per 22 km² in 1900, as the native population continued a decline that must have begun with the arrival of the Russians, and—of local importance—as settle-

**TABLE 12.2**
**Late-nineteenth-century Native Population of Some Settlements of the Alaska Peninsula[a]**

|                     | 1880 | 1890 | 1900 |
|---------------------|------|------|------|
| Pacific drainage    |      |      |      |
| Douglas             | 40   | 82   | 70   |
| Kukak               | 37   | —    | —    |
| Katmai              | 218  | 131  | 62   |
| Total               | 295  | 213  | 132  |
| Bering Sea drainage |      |      |      |
| Pauwik              | 192  | 93   | 92   |
| Naknek              | —    | —    | 39   |
| Savonoski           | 162  | 94   | 100  |
| Total               | 354  | 187  | 231  |
| Grand total         | 649  | 400  | 363  |

[a]From Petroff (1884), Porter (1893), and U.S. Census (1900).

ment distribution was drastically affected by the attraction of commercial fish processing stations that were being established at various locations, many of them outside the area just referred to. Perhaps a better idea of the density within the entire sustaining area can be gained by considering a larger area of the peninsula. The region from Iliamna Lake on the north to Chignik on the south, a distance of about 400 km, contains a total area of about 53,000 km². The roughly 1230 individuals of the native population of 1880 constituted a density of about one person in 32 km², the 1000 individuals of 1900 a density of one in 53 km².

## The Eruption of 1912

To the thinness of the population can be partly attributed the fact that no lives were lost as a direct result of the Katmai eruption of 6 June 1912. But at least equally important, a large proportion of even the few people resident in the vicinity were induced to leave their homes ahead of the real disaster.

Earthquakes are said to have begun some days before the explosion, perhaps by 1 June, and shocks were severe by 4 and 5 June (Griggs 1922:19). On 4 June, the two families (a total of six people) then staying in Katmai village fled by boat southwestward along the peninsula coast. The majority of the Katmai people, however, were already camped at a fishing station at Kaflia Bay (Griggs 1922:17).

According to one recently interviewed survivor (Toleffson 1975) the people of both Katmai and Douglas had for several years gathered at Kaflia Bay to catch and smoke salmon for sale to a Kodiak entrepreneur, who also ran a small store at the site. This report is rendered particularly plausible by the lack of any evidence in Alaska Commercial Company records (Oswalt 1967a) that the early trading posts at Katmai and Douglas continued in operation much after 1900; indeed, it seems possible that these two settlements were actually in the informal process of being relocated at Kaflia Bay.

Apparently no attempt was made there to flee the initial shocks, and when the eruption came, the Katmai people were still camped south of the bay and the Douglas people north of it. Darkness from the rain of ash was said to have lasted for 48 hours, until the pumice became a meter deep and the streams and springs were clogged; the position would have been deadly had not the exhortations of an old man been heeded, and water collected in every available pot as the ash fall began.

When daylight finally broke and the plight became evident, a trio of three-man kayaks was sent across Shelikof Strait to the Kodiak settlements for help. Nearly a week after the eruption, on 12 June, the U.S. Revenue Cutter Service responded with an emergency run that picked up the total community of about 100 people and transported them to Afognak Island of the Kodiak group (Hussey 1971:356–357, with references to U.S. Revenue Cutter Service reports; Toleffson 1975).

In the interior, most if not all of the people of Savonoski—only 34 km from

Mount Katmai itself—had moved to the mouth of the Naknek River by the time of the major explosions. According to one account, this was through fear of the earth tremors of the days before 6 June (Griggs 1922:17), although native informants interviewed more recently around modern Naknek have indicated that by the day of the eruption nearly all of the Savonoski people had moved to that area in search of seasonal employment in the commercial fishery, with only two families actually maintaining interior residence at the time of the volcanic event (Davis 1961).

If the settlements closest to Mount Katmai were already largely abandoned, it is not particularly surprising that there were no casualties resulting directly from the eruption.

## Aftermath and Recovery

Because of the obvious impossibility of living in any village on the peninsula coast of Shelikof Strait in the days after the eruption, in July a total of 92 of the survivors of the settlements of Katmai and Douglas were resettled at Ivanof Bay, on the peninsula coast southwest of Chignik, and in August were moved again at their own request to a final site at modern Perryville, where they were left with supplies, building materials, and the expectation that they would be thenceforth self-supporting. The settlement so established has endured.

On the other side of the peninsula, the displaced Savonoski people first scattered to stay with various relatives and contacts in the vicinity of present Naknek, South Naknek, and Levelock, and in the fall of 1912, after the Russian priest from the mission at Nushagak met with the leaders of both Savonoski and the coastal villages to determine upon a site, they established the new settlement of Savonoski only some 13 km upstream from the mouth of the Naknek River (Dumond 1974). Whether either of the two communities of refugees would have tried to return to their pre-eruption homesites seems doubtful, given the successful establishment of the new Perryville and the availability of at least some jobs in the growing salmon packing industry at the mouth of the Naknek River.

The likelihood of a return of former residents to the old Savonoski was lessened even further by the traumatic impact of the influenza epidemic of 1919. The disease raged for about 4 weeks in the Naknek settlements, leaving more than 60 natives dead (Heinbockel 1919). Considering the figures in Table 12.2, this must have been well over one-fourth—perhaps even one-third—of the native population of the Naknek River. Savonoski, according to one account, numbered 54 people at the onset of the sickness (Davis 1954:71); the total number of its dead was not reported, but 13 were said to have succumbed well before the disease had run its course (Heinbockel 1919). There are no similar direct accounts of influenza losses at Perryville, but for one cause or another the population there was reported in the 1920 census to have declined to 85.

Following the eruption the landscape was desolate. Even in the broad areas only moderately affected by ejecta, such as parts of Kodiak Island and around the eastern edge of Naknek Lake, the ash was said to lie 15 cm deep and to have destroyed 70% of the vegetation (Ball 1914:63–64). Although two families reportedly returned almost immediately to the inland Savonoski settlement and tried to live there, before a year was out they found the dust and residual heat impossible to stand (Davis 1961). The Pacific coastal area directly in the path of the major eastward-moving cloud of ejecta was even worse. When in 1914 two Alaskans visited Katmai Bay, they found what seemed a plain, the former bay still solid with floating pumice. Later in 1915 when a National Geographic Society party entered the region, there were few living plants in Katmai Valley except some herbaceous varieties surviving in locations where the wind had swept away the pumice; some balsam poplars had also survived, but the profuse alder and birch cover of the valley was dead (Griggs 1922:82–83). Later years would bring only slight respite to the valley, as the load of pumice steadily borne downward by the river choked the valley lowlands, raising the water table well above the old ground surface at Katmai village, until by 1953 excavations in the pumice-covered structures were impossible (Davis 1954:38).

Yet in all but those areas closest to the eruption—the Valley of Ten Thousand Smokes to the west of the mountain ridge (the real seat of the activity) and Katmai Valley and some outliers to the east—vegetational recovery was remarkably swift. In areas where 15 cm of ash hid virtually all low-lying vegetation, the first rains created cracks in the surface, through which woody plants immediately emerged (Ball 1914:63). People moved in soon after. Even on the heavily affected coast there were trapping cabins in use by 1914 (Davis 1954:69); in 1923 the commercial digging and canning of clams began around former Douglas (Hussey 1971:423). Inland, despite unsuccessful attempts to colonize immediately, by 1918 former residents of Savonoski were making annual bear hunts to the vicinity of the abandoned village, although finding it still too desolate for comfortable permanent habitation (Griggs 1922:17).

That coastal animals and the larger land mammals may have survived with relatively little mortality, even in the great area involved, is not entirely surprising, but what of other fauna? Again, information suggests that mortality was less than might be expected. Among small land mammals and birds it was clearly heavy, and among fish ascending the rivers, finding themselves suddenly battling thick loads of ash that shortly turned to mud, the kill was almost complete (Ball 1914). But the major part of the important run of sockeye salmon had not yet entered the streams at the time of the eruption.

In the Kodiak Island group, the figures for the commercial salmon catch in one area heavily affected by the ash fall reveal that there was no noticeable drop in the catch in the years 1912 through 1914, but rather that a substantial decline—to less than one-half of normal—did occur between 1915 and 1920, as an effect of the smallness of the salmon population that was able to spawn in the years immediately after the eruption. Recovery thereafter, however, was virtually instantaneous (Eicher and Rounsefell 1957).

The situation just described is a case in which a number of adjacent streams were affected equally by the ash. Around the mouth of the Naknek River, on the other hand, during normal spawning seasons there are more fish arriving in search of the mouth of the Kvichak River, only 15 km away, than those seeking the Naknek, and typically many of these enter and school about the Naknek River mouth before finding their way onward to their destination. The Kvichak drains an area that was not significantly affected by the Katmai eruption. Therefore, although any inland fishermen higher up the Naknek drainage system must have experienced a severe shortage of fish in some years following 1912, it is likely that native fishermen at the very mouth of the Naknek River would have noticed no particular diminution in the numbers of arriving fish. Indeed, figures for the commercial catch of sockeye salmon in the Naknek-Kvichak district of Bristol Bay reflect no sustained slump that is apparently attributable to results of the eruption (Alaska Department of Fish and Game n.d.). Although 1915 was not a good year, it was preceded and followed by years of high catch. The first sustained decline after 1912 did not begin until 1919, and although during the ensuing years Fish and Wildlife Service personnel were busy in the Naknek system trying to reduce predation on young sockeye and to improve salmon spawning, they made only the briefest of references either to the presence of the noticeable and recent volcanic ash deposit, or to possible effects of the eruption on the fish runs (Bower 1921:31–32; 1922:16–17). One can only conclude that by then, at least, the effects of the ash fall were considerably ameliorated.

So far as saltwater fishes are concerned, although a heavy deposition of ash, as well as certain other accompaniments of eruptions (such as shock waves and underwater landslides) are reported to take it least a sporadic toll of some forms such as cod and rockfish (Hanna 1971), it is doubtful that the effect upon an area-wide spawning population such as that around the north Pacific would be great enough to be noticeable to the eyes of native fishermen.

In 1918 a Presidential proclamation established Katmai National Monument, to consist of the vicinity of Katmai Volcano with the flanking valleys of the Katmai River and the River Lethe—the latter newly designated the Valley of Ten Thousand Smokes. In 1931 a second proclamation extended the monument boundaries to include the area north of Douglas village to Cape Douglas, as well as most of the lake system in the Naknek drainage. It was 1940, however, before the National Park Service made even a rudimentary on-ground examination of the area, by which time illicit trappers' cabins were widespread, and it was not until 1950 that a summer ranger station was established within the monument (Hussey 1971:422–429). In 1953 and 1954, the Park Service's Katmai Project of research and description provided an inventory of the monument territory; the biological survey that was a part of the study (Cahalane 1959) makes it clear that with the exception of the relatively small areas of "total volcanism"—that is, within the Valley of Ten Thousand Smokes itself—and a very few nearby areas, recovery from the effects of the eruption were complete and in some places had been so for many years.

It thus seems clear that the regions of the coast on the one hand and the lakes on the other—both of them 30 km and more from the center of the eruption, and both of them the locations of abandoned villages—were with few or no exceptions completely habitable within 20 years after the eruption, and much of their areas must have been habitable at least 10 years earlier than that. By 1932 at the latest, the pre-eruption lifeway could have been pursued with little if any inconvenience stemming from the volcanic activity itself.

## Process and Volcanic Event at Brooks River

This returns us to a consideration of the volcanic ash deposits set out in Table 12.1, to the relationships portrayed in Figure 12.2, and in particular to the probable impact of Volcanic Ashes C, F, and G upon people inhabiting the Naknek drainage.

Given aboriginal conditions, among which was the absence of rescue capabilities such as those of the Revenue Cutter Service, an eruption like that of Katmai Volcano, or even one of considerably less magnitude, might well have had lethal effects. Given also the apparent fact that the aboriginal population of the region was greater than the native population of the late nineteenth century—perhaps twice as great (Dumond 1973)—the number of people displaced by a single eruption could conceivably have been fairly substantial. With densities approximately double those mentioned earlier for the entire northern Alaska Peninsula in 1880, a terrestrial region the same size as the total area that was covered by Katmai ash to depths of 15 cm and more (Figure 12.1) would have included more than 1000 inhabitants. Nevertheless, whether even this amount of ash would have displaced the entire human population is extremely doubtful. The Katmai eruption is not reported to have occasioned any significant relocation of human population on Kodiak Island, for instance, even though some places were blanketed with its ash to a depth of 30 cm.

Even taking the most cataclysmic view possible, therefore, from analogy with the Katmai eruption it seems certain that the fall of Ashes C, F, and G at Brooks River—all of them now much thinner than the ash of 1912—could each have occasioned the abandonment of that inland locality for no more than 2 decades at the very most, and should each have been sufficient to interrupt seriously the customary subsistence use of the region for less than 10 years. Meanwhile, the lower portions of the Naknek drainage were substantially or even totally unaffected by those ashes. Indeed, only two ash deposits are visible in cuts at the mouth of the Naknek River—one of these the Katmai ash, the other an earlier deposit that on the basis of position and apparent date must be Ash C. Neither of these is now more than a centimeter or two in thickness. Although the known Katmai eruption caused some local inconvenience on the coast of Bristol Bay, it occasioned no disruption at all of the coastal pattern of subsistence and settlement. Thus one is led to presume that even the heaviest prehistoric deposits of volcanic material in evidence in the upper Naknek drainage would have caused no more than a temporary shift of population to lower portions of the Naknek River system—to areas that were almost certainly in seasonal use by the same people.

To consider the three crucial ash deposits in turn, and to reiterate some comments made earlier, the fall of Ash G coincided with the only case of complete population replacement in the prehistory of the Naknek drainage. It is just possible, although highly speculative, that the associated volcanic eruption—an eruption that must have occurred somewhere on the high mountain backbone of the Alaska Peninsula—effectively but briefly blocked the path of Pacific coastal people who were one of two groups who had previously hunted caribou in the upper Naknek drainage, and so may have served to ease the occupation of that area by new people from the north. That it had any other repercussion on the specific course of evolution of local culture, however, is not suggested at all by any evidence at hand, and the advent of the new foreigners was earlier indicated to have been but a single detail of an important and widespread development in Alaskan prehistory, in which the localized effect of a volcanic eruption could scarcely have been significant. The circumstances surrounding the deposition of Ash G therefore throw no light upon the question posed at the outset of this discussion: Did the series of volcanic events have any perceptible impact upon the evolution of local aboriginal culture?

As regards the fall of Ash F, it seems evident that even at Brooks River, where the ash is so clearly present, the long period without occupation that coincides with the volcanic deposition far exceeds in length any such period that could reasonably be attributed to the ashfall. Furthermore, as mentioned earlier, the apparent abandonment of the Naknek drainage at about this time—either before or after the actual deposition of Ash F—seems to be at least roughly matched by similar hiatuses in occupation elsewhere. The abandonment of the Brooks River area around the time of the volcanic eruption, then, appears to be no more than another coincidence.

Concerning Ash C, on the other hand, there is no archaeologically perceptible evidence of any period of abandonment at all, and presumption of the existence of such a period must depend solely upon the presence of the volcanic ash and upon analogy with events of 1912. Furthermore, since at the mouth of the Naknek River there is clear evidence of Brooks River Camp phase occupation stratigraphically beneath the presumed Volcanic Ash C, and even clearer evidence of Brooks River Bluffs phase occupation immediately above the same ash, it is reasonable to think that the transition from Camp to Bluffs phase defined at Brooks River occurred also in the lower drainage, where it was also probably equally rapid and abrupt. Although the archaeological sample from the mouth of the river is not adequate now to demonstrate the truth of this latter suggestion as fully as might be desired, there is certainly no evidence of any more leisurely transition from one phase to the other that might have coincided with a period in which the upper drainage was not occupied, nor is there any evidence at all of any significant shift in any subsistence or settlement pattern from one phase to the other.

The upshot is that one must conclude that the transition from the Camp phase to the Bluffs phase, like the earlier transitions in evidence in the Naknek drainage, resulted from factors largely or completely independent of any coincidental volcanic events.

## SUMMARY CONCLUSIONS

In the foregoing pages two separate questions or sets of questions have been addressed, the first of them purely archaeological—regarding the usefulness of volcanic ash deposits in the dating of one particular sequence—and the second of significance to prehistory—regarding the impact of a series of volcanic events in the upper Naknek drainage upon the culture history of its inhabitants.

In regard to the first, it must be said that the presence of a relatively large number of volcanic ash deposits in a single restricted area is extremely helpful to the field archaeologist, once the stratigraphy and certain temporal factors are understood, permitting the identification of occupations encountered in limited tests with an efficiency and rapidity that can scarcely be achieved in any other way. This is, however, a seat-of-the-pants, impressionistic use of the ash stratigraphy, and estimates of dates so derived must be confirmed by artifact typology or by radiometric determinations. Less enthusiasm appears indicated for the results of the more extensive laboratory analysis of deposits necessary to achieve even a rudimentary grasp of characteristics of the deposits that might permit the eventual and positive identification of each ash no matter in what situation it appears. Attempts to discriminate the 10 major deposits of the upper Naknek drainage so precisely as to permit their use as definitive dating devices, rather than as convenient field supplements to other techniques, were not successful within limits provided by considerations of economy of either time or money. To generalize, this can be expected to be the case in at least most situations in which multiple deposits derive from only a few sources.

In regard to the second set of questions, it has been indicated that in the upper Naknek drainage all major cultural changes seem uniformly to have resulted from processes of development that bore no direct relation to specific volcanic events. This is not to say that the eruptions indicated by the 10 major volcanic ash deposits at Brooks River were of no consequence to the lives of the particular people who experienced them, for all of them were potentially traumatic. Instead, it is to suggest that volcanic events of the magnitude of those in evidence in the upper Naknek drainage, even up to that of the 1912 eruption of Mount Katmai, are simply one among many classes of environmental phenomena (together with all other factors affecting the supply of faunal and floral resources) that bring about intermittent and relatively short-range and short-term displacements of human occupants—displacements that are often almost impossible to recognize in the archaeological record of prehistoric hunters and gatherers, that are generally even more nearly impossible to account for if recognized, and that affect material culture—if at all—only so indirectly that causal connections are imperceptible.

To generalize again, in the evolution of culture among relatively mobile people of low population density, the common recurrence of unsettling volcanic disturbances may constitute an important factor in the determination of overall strategies of survival, subsistence, and settlement. But in regard to northern peoples in particular, it seems highly unlikely that any single volcanic event would have so completely destroyed the resources of the entire extensive territory of any single people as to render their flexible social organization—with its mechanisms permitting the peaceful relocation of families with relatives and partners—incapable of dealing with the emergency.

Is this a general negation of the importance of sudden and unforeseen events in human life? The answer is no; the conclusions suggested here cannot be generalized that far.

In the North, volcanic eruptions must always have been far less disruptive, say, than major cyclical fluctuations in the sizes of regional caribou populations. It would be folly to suggest that a sharp reduction in caribou strength would not have immediate effects upon life among northern hunters, or that it might not well have longer-term repercussions in patterned changes in their customary behavior. Thus the ultimate importance of certain unforeseen events cannot be denied wholesale.

The conclusion here, then, is a more restricted one: As the culture history of northern hunting peoples is viewed through a glass as fogged as that available to prehistorians, **individual volcanic events** of the magnitude of those evidenced by the specific ash deposits at Brooks River tend to be simply and consistently irrelevant.

## ACKNOWLEDGMENTS

Material reported here was derived from field and laboratory work between 1960 and 1975, which was variously supported by grants from the National Science Foundation and from the University of Oregon, by contracts with the National Park Service, and through material assistance from the National Marine Fisheries Service (formerly the Bureau of Commercial Fisheries). I am grateful for the library aid of G. H. Clark in historical matters, and for the laboratory assistance of Michael Nowak—who acted under advice from L. R. Kittleman—in work with the volcanic ashes. I thank Mike Toleffson for access to unpublished material, William B. Workman for comments upon an earlier draft of this paper, and—in particular—the people of the Naknek region for their consistent and generous cooperation in all things over the years. The map was drawn by Carol Steichen Dumond.

## REFERENCES

Anderson, Douglas D.
    1968   A Stone Age campsite at the gateway to America. *Scientific American* 218(6):24–33.

Alaska Department of Fish and Game
   n.d.   Unpublished compilation of commercial catch of Bristol Bay sockeye salmon, 1893–1970,
          based on various sources. On file in the Commercial Fisheries Division, Alaska Depart-
          ment of Fish and Game, Anchorage.
Alaska Russian Church
   1816–  Vital statistics, Kodiak, Nushagak, Unalaska missions.
   1936   Microfilmed portion of Russian Orthodox Greek Catholic Church of North America,
          Diocese of Alaska, Alaska Russian Church Archives. Library of Congress MS 64-1221.
Ball, Edward M.
   1914   Untitled report on the effect of the eruption of Katmai Volcano on fisheries, animals, and
          plant life of the Afognak Island Reservation. In Barton W. Evermann, *Alaska fisheries and
          fur seal industries in 1913*. U.S. Bureau of Fisheries Document No. 797.
Bell, Barbara
   1971   The Dark Ages in ancient history. *American Journal of Archaeology* 75(1):1–26.
Bower, Ward T.
   1921   Alaska fishery and fur-seal industries in 1920. U.S. Bureau of Fisheries Document No.
          909.
   1922   Alaska fishery and fur-seal industries in 1921. U.S. Bureau of Fisheries Document No.
          933.
Cahalane, Victor H.
   1959   A biological survey of Katmai National Monument. Smithsonian Miscellaneous Collec-
          tions 138(5).
Clark, Donald W.
   1975   Technological continuity and change within a persistent maritime adaptation: Kodiak
          Island, Alaska. In *Prehistoric maritime adaptations of the circumpolar zone*, edited by W.
          Fitzhugh. The Hague: Mouton.
Curtis, G. H.
   1955   Importance of Novarupta during eruption of Mount Katmai, Alaska, in 1912 [abstract].
          *Geological Society of America Bulletin* 66.1547.
Davis, Wilbur A.
   1954   Archaeological investigations of inland and coastal sites of the Katmai National Monu-
          ment, Alaska. Report to the U.S. National Park Service. Archives of Archaeology No. 4.
   1961   Tape recordings of eyewitness accounts of Mount Katmai eruption of 6 June, 1912. Tape
          and transcript in the Department of Anthropology, University of Oregon.
Dumond, Don E.
   1971   Archaeology in the Katmai region, southwestern Alaska. *University of Oregon An-
          thropological Papers* No. 2.
   1973   Late aboriginal population of the Alaska Peninsula. Paper presented at the annual
          meeting of the American Anthropological Association.
   1974   Field notes of interviews in the Naknek vicinity. On file in the Department of Anthropol-
          ogy, University of Oregon.
   1977   *The Eskimos and Aleuts*. London: Thames and Hudson.
Dumond, Don E., Winfield Henn, and Robert Stuckenrath
   1976   Archaeology and prehistory on the Alaska Peninsula. *Anthropological Papers of the
          University of Alaska* 18(1):17–29.
Eicher, George J., Jr., and George A. Rounsefell
   1957   Effects of lake fertilization by volcanic activity on abundance of salmon. *Limnology and
          Oceanography* 2(2):70–78.
Evans, Sir Arthur J.
   1921–  *The palace of Minos* (4 Vols.). London: Macmillan.
   1935
Griggs, Robert F.
   1922   *The Valley of Ten Thousand Smokes*. Washington, D.C.: National Geographic Society.
Hanna, G. Dallas
   1971   Introduction: Biological effects of the earthquake as observed in 1965. In *The great Alaska*

*earthquake of 1964, Biology*, by Committee on the Alaska Earthquake, Division of Earth Sciences, National Research Council. Washington, D.C.: National Academy of Sciences.

Heinbockel, J. F.
   1919   Report on 1919 influenza epidemic: Alaska Packers Association, Naknek Station. Duplicated report, Box 468, 9-1-71, RG 126, Office of Territories Classified Files, 1907–1951. U.S. National Archives, Washington, D.C.

Hussey, John A.
   1971   *Embattled Katmai: A history of Katmai National Monument*. National Park Service Historic Resource Survey. San Francisco: Western Service Center, NPS.

Krauss, Michael E.
   1974   *Native peoples and languages of Alaska* [map]. Fairbanks: Alaska Native Language Center, University of Alaska.

Nowak, Michael
   1968   Archaeological dating by means of volanic ash strata. Doctoral dissertation, Dept. of Anthropology, University of Oregon. University Microfilms 69-6654.

Oswalt, Wendell H.
   1967a   *Alaska Commercial Company records: 1868–1911*. College, Alaska: University of Alaska Library.

   1967b   *Alaskan Eskimos*. San Francisco: Chandler.

Petroff, Ivan
   1884   *Report on population, industries, and resources of Alaska*. Washington, D.C.: Department of the Interior, Census Office.

Porter, Robert P.
   1893   *Report on population and resources of Alaska at the eleventh census: 1890*. Washington, D.C.: Department of the Interior, Census Office.

Toleffson, Mike
   1975   Taped interview with Father Harry Kiakokonok, April 29, 1975. Tape and transcript at Katmai National Monument, King Salmon, Alaska.

U.S. Census
   1900   Enumeration records 17, 18 (Vol. 6), and 20 (Vol. 7), Census of 1900. U.S. National Archives, Washington, D.C.

Willey, Gordon R., and Philip Phillips
   1958   *Method and theory in American archaeology*. Chicago: University of Chicago Press.

Wrangell, Ferdinand P. von
   1839   *Statische und etnographische Nachrichten über die russischen Besitzungen an der Nordwestküste von America*, edited by K. E. von Baer. St. Petersburg: Kaiserlich Akademie der Wissenschaften.

# 13

# Pollen Influx and the Deposition of Mazama and Glacier Peak Tephra

ERIC BLINMAN
PETER J. MEHRINGER, JR.
JOHN C. SHEPPARD

## INTRODUCTION

White layers of volcanic ash stand out dramatically against landscapes and sediments and serve as mute evidence of the spectacular volcanic activity that has characterized many parts of the earth through the late Quaternary. As one of the tools of the Quaternary stratigrapher, these ash layers have been used to correlate events in time over hundreds of kilometers. Additionally, the destructive potential of ashfalls has prompted speculations as to their ecological consequences.

The importance of these speculations to archaeologists lies in the correlation of ashfalls with changes in the archaeological record. Such correlations often result in hypotheses of causation that are usually based on inferred damage to prehistoric resources (Bedwell and Cressman 1971; Workman 1974). Although analogies are often drawn from modern ashfalls and their effects, the information necessary to characterize ancient ashfalls, and to test hypotheses of their influence on man and nature, is rarely available.

Studies of modern ashfalls demonstrate the complex relationship between ashfalls and their effects. Biologists have correlated ecological impact with the thickness, duration, and season of ashfall (Griggs 1915:198, 201; Eggler 1948:435). In addition, they have noted the importance of the redeposition of ash, especially for aquatic ecosystems (Evermann 1914; Eicher and Rounsefell 1957:72). Unfortunately, these variables are difficult to measure for ancient ashfalls. The thickness of an ash layer, as measured in the sedimentary record,

*VOLCANIC ACTIVITY AND HUMAN ECOLOGY*

may have been modified by redeposition or erosion. Although volcanic ashes may be radiocarbon-dated, the uncertainties of these measurements usually far exceed the duration of the ashfall. Tree-rings have been used to date eruptions and they can also indicate whether the event occurred during or after the growing season (Lawrence 1954; Smiley 1958:190; Breternitz 1967:73), but this technique is temporally and geographically limited. Other dating methods, such as varves (Rymer and Sims 1976:9–12), may have sufficient resolution to determine the season and duration of ashfall, but they are either restricted in application or are untried.

Pollen analysis has been used in the study of postashfall forest succession (Hansen 1947:99–103) but has only recently been developed as a method of studying the deposition of ancient ashfalls (Mehringer, Arno, and Petersen 1977; Mehringer, Blinman, and Petersen 1977; Blinman 1978). The number of pollen grains in an ash may reflect the duration of its fall and may also be useful in distinguishing ashfall layers from redeposited ash. The sequence of pollen types within ashes may indicate the season of ashfall and the effects of ashfall on local and regional pollen production.

In this chapter we report the pollen content of Glacier Peak or Mazama ash layers from three localities in the Pacific Northwest (Figure 13.1). Layers of both ashes from Lost Trail Pass Bog in the Bitterroot Mountains of Montana were reanalyzed to verify and expand the conclusions of a previous study (Mehringer, Blinman, and Petersen 1977). We also studied the pollen content of Mazama ash and adjacent sediments from Wildcat Lake in the channeled scablands of eastern Washington and Wildhorse Lake on Steens Mountain in Southeastern Oregon.

## POLLEN INFLUX

Pollen influx is an estimate of pollen deposition per weight or volume of sediment. Vegetation changes through time may cause variation in the influx of pollen types and also in total pollen influx. Because annual pollen influx is a summation of seasonal pollen production and dispersal, it is dominated by different pollen types through the year. Although weather may delay or advance flowering times, the relative sequence of flowering is usually predictable (Tippett 1964:1699).

Because factors controlling pollen production, dispersal, and deposition vary, we have used averages to quantify pollen influx. Pollen production is influenced by weather, and pollen dispersal is controlled by water, wind, and vegetation density (Tauber 1977). Sedimentation rates vary with the morphology of a lake basin (Lehman 1975), and water circulation in a lake may erode and redeposit pollen from bottom sediments (M. Davis 1973). An example of variation in annual pollen influx can be seen as relative percentage variations in Tippett (1964:Figures 2, 3, 4).

Estimates of past pollen influx are derived from the pollen concentration

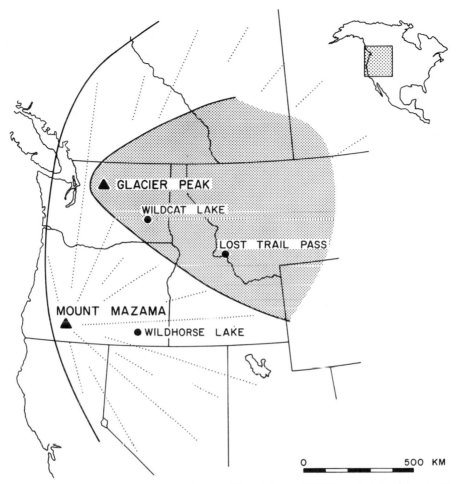

FIGURE 13.1. Locations of Lost Trail Pass, Wildcat Lake, and Wildhorse Lake and the distributions of Mazama and Glacier Peak tephra. (After Lemke et al. 1975, Figure 1.)

and the deposition rate of lake sediment. A deposition rate curve must be
established by radiocarbon dating, varves, or correlation with dated events.
Pollen per volume of sediment is usually determined by the ratio of fossil pollen
to introduced tracers. The pollen concentration of a sample (pollen per cubic
centimeter) is multiplied by the deposition rate (centimeters per year), and the
product is the estimated annual pollen influx (pollen per square centimeter per
year). Pollen influx estimates may be calculated from the average pollen
concentration and average deposition rate for each pollen sample or each
 pollen zone (M. Davis 1969:Figure 6; Mehringer, Arno, and Petersen
1977:358).
    The duration of past depositional events can be estimated by comparing
the pollen content of the deposit with the pollen influx of adjacent sediments

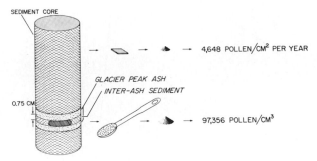

97,356 POLLEN/CM³ x 0.75 CM/DEPOSITIONAL UNIT = 73,017 POLLEN/CM² PER DEPOSITIONAL UNIT

$$\frac{73,017 \text{ POLLEN/CM}^2 \text{ PER DEPOSITIONAL UNIT}}{4,648 \text{ POLLEN/CM}^2 \text{ PER YEAR}} = 15.7 \text{ YEARS/DEPOSITIONAL UNIT}$$

a

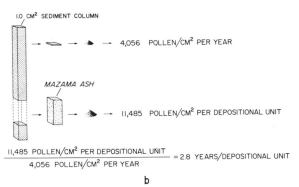

$$\frac{11,485 \text{ POLLEN/CM}^2 \text{ PER DEPOSITIONAL UNIT}}{4,056 \text{ POLLEN/CM}^2 \text{ PER YEAR}} = 2.8 \text{ YEARS/DEPOSITIONAL UNIT}$$

b

**FIGURE 13.2. Calculation of duration of deposition by comparing the pollen content of a depositional unit with adjacent pollen influx: (a) sample taken by excavating sediment and packing it into a spoon; (b) sample taken as part of a sediment column. (After Mehringer, Blinman, and Petersen, _Science_, Vol. 198, "Pollen influx and volcanic ash." Copyright 1977 by the American Association for the Advancement of Science.)**

(Figure 13.2). Pollen content of the unit must be expressed as pollen deposited through a surface of 1 cm² (pollen per square centimeter) and, when divided by the adjacent influx value (pollen per square centimeter per year), the quotient is an estimate of the duration in years. Subsamples may be used to refine the estimate if they reveal clues to the seasonal pattern of pollen deposition or major deviations from the "normal" pollen influx.

When an ash falls into a lake, the shards settle to the bottom and usually form a graded lamina (Reineck and Singh 1975:104) for each billow of the ash cloud. The pollen and other microfossils in the water settle with the shards and are incorporated in the laminae. The pollen types within the laminae may reflect the season of ashfall and the number of pollen will reflect its duration. For example, an ashfall of a day's duration will incorporate little pollen, whereas sporadic ashfall over several months will incorporate considerable pollen. If an ashfall selectively discourages or encourages subsequent flowering, the altered pollen production and deposition may also be detectable. Secondary deposits of ash, eroded from the surrounding landscape into a lake, may be distinguished from airfall deposits by relatively higher pollen content (reflecting a slower deposition rate) and the presence of nonpyroclastic detrital materials. Also, in our experience, redeposited ash usually lacks the distinctive graded laminae that characterize primary ashfalls incorporated into lake sediments.

## METHODS

Sediments were obtained with a chain-hoist-operated piston corer (Cushing and Wright 1965) using 10-cm diameter plastic pipe as the core barrel. Cores containing volcanic ash layers of less than 10-cm thickness were returned from the field in their plastic barrels without special handling. Cores containing ash layers anticipated to be thicker than 10-cm were frozen before transport from the field. If cores containing thick ash layers were not frozen, vibration during transport caused the ash shards to become suspended in their interstitial water. Thin ash layers were rarely disturbed by transport.

In the laboratory, the sediments were exposed by making two shallow (<5 mm) cuts the length of the core and removing half of the plastic barrel. Distorted and contaminated sediment from the outer layer of the core was discarded. A surface of the undisturbed sediment was faced, and the ashes and their surrounding sediments were examined, described, and photographed. Frozen cores were opened, allowed to thaw, and then prepared for sampling.

Samples were taken by two different techniques (Figure 13.2). The ash layers from Lost Trail Pass Bog and Wildhorse Lake were sampled by excavating sediment with a spatula and packing it into a 2-cm³ spoon. Two samples from Wildcat Lake were also taken in this manner, but the remainder were taken as portions of a sediment column with a 1-cm² cross section. The column was isolated using single-edged industrial razor blades, and once isolated, the column was divided into ≃2-cm³ samples at either arbitrary or natural stratigraphic boundaries.

The two sampling techniques differ in their utility and precision. Excavation and packing of the sample into a spoon changes the bulk density of the ash and may distort the actual sample volume. It also requires an estimate of the height of the sampled unit; this estimation may be difficult if the upper and lower boundaries are not planar. Also, sampling must be carried out so that no portion of the unit is over- or underrepresented. On the positive side, this technique ensures an adequate sample from very thin units.

Column sampling results in the waste of portions of the core during preparation of the column, and column samples cannot be used for very thin units because the small sample may contain too few pollen for analysis. However, the consistent cross-sectional area of the column samples eliminates the need for height estimates. Also, there is no distortion of sediment volume during sampling, and the column samples facilitate the subsampling of large units by arbitrary intervals. Thus, the choice of techniques is guided by the amount of material available for sampling and the thickness of the units to be sampled.

Laboratory processing began with the addition of calibrated tablets of *Lycopodium* spore tracers (Stockmarr 1971, 1973). Four tablets (50,000 *Lycopodium*) were added to samples that were predominantly ash, whereas eight tablets (100,000 *Lycopodium*) were added to all other samples. Silica was removed by HF, and organic material was removed by acetolysis for 1 minute. The samples were stained and mounted in silicone oil.

Depending upon the pollen-to-*Lycopodium* ratio, samples were counted at a magnification of either 200× or 400×; all pollen identifications were made at 400×. With two exceptions, at least 200 pollen were counted for each ash sample, and at least 300 pollen were counted for all other samples. Population estimates were calculated from the ratio of pollen to *Lycopodium* tracers. Confidence intervals (95%) were calculated for each estimate of pollen abundance following Maher (1972:89) and Mosimann (1965:659), and by the error propagation method (Crandell and Seabloom 1970). Errors for total pollen per square centimeter range from 11.9 to 15.1% of the estimates.

Pollen influx estimates for sediments adjacent to the Lost Trail Pass Bog and Wildcat Lake ashes were derived from $^{14}$C dates and pollen analysis of the cores. Pollen zones were defined, and average pollen influx was calculated for each zone. The pollen content of ashes from Lost Trail Pass Bog is compared with average pollen influx for the inclusive zone. Although pollen samples and radiocarbon dates were taken within and below Mazama ash from Wildcat Lake, the dating is inadequate for pollen influx calculation and comparisons are limited to pollen zones above Mazama ash. Pollen influx estimates are not yet available for the Wildhorse Lake cores.

## GLACIER PEAK ASHFALLS

### Lost Trail Pass Bog

Lost Trail Pass Bog is located near the Montana–Idaho border, at 2152 m elevation on the crest of the Bitterroot Range. It is about 140 km south of Missoula, Montana, and 60 km north of Salmon, Idaho. The 2-ha bog and meadow occur in a small forested depression left by glacial retreat (Figure 13.3).

Moist forest sites around the bog are in the *Abies lasiocarpa/Menziesia ferruginea* habitat type (Pfister *et al.* 1977). They are dominated by 300–400-year-old *Picea engelmannii* (Engelmann spruce) with abundant smaller *Abies lasiocarpa* (subalpine fir) and an occasional *Pinus contorta* (lodgepole pine).

FIGURE 13.3. Lost Trail Pass Bog, Bitterroot Mountains, Montana. (Photo by Jill Williams, July 1978, courtesy U.S.D.A. Forest Service.)

Scattered mature *Pinus albicaulis* (whitebark pine) grow around the bog. However, Lost Trail Pass Bog is at the lower limits of *Pinus albicaulis*, which becomes a major forest component in stands above 2300 m.

*Pseudotsuga menziesii* (Douglas-fir) is near its upper elevational limit in the vicinity of Lost Trail Pass Bog but is a major component of most stands below 1950 m (within 5 km of the bog). *Pinus ponderosa* (ponderosa pine) is a major forest component on south and west exposures below 1800 m, but these stands are about 3 km farther away than those dominated by *Pseudotsuga*.

Densely forested terrain extends several dozen kilometers in most directions from Lost Trail Pass Bog. Only downwind to the east (beginning about 12 km away in the Big Hole Valley) and far to the south (beginning about 50 km away near Salmon, Idaho) are there considerable expanses of *Artemisia tridentata* steppe and *Agropyron spicatum/Festuca idahoensis* grasslands. Sizable grasslands also occur as close as 30 km to the north in the Bitterroot Valley. However, prevailing winds are from the southwest, west, and northwest and coniferous forest is the dominant vegetation for 150 km in those directions (Mehringer, Arno, and Petersen 1977:Figures 3, 4).

Sediments from the bog were collected to reconstruct postglacial bog, forest, and fire history. Glacial ice withdrew from the basin of Lost Trail Pass Bog by 12,000 B.P. and left a moraine-dammed lake. For 400–500 years sagebrush steppe dominated the landscape. During a brief transition period ending by 11,500 B.P., whitebark and possibly some lodgepole pine replaced the open vegetation. Two distinct layers of Glacier Peak volcanic ash were deposited about 11,250 B.P. Forests dominated by whitebark pine prevailed for the next 3000–4000 years. During this time climatic conditions were probably slightly cooler than today.

By 7000 years ago, Douglas-fir was dominant with lodgepole and possibly some ponderosa pine. The climate at this time was warmer but not necessarily drier than the present. About 6700 B.P., 7 cm of Mazama ash were deposited in the lake. Peat deposition began about 5000 B.P., and microfossils of aquatic, fen, and bog genera appeared in significant numbers for the first time. With a return to a cooler climate, by about 4000 B.P., the Douglas-fir were replaced by pines, and whitebark pine was locally more important than during the previous 4000–5000 years. After 4000 B.P. there were no fluctuations in pollen content that suggested important changes in forest composition.

### Glacier Peak Ash

Glacier Peak ash at Lost Trail Pass Bog (Figure 13.4) consists of two bands separated by 7.5 mm of mixed ash and lake sediment. The lower of these (12 mm thick) contains graded laminae of volcanic ash above and below a thin (<0.5 mm) organic lamina. The upper band (23 mm thick) contains two sets of graded laminae that are separated by 8 mm of convoluted ash. The two bands of ash are identical in chemical and petrographic properties and were correlated with the youngest of three Glacier Peak ash layers by Smith *et al.* (1977b:203; layer B, Porter 1978:37). Lake sediments directly below and above the bands provided $^{14}$C dates of 11,200 ± 100 (WSU1548) and 11,300 ± 230 B.P. (WSU1554) (Mehringer, Arno, and Petersen 1977:Table 3).

Five pollen samples from the Glacier Peak ash sequence (two from the

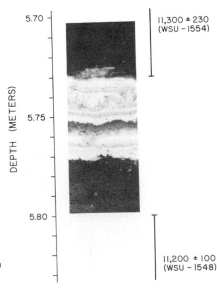

**FIGURE 13.4. Glacier Peak ashes and radiocarbon dates (years B.P.), Lost Trail Pass Bog, Montana.**

lower ash and three from the upper ash) were analyzed to characterize the
duration, timing, and possible effects of the ashfalls. We also wished to assess
the possibility that the upper ash resulted from redeposition of the lower ash.
One sample was taken from between the ashes to estimate the time separating
them (Mehringer, Blinman, and Petersen 1977). All samples were taken by
excavating sediment and packing it into a 2-cm³ spoon.

The total terrestrial pollen contents of the Glacier Peak ash samples are
given in Table 13.1 and Figure 13.5. Pollen years are calculated by comparison
with average pollen influx for the pollen zone encompassing the ashes (Zone 2;
Mehringer, Arno, and Petersen 1977:Table 5). The pollen content of the two
samples from the lower ash represents less than 0.2 year of average pollen
influx. Despite this short period, there is no clear indication of the season of
deposition.

The lower Glacier Peak ash is overlain by organic sediment that contains
volcanic ash derived from the mixing of ash in lake sediments and redeposition
of ash from the surrounding landscape. The effects of redeposition can be seen
in Figure 13.4, where a portion of the lower band has been disturbed and
diffuse white ash is visible in the lower half of the overlying sediment. Weight
loss after combustion at 600°C (a measure of organic carbon) reveals only 7%
organic carbon as compared with 25% for preash sediments (Mehringer, Arno,
and Petersen 1977:Fig. 7). The total pollen per square centimeter for this
sample is the equivalent of 15.7 years of average pollen influx (Figure 13.2).
When possible errors of sampling and population estimation are considered,
10–25 years is reasonable for the Glacier Peak interval (Mehringer, Blinman,
and Petersen 1977:Note 10).

Organic sedimentation was interrupted by deposition of the upper Glacier
Peak ash. Sample 4 contains about 0.3 year of average pollen influx and shows
no evidence of seasonality. Sample 5 was taken from the deformed ash and
contains only 0.03 year of average pollen influx. *Pinus* pollen is clearly underrepresented, whereas *Alnus*, *Artemisia*, Chenopodiineae, and other pollen
(dominated by *Salix* and *Arceuthobium*) are overrepresented. *Alnus*, *Salix*, and
*Arceuthobium americanum* (Mehringer, Arno, and Petersen 1977:363) bloom
in the early spring, whereas *Artemisia* and Chenopodiineae pollen are produced in the summer and autumn but may be recirculated from lower elevations during the winter (Mehringer, Blinman, and Petersen 1977:260). The
relative abundance of early spring pollen and the scarcity of *Pinus* pollen
indicates that deposition occurred before late spring. Sample 6 contains about
0.3 year of average influx and although *Picea* pollen is overrepresented, there
is no clear evidence of the season of ashfall.

Because the lower Glacier Peak ash contains little pollen, its deposition
must have been extremely rapid. Pollen content of Samples 1 and 2 represents
only a fraction of a year. Such rapid deposition, combined with graded
laminae, is consistent with an origin via airfall into standing water. The thin
organic lamina through the center of the band was not sampled, but it probably
represents a short hiatus in ashfall. Considering its thickness in comparison

**TABLE 13.1**

**Estimates of Pollen Influx and Duration for Selected Pollen Types and Total Terrestrial Pollen for Samples of Glacier Peak Ash and Interash Sediment, Lost Trail Pass Bog, Montana**

| Sample number | Pinus | | Picea | | Alnus | | Artemisia | | Chenopodiineae | | Gramineae | | Other | | Total | |
|---|---|---|---|---|---|---|---|---|---|---|---|---|---|---|---|---|
| | Pollen | Years | Pollen | Years | Pollen | Years | Pollen | Years | Pollen | Years | Pollen | Years | Pollen | Years | Pollen | Years |
| 6 | 908 | 0.26 | 169 | 1.11 | 72 | 0.52 | 77 | 0.23 | 36 | 0.44 | 15 | 0.16 | 92 | 0.27 | 1,370 | 0.29 |
| 5 | 40 | 0.01 | 4 | 0.03 | 18 | 0.13 | 18 | 0.05 | 14 | 0.17 | 3 | 0.03 | 40 | 0.12 | 137 | 0.03 |
| 4 | 928 | 0.26 | 37 | 0.24 | 70 | 0.51 | 94 | 0.29 | 58 | 0.71 | 58 | 0.62 | 74 | 0.22 | 1,322 | 0.28 |
| 3 | 45,072 | 12.81 | 3,967 | 26.10 | 3,435 | 24.89 | 7,392 | 22.47 | 1,082 | 13.19 | 2,704 | 28.77 | 9,375 | 27.98 | 73,017 | 15.72 |
| 2 | 47 | 0.01 | 4 | 0.03 | 9 | 0.06 | 11 | 0.03 | 10 | 0.12 | 4 | 0.04 | 21 | 0.06 | 106 | 0.02 |
| 1 | 380 | 0.11 | 37 | 0.24 | 31 | 0.22 | 60 | 0.18 | 23 | 0.28 | 17 | 0.18 | 23 | 0.07 | 571 | 0.12 |
| Zone average | 3,518 | 1.00 | 152 | 1.00 | 138 | 1.00 | 329 | 1.00 | 82 | 1.00 | 94 | 1.00 | 335 | 1.00 | 4,648 | 1.00 |

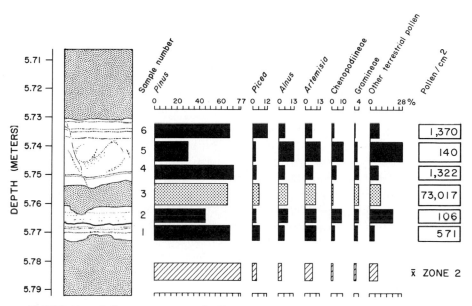

**FIGURE 13.5.** Pollen percentages and pollen influx for Glacier Peak ash and interash sediment, Lost Trail Pass Bog, Montana. Areas of stippling represent lake sediments adjacent to ash layers.

with the lake sediment between the bands, it could have been deposited in 1 year or less.

The lake sediment separating the ashes represents a major hiatus in ashfall lasting 10–25 years. Some ash was mixed with the lake sediment, but there is no stratigraphic evidence for deposition from another ashfall. The pollen spectra (Figure 13.5) show no evidence of ecological changes resulting from the first ashfall, but short-term effects immediately after the ashfall could have been masked by later pollen production.

Less than 1 year of pollen influx is present in the upper ash. We interpret its pollen content and the graded laminae of Samples 4 and 6 as evidence of ashfall deposition. Sample 5 was deposited rapidly, probably in the early spring, and although it lacks graded laminae, we feel that it resulted from ashfall. Its deformation could have been caused by disturbance before the deposition of Sample 6 or from special circumstances attending its deposition; perhaps it resulted from secondary deposition following ashfall onto snow or ice covering the lake surface. In any case, its deposition was both preceded and followed by ashfall.

We conclude that the pollen content and stratigraphy of the two Glacier Peak ashes at Lost Trail Pass Bog were the result of two distinct ashfalls. The ashfalls were rapid (less than a year for the first and probably less than a year for the second) and were separated by 10 to 25 years. The season of ashfall is unclear for both ashes, and the ashfalls had no apparent effect on the pollen production of surrounding vegetation.

## MAZAMA ASHFALLS

Ash produced by eruptions of Mount Mazama (Crater Lake, Oregon) is widely distributed over western North America (Figure 13.1). In a previous study we analyzed the pollen from a 1-cm² column of a Mazama ash layer from Lost Trail Pass Bog (Mehringer, Blinman, and Petersen 1977). The ash contained less than 3 years of average pollen influx, and pollen spectra within the ash indicated that the lower 4.6 cm of ash fell rapidly beginning in the autumn. The ash also caused a drastic but temporary decrease in *Botryococcus* and *Pediastrum* algae in the lake and may have increased the vigor of steppe vegetation in the region, perhaps through mulching. In this study, Mazama ash from Lost Trail Pass Bog was resampled by a different technique to compare the sampling methods and to reexamine our previous conclusions. In addition, analyses of cores containing the Mazama ash sequence from Wildcat Lake, Washington, and Wildhorse Lake, Oregon provided additional information on variation in the regional and chronological details of Mazama ashfall.

### Lost Trail Pass Bog

Mazama ash within the sediments of Lost Trail Pass Bog lies between 5.055 and 5.125 m depth. The single 7-cm-thick ash is composed of at least 23 graded laminae, and approximately 3 cm of mixed ash and organic sediments overlie the ash. Two very thin organic laminae in the upper part of the ash are noticeable as slightly darker boundaries between ash laminae (Figure 13.6).

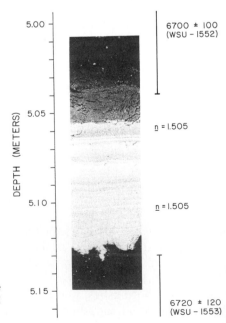

FIGURE 13.6 Mazama ash layer modal refractive indices and radiocarbon dates (years B.P.), Lost Trail Pass Bog, Montana.

The age and petrographic and morphologic characters of the ash are consistent with those of ash from Mount Mazama (Kittleman 1973:2957). Relative phenocryst abundance varies from lamina to lamina, but the modal refractive index is 1.505 ± 0.002 for glass from both upper and lower portions of the ash. Radiocarbon dates on organic sediments below and above the ash are 6720 ± 120 (WSU1553) and 6700 ± 100 B.P. (WSU1552) (Mehringer, Arno, and Petersen 1977:Table 3).

We defined sample boundaries by characters of the ash stratigraphy and collected samples from the core by using the excavation and spoon-packing technique. The bulk density of a few samples appeared to change slightly during excavation and packing, but the overall effect was negligible. Seven samples were taken from the ash and one was taken less than 5 mm below the ash to characterize pre-ash influx. Three samples were taken from the mixed ash and organic sediments above the Mazama ash.

Lake sediments immediately below the ash (Sample 1, Figure 13.7) contain a lower relative frequency of *Pinus* pollen and higher frequencies of Gramineae, Chenopodiineae, and *Artemisia* than the zone average. These departures from the average are also found in other samples from the lower part of the zone, and we assume them to be characteristic of the pollen influx at that time. Acid-resistant algae (*Botryococcus* and *Pediastrum*) are highly variable throughout the zone, but compared with the total pollen duration

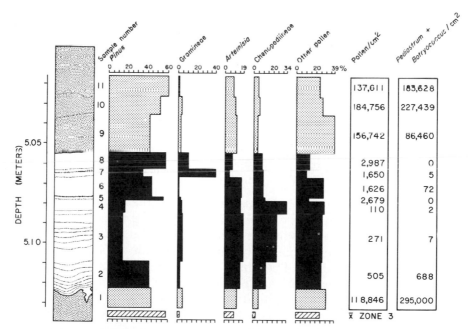

FIGURE 13.7. Pollen percentages, pollen influx, and acid-resistant algae influx for Mazama ash and adjacent sediment, Lost Trail Pass Bog, Montana. Areas of stippling represent lake sediments above and below Mazama ash.

estimate for Sample 1 (Table 13.2), the estimated preash algae influx is 10,000 per cm² per year.

Samples 2, 3, and 4 are taken from the lower 4.9 cm of ash and contain total pollen equivalent to only 0.2 year of average pollen influx (Figure 13.7; Table 13.2). *Pinus* pollen is underrepresented (only 0.1 year), whereas Chenopodiineae and *Artemisia* influx are close to 1.3 and 0.4 years respectively. The frequencies of these pollen types are consistent with autumn deposition, as are the numbers of *Pediastrum* and *Botryococcus* (8% of their estimated preash influx).

The pollen content of the next four samples rises significantly. Sample 5 is 2 mm high, includes the first of the dark (presumably organic) laminae, and contains pollen equivalent to 0.7 year of average pollen influx. Compared with preash frequencies, *Pinus* pollen is slightly more abundant than expected and Gramineae and *Artemisia* are less abundant. Chenopodiineae pollen is also more abundant and acid-resistant algae are absent. The *Pinus* and *Artemisia* frequencies and the dark lamina suggest spring or early summer deposition during a hiatus in ashfall.

Sample 6 has a total pollen content equivalent to 0.4 year of influx. *Pinus* and Gramineae pollen frequencies decrease and *Artemisia* pollen increases slightly relative to the previous sample. Chenopodiineae remain unusually abundant and algae are present though scarce. The season of deposition for this sample appears to be summer or early autumn.

The second dark lamina is included in Sample 7. About 0.4 year of average pollen influx is present and algae are scarce. *Pinus* and *Artemisia* pollen are equivalent to 0.2 year, Chenopodiineae pollen is equivalent to 1.3 years, and Gramineae pollen represents 6.7 years of average influx. This overwhelming abundance of Gramineae pollen was observed in the column sample (Mehringer, Blinman, and Petersen 1977:Fig. 5, Table 1) and was confirmed by two duplicate samples in this study. Therefore, contamination of the sample or inclusion of a flower is unlikely. Despite the abundance of grass pollen, season of deposition is uncertain because other spring and summer indicators are not overrepresented. The coincidence of large numbers of pollen with the dark (organic) lamina suggests a hiatus in ashfall of a fraction of a year.

Sample 8, the uppermost sample within the ash, contains the equivalent of 0.7 year of average pollen influx. *Pinus* pollen is relatively abundant, *Artemisia* pollen is scarce, and Chenopodiineae pollen is still unusually abundant. Numbers of Gramineae pollen decrease by half but continue to be very abundant (3.5 years of average influx). The *Pinus* and *Artemisia* frequencies compare well with expected frequencies for summer deposition.

The cessation of ashfall and return to organic deposition is marked by the extremely high pollen content of Samples 9, 10, and 11. Sample 9 represents the equivalent of over 35 years of average pollen influx, and pollen percentages are similar to the preash sample. Gramineae pollen declines to less than its preash frequency, and *Pediastrum* and *Botryococcus* amount to only 22% of their

**TABLE 13.2**

**Estimates of Pollen Influx and Duration for Selected Pollen Types and Total Terrestrial Pollen for Samples of Mazama Ash and Adjacent Sediment, Lost Trail Pass Bog, Montana**

| Sample number | Pinus Pollen | Pinus Years | Gramineae Pollen | Gramineae Years | Artemisia Pollen | Artemisia Years | Chenopodiineae Pollen | Chenopodiineae Years | Other pollen Pollen | Other pollen Years | Total Pollen | Total Years |
|---|---|---|---|---|---|---|---|---|---|---|---|---|
| 11 | 84,513 | 35.30 | 1,911 | 19.50 | 14,401 | 34.67 | 7,979 | 63.83 | 42,896 | 41.93 | 137,611 | 33.93 |
| 10 | 96,951 | 40.50 | 7,317 | 74.66 | 20,122 | 48.37 | 11,188 | 89.51 | 70,749 | 69.16 | 184,756 | 45.55 |
| 9 | 65,730 | 27.46 | 4,045 | 41.27 | 19,213 | 46.19 | 6,573 | 52.58 | 61,180 | 59.80 | 156,742 | 38.64 |
| 8 | 1,736 | 0.72 | 343 | 3.50 | 242 | 0.58 | 283 | 2.26 | 383 | 0.37 | 2,987 | 0.74 |
| 7 | 554 | 0.23 | 661 | 6.75 | 37 | 0.21 | 164 | 1.31 | 36 | 0.04 | 1,650 | 0.41 |
| 6 | 721 | 0.30 | 16 | 0.16 | 230 | 0.67 | 168 | 1.35 | 441 | 0.43 | 1,626 | 0.40 |
| 5 | 1,504 | 0.63 | 88 | 0.90 | 430 | 1.03 | 322 | 2.58 | 53 | 0.05 | 2,679 | 0.66 |
| 4 | 19 | 0.01 | 4 | 0.04 | 17 | 0.04 | 37 | 0.30 | 77 | 0.08 | 110 | 0.03 |
| 3 | 41 | 0.02 | 14 | 0.14 | 52 | 0.13 | 66 | 0.53 | 99 | 0.10 | 271 | 0.07 |
| 2 | 212 | 0.09 | 15 | 0.15 | 30 | 0.19 | 65 | 0.52 | 53 | 0.05 | 505 | 0.17 |
| 1 | 52,692 | 22.01 | 6,538 | 66.72 | 15,335 | 36.98 | 6,538 | 52.31 | 37,692 | 36.84 | 118,846 | 29.30 |
| Total in ash | 4,787 | 2.00 | 1,141 | 11.64 | 1,138 | 2.86 | 1,105 | 8.84 | 1,142 | 1.12 | 9,828 | 2.42 |
| Zone average | 2,394 | 1.00 | 98 | 1.00 | 4.6 | 1.00 | 125 | 1.00 | 1,023 | 1.00 | 4,056 | 1.00 |

407

estimated preash influx. Redeposited ash is indicated by the light-colored sediments of Samples 9 and 10 (Figure 13.6); apparent redeposition had ceased before Sample 11 was deposited.

In summary, the total pollen content of the Mazama ash at Lost Trail Pass Bog is the equivalent of 2.4 years of average influx. The ashfall began in the autumn with the rapid fall of 4.9 cm of ash; after its deposition, no ash fell on the lake through the next spring and summer. This hiatus was ended by another ashfall that deposited slightly more than 1 cm of ash before the following winter. During the second spring or summer the final 1 cm of primary ash fell on the lake. Subsequent redeposition of ash lasted approximately 80 years.

Some ecological consequences of the ashfall are evident in the Gramineae, algae, and possibly Chenopodiineae records. The increase in Gramineae pollen may reflect a response comparable with those observed after the historic eruptions and ashfalls of Parícutin, Katmai and Taal volcanoes. Two and one-half years after the beginning of the Parícutin eruptions, grasses had pushed through the ash and spread laterally, covering as much as one-half of the surface area of ash layers less than 30 cm thick (Eggler 1948:432). On Kodiak Island, grass grew through cracks in the Katmai ash layer in areas where grass was part of the preash vegetation (Martin 1913:143); 3 years after the ashfall, grasses had covered devastated hillsides and had exceeded their preash luxuriance (Griggs 1918:4–7). On Taal Volcano in the Philippines, grasses appeared 2 years after the eruption and "with no opposition, rapidly became established and spread in all directions [Gates 1914:395]." The conditions created by the initial Mazama ashfall may have selectively encouraged the growth of grasses for at least several seasons. The productivities of *Pediastrum* and *Botryococcus* were reduced markedly by the ashfall and stayed at low levels for at least the next 40 years. The relative abundance of Chenopodiineae pollen throughout the ash is more than twice its preash frequency. This may be the result of a mulching effect of the ash on steppe genera (Mehringer, Blinman, and Petersen 1977:260) or may have resulted from a short-term climatic trend that enhanced Chenopodiineae pollen production.

Despite the difference in sampling technique, these results are remarkably similar to those of Mehringer, Blinman, and Petersen (1977). Total pollen content estimates for the ash layer are close (9,828 and 11,485 pollen/cm²) and overlap at the 95% confidence level. Slight differences in the patterns of pollen frequencies through the ash may be explained by the different sampling locations.

## Wildcat Lake

Wildcat Lake is located at 335 m elevation near the eastern edge of the channeled scablands (Figures 13.1 and 13.8). The lake lies in one of the numerous depressions excavated when the disintegrating ice dam of Glacial

FIGURE 13.8. Wildcat Lake, Washington. (Photo by David Kolva, 2 May 1974.)

Lake Missoula released waters of the Spokane flood. The evidence for this late Quaternary deluge is unrivaled in all of geologic history (Bretz 1969; Baker 1973).

The lake lies within *Agropyron–Festuca* grassland. This vegetation zone is bounded on the west by the warmer and drier *Artemisia tridentata–Festuca idahoensis* zone. The cooler *Artemisia tripartita–Festuca idahoensis* zone lies to the north, and the cooler and more moist *Festuca–Symphoricarpos* and *Festuca–Rosa* zones occur to the east (Daubenmire 1970). As available moisture increases, the steppe vegetation of eastern Washington grades into ponderosa pine forest. The nearest stands of ponderosa pine are in the Blue Mountains, about 40 km to the southeast, and about 55 km to the east, near Colfax, Washington.

Bretz *et al.* (1956:1017) described the Wildcat Lake basin as a great empty hole, but beneath 3 to 5 m of water lie at least 18.45 m of fossil-rich sediments that span the last 7000 [14]C years. These sediments include three Mount St. Helens ashes and a depositional record of Mazama ashfalls and redeposited ash (Figure 13.9). Presumably, the deeper unsampled lake sediments began accumulating after the last scabland flood of about 13,000 B.P. (Mullineaux *et al.* 1978).

Pollen from the 11.34 m of sediment above redeposited Mazama ash has been analyzed (Davis 1975; Kolva 1975; Mehringer, unpublished data), and the pollen record of the upper 4.00 m (the last 1000 [14]C years) has been published (O. Davis *et al.* 1977). Ten [14]C dates establish a pollen chronology that extends back to 5400 B.P., and 131 pollen samples have been used to define seven pollen zones. Average pollen influx values have been calculated for total pollen and selected pollen types for each zone.

For the past 5400 ¹⁴C years the pollen of Wildcat Lake sediments has been dominated by grass and conifers (primarily pine with a few percent spruce, fir, hemlock, and Douglas-fir). Conifer pollen is transported from forests encircling the Columbia Basin, and the pollen record is relatively stable and indicates only minor changes attributed to variation in steppe vegetation. The only major changes in the sedimentary, pollen and algae records followed the introduction of livestock and agriculture during the last 130 years (O. Davis *et al.* 1977).

Between 11.34 and 18.45 m, Wildcat Lake sediments are predominantly volcanic ash. Cores of sediment between 11.34 and 15.00 m consist of a massive gray ash mixed with nonash detrital materials. These cores were not frozen and were subsequently disturbed during transport. The core from 15.00 to 17.20 m was frozen before transport and suffered only slight disturbance. Sediment between 15.00 and 16.82 m is a gray ash mixed with nonash detrital minerals; the only bedding occurs as several thin lenses of organic sediment at 16.65 m. Thin organic lenses appear between 16.80 and 16.96 m (Figure 13.10) and increase in frequency downward. At 16.97 m a thin layer (2.5 mm) of white ash

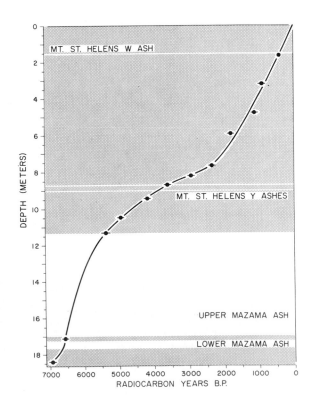

FIGURE  13.9. Deposition rate curve for lake sediments and volcanic ashes, Wildcat Lake, Washington. Nearly one-third of the sediments deposited in the last 7000 radiocarbon years resulted from redeposition of volcanic ash in the 1200 years following the second Mazama ashfall.

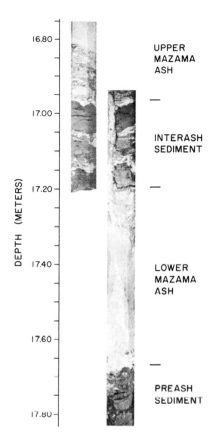

16.80

UPPER
MAZAMA
ASH

17.00 —

INTERASH
SEDIMENT

17.20 —

DEPTH (METERS)

17.40 —

LOWER
MAZAMA
ASH

17.60 —

PREASH
SEDIMENT

17.80 —

FIGURE 13.10. Mazama ash sequence, Wildcat Lake, Washington.

abruptly overlies organic sediment. The organic sediment extends from 16.97 to 17.20 m and encloses lenses of volcanic ash at 17.06 and 17.14 m.

The core containing the sediments from 16.00 to 18.45 m was not frozen and ash layers within it suffered some damage; portions of the core with lenses or laminae of organic sediment were only slightly disturbed. The stratigraphy between 16.80 and 17.20 m corresponds to that of the previous core. At 17.20 m there is an abrupt (1–2 mm) boundary between the overlying organic sediment and underlying ash with organic lenses. These lenses decrease in frequency downward to gray ash at 17.32 m. From 17.32 m to 17.67 m the ash was disturbed. A portion of the ash at 17.50 m is white (Figure 13.10) and may be the displaced remnant of a white layer similar to that at 16.97 m. The sediment below 17.67 m is organic, horizontally laminated, and lacks lenses or laminae of volcanic ash.

Petrographic and morphologic characters of ash samples below 11.34 are consistent with those of Mazama ash (H. W. Smith, pers. comm.). Modal refractive index measurements of glass shards are also consistent with Mazama ash, but there are variations among samples (Figure 13.11). An ash sample from 17.65 m has a modal refractive index of 1.508 ± 0.002, whereas ash from

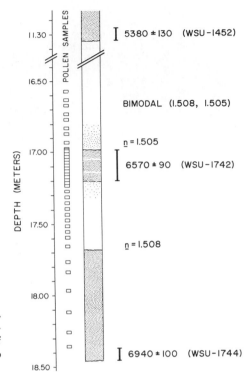

**FIGURE 13.11.** Mazama ash stratigraphy, radiocarbon dates (years B.P.), and modal refractive indices, Wildcat Lake, Washington. Areas of stippling represent lake sediments adjacent to Mazama ash layers.

16.97 has a modal refractive index of 1.505 ± 0.002; ash shards from 16.55 m were bimodal with indices of both 1.508 and 1.505. Weight percentages of major elements of all Mazama ash layers described in this study were determined by electron microprobe analysis. The samples are nearly identical (Figure 13.12) and fall within the range of other reported analyses of Mazama ash (Smith *et al*. 1977a:Figure 5).

Preash sediment at 18.40 m was radiocarbon-dated at 6940 ± 120 B.P. (WSU1553) and interash sediment between 17.00 and 17.20 m was dated at 6750 ± 90 B.P. (WSU1742) (Figures 13.9 and 13.11). These ages are consistent with other ages reported for Mazama ash (Kittleman 1973:Table 1). The 5380 ± 130 B.P. (WSU1452) age for sediments above the Mazama ash dates the end of ash redeposition at Wildcat Lake.

Thirty-nine pollen samples were taken between 16.56 and 18.36 m (Figure 13.13). Six samples of preash sediment characterize pollen deposition before the initial ashfall. Twelve samples were taken from the lower ash, the interash sediment was continuously sampled (12 samples), and 9 samples were taken from the upper Mazama ash. All samples except 16.96–16.97 m and 16.97–16.98 m were part of a 1-cm² column.

Because we lack adequate chronological controls below Mazama ash, the basis for estimating pollen years within the ash sequence must come from pollen influx values of the last 5400 ¹⁴C years. Two obvious choices occurred to

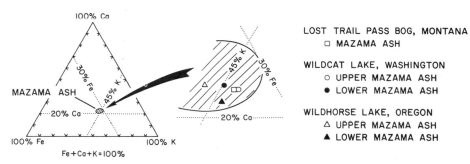

**FIGURE 13.12.** Electron microprobe weight percentages for Ca, K, and Fe (sum = 100%) for Mazama ashes from Lost Trail Pass Bog, Montana, Wildcat Lake, Washington, and Wildhorse Lake, Oregon. The shaded oval represents the range of other samples of Mazama tephra (Smith *et al.* 1977a:Figure 5).

us; we could either use the influx average for all six pregrazing zones, or the influx average for Zone 1, which immediately overlies redeposited Mazama ash and spans the next 1200 years. Excluding the recent period of overgrazing, the average pollen influx of six pollen zones dating from 5400 to 130 B.P. (Zones 1 through 6) is 5153 pollen/cm² · year with a range of about 3000 to 6100 pollen/cm² · year. Zone 1 average pollen influx (5434 pollen/cm² · year) is close to the combined average, and its pollen percentages are most like those of preash sediment and the Mazama ash sequence. This is most apparent in small *Pinus* (<35%), large *Artemisia* (>8%), and large other Compositae pollen percentages. When compared with pollen percentages of younger Wildcat Lake sediments (4200–130 B.P.), these differences may indicate generally drier conditions from before the fall of Mazama ash through the end of Zone 1 time. Therefore, because of its similarity in pollen percentages and proximity to the Mazama ash sequence, we have selected the Zone 1 pollen influx value for estimating pollen years.

Average pollen content of preash samples is equivalent to 14 years of Zone 1 pollen influx (Table 13.3). *Pediastrum* and *Botryococcus* influx is variable, with an average estimated influx of 750 per cm² per year. Pollen percentage variations between both preash and ash samples are minor (Figure 13.13) and are less than those occurring within single pollen zones of the past 5380 ¹⁴C years. Therefore, season of deposition cannot be derived from these data, and there is no indication of selective effects of ashfall on pollen production. However, pollen influx does reveal some details of depositional history.

Below 17.32 m the lower Mazama ash was disturbed, but all samples contain the equivalent of less than 1 year of Zone 1 pollen influx. The four samples taken from or including portions of the white patch of ash (17.42–17.56 m) have the lowest pollen content of any lower ash samples, and *Pediastrum* and *Botryococcus* are either absent or rare. Above 17.32 m, the ash contains lenses of organic sediment and suffered little disturbance. Samples from this portion of the ash contain from 1 to 3 years of pollen influx, and acid-resistant algae influx returns to preash levels.

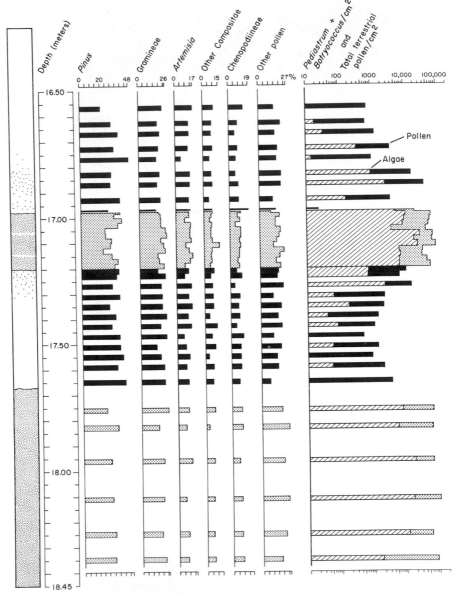

**FIGURE 13.13.** Pollen percentages and logarithmic plot of acid-resistant algae and pollen influx for Mazama ashes and adjacent sediment, Wildcat Lake, Washington. Areas of stippling represent lake sediments adjacent to Mazama ash layers.

**TABLE 13.3**

Estimates of Terrestrial Pollen/cm³, Terrestrial Pollen/cm², Pollen Years, and *Botryococcus* and *Pediastrum*/cm² for 39 2-cm³ Samples of pre-Mazama Ash Sediment, Lower Mazama Ash, Interash Sediment, and Upper Mazama Ash, Wildcat Lake, Washington

| Depositional unit | Sample depth (meters) | Pollen/cm³ | Pollen/cm² | Pollen years | *Botryococcus* and *Pediastrum*/cm² |
|---|---|---|---|---|---|
| Upper ash | 16.56–16.58 | 366 | 731 | 0.134 | — |
| | 16.62–16.64 | 329 | 658 | 0.121 | 19 |
| | 16.66–16.68 | 628 | 1,256 | 0.231 | 34 |
| | 16.72–16.74 | 1,814 | 3,629 | 0.668 | 358 |
| | 16.76–16.78 | 510 | 1,019 | 0.187 | 15 |
| | 16.82–16.84 | 8,392 | 16,784 | 3.089 | 966 |
| | 16.86–16.88 | 20,602 | 41,205 | 7.583 | 2,676 |
| | 16.92–16.94 | 1,851 | 3,702 | 0.681 | 171 |
| | 16.97[a] | 96 | 24 | 0.004 | 3 |
| Interash | 16.98[a] | 40,698 | 20,349 | 3.745 | 5,107 |
| | 16.98–17.00 | 32,874 | 65,748 | 12.099 | 11,024 |
| | 17.00–17.02 | 30,036 | 60,073 | 11.055 | 9,524 |
| | 17.02–17.04 | 28,142 | 56,283 | 10.358 | 8,673 |
| | 17.04–17.06 | 23,166 | 46,331 | 8.526 | 3,885 |
| | 17.06–17.08 | 36,936 | 73,872 | 13.594 | 30,166 |
| | 17.08–17.10 | 35,618 | 71,236 | 13.109 | 20,000 |
| | 17.10–17.12 | 45,738 | 91,477 | 16.834 | 26,420 |
| | 17.12–17.14 | 18,674 | 37,347 | 6.873 | 6,801 |
| | 17.14–17.16 | 17,742 | 35,484 | 6.530 | 7,527 |
| | 17.16–17.18 | 25,502 | 51,003 | 9.386 | 6,877 |
| | 17.18–17.20 | 31,940 | 63,879 | 11.755 | 6,939 |
| | Mean | 30,589 | 56,090 | 10.322 | 11,912 |
| | Total[b] | — | 673,082 | 123.865 | 142,943 |
| Lower ash | 17.20–17.22 | 5,506 | 11,013 | 2.027 | 787 |
| | 17.22–17.24 | 3,438 | 6,876 | 1.265 | 754 |
| | 17.26–17.28 | 8,060 | 16,121 | 2.967 | 2,484 |
| | 17.30–17.32 | 1,173 | 2,346 | 0.432 | 67 |
| | 17.34–17.36 | 1,110 | 2,221 | 0.409 | 196 |
| | 17.38–17.40 | 731 | 1,462 | 0.269 | 43 |
| | 17.42–17.44 | 567 | 1,134 | 0.209 | 85 |
| | 17.46 17.48 | 261 | 522 | 0.096 | — |
| | 17.50 17.52 | 726 | 1,452 | 0.267 | 61 |
| | 17.54–17.56 | 466 | 931 | 0.171 | — |
| | 17.58–17.60 | 1,076 | 2,153 | 0.396 | 57 |
| | 17.64–17.66 | 1,879 | 3,759 | 0.692 | — |
| | Mean | 2,083 | 4,165 | 0.767 | 378 |
| | Total[c] | — | 95,814 | 17.632 | 8,690 |
| e-ash | 17.75–17.77 | 34,488 | 68,977 | 12.694 | 7,594 |
| | 17.82–17.84 | 33,166 | 66,332 | 12.207 | 5,612 |
| | 17.95–17.97 | 33,922 | 67,844 | 12.485 | 18,551 |
| | 18.10–18.12 | 54,648 | 109,295 | 20.113 | 17,308 |
| | 18.24–18.26 | 30,187 | 60,374 | 11.110 | 11,905 |
| | 18.34–18.36 | 44,603 | 89,206 | 16.416 | 1,746 |
| | Mean | 38,502 | 77,004 | 14.171 | 10,453 |

[a]Samples were taken by excavating and packing sediment into a 2-cm³ spoon.
[b]Interash sediments were completely sampled, and total terrestrial pollen, algae, and total pollen years are the sum of mates for all samples of the unit.
[c]Totals for the entire lower Mazama ash unit are estimated from the average pollen and algae content of 12 2-cm-high ples taken from the 46-cm thick unit (Figure 13.10).

The white patch of ash is probably a remnant of the original ashfall deposit, whereas adjacent ash contains more pollen and represents less rapid redeposition. The upper three samples resulted from a slowing of redeposition until organic lake sedimentation resumed. We estimated the total pollen per square centimeter of the lower Mazama ash from the average pollen content of 12 samples. This estimate (95,814 pollen/cm²) is equivalent to 17.6 years of Zone 1 influx (Table 13.3).

The average sample of interash sediment contains pollen equivalent to about 10 years of Zone 1 pollen influx. Compared with the average pollen content, *Pediastrum* and *Botryococcus* influx (1150 algae/cm² · year) is greater than before the ashfall. Some redeposition of ash undoubtedly continued, although the two ash lenses could also represent minor primary ashfalls. A 1-cm² column of the 23 cm of interash sediment was continuously sampled and the total pollen content (673,082 pollen/cm²) is equivalent to 123.9 years of Zone 1 influx; the 95% confidence interval for this estimate is 116.5–131.3 years. The combined pollen year estimates for the lower Mazama ash and these interash lake sediments are about 142 pollen years.

Normal lake deposition came to an abrupt end when a second Mazama ashfall deposited 2.5 mm of white ash. This portion of the upper Mazama ash contains only 24 pollen/cm², and its lower refractive index ($n = 1.505$) precludes the possibility that it resulted from redeposition of the lower Mazama ash. Pollen content of the upper Mazama ash increases through the next two samples but then declines to just over 0.1 year of Zone 1 pollen influx. The numbers of *Pediastrum* and *Botryococcus* also increase and then decline upwards. These trends in microfossil abundance correlate roughly with the occurrence of organic lenses. The stratigraphy and bimodal refractive indices from this portion of the lower ash suggest that it is redeposited, even though its pollen content is low.

In summary, sediments from Wildcat Lake include evidence for two Mazama ashfalls. The first ashfall had a modal refractive index of 1.508 and may have deposited 1 cm of ash. Redeposition of this ash lasted about 17 years, until organic sedimentation resumed in the lake. About 142 years after the first ashfall, a second ashfall deposited 2.5 mm of ash with a modal refractive index of 1.505. Redeposition after the second ashfall included material from both ashes and dominated lake sedimentation until 5380 ¹⁴C years ago. During this time close to 5.7 m of ash were deposited in the lake (Figure 13.9). There is no pollen evidence for the season of either ashfall or for any effect of the ashfalls on vegetation.

### Wildhorse Lake

Wildhorse Lake is located on Steens Mountain in southeastern Oregon, 290 km east of Mount Mazama (Figures 13.1 and 13.14). The 13-m-deep lake lies within a cirque at 2600 m elevation and is surrounded by subalpine grassland (Mairs 1977). In 1977, 4.2 m of sediment containing six tephra layers

**FIGURE 13.14.** Wildhorse Lake, Steens Mountain, Oregon. (Photo by Peter J. Mehringer, Jr., October 1975.)

were recovered from the lake. Two of these layers, separated by 6 cm of organic lake sediment, are identified as Mazama ash (Figure 13.12). An apparently identical Mazama ash sequence has also been recovered from Fish Lake, 13 km to the northwest. Interash sediment from this lake has a radiocarbon age of 6765 ± 70 years B.P. (WSU2035).

The lower Mazama ash from Wildhorse Lake is 1 mm thick (Figure 13.15). Six centimeters of horizontally laminated organic sediment overlie this ash and separate it from the upper Mazama ash. This second layer consists of about 2 cm of white ash, with alternating coarse and fine laminae, overlain by about 5 cm of gray ash. Several thin laminae are present in the gray ash, and its color darkens upward. The overlying organic sediment is separated from the gray ash by an abrupt boundary.

Samples of the upper ash and immediately postash sediment were taken using the excavation and spoon-packing technique. Interash and preash sediments were sampled by removing 2-cm$^3$ blocks with a sampling frame (Kolva 1975:16–18). The lower Mazama ash was not sampled because of its thinness and the likelihood of contamination from adjacent sediments. Because pollen influx estimates are not yet available for Wildhorse Lake, only estimates of total pollen, *Isoetes* microspores, and *Pediastrum* + *Botryococcus* per cm$^2$ have been calculated and are presented in Figure 13.15. (*Isoetes* is a rooted aquatic plant.)

Pollen content of preash and interash samples averages about 155,000 pollen/cm$^2$. This contrasts markedly with the first two samples of the upper Mazama ash whose pollen contents average only about 2270 pollen/cm$^2$. In the gray ash immediately above these samples, pollen content increases to nearly 50,000 pollen/cm$^2$. This increase continues upward through the gray ash until

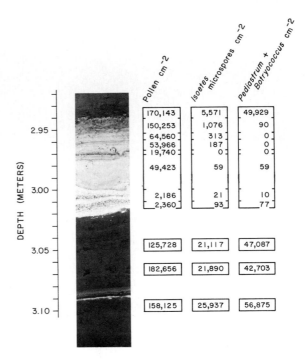

**FIGURE 13.15. Mazama ashes and pollen, *Isoetes*, and acid-resistant algae per cm², Wildhorse Lake, Oregon.**

the pollen content of the last ash sample falls within the preash range. Frequencies of *Isoetes* microspores follow a similar trend, but instead of the lowest value coinciding with the white ash, it coincides with the middle of the gray ash. *Isoetes* then increases upwards, but even above the ash it amounts to only 25% of the preash and interash estimates. Deposition of acid-resistant algae declines abruptly from an average of about 49,000/cm² in preash and interash samples to 0–90/cm² within both the white and gray ash.

Despite the lack of annual pollen influx estimates, we can use these microfossil and stratigraphic data to support inferences of relative deposition rates and effects of the upper Mazama ash on the aquatic environment at Wildhorse Lake. The lower 2 cm of the upper ash are white, distinctly laminated, and contain little pollen relative to adjacent ash and organic sediment. Overlying gray ash contains considerably more pollen and few distinct laminae, and its decrease in color value upward coincides with increasing pollen content. We interpret this sequence as representing an initial ashfall of about 2 cm followed by a prolonged period of redeposition. *Isoetes* was severely affected by the ashfall and initial erosion of ash into the lake and, although *Isoetes* frequencies increase through the redeposited ash, *Isoetes* had not reached its preash abundance by the end of active redeposition. In contrast, *Botryococcus* and *Pediastrum* were drastically reduced for the duration of redeposition but regained their preash abundance immediately thereafter.

Thus, Wildhorse Lake contains a record of two distinct Mazama ashfalls.

The first deposited about 1 mm of ash and probably had little direct effect on local environments. After its fall, organic deposition resumed until a second Mazama ashfall deposited about 2 cm of primary ash; a prolonged period of redeposition followed. Without comparative pollen influx data from adjacent sediments we cannot estimate the duration, season, and possible effects of the second ashfall on terrestrial environments. However, the abundances of *Isoetes* and acid-resistant algae indicate that at least some aquatic organisms were severely affected by the ashfall and its subsquent redeposition.

## DISCUSSION AND CONCLUSION

### The Method

Interpretation of ecological effects of ancient ashfalls requires information on the thickness, duration, and season of each event. These data may be provided by the stratigraphy and pollen analysis of ash layers and by comparing their pollen content with the estimated annual pollen influx of adjacent sediments. The amount of pollen in ashes may be used to estimate rates of deposition and, when combined with stratigraphy, to distinguish primary ashfalls from redeposited ash. In addition, types of pollen present in ashes may provide clues to the seasons of ashfalls and to the effects of ashfalls on vegetation.

Although the underlying concepts are simple, this application of pollen analysis has several limitations. Pollen production and pollen influx vary from year to year. Ideally, influx estimates needed for studies of ashfall deposition could be obtained from adjacent varved sediments where annual variation in pollen influx could be precisely determined. Lacking varved sediments, we have relied on average pollen influx estimates derived from analysis of many pollen samples and many radiocarbon dates; these values are the best estimates presently available.

Deposition of pollen is not uniform throughout the year. In the Pacific Northwest, most pollen is released during spring and early summer. Thus, one-half year of pollen influx could be deposited in 1 month or 2. For this reason, the durations represented by fractions of a year of pollen influx vary from month to month. Lakes subject to seasonal turnover may have different patterns of pollen deposition because of resuspension of pollen from bottom sediments (M. Davis 1973). It is even conceivable that a tephra fall during lake turnover would incorporate more than 1 year's pollen rather than a fraction of a pollen year.

Ashfall could perhaps effect pollen influx by "sweeping" suspended pollen from the water (Mehringer, Blinman, and Petersen 1977:260). This would create a longer apparent duration for initial pulses in the same season. This phenomenon may explain the slightly greater numbers of pollen in the lower third (Sample 2) of the initial 4.9 cm of Mazama ash at Lost Trail Pass. Ash layers could also seal bottom sediments, limit resuspension of pollen during lake turnover, and thereby reduce pollen influx to deep-water sediments for a few years.

Redeposition of ash is controlled by local factors such as vegetation and topography. As these factors vary from one locality to another, the criteria necessary to distinguish primary ashfall from redeposited ash may also change. Although we expect ashfalls to incorporate little pollen, because of their extremely rapid deposition, rapidly redeposited ash may contain as little or less pollen. Lost Trail Pass and Wildhorse Lake primary ashfalls were identified by pollen content, graded bedding, and color; whereas Wildcat Lake primary ashfalls were distinguished from redeposited ash by color, refractive index, stratigraphy, and pollen content. Pollen abundance in some redeposited ash at Wildcat Lake was less than that of ashfall at Lost Trail Pass. Although we did not study nonpyroclastic detrital minerals, our observations indicate they may also be useful in distinguishing ashfall from redeposited ash.

Seasonal interpretations from pollen content are limited by the scarcity of comparative data and the unique characteristics of the pollen rain at each site. *Artemisia* pollen is released in the autumn and carried to Lost Trail Pass from as far as the Snake River Plain. This accentuates the contrast with spring and summer pollen production dominated by pine and grass. At Wildcat Lake, much of the pollen is transported from long distances, and in such semi-arid environments, pollen may be repeatedly deposited, eroded and resuspended from the landscape. This recirculation reduces the seasonal contrast and thereby complicates interpretation. Uncommon pollen types may be valuable in seasonal interpretations, and we recorded them during analysis. *Salix* and *Arceuthobium* were helpful in interpreting the season of one sample from Lost Trail Pass, but in all other cases the rare pollen types were apparently uninformative. Finally, mixing at the mud-water interface by organisms or annual turnover may obscure seasonal data.

The few pollen in ash samples may cause special problems in analysis. During sampling and extraction, care must be taken to avoid contamination, especially when 1 cm³ of ash may contain only 100 to 500 pollen. Extraction procedures must also be altered to minimize pollen loss on stirring rods and glassware. Before counting, we decided that 200 pollen would be the minimum acceptable pollen sum; to reach this goal, we counted thousands of *Lycopodium* tracers for each sample. In two cases samples did not contain 200 pollen following extraction, and in another case, we extracted a second sample and counted 49,388 *Lycopodium* tracers to achieve a pollen sum of 219.

Despite these complications, pollen analysis can provide valuable insights about the deposition of volcanic ashes. Knowledge of lake history and the studies needed to establish pollen influx estimates can eliminate many possible problems of interpretation. Similarly, supporting data from stratigraphy and ash characteristics can increase the probability of correct interpretation.

## The Ashfalls

Our study of the Glacier Peak ash layers from Lost Trail Pass Bog reveals that two distinct ashfalls were separated by 10–25 years. The first ashfall rapidly deposited about 1 cm of ash, but the season of fall is unknown. The second ashfall lasted somewhat longer, perhaps through a winter and into early spring, and deposited a little over 2 cm of ash. There is no evidence that either ashfall affected pollen production, and such thin ashfalls would have had little effect on vegetation. Although both ashfalls could have elicited cultural responses, we do not believe that important subsistence resources were significantly affected in this portion of the Bitterroot Range and adjacent areas.

One 7-cm-thick Mazama ash layer was found in Lost Trail Pass Bog. Its pollen content indicates that the lower 4.9 cm fell in the autumn. No new ash fell until after the next spring and early summer, when just over 1 cm of ash entered the lake. During the second spring and summer, an unusual number of grass pollen was deposited along with the final 1 cm of ash. The abundance of grass pollen suggests that herbaceous vegetation was affected by the initial ashfall. An increase in grass vigor followed, but the effect was short lived (possibly as short as 2 years). Grazing and browsing animals may have been affected the first year, but it is likely that grazing conditions were average or better than average the second year. Although accounting for less than 3 cm of sediment, obvious redeposition lasted about 80 years in the forested mountains. Its influences, if significant, were probably limited to aquatic environments. The effects of redeposition in lowlands around Lost Trail Pass are unknown.

About 400 km away at Wildcat Lake, the record of Mazama ashfalls is different. Sediments contain at least two Mazama ashes separated by 70 cm of redeposited ash and lake sediment. The first ashfall was about 1 cm thick, and during the following 15–20 years, about 45 cm of ash were redeposited. About 125–160 years after the first ashfall, a fraction of a centimeter of a second ash fell and was followed by 1200 years (5.4 m) of redeposition. Fossil pollen reveals no clear indication of the season of either ashfall or of effects on terrestrial vegetation. Given the thickness of these primary ash layers, it is unlikely that initial ashfall had major influences on vegetation. However, because of the volume and mobility of ash on the landscape, redeposited ash must have altered both terrestrial and aquatic environments in certain areas of eastern Washington. We do not know the extent of these inferred changes or their potential effects on prehistoric resources.

Sediments at Wildhorse Lake, Oregon, also contain multiple Mazama ash layers. The first ashfall was only a millimeter thick; this ash was overlain by 6 cm of organic sediment. A second ashfall deposited about 2 cm of ash that was overlain by about 5 cm of redeposited ash. Excepting the aquatic environments, neither of these two Mazama ashfalls is likely to have had major local effects. Specific interpretations of duration, seasonality, and effect await comparative data from estimates of pollen influx to adjacent sediments.

Speculations about the effects of Mazama ash on man in the interior Northwest have varied from catastrophe to no appreciable effect. Malde (1964:10–11) presented a view wherein terrestrial and aquatic resources were so seriously damaged that people were forced to move from affected areas. Bedwell and Cressman (1971) accounted for gaps in archaeological records by assuming abandonment of parts of southeastern Oregon following eruption of Mount Mazama. Other archaeologists, working in southeastern Washington, saw no evidence for such catastrophes (Bense 1972; Brauner 1976:306).

Our studies support the latter interpretation but also point out the danger of extrapolating ashfall data from one locality to another. The existence of multiple Mazama ashfalls with quite different thicknesses, durations, and intensities of redeposition requires an independent assessment of ashfalls and their effects for each locality studied. Also, the pollen record from Lost Trail Pass Bog indicates the onset of warmer climatic conditions shortly before the Mazama ashfall (Mehringer, Arno and Petersen 1977:366–367). This near coincidence of climatic change and ashfall must also be considered in evaluating the effects of the Mazama eruption.

Ashfalls undoubtedly affected prehistoric resources in some areas of the Pacific Northwest, but social responses may have been of equal or greater importance. Teit (1930:291–292) recorded a Nespelim account of a late eighteenth-century ashfall in northeastern Washington. The fall of "dry snow" caused much alarm, and people sought an explanation through prayer and dancing. They neglected collecting and storing food and, during the long and severe winter that followed, many starved and some died. This ashfall was probably what is now known as Mount St. Helens pyroclastic layer T (Okazaki *et al.* 1972), which resulted from an eruption in 1800 (Lawrence 1954). Neither Teit's nor Ray's (1932:108) informants recall any damage from the ashfall, which was probably less than 1 cm thick (Okazaki *et al.* 1972:83). Thus, a major but short-lived culturally induced catastrophe occurred without significant environmental disturbance.

Although analogies may be drawn from modern eruptions and human responses, the nature of past ashfalls and their effects must ultimately be determined from regional studies of each past event. Volcanic activity was undoubtedly important in shaping the prehistory of the interior Northwest, but there are as yet no clear archaeological examples of human response. Details of volcanic ashfalls, as revealed through their pollen content, should help us ask productive questions of the archaeological record.

## ACKNOWLEDGMENTS

This research was supported in part by a USDA Forest Service and Washington State University Cooperative Research Agreement through the Fire in Multiple-Use Management Research Development, and Application Program of the Intermountain Forest and Range Experiment Station, Ogden, Utah, National Science Foundation Grant BNS 77-12556, and by the Donors of the Petroleum Research Fund, administered by the American Chemical Society.

We thank H. W. Smith, R. Okazaki, and C. Knowles for volcanic ash identification, K. L. Petersen, and P. E. Wigand for field and laboratory assistance, and M. A. Mehringer, and K. A. Nelson for aid in manuscript preparation.

# REFERENCES

Baker, V. R.
   1973  Paleohydrology and sedimentology of Lake Missoula flooding in eastern Washington. Geological Society of America Special Paper 144.
Bedwell, S. F., and L. S. Cressman
   1971  Fort Rock report: Prehistory and environment of the pluvial Fort Rock area of south-central Oregon. In *Great Basin anthropological conference 1970, selected papers*, edited by C. M. Aikens. University of Oregon Anthropological Papers 1:1–25.
Bense, J. A.
   1972  The Cascade phase: A study in the effect of the altithermal on a cultural system. Doctoral dissertation, Department of Anthropology, Washington State University.
Blinman, E.
   1978  Pollen analysis of Glacier Peak and Mazama volcanic ashes. Master's thesis, Department of Anthropology, Washington State University.
Brauner, D. R.
   1976  Alpowai: The culture history of the Alpowa locality. Doctoral dissertation, Department of Anthropology, Washington State University.
Breternitz, D. A.
   1967  The eruption(s) of Sunset Crater: Dating and effects. *Plateau* 40:72–76.
Bretz, J H.
   1969  The Lake Missoula floods and the channeled scabland. *Journal of Geology* 77:505–543.
Bretz, J H., H. T. U. Smith, and G. E. Neff
   1956  Channeled scabland of Washington: New data and interpretations. *Geological Society of America Bulletin* 67:957–1047.
Crandell, K. C., and R. W. Seabloom
   1970  *Engineering fundamentals in measurements, probability, statistics, and dimensions*. New York: McGraw-Hill.
Cushing, E. J., and H. E. Wright, Jr.
   1965  Hand-operated piston corers for lake sediments. *Ecology* 46:380–384.
Daubenmire, R.
   1970  Steppe vegetation of Washington. *Washington Agricultural Experiment Station Technical Bulletin* 62.
Davis, M. B.
   1969  Palynology and environmental history during the Quaternary Period. *American Scientist* 57:317–332.
   1973  Redeposition of pollen grains in lake sediment. *Limnology and Oceanography* 18:44–52.
Davis, O. K.
   1975  Pollen analysis of Wildcat Lake, Whitman County, Washington: The introduction of grazing. Master's thesis, Department of Botany, Washington State University.
Davis, O. K., D. A. Kolva, and P. J. Mehringer, Jr.
   1977  Pollen analysis of Wildcat Lake, Whitman County, Washington: The last 1000 years. *Northwest Science* 51:13–30.
Eggler, W. A.
   1948  Plant communities in the vicinity of the volcano El Parícutin, Mexico, after two and a half years of eruption. *Ecology* 29:415–436.
Eicher, G. J., Jr., and G. A. Rounsefell
   1957  Effects of lake fertilization by volcanic activity on abundance of salmon. *Limnology and Oceanography* 2:70–76.

Evermann, B. W.
  1914  *Alaska fisheries and fur seal industries in 1913.* Washington, D.C., U.S. Fish and Wildlife
         Services, Bureau of Fisheries.
Gates, F. C.
  1914  The pioneer vegetation of Taal Volcano. *The Philippine Journal of Science, C., Botany*
         9:391–434.
Griggs, R. F.
  1915  The effect of the eruption of Katmai on land vegetation. *American Geographical Society
         Bulletin* 47:193–203.
  1918  The recovery of vegetation at Kodiak. *The Ohio Journal of Science* 19:1–57.
Hansen, H. P.
  1947  Postglacial forest succession, climate and chronology in the Pacific Northwest. *Transac-
         tions of the American Philosophical Society* n.s. 37(1):1–130.
Kittleman, L. R.
  1973  Mineralogy, correlation, and grain-size distribution of Mazama tephra and other postgla-
         cial pyroclastic layers, Pacific Northwest. *Geological Society of America Bulletin*
         84:2957–2980.
Kolva, D. A.
  1975  Exploratory palynology of a scabland lake, Whitman County, Washington. Master's
         thesis, Department of Anthropology, Washington State University.
Lawrence, D. B.
  1954  Diagrammatic history of the northeast slope of Mt. St. Helens, Washington. *Mazama*
         36:41–44.
Lehman, J. T.
  1975  Reconstructing the rate of accumulation of lake sediment: The effect of sediment focus-
         ing. *Quaternary Research* 5:541–550.
Lemke, R. W., M. R. Mudge, R. E. Wilcox, and H. A. Powers
  1975  Geologic setting of the Glacier Peak and Mazama ash-bed markers in west-central
         Montana. *U.S. Geological Survey Bulletin* 1395-H.
Maher, L. J., Jr.
  1972  Nomograms for computing 0.95 confidence limits of pollen data. *Review of Palaeobotany
         and Palynology* 13:85–93.
Mairs, J. W.
  1977  Plant communities of the Steens Mountain subalpine grassland and their relationship to
         certain environmental elements. Doctoral dissertation, Department of Geography, Ore-
         gon State University.
Malde, H. E.
  1964  The ecologic significance of some unfamiliar geologic processes. In *The reconstruction of
         past environments*, edited by J. J. Hester and J. Schoenwetter. *Proceedings of the Fort
         Burgwin Conference on Paleoecology* 3:7–15.
Martin, G. C.
  1913  The recent eruption of Katmai Volcano in Alaska. *National Geographic Magazine*
         24:131–181.
Mehringer, P. J., Jr., S. F. Arno, and K. L. Petersen
  1977  Postglacial history of Lost Trail Pass Bog, Bitterroot Mountains, Montana. *Arctic and
         Alpine Research* 9:345–368.
Mehringer, P. J., Jr., E. Blinman, and K. L. Petersen
  1977  Pollen influx and volcanic ash. *Science* 198:257–261.
Mosimann, J. E.
  1965  Statistical methods for the pollen analyst: Multinomial and negative multinomial tech-
         niques. In *Handbook of paleontological techniques*, edited by B. Kummell and D. Raup.
         San Francisco: Freeman.
Mullineaux, D. R., R. E. Wilcox, W. F. Ebaugh, R. Fryxell, and M. Meyer
  1978  Age of the last major scabland flood of the Columbia Plateau in eastern Washington.
         *Quaternary Research* 10:171–180.

Okazaki, R., H. W. Smith, R. A. Gilkeson, and J. Franklin
   1972   Correlation of West Blacktail Ash with Pyroclastic Layer T from the 1800 A.D. eruption of
          Mount St. Helens. *Northwest Science* 46:77–89.
Pfister, R. D., B. L. Kovlachik, S. F. Arno, and R. C. Presby
   1977   Forest habitat types of Montana. U.S. Department of Agriculture Forest Service General
          Technical Report INT-34.
Porter, S. C.
   1978   Glacier Peak tephra in the North Cascade Range, Washington: Stratigraphy, distribution,
          and relationship to late-glacial events. *Quaternary Research* 10:30–41.
Ray, V. F.
   1932   The Sanpoil and Nespelem: Salishan peoples of northeastern Washington. University of
          Washington Publications in Anthropology, 5.
Reineck, H.-E., and I. B. Singh
   1975   *Depositional sedimentary environments*. New York: Springer-Verlag.
Rymer, M. J., and J. D. Sims
   1976   Preliminary survey of modern glaciolacustrine sediments for earthquake-induced defor-
          mational structures, south-central Alaska. U.S. Geological Survey Open File Report
          76-373.
Smiley, T. L.
   1958   The geology and dating of Sunset Crater, Flagstaff, Arizona. In *Guidebook of the Black
          Mesa Basin, northeastern Arizona*. New Mexico Geological Society, Ninth Field Confer-
          ence.
Smith, H. W., R. Okazaki, and C. R. Knowles
   1977a  Electron microprobe analysis of glass shards from tephra assigned to Set W, Mount St.
          Helens, Washington. *Quaternary Research* 7:207–217.
   1977b  Electron microprobe data for tephra attributed to Glacier Peak, Washington. *Quater-
          nary Research* 7:197–206.
Stockmarr, J.
   1971   Tablets with spores used in absolute pollen analysis. *Pollen et Spores* 13:615–621.
   1973   Determination of spore concentration with an electronic particle counter. *Geological
          Survey of Denmark Yearbook* 1972:87–89.
Tauber, H.
   1977   Investigations of aerial pollen transport in a forested area. *Dansk Botanisk Arkiv* 32:3–121.
Teit, J. A.
   1930   The Salishan tribes of the western plateaus. Bureau of American Ethnology Annual
          Report, 1927–1928.
Tippett, R.
   1964   An investigation into the nature of the layering of deep-water sediments in two eastern
          Ontario lakes. *Canadian Journal of Botany* 42:1693–1708.
Workman, W. B.
   1974   The cultural significance of a volcanic ash which fell in the Upper Yukon Basin about
          1400 years ago. In *International conference on the prehistory and paleoecology of western
          North American Arctic and Subarctic*, edited by S. Raymond and P. Schledermann.
          Calgary: The University of Calgary Archaeological Association.

# 14

# Mount Mazama, Climatic Change, and Fort Rock Basin Archaeofaunas

DONALD K. GRAYSON

Prior to its eruption and subsequent collapse, Mount Mazama formed an imposing part of southwestern Oregon's Cascade Mountains. Reaching an elevation estimated to have been somewhat less than 3700 m, Mount Mazama erupted at about 7000 B.P., spewing some 30 km³ of liquid magma, entrained crystals, and lithic fragments into the air. Though pumice flows directly affected only immediately adjacent areas, airborne volcanic debris was carried toward the northeast, and the resulting tephra fall affected vast portions of western North America north and east of Mount Mazama. The eruption of Mount Mazama led to the formation of a classic caldera, in which Crater Lake now sits, and subsequent volcanic activity produced a small cinder cone (Wizard Island) within the caldera (Williams 1942, 1969; Williams and Goles 1968).

It has often been suggested that the Mazama eruption must have had a major impact upon the flora and fauna of adjacent regions and that local human settlements must have been similarly affected (Hansen 1942a,b, 1947a; Williams 1942; Malde 1964; Bedwell 1971, 1973). Given these speculations, it is surprising that, aside from the pioneering palynological studies by Hansen (1942a,b, 1947a,b), there have been few attempts to gauge the impacts of the Mazama eruption upon the flora and fauna of immediately adjacent areas, and that there have been no archaeological programs designed to determine the impacts of the eruption upon human settlements here. In fact, aside from having some knowledge of the distribution and nature of Mazama tephra, we know little about the effects of the eruption on eastern Oregon, one of the areas most heavily affected.

427

Of the region to the immediate north and east of Mount Mazama, only the Fort Rock Basin has been the locus of research providing a sequence of archaeological remains that spans most of the Holocene (Cressman *et al.* 1940; Cressman 1942; Bedwell 1969, 1973; Bedwell and Cressman 1971). In addition, the archaeological work that Bedwell (1969, 1973) conducted here during the 1960s provided a series of vertebrate archaeofaunas dating to between about 11,000 and 3000 B.P. Even though the sites that Bedwell excavated apparently lacked deposits immediately postdating the Mazama eruption, I was hopeful that the essentially unanalyzed faunas collected from these sites would provide insight into the impact of Mount Mazama on the fauna of the Fort Rock Basin. Given that human settlements here at the time of the Mazama eruption must in part have been keyed to the distribution and abundance of vertebrates, I was also hopeful that analysis of these faunas would provide insight into the possible effects of the eruption on human subsistence and settlements systems in the Fort Rock Basin. These faunas are the subject of this paper.

## THE FORT ROCK BASIN

The Fort Rock Basin is the northernmost drainage system within the Northern Great Basin. Some 3900 km² in area, the basin is both high, with valley floors at about 1310 m and peaks reaching 1830 m, and arid, with less than 25.4 cm of precipitation a year. During the Pleistocene, the basin was intermittently occupied by a pluvial lake whose maximum depth may have reached 60 m (Phillips and Van Denburgh 1971); there is some radiocarbon evidence to indicate that this lake had retreated to an elevation of about 1340 m by 13,000 B.P. (Bedwell 1973). The subsequent history of the lake is unknown. Today, Silver Lake and Paulina Marsh, in the southwestern corner of the basin, are the only appreciable natural bodies of standing water here, and even they are often dry.

The vegetation of the Fort Rock Basin is dominated by shrubs, especially on soils developed in lacustrine deposits. The big sagebrush (*Artemisia tridentata*) community is found on nonalkaline soils with a high sand content; minor shrubs in this community include rabbitbrush (*Chrysothamnus* spp.), smooth horsebrush (*Tetradymia glabrata*), and low sagebrush (*Artemisia arbuscula*). On more finely textured soils developed in alkaline playas, a silver sagebrush (*Artemisia cana*) community is found. *Artemisia tridentata* and *Chrysothamnus* spp. are dominant shrubs, but the alkaline nature of the soils here is indicated by the presence of *Artemisia cana*, black greasewood (*Sarcobatus vermiculatus*), alkali muhlenbergia (*Muhlenbergia asperifolia*), and thickspike wheatgrass (*Agropyron dasytachum*). Plant communities lacking a major shrub component are rare in the basin; where they occur, they may be the result of burning. On more mesic habitats, and typically on escarpments and rock outcrops, western junipers (*Juniperus occidentalis*) are common, variably associated with *Artemisia tridentata* and A. *arbuscula* as well as with various perennial grasses (Franklin and Dyrness 1973; Moir *et al.* 1973).

Archaeology in the Fort Rock Basin began in the nineteenth century (Cope 1878, 1889), but it was the work of Luther Cressman at Fort Rock Cave in the 1930s (Cressman *et al.* 1940; Cressman 1942) that initiated professional studies of the region's prehistory. After a long lapse that saw little professional work, the late S. F. Bedwell, working with Cressman, began further studies of the prehistory of the area in the late 1960s. Bedwell and Cressman reexcavated portions of Fort Rock Cave in 1966 and 1967, and Bedwell excavated the Connley Caves, a series of rockshelters with deep archaeological deposits, in 1967.

Bedwell was successful in obtaining a large number of radiocarbon dates for some of the sites he excavated: He obtained 21 $^{14}$C dates for the Connley Caves and 4 for Fort Rock Cave. The archaeological sequence obtained by Bedwell is not markedly different from other long sequences in the Great Basin. The cultural record begins by 11,000 B.P. and continues to 3000 B.P.; the later record of the sites had been severely truncated by the depredations of looters. The earlier occupations show similarities to Windust (Rice 1972) and San Dieguito (Warren 1967) materials, whereas later complexes mirror established Great Basin cultural sequences (Bedwell 1969, 1973; Hester 1973; Aikens 1978).

The distribution of radiocarbon dates through time at the Connley Caves is of interest, since none of the dates fall between 7000 and 5000 B.P. (see p. 430). Bedwell argued that this gap in radiocarbon determinations implied an equally long gap in human occupation of the sites. The gap, he argued, could be accounted for either by unfavorable middle Holocene climatic conditions (the Altithermal) or by the effects of the eruption of Mount Mazama. He suggested that though the initial abandonment of the sites might be explained by the Mazama eruption, the entire 2000-year interval of apparent nonoccupation could not be accounted for in this fashion. Therefore, he concluded, middle Holocene heat and aridity account for the lengthy abandonment, even if initiated by the Mazama eruption, and reoccupation of the area did not occur until the end of this climatic episode.

Unfortunately, though Bedwell could assume that the eruption of Mount Mazama had an impact on the Fort Rock Basin, at the time he wrote extremely little was known about the nature of Holocene climatic change in the Northern Great Basin (Grayson 1977b). In addition, the potentially most interesting bits of paleoenvironmental data that Bedwell presented are now known to have been unreliable: the identification of piñon pine (*Pinus monophylla* or *edulis*) and turkeys (*Meleagris gallopavo*) in the Fort Rock Basin prior to 7000 B.P. The wood interpreted as piñon pine was actually securely identified to the genus level only, whereas the bones identified as turkey are, in fact, from sage grouse (*Centrocercus urophasianus*) (Grayson 1977c). In short, because so little is known of changing Holocene climates in the Fort Rock Basin, the scenario provided by Bedwell to account for the archaeological data that he recovered here must be seen as essentially un-tested, although certainly not unreasonable, speculation.

## THE CONNLEY CAVES ARCHAEOFAUNAS

Bedwell recovered sizable faunas from both the Connley Caves and Fort Rock Cave. Unfortunately, the bulk of the fauna from Fort Rock Cave lacked adequate provenience: The 160 elements with adequate provenience that I was able to identify do not provide an adequate sample from which to generalize about the nature of prehistoric Fort Rock Basin environments (Table 14.1). The Connley Caves archaeofauna does, however, allow insight into the nature of these environments.

The Connley Caves are a series of contiguous rockshelters located in volcanic rocks of the Pliocene Fort Rock Formation (Hampton 1964; Bedwell 1969, 1973). Located approximately 16 km southeast of Fort Rock Cave, these sites are about 1.5 km northeast of the fluctuating edge of Paulina Marsh (Figures 14.1 and 14.2). This marsh is fed by the only perennial streams within the Fort Rock Basin and constitutes the basin's largest natural body of standing water.

The vegetation surrounding the Connley Caves is characteristic of that found on escarpments and rock outcrops within the Fort Rock Basin. Domi-nant shrubs are big sagebrush, gray rabbitbrush (*Chrysothamnus nauseosus*) and green rabbitbrush (*C. viscidiflorus*). The dominant grasses are Sandberg's bluegrass (*Poa sandbergii*) and bottlebrush squirreltail (*Sitanion hystrix*). West-ern junipers are scattered throughout the area.

Prior to excavation, Bedwell designated these sites Connley Caves Num-bers 1 through 6 (35 LK 50/1–6). Connley Cave Number 2 subsequently proved to lack evidence of human occupation. Of the remaining rockshelters, Bedwell recovered faunal remains from all but Connley Cave Number 1.

Bedwell (1969, 1973) has described the archaeology of the Connley Caves in detail, and that information will not be repeated here. Excluding the Mazama ash, Bedwell defined four strata for these sites, and argued that each stratum was present in each site. I have assigned the following ages to these strata based upon the radiocarbon dates available for each: Stratum 1, 3000–3400 B.P.; Stratum 2, 3400–4400 B.P.; Stratum 3, 7200–9500 B.P.; and, Stratum 4, 9500–11,200 B.P. (Grayson 1977a,c). Problems of overlap among radiocarbon determinations make the dated boundaries between Strata 1 and 2 on the one hand, and especially Strata 3 and 4 on the other, little more than very rough estimates (a full list of radiocarbon dates from the Connley Caves is presented in Bedwell [1973]).

Recovery of the Connley Caves fauna proceeded by excavating the sites by 10-cm levels within each major stratum; deposits shoveled from the sites were passed through ¼-in. screens and faunal materials retrieved from those screens.

**TABLE 14.1**
**Numbers of Identified Elements by Stratum at Fort Rock Cave**[a]

| | Fort Rock Cave Stratum | | | |
|---|---|---|---|---|
| Taxon | Surface | 1 | 2 | 3 |
| Pygmy rabbit, *Sylvilagus idahoensis* | | 4 | 6 | 1 |
| Nuttall's cottontail, *S.* cf. *nuttallii* | | 8 | 22 | 1 |
| Rabbits, *Sylvilagus* spp. | | 1 | 5 | |
| Hares, *Lepus* spp. | | 36 | 38 | 2 |
| Yellow-bellied marmot, *Marmota flaviventris* | | 1 | | |
| Townsend's ground squirrel, *Spermophilus townsendii* | | 1 | | |
| Ground squirrels, *Spermophilus* sp. | | 7 | | |
| Northern pocket gopher, *Thomomys talpoides* | | 6 | | |
| Ord's kangaroo rat, *Dipodomys ordii* | | 2 | | |
| Bushy-tailed wood rat, *Neotoma cinerea* | 2 | 2 | | |
| Elk, *Cervus canadensis* | | 1 | | |
| Pronghorn, *Antilocapra americana* | | 2 | | |
| Bison, *Bison* sp. | | 5 | | |
| Mountain Sheep, *Ovis canadensis* | | | 2 | 4 |
| Mallard, *Anas platyrhynchos* | | 1 | | |

Totals. Surface:  2
Stratum 1:  77
Stratum 2:  73
Stratum 3:   8
───
160

[a]Strata are those provided by Bedwell (1969, 1973, field notes). Four radiocarbon dates are available for these strata: Stratum 3, 13,200 ± 720; Stratum 2, 10,200 ± 230, 8550 ± 150, and 4450 ± 100. All dates are in radiocarbon years B.P. The problematic date for Stratum 3 was obtained from the same level that provided the identified bone indicated in the table. However, the nature of the association between the date and the identified bone cannot be evaluated.

**FIGURE 14.1. The setting of the Connley Caves. Arrow indicates location of the Connley Caves themselves.**

## Quantification of Taxonomic Abundance

Paleoenvironmental information may be extracted from vertebrate archaeofaunas in a number of ways. Analysis of incremental structures (e.g., Casteel *et al.* 1977), changes in body size of a given species through time (e.g., Davis 1977), and marked changes in the geographic distribution of taxa (e.g., Anderson 1968; Graham 1976) may all provide information on the nature of the environment at the time of deposition of the fauna being studied. In addition, changes in the relative abundances of taxa within archaeofaunas through time may also provide paleoenvironmental information. Because changes in taxonomic abundance through time seem to promise much precision in reconstructing past environments, a great deal of energy has been expended in tabulating and studying such abundances (e.g., Butler 1969; Grayson 1976, 1977b). However, before much faith can be placed in the results of these studies, two major difficulties must be resolved.

First, the nature of the relationship between the animals in the archaeological site and those which existed in the area surrounding the site must be understood. From the point of view of the vertebrate faunal analyst interested in analyzing past environments, an archaeological site represents a point in the environment at which remains of animals occupying the area surrounding the site tended to accumulate. In such complex settings as

FIGURE 14.2. Paulina Marsh as seen from immediately above the Connley Caves. Arrow indicates location of the marsh.

rockshelters, the precise mechanisms by which such accumulations occur are rarely, if ever, known. A range of predators and scavengers—often including humans—may deposit vertebrates in rockshelters, augmenting the set of animals that entered the site and died without the intervention of other mechanisms. Unlike the faunal analyst who is interested in using vertebrate archaeofaunas to study past human economies, the faunal analyst interested in past environments is not interested in the faunal accumulation in the archaeological site per se, but is instead interested in what the accumulation reveals about the nature of the vertebrate communities surrounding the site of deposition. As a result, analysts conducting these kinds of studies must view the faunal accumulation in the site as a sample of the vertebrates that surrounded that site. Since the relationship between the accumulation (the faunal sample) and the vertebrates that lived in the area surrounding the site at the time of accumulation (the target faunal population) is never known, the vertebrate faunal analyst is at a decided disadvantage in attempting to use that sample to infer aspects of the target population unless all that is sought is information on what was present in that population. That is, the analyst can be reasonably certain that animals present in the faunal accumulation were, in fact, present in the area surrounding the site. He or she can be decidedly less certain that animals absent in the accumulation were correspondingly absent in that area. Moreover, the analyst is on extremely shaky ground when changing abundances of taxa become the target of investigation. Significant changes in abundances may reflect changing abundances of animals in the surrounding environment, but they may also represent changing use of the point of accumulation by one or more predators or scavengers, or, more generally, changes in any of the mechanisms that caused the accumulation of the animals in the deposit. If one is interested in using the abundances of vertebrate taxa in an archaeological fauna to infer changing climates, then it becomes essential to filter out those changes in vertebrate abundances that are due to all other causes. Unfortunately, it is impossible to accomplish this given current knowledge of the mechanisms by which archaeofaunas accumulate.

In short, when paleoenvironmental inferences are the analytic goal, vertebrate archaeofaunas must be seen as poorly controlled samples of the vertebrates that existed in the environment about which inferences are to be made. Because the relationship between the sample and the target population is unknown, it seems unduly optimistic to quantify taxonomic abundances within vertebrate archaeofaunas with the aim of treating differences through time in those abundances as reflecting the influence of known environmental causes.

This does not necessarily mean that all attempts to infer the nature of past environments from changing abundances of taxa within vertebrate archaeofaunas should be abandoned until more is known about the ways in which such faunas accumulate. Instead, steps must be taken to increase the likelihood that the faunal changes under study are, in fact, reflecting known environmental causes. One way of accomplishing this is, of course, to examine suites of taxa, each component of which can provide information on the environmental variable of interest. If all the taxa used as paleoenvironmental indicators all change in abundance in equivalent directions, the chances that the variable of interest is, in fact, the cause of these changes increases. This is not to say that proceeding in this fashion will necessarily provide information on the variable of interest. If, for instance, the taxa involved are rodents of either mesic or xeric habitats, changes in abundances of either set of taxa could reflect either changes in abundance of mesic or xeric habitats surrounding the point of accumulation or a change in accumulation mechanism from one that preferentially sampled one of these habitats to one that preferentially sampled the other. One way of avoiding this difficulty is to use a widely varied range of taxa as indicators of the paleoenvironmental attribute of interest so as to decrease the chances that changes in abundance of one of those taxa will be dependent upon the same mechanism of accumulation as changes in abundance of the other taxa under study. For instance, animals that are widely different in size are less likely to be sampled by a single accumulation mechanism than those that are similar in size; as a result, the more varied the size of the taxa used as paleoenvironmental indicators, the less the chances that changes in accumulation mechanism can account for correlated changes in the abundances of those taxa. Clearly, even this approach will not solve the problem. However, the problem is not likely to be fully solved until a great deal more is known about the ways in which vertebrate faunas accumulate in such complex settings as rockshelters. The approach I have just suggested will at least increase the likelihood that changes in the abundance of taxa used as indicators of specific attributes of past environments will, in fact, be informing on those attributes. The only other option would appear to be to ignore the information that the changing abundances of taxa within vertebrate archaeofaunas might provide concerning past environments, and to treat these taxa as attributes that may only be present or absent.

The second major difficulty encountered when analyzing the abundances of taxa within vertebrate archaeofaunas involves precisely how those abundances should be measured. This problem is also a long way from being solved, although the pitfalls of various measures are becoming well described (see, for instance, Casteel 1978, n.d.; Grayson 1973, 1978, and references in those papers). The values provided by the unit favored by archaeologists—the minimum number of individuals—vary with the way in which the faunal material is divided into the smaller aggregates that in turn form the basis of minimum number definition, and are, in addition, a function of the number of identified elements per taxon (Casteel n.d.; Grayson 1973, 1978). It would seem that minimum numbers cannot tell us very much about taxonomic abundances, but what they can tell us is in general also supplied by simple element counts. The argument that more can be done with minimum numbers than with other abundance measures (Chaplin 1971), since they are felt to indicate relative abundances of actual numbers of individual animals per taxon, cannot stand in the face of the facts that the relationship between minimum and actual numbers of animals per taxon is never known, and that the relationship between these two figures must vary among taxa. Element counts, however, possess the unfortunate attribute of interdependence, as I have pointed out many times (Grayson 1973, 1974, 1977b). Even given that minimum number values are a function of numbers of identified elements per taxon, the increase in sample size that occurs as the bones of a single animal deposited in an archaeological site are increasingly fragmented will cause interpretive difficulties even if fragmentation is equally distributed across all taxa: Differences in taxonomic abundance that are statistically insignificant when examined using the relatively small numbers provided by minimum numbers of individuals may well be statistically significant when examined using the larger numbers provided by element counts. I have no universal solution for these problems. In the current context, however, a solution is possible because of the general nature of the questions being asked of the faunal data. In this chapter, I shall primarily be interested in comparing abundances of taxa before and after the Mazama eruption, and will include in my discussion only those taxa whose abundances across this boundary change significantly in terms of both measures of abundance. I use minimum number values only to provide a control for the effects of interdependence. Given my approach, the results are the same no matter which measure of abundance is employed.

Connley Caves 3, 4, 5 and 6 yielded 2114 bones, bone fragments, and teeth that I was able to identify. These elements pertain to at least 22 genera of mammals and 10 genera of birds. In addition, 9 elements of fish were recovered that have yet to be identified. The number of identified elements per taxon (E) and the minimum number of individuals per taxon (MNI) are presented in Tables 14.2–14.5.

The calculation of E values for these data was straightforward: These values simply refer to the number of identified elements per taxon. Minimum numbers were calculated as follows. First, all faunal materials were separated by site, then subdivided by major natural stratum within each site, then, since each natural stratum represents between about 400 and 2300 years of deposition, by 10-cm levels within those strata. The operational definition of minimum numbers (Grayson 1973) was then applied separately to each of the clusters of faunal material so defined, and the numbers for separate 10-cm levels within a stratum combined to provide MNI values for each species within each stratum. In proceeding in that fashion I assumed that individual animals had not been distributed across sites or across arbitrary levels; to the extent that they were so distributed, the minimum numbers reported are interdependent and inflated. Minimum numbers are not provided for the Anatinae, since the elements assigned to this subfamily may pertain to individuals already counted among other, lower taxa of ducks.

## The Connley Caves Birds

Although the sample of birds from the Connley Caves is small (563 identified elements and 195 individuals), the sample is nonetheless interesting. Of the 563 elements and 195 individuals, 537 elements and 174 individuals were found in Strata 3 and 4, whereas only 26 elements and 21 individuals were recovered from Strata 1 and 2. A large proportion of this decrease may be attributed to Bedwell's (1973) excavation strategy, which properly emphasized the excavation of deeper, less disturbed deposits within the sites. In order to reach these deposits in Connley Caves 3 and 4, "a backhoe was employed to remove the fill down to Mazama ash [p. 19]"—that is, to remove disturbed portions of Strata 1 and 2. As a result, the reduction in volume of deposits excavated above Stratum 3 may account for much of the post-Mazama reduction in size of the avian faunal sample.

However, not all of the reduction may be explained in this fashion. Though mammals undergo a similar reduction (from 1275 elements and 298 individuals in Strata 3 and 4 to 277 elements and 169 individuals in Strata 1 and 2), for both elements ($\chi^2 = 43.44$, $p < .01$) and individuals ($\chi^2 = 60.51$, $p < .01$) this decrease is significantly less than that which affects the birds. Clearly, the decrease in the avian component of the archaeofauna is not due solely to excavation procedures.

All of the birds are affected by this reduction; only a few sage grouse, scattered waterbirds, and an occasional raptor are present in the upper deposits of the Connley Caves. I assume that this reduction relates to the availability of the birds, not to changes in the mechanisms that accumulated the fauna. That is, I assume that the reduction in the numbers of birds in the sites is a direct reflection of a reduction of birds in the surrounding environment.

Of the 20 genera of birds identified from the Connley Caves, 12 require significant amounts of open water for survival (*Podiceps, Aechmophorus, Podilymbus, Anas, Aythya, Bucephala, Oxyura, Fulica, Numenius, Recurvirostra, Larus,* and *Sterna*). Reduction of Paulina Marsh after 7000 B.P. would most certainly adversely affect populations of these birds. Of the remaining genera, only sage grouse were abundant in Strata 3 and 4. Could reduction of Paulina Marsh account for the decrease of this bird as well?

Sage grouse do not require continuous supplies of open water to survive. Nonetheless, these birds "attain their highest population densities in those areas which contain abundant and well-distributed surface water supplies [Patterson 1952:208]." Reduction in the amount of open water available in any given region in which sage grouse are present reduces grouse populations in marginal areas and concentrates the birds in regions that still maintain such water (Patterson 1952). Continued drought would not be likely to extirpate sage grouse from large expanses of their habitat but would greatly reduce their numbers. That is, reduction of Paulina Marsh would have much the same effect upon sage grouse as it would upon waterbirds: Populations of both would decrease in size.

**TABLE 14.2**
**Numbers of Identified Elements for the Connley Caves Mammals**

| | Connley cave number | | | | | | | | | | | | | | | |
|---|---|---|---|---|---|---|---|---|---|---|---|---|---|---|---|---|
| | 3 | | | | 4 | | | | 5 | | | | 6 | | | |
| | Stratum | | | | Stratum | | | | Stratum | | | | Stratum | | | |
| Taxon | 1 | 2 | 3 | 4 | 1 | 2 | 3 | 4 | 1 | 2 | 3 | 4 | 1 | 2 | 3 | 4 |
| Broad-footed mole, *Scapanus latimanus* | | | | | 1 | | | | | | | | | | | |
| Pika, *Ochotona princeps* | | | | | | | 7 | 6 | | | 3 | | | | | |
| Pygmy rabbit, *Sylvilagus idahoensis* | 1 | | | | | 5 | 43 | 34 | | | 13 | | | | | |
| Nuttall's cottontail, *S.* cf. *nuttallii* | 7 | | | | 10 | | 75 | 27 | | 3 | 44 | | 1 | 34 | | 5 |
| Rabbits, *Sylvilagus* sp. | | | | | 1 | 2 | 8 | 12 | | | 5 | | | 1 | | |
| White-tailed and black-tailed jackrabbits, *Lepus* spp. | 34 | 1 | | | 7 | 3 | 256 | 388 | 9 | 1 | 114 | | 4 | 88 | | 4 |
| Yellow-bellied marmot, *Marmota flaviventris* | 5 | | | | 3 | 2 | 3 | | | | 1 | | | | | |
| Belding's ground squirrel, *Spermophilus beldingi* | | | | | | | | 1 | | | | | | | | |
| Ground squirrel, *Spermophilus* sp. | | | | | 1 | | | | | | | | | | | |
| Northern pocket gopher, *Thomomys talpoides* | | | | | 4 | | 31 | 8 | 3 | 1 | 15 | | | 6 | | |
| Ord's kangaroo rat, *Dipodomys ordii* | | | | | | 1 | 4 | 2 | | | 2 | | | | | |

| Bushy-tailed wood rat, *Neotoma cinerea* | 2 | 3 | 3 | 40 | 9 | 3 | 8 | 3 |
|---|---|---|---|---|---|---|---|---|
| Voles, Microtinae sp. | | | 3 | 1 | 3 | 3 | | 1 |
| Porcupine, *Erethizon dorsatum* | 1 | | | 1 | 2 | | | |
| Coyote, *Canis* cf. *latrans* | | | 1 | 1 | 4 | 1 | | |
| Red fox, *Vulpes fulva* | | | | 2 | 2 | | | |
| Long-tailed weasel, *Mustela frenata* | | | | 2 | | | | |
| Mink, *M. vison* | 2 | | | | | | | |
| Wolverine, *Gulo luscus* | | | | | 1 | | | |
| Badger, *Taxidea taxus* | | | | | | | 1 | 1 |
| Spotted skunk, *Spilogale putorius* | | | | | | | | |
| Bobcat, *Lynx* cf. *rufus* | | | | 1 | 3 | | | |
| Elk, *Cervus canadensis* | | | | 8 | 13 | | 3 | |
| Black-tailed deer, *Odocoileus* cf. *hemionus* | | | | 10 | 2 | | 1 | 1 |
| Pronghorn, *Antilocapra americana* | | | | | | | | |
| Bison, *Bison* cf. *bison* | 1 | | | 8 | 19 | | 3 | 3 |
| Mountain sheep, *Ovis canadensis* | 1 | | | 1 | | | 12 | |

Totals:  Stratum 1: 422  
Stratum 2: 455  
Stratum 3: 128  
Stratum 4: 547  
‾‾‾‾  
1552

# TABLE 14.3
## Minimum Numbers of Individuals for the Connley Caves Mammals

| Taxon | 3 | | | | 4 | | | | 5 | | | | 6 | | | |
|---|---|---|---|---|---|---|---|---|---|---|---|---|---|---|---|---|
| | Stratum | | | | Stratum | | | | Stratum | | | | Stratum | | | |
| | 1 | 2 | 3 | 4 | 1 | 2 | 3 | 4 | 1 | 2 | 3 | 4 | 1 | 2 | 3 | 4 |
| Broad-footed mole, *Scapanus latimanus* | | | | | 1 | | | | | | | | | | | |
| Pika, *Ochotona princeps* | | | | | | | | | | | 2 | | | | | |
| Pygmy rabbit, *Sylvilagus idahoensis* | 1 | | | | 7 | | 3 | 6 | | | | | | | | |
| Nuttall's cottontail, *S.* cf. *nuttallii* | 6 | | | | 1 | 3 | 11 | 12 | | 1 | 10 | | | | | 2 |
| Rabbits, *Sylvilagus* sp. | | | | | | 1 | 16 | 12 | | | 4 | | | 1 | | |
| White-tailed and black-tailed jackrabbits, *Lepus* spp. | 13 | | 1 | | 7 | 3 | 25 | 37 | 1 | 1 | 18 | | 1 | 23 | | 4 |
| Yellow-bellied marmot, *Marmota flaviventris* | 1 | | | | | | 2 | | | | 1 | | | | | |
| Belding's ground squirrel, *Spermophilus beldingi* | | | | | | | 1 | | | | | | | | | |
| Ground squirrel, *Spermophilus* sp. | | | | | | | | | 1 | | | | | | | |
| Northern pocket gopher, *Thomomys talpoides* | | | | | 3 | 16 | 4 | | | 1 | 7 | | | 6 | | |
| Ord's kangaroo rat, *Dipodomys ordii* | | | | | | 4 | 2 | | | | | | | | | |

bushy-tailed wood rat,

| Species | | | | | | | | |
|---|---|---|---|---|---|---|---|---|
| Neotoma cinerea | 1 | 2 | 2 | 12 | 7 | 3 | 4 | 3 |
| Voles, Microtinae sp. | | | | 1 | 2 | | | |
| Porcupine, Erethizon dorsatum | 1 | | | 1 | 1 | | | |
| Coyote, Canis cf. latrans | | | | 1 | 4 | 1 | | |
| Red fox, Vulpes fulva | | | | 2 | 1 | | | |
| Long-tailed weasel, Mustela frenata | | | | 2 | | | | |
| Mink, M. vison | 2 | | | | | | | |
| Wolverine, Gulo luscus | | | | | 1 | | | |
| Badger, Taxidea taxus | | 1 | | | | | 1 | 1 |
| Spotted skunk, Spilogale putorius | | | | | | | | |
| Bobcat, Lynx cf. rufus | | | | 1 | 1 | | 1 | |
| Elk, Cervus canadensis | | | | 3 | 6 | | 2 | 1 |
| Black-tailed deer, Odocoileus cf. hemionus | | | | 3 | 2 | | 1 | 1 |
| Pronghorn, Antilocapra americana | 1 | | | 1 | | | 2 | 1 |
| Bison, Bison cf. bison | 1 | | | 4 | 5 | | 6 | |
| Mountain sheep, Ovis canadensis | | | | 1 | | | | |

Totals:  Stratum 1:  73
         Stratum 2:  66
         Stratum 3:  179
         Stratum 4:  119
                     437

441

**TABLE 14.4**
**Numbers of Identified Elements for the Connley Caves Birds**

| | Connley cave number | | | | | | | | | | | | | | | |
| | 3 | | | | 4 | | | | 5 | | | | 6 | | | |
| | Stratum | | | | Stratum | | | | Stratum | | | | Stratum | | | |
| Taxon | 1 | 2 | 3 | 4 | 1 | 2 | 3 | 4 | 1 | 2 | 3 | 4 | 1 | 2 | 3 | 4 |
| Horned grebe, *Podiceps auritus* | | | | | | | 1 | | | | | | | | | |
| Eared grebe, *P. nigricollis* | | | | | | | 1 | | | | | | | | | |
| Horned grebe and/or eared grebe, *Podiceps* spp. | | | | | | 5 | | 2 | | | 2 | | | 1 | | |
| Western grebe, *Aechmophorus occidentalis* | | | | | | | 1 | | | | 1 | | | | | |
| Pied-billed grebe, *Podilymbus podiceps* | | | | | | | | 1 | | | | | | | | |
| Grebes, *Podiceps* and/or *Podilymbus* | | | | | | 2 | 2 | 3 | | | 2 | | | | | |
| Goose spp., *Anser* and/or *Branta* | | | | | | 1 | | | | | 3 | | | | | |
| Mallard, *Anas platyrhynchos* | | | | | | 1 | 3 | 2 | | | | | | 1 | | |
| Gadwall, pintail, American wigeon, and/or northern shoveler, *Anas* spp. | | | | | | | 22 | 3 | | | 6 | | | 2 | | |
| Green-winged teal, blue-winged teal, and/or cinnamon teal, *Anas* spp. | | | | | 1 | | 39 | 15 | | | 15 | | | 1 | | |
| Redhead, ring-necked duck, greater scaup, and/or lesser scaup, *Aythya* spp. | | | | | 1 | 1 | | 4 | | | | | | | | |
| Bufflehead, *Bucephala albeola* | | | | | | | 1 | | | | | | | | | |
| Ruddy duck, *Oxyura jamaicensis* | | | | | | | 1 | | | | 1 | | | 1 | | |

| Species | Stratum 1 | Stratum 2 | Stratum 3 | Stratum 4 |
|---|---|---|---|---|
| Ducks, Anatini | 1 | 17 | 45 | 13 |
| Red-tailed hawk, *Buteo* cf. *jamaicensis* | | 1 | 1 | |
| Rough-legged hawk, *B. lagopus* | 1 | | 1 | |
| Hawk, *Buteo* sp. | | | 1 | |
| American kestrel, *Falco sparverius* | | 1 | 1 | |
| Sharp-tailed grouse, *Pediocetes phasianellus* | | | 1 | 1 |
| Sage grouse, *Centrocercus urophasianus* | 3 | 35 | 197 | 50 |
| American coot, *Fulica americana* | | 2 | 8 | 1 |
| Long-billed curlew, *Numenius americanus* | | | 1 | |
| American avocet, *Recurvirostra americana* | | | 3 | |
| Bonaparte's gull, *Larus philadelphia* | | | 1 | |
| Ring-billed gull, *L. delawarensis* | | | 1 | |
| Tern, *Sterna* sp. | | | 1 | |
| Great horned owl, *Bubo virginianus* | | | 5 | |
| Short-eared owl, *Asio flammeus* | | 1 | 2 | 1 |
| Short-eared owl, *Asio* cf. *flammeus* | | | 1 | |
| Owl, *Asio* sp. | | | | 1 |
| Common raven, *Corvus corax* | | 1 | | |
| Common crow, *C. brachyrhynchos* | | 1 | | |
| Meadowlark, cf. *Sturnella* sp. | | 1 | | |
| Meadowlark or blackbird, Icteridae gen. et sp. indet. | 3 | 2 | 1 | |
| Songbirds, Passeriformes | | 2 | 1 | |

Totals:  Stratum 1: 8  
Stratum 2: 18  
Stratum 3: 439  
Stratum 4: 98  
563

443

**TABLE 14.5**
**Minimum Numbers of Individuals for the Connley Caves Birds**

| | Connley cave number | | | | | | | | | | | | | | |
|---|---|---|---|---|---|---|---|---|---|---|---|---|---|---|---|
| | 3 | | | | 4 | | | | 5 | | | | 6 | | |
| | Stratum | | | | Stratum | | | | Stratum | | | | Stratum | | |
| Taxon | 1 | 2 | 3 | 4 | 1 | 2 | 3 | 4 | 1 | 2 | 3 | 4 | 2 | 3 | 4 |
| Horned grebe, *Podiceps auritus* | | | | | | | 1 | | | | | | | | |
| Eared grebe, *P. nigricollis* | | | | | | | 1 | | | | | | | | |
| Horned grebe and/or eared grebe, *Podiceps* spp. | | | | | | | 4 | 1 | | | 2 | | | 1 | |
| Western grebe, *Aechmophorus occidentalis* | | | | | | | 1 | | | | 1 | | | | |
| Pied-billed grebe, *Podilymbus podiceps* | | | | | | | 1 | | | | | | | | |
| Grebes, *Podiceps* and/or *Podilymbus* | | | | | | 2 | 2 | 3 | | 2 | 1 | | | | |
| Goose spp. *Anser* and/or *Branta* | | | | | | | 1 | | | | | | | | |
| Mallard, *Anas platyrhynchos* | | | | | | 2 | 2 | | | | 1 | | | 1 | |
| Gadwall, pintail, American wigeon, and/or northern shoveler, *Anas* spp. | | | | | | | 8 | 3 | | | 4 | | | 2 | |
| Green-winged teal, blue-winged teal, and/or cinnamon teal, *Anas* spp. | | | | | | 1 | 14 | 7 | 1 | 1 | 10 | | | | |
| Redhead, ring-necked duck, greater scaup, and/or lesser scaup, *Aythya* spp. | | | | | | | 1 | | | | | | | 1 | |
| Bufflehead, *Bucephala albeola* | | | | | | | 1 | | | | | | | | |

| Species | Stratum 1 | Stratum 2 | Stratum 3 | Stratum 4 |
|---|---|---|---|---|
| Ruddy duck, *Oxyura jamaicensis* | 1 | | | 1 |
| Ducks, Anatini (see text) | | 1 | | |
| Red-tailed hawk, *Buteo* cf. *jamaicensis* | 1 | | 1 | 1 |
| Rough-legged hawk, *B. lagopus* | | | | 1 |
| Hawk, *Buteo* sp. | | | 1 | 1 |
| American kestrel, *Falco sparverius* | | | 1 | 1 |
| Sharp-tailed grouse, *Pediocetes phasianellus* | | | | 1 |
| Sage grouse, *Centrocercus urophasianus* | 3 | 12 | 21 | 28 |
| American coot, *Fulica americana* | | 2 | 1 | 4 |
| Long-billed curlew, *Numenius americanus* | | | | 1 |
| American avocet, *Recurvirostra americana* | | | | 3 |
| Bonaparte's gull, *Larus philadelphia* | | | | 1 |
| Ring-billed gull, *L. delawarensis* | | | | 1 |
| Tern, *Sterna* sp. | | | | 1 |
| Great horned owl, *Bubo virginianus* | | | 4 | |
| Short-eared owl, *Asio flammeus* | | 1 | 1 | 2 |
| Short-eared owl, *Asio* cf. *flammeus* | | | | 1 |
| Owl, *Asio* sp. | | | | 1 |
| Common raven, *Corvus corax* | | 1 | | |
| Common crow, *C. brachyrhynchos* | | 2 | | 1 |
| Meadowlark, cf. *Sturnella* sp. | | 1 | | |
| Meadowlark or blackbird, Icteridae gen. et sp. indet. | 2 | 2 | | 1 |
| Songbirds, Passeriformes | | 1 | | 1 |

Totals: Stratum 1: 7  
Stratum 2: 14  
Stratum 3: 128  
Stratum 4: 46  
195

Thus, the great reduction in numbers of birds that is characteristic of post-Mazama levels of the Connley Caves can be accounted for by a major reduction in size of the marsh between 7000 and 5000 B.P. The continued presence of small numbers of waterbirds in post-Mazama levels of these sites suggests that open water continued to exist in the area, at least seasonally.

It is hard to resist providing a bit of speculative reconstruction by noting that the sex ratios characteristic of the Connley Caves sage grouse are not inconsistent with this view. In an earlier paper (Grayson 1977c), I distinguished male and female grouse elements from the Connley Caves on the basis of bivariate plots of measurable limb elements. The measurable elements from the combined Strata 3 and 4 sage grouse sample indicated the presence of 28 elements from male and 48 elements from female sage grouse. Mayr (1939) noted long ago that sex ratios in polygynous or promiscuous species of birds were unbalanced in the favor of females, a finding confirmed by subsequent research (Trivers 1972). Sage grouse are no exception to this rule: Assuming that element counts are here accurately reflecting ratios of male and female birds, the pre-Mazama Connley Caves ratio of 1 male per 1.7 female sage grouse is not significantly different from sex ratios reported for modern sage grouse populations (Dalke et al. 1960). Given the patterned aggregation and disaggregation of the sexes that is characteristic of sage grouse social organization (Patterson 1952; Dalke et al. 1960, 1963; Wiley 1973), and the fact that immature sage grouse are virtually lacking in the Connley Caves avifauna (one element in Stratum 3), three patterns of sage grouse predation by the occupants of the Connley Caves might account for the male: female and mature: immature ratios characteristic of the pre-Mazama Connley Caves sage grouse: (a) predation on separate male and female flocks in the uplands during the winter months; (b) predation on male flocks and dispersed females in the uplands during late spring; (c) predation on all individuals as they came to drink at Paulina Marsh from late fall to late spring.

Of these alternatives, the first two require the assumption that separated male and female flocks, or separated male flocks and dispersed females, were preyed upon in equal proportions, such as to preserve the sex ratio of the entire population in the sample of birds brought back to the Connley Caves. The third alternative simply requires the assumption that all birds were preyed upon as they came to drink at Paulina Marsh. Because many factors relating to the inconspicuous nature of female sage grouse (Patterson 1952) would make males much more prone to predation in the first two settings and would, therefore, make it highly unlikely that population sex ratios would be retained in the sample of birds retrieved in these settings, it is most reasonable to hypothesize that the Connley Caves sage grouse accumulated as a result of predation on birds that were utilizing Paulina Marsh as a water source.

Fortunately, even if this extremely speculative (and barely testable) exercise in reconstruction is incorrect, it remains true that the Connley Caves birds whose abundances decreased dramatically after the eruption of Mount Mazama were birds whose abundances would have been negatively affected by decreases in the availability of shallow water habitat in the surrounding area. The reduction of Paulina Marsh sometime after the eruption of Mount Mazama can, therefore, account for the decreased abundance of birds in post-Mazama deposits at the Connley Caves.

## The Connley Caves Mammals

The Connley Caves birds suggest that significant environmental change, involving the reduction of Paulina Marsh, occurred in the Fort Rock Basin between 7000 and 5000 B.P. The sample of mammals from the Connley Caves is consistent with these implications. There are three aspects of this sample that I wish to examine here.

### PIKAS IN THE FORT ROCK BASIN

Although they have not been reported for the modern Fort Rock Basin, pikas (*Ochotona princeps*) were present in the pre-Mazama levels of the Connley Caves (Grayson 1977a). Because pikas are primarily mountain animals and are not known to occur in the Fort Rock Basin today, their presence at the Connley Caves between 11,000 and 7000 B.P. is noteworthy.

The record for pikas at these sites is not abundant: Sixteen elements from a minimum of 11 individuals were recovered from Strata 3 and 4. When compared to the distribution of all other mammals, the decrease in both elements ($\chi^2 = 3.52$, $p < .10$) and individuals ($\chi^2 = 5.26$, $p < .05$) is significant. More importantly, Bedwell's excavations removed nearly all of the undisturbed post-Mazama deposits in the Connley Caves; as a result, it seems unlikely that pikas were, in fact, represented in the upper strata of these sites.

As I have noted elsewhere (Grayson 1977a), pikas in the Great Basin tend to occur as isolated populations on mountainous uplands (Hall 1946; Durrant 1952; Durrant et al. 1955); the animals of one mountain range separated from those of another by "a vast sea of sagebrush desert [Brown 1971:467]." In the Northern Great Basin and adjacent areas, pikas are found primarily in such rugged uplands as the Steens and Warner Mountains. Though pikas have been reported to the west, south, and east of the Fort Rock Basin, there are no records for the basin itself (B. J. Verts, personal communication). A recent survey of the area indicates that although talus suitable for pikas is present, sufficient succulent vegetation is lacking (C. Maser, personal communication). Thus, today's Fort Rock Basin seems too zeric to support these animals.

It is possible that early Holocene pikas in the Fort Rock Basin succumbed to decreasing effective precipitation at or sometime after 7000 B.P., and that conditions in the basin since that time have been too xeric to allow recolonization. This hypothesis is in accord with the remainder of the Connley Caves fauna, as well as with climatic reconstructions for adjacent areas (e.g., Butler 1969, 1972a,b; Grayson 1976, 1977b). However, even if this is correct, the initial factor in the local extinction of these mammals may have been the eruption of Mount Mazama at about 7000 B.P. Though relatively little is known of the impact of this eruption on the Fort Rock Basin, the eruption did cause the deposition of approximately 15 cm of ash in the basin (Bedwell 1973). This event would have been very significant to the pika populations here. Not only would the talus upon which these animals depend for shelter have been at least partly choked with ash, but the local vegetation—the food source for pikas— would have been decimated by the ashfall (Hansen 1942a, 1947a). In addition, the exceedingly abrasive effect of pumice on the teeth of these small herbivores may have been extremely deleterious, an effect documented for historic lagomorph populations (Burt 1960). With or without climatic change, the deposition of Mazama ash in the Fort Rock Basin would have caused great damage to the local pika populations, and may have caused the initial disappearance of these animals here. An overlapping or subsequent shift toward an increasingly xeric environment, as is indicated by other aspects of the Connley Caves fauna, would have prevented pikas from reinvading after the ashfall. Since pikas have restricted dispersal abilities, especially at lower elevations (Smith 1974a,b), reoccupation of the basin by pikas may not have occurred after this ashfall even if the shift toward an increasingly xeric environment began long after the eruption.

Thus, it is reasonable to suggest that pikas were eliminated from the Fort Rock Basin as the result of the superposition of a catastrophic event upon unfavorable climatic change approximately 7000 years ago. Even if climatic change alone accounts for the extinction of pikas within the Fort Rock Basin, these data provide support for the argument that the distribution of boreal mammals on mountainous islands within the Great Basin is to be accounted for not by equilibria between rates of extinction and rates of colonization (MacArthur and Wilson 1967), but instead by initial colonization during the Pleistocene followed by subsequent extinctions of geographically intermediate populations. The Connley Caves, in fact, seem to provide insight into the process of extinction of one such geographically intermediate population.

### LEPUS TOWNSENDII AND LEPUS CALIFORNICUS

One of the more interesting differences between pre- and post-Mazama Connley Caves faunas involves the apparent shift in the species of *Lepus* that are represented in the rockshelters. Because I have discussed this shift elsewhere (Grayson 1977a), I will simply summarize my arguments here.

Two species of *Lepus* are found in the lower elevations of the modern Fort Rock Basin—black-tailed jackrabbits (*L. californicus*) and white-tailed jackrabbits (*L. townsendii*). Of these species, white-tailed jackrabbits are more northerly in distribution, tend to occupy higher elevations and grassier habitats, and are better adapted to the colder climates of higher or more northerly regions. In contrast, black-tails are more southerly, prefer lower elevations and shrubbier habitats, and lack some of the adaptations of white-tailed jackrabbits to colder climates (Anthony 1913; Orr 1940; Sevareid 1950; Hansen and Flinders 1969; Flinders and Hansen 1972, 1973, 1975).

I have argued that the Connley Caves birds imply that the amount of open-water habitat in the Fort Rock Basin decreased after about 7000 B.P. Though this hypothesized decrease could have been caused by local tectonic activity or by alterations in drainage resulting from the deposition of Mazama tephra in the Fort Rock Basin, it could also have resulted from changes in effective precipitation during the middle Holocene. If the hypothesized reduction in abundance of open-water habitat did occur, and if the cause of this reduction was a shift in effective precipitation, then the relative abundances of *L. californicus* and *L. townsendii* should reflect this shift. *L. townsendii* may be predicted to have been the abundant hare in the cooler and/or moister environments of the early Holocene in the lower elevations of the Fort Rock Basin, whereas *L. californicus* may be predicted to have been the abundant hare after this time.

Unfortunately, black-tailed and white-tailed jackrabbits are difficult to distinguish osteologically: Even complete skulls may be difficult to distinguish (Hoffman and Pattie 1968). Fortunately, *L. townsendii* and *L. californicus* obey Bergmann's rule—the more northerly *L. townsendii* tends to be larger than the more southerly *L. californicus*. Although the size overlap is considerable, white-tailed jackrabbit bones and teeth are, on the average, larger than black-tailed jackrabbit bones and teeth. These differences may allow the species composition of any set of elements that are either *L. townsendii* or *L. californicus* to be assessed.

Using the alveolar length of the mandibular toothrow, I have demonstrated (Grayson 1977a) that

1. *Lepus* mandibles in Strata 3 and 4 of the Connley Caves are significantly larger than those of Strata 1 and 2 ($t = -5.07$, $p < .01$);
2. *Lepus* mandibles from the combined Connley Caves Strata 1 and 2 samples are significantly smaller than those of modern *L. townsendii* ($t = -2.10$, $p < .05$), but are not significantly different from modern *L. californicus* ($t = 0.98$, $p > .20$);
3. *Lepus* mandibles from the combined Connley Caves Strata 3 and 4 samples are significantly larger than those of modern *L. californicus* ($t = 7.07$, $p < .01$) but are not significantly different from modern *L. townsendii* ($t = 1.53$, $p > .10$).

Details of this analysis are presented in Grayson (1977a). Given these similarities and differences, it is reasonable to conclude that the hares beneath Mazama ash at the Connley Caves are predominantly *L. townsendii*, whereas those above the ash are predominantly *L. californicus*.

It is conceivable that the shift in mandible size through time is chronoclinal, reflecting evolutionary change within local populations of *Lepus*, rather than taxonomic, reflecting changing distributions of species. However, the relatively abrupt change in mandible size between pre- and post-Mazama strata coupled with the lack of evidence for change within Strata 3 and 4 on the one hand, and Strata 1 and 2 on the other, strongly suggests that the shift is due to changing species composition within the local leporid community.

These findings support the hypothesis of a shift toward decreased effective precipitation in the Fort Rock Basin during the middle Holocene, and are congruent with the implications of the Connley Caves avifauna concerning environmental change in this area.

OTHER FAUNAL SHIFTS

In addition to the changes in composition of the Connley Caves lagomorph sample, the pre- and post-Mazama mammalian faunas from these sites also differ in other ways. As measured by both numbers of identified elements and minimum numbers of individuals, the abundances of four other taxa (*Sylvilagus idahoensis*, *Lepus* spp., *Cervus canadensis*, and *Bison bison*) decrease significantly across the Mazama boundary when compared with the abundance of all other mammals, whereas the abundance of one taxon (*Sylvilagus nuttallii*) increases significantly (see Table 14.6).

Unfortunately, it is impossible to say with certainty whether these differences are reflecting changes in the mechanisms that accumulated the Connley Caves fauna, or whether they reflect instead changing absolute abundances of the animals in the area surrounding the sites. Clearly, all of these animals may have been present because of the hunting activities of the human occupants of the Connley Caves, and the changing abundances through time could reflect changes in preferred prey by those occupants. However, since all of these shifts can be explained by reference to a single climatic variable (with one possible exception), it seems most reasonable to assume that these changes are, in fact, reflecting abundances of the animals in the area surrounding the site, rather than changes in accumulation mechanisms.

With the possible exception of the decreased abundance of *Lepus* spp., these shifts can be explained by positing a change in the nature of the vegetation of the Fort Rock Basin from one that, prior to about 7000 B.P., was characterized by a higher percentage of herbaceous vegetation to one which after that time was characterized by a higher percentage of shrubs. Such a shift would account for the decrease of the two large grazing ungulates, elk and

**TABLE 14.6**
**Significant Differences in Abundances of Mammalian Taxa in Pre- and Post-Mazama Faunas at the Connley Caves**

| Taxon | $E^a$ | | $MNI^b$ | |
|---|---|---|---|---|
| | $\chi^2$ | $p$ | $\chi^2$ | $p$ |
| *Sylvilagus idahoensis* | 16.38 | <.001 | 10.91 | <.001 |
| *S. nuttallii* | 37.03 | <.001 | 11.43 | <.001 |
| *Lepus* spp. | 4.20 | <.05 | 4.99 | <.05 |
| *Cervus canadensis* | 3.32 | <.10 | 3.14 | <.10 |
| *Bison* sp. | 6.60 | <.02 | 5.00 | <.05 |

[a]E = numbers of identified elements.
[b]MNI = minimum numbers of individuals.

bison. The decrease in abundance of both white-tailed jackrabbits and pikas is fully consistent with such a change. If this shift was caused by decreasing effective precipitation, as seems likely, then the marked decline in the abundance of pygmy cottontails is also consistent with this interpretation. Though these animals are poorly known, they do seem most abundant in tall, decadent stands of big sagebrush; any factor that caused the loss of such stands—as decreased effective precipitation would do—would at the same time reduce pygmy cottontail population sizes. Finally, the increase in shrubbiness of the vegetation would also increase habitat for Nuttall's cottontails, animals whose abundances are greatest in such shrubby habitats. The only shift in taxonomic abundance that might not be accounted for by positing a decrease in herbaceous, and an increase in shrubby, vegetation between about 7000 and 5000 B.P. is the overall decrease in abundance of *Lepus* spp. I am not aware of any data that would either support or refute the notion that *L. townsendii* is more abundant in grassy habitats than *L. californicus* is in shrubby ones.

## PALEOENVIRONMENTAL OVERVIEW

The avian and mammalian faunas provided by the Connley Caves provide an internally consistent picture of the differences between pre- and post-Mazama environments surrounding these sites. These faunas suggest that Paulina Marsh retreated extensively between about 7000 and 5000 B.P., and suggest also that the herbaceous component of the local vegetation decreased as the shrub component increased during the same time interval. All aspects of the faunal data consistently imply such changes.

Given these hypothesized changes, it must be asked which of the two most likely possible causes best explain them—the effects of the eruption of Mount Mazama, or climatic change. If nearby areas that were not affected to any great extent by the eruption of Mount Mazama underwent similar changes, then it is reasonable to infer that Mount Mazama need not be considered the cause of these changes. A brief review of paleoenvironmental studies from adjacent regions indicates that this does, in fact, appear to be the case.

Mehringer (1977) has noted that precipitation–evaporation ratios like those of the present did not become established in the Great Basin until about 7500 B.P. The precise date at which modern precipitation–evaporation ratios were reached in the Great Basin may be expected to have varied from place to place as might the effects of this event. In the southern Plateau, immediately adjacent to the Northern Great Basin, these modern ratios seem to have been reached at about 7000 B.P.

Arguing primarily on the basis of faunal data from the Wasden Site (Owl Cave) on the Snake River Plain of southern Idaho, Butler (1969, 1972a,b) concluded that relatively cool and moist conditions characterized this area prior to 7000 B.P., whereas after this time and continuing to at least 2000 B.P., conditions similar to those of today prevailed. Though Butler's argument is based primarily on vertebrate faunal data, it is also supported by other lines of evidence from Owl Cave (Dort 1968). Butler (1972) argued that, as a result of this climatic shift, "essentially two things happened 7,000 years ago: 1) a general reduction in vegetation cover and 2) a greater reduction in grasses and forbs than in sagebrush," and that "prior to 7,000 years ago there appears to have been a greater abundance of grasses and forbs in the sagebrush-grass biotic community than at any time since . . . [pp. 52–53]."

Analysis of the mammalian fauna from Dirty Shame Rockshelter, a deep, stratified rockshelter in the Owyhee drainage in far southeastern Oregon, suggests a nearly identical sequence of paleoenvironmental change. The analyzed fauna covers much of the Holocene, with the exception of an interval of approximately 3000 years between about 5900 and 2700 B.P. Shifts in abundance of "mesic" and "xeric" rodents through time suggests that relatively cool and/or moist climates characterized the region surrounding the site between 9500 and 6800 B.P., and that warmer and/or drier climates have prevailed in that area since that time (Grayson 1977b).

Both of these studies are, of course, faunal, and must be treated with caution because of the difficulties inherent in using such data in making paleoenvironmental inferences, as discussed above. However, each is fully consistent with a wide range of data indicating that precipitation–evaporation ratios like those of the present did not become established in the Great Basin until about 7500 B.P. These similarities are not likely to be due to coincidence.

In short, changes similar to those I have hypothesized for the Fort Rock Basin on the basis of the Connley Caves archaeofauna seem to have occurred as well in the southern Plateau of southeastern Oregon and adjacent Idaho. It is reasonable to conclude that the shifts hypothesized for the Fort Rock Basin would have occurred even if Mount Mazama had not erupted some 7000 years ago.

## CONCLUSIONS

Analysis of the Connley Caves vertebrates suggests that two major paleoenvironmental changes occurred between 7000 and 5000 B.P. in the Fort Rock Basin. First, Paulina Marsh, today the largest natural body of standing water in the basin, was greatly reduced in size. Second, the nature of the vegetation changed from a flora characterized by a relatively high percentage of herbaceous vegetation to one characterized by a relatively high percentage of shrubby vegetation. A shift toward decreased effective precipitation between 7000 and 5000 B.P. may account for both of these changes. Similar shifts have been documented for the southern Plateau of southeastern Oregon and adjacent Idaho, areas that were not greatly affected by the Mazama eruption. In both of these nearby areas, a shift toward decreased effective precipitation seems to have occurred at about 7000 B.P. Since such a shift can account for the changes that occurred in the Connley Caves vertebrate faunas between

7000 and 5000 B.P., there is no need to call upon the eruption of Mount Mazama as a cause of these changes. Only the disappearance of pikas in the Fort Rock Basin can be linked to the eruption of Mount Mazama in any meaningful fashion, and even here it is possible that climatic change alone may account for the local extinction of this animal.

Unfortunately, my analysis of the Connley Caves archaeofauna answers relatively few questions concerning the impact of the Mazama eruption upon vertebrate faunas in the Fort Rock Basin. The nature of short-term impacts on these faunas cannot be addressed through the Connley Caves data since there are no detailed pre- and post-Mazama faunas available from these sites. Thus, though it would appear that significant long-term impact may be eliminated, significant short-term impact most certainly cannot. Indeed, studies of the effects of modern volcanism on terrestrial vertebrates suggests that the short-term impacts of the Mazama eruption upon local terrestrial vertebrate communities may have been considerable (e.g., Burt 1960; Thorarinsson, Chapter 5 of this volume). This issue cannot be addressed given currently available information. Excavation of mid-Holocene vertebrate faunas with painstaking attention paid to problems of chronology, stratigraphy, and the recovery of small vertebrates is required before these fascinating problems can be empirically approached.

It is also true that the results of the Connley Caves study allow little to be said about the nature of the relationship between the Mazama eruption and human settlement systems in the Fort Rock Basin. Though it appears that long-term effects of the Mazama eruption upon nonhuman vertebrates here could have been no greater than those which would have been induced by mid-Holocene climatic change, long-term impacts upon human settlements could have resulted either from transient impacts upon terrestrial vertebrates, or from transient or lasting impacts upon other segments of the environment. Detailed regional paleoenvironmental and archaeological studies are needed before this issue can be resolved.

Thus, I end with an unfulfilling conclusion. The data from the Connley Caves do not support the notion that the eruption of Mount Mazama had any great long-lasting effect upon the terrestrial vertebrate fauna of the Fort Rock Basin: The known faunal changes that occurred here between 7000 and 5000 B.P. would have occurred, it would seem, whether or not that eruption had occurred. However, the nature of short-term effects of this eruption on the vertebrates of the region remains totally unknown, as does the impact of the eruption on human settlement systems here.

## ACKNOWLEDGMENTS

I thank M. L. Johnson and E. Kritzman (Puget Sound Museum of Natural History, University of Puget Sound), M. Foster and J. L. Patton (Museum of Vertebrate Zoology, University of California), and S. A. Rohwer (Thomas Burke Memorial Washington State Museum, University of

Washington) for the use of skeletal collections under their care. I gratefully acknowledge the assistance of C. M. Aikens, D. E. Dumond, B. Grayson, R. M. Hansen, R. Hevly, D. Hill, L. Kittleman, C. Maser, and B. J. Verts during the preparation of this chapter. The analysis of the Connley Caves vertebrates was supported by the American Philosophical Society. This chapter was prepared for and presented at the Forty-second Annual Meeting of the Society for American Archaeology (1977).

## REFERENCES

Aikens, C. M.
   1978   The far west. In *The ancient Native Americans*, edited by J. D. Jennings. San Francisco: W. H. Freeman.
Anderson, E.
   1968   Fauna of the Little Box Elder Cave. Converse County, Wyoming. *University of Colorado Studies in Earth Sciences* 6:1–59.
Anthony, H. E.
   1913   Mammals of northern Malheur County, Oregon. *Bulletin of American Museum of Natural History* 32:1–27.
Bedwell, S. F.
   1969   Prehistory and environment of the pluvial Fort Rock area of south central Oregon. Doctoral dissertation. Department of Anthropology, University of Oregon.
   1971   New evidence for the presence of turkey in the early postglacial period of the Northern Great Basin. *Great Basin Naturalist* 31:48–49.
   1973   *Fort Rock Basin: Prehistory and environment*. Eugene: University of Oregon Books.
Bedwell, S. F., and L. S. Cressman
   1971   Fort Rock report: Prehistory and environment of the pluvial Fort Rock area of south-central Oregon. In *Great Basin anthropological conference 1970: Selected papers*, edited by C. M. Aikens. *University of Oregon Anthropological Papers* 1:1–26.
Brown, J. H.
   1971   Mammals on mountaintops: Nonequilibrium insular biogeography. *American Naturalist* 105:467–478.
Burt, W. H.
   1960   Some effects of Volcan Parícutin on vertebrates. *Occasional Papers of the Museum of Zoology, University of Michigan* 620:1–24.
Butler, B. R.
   1969   More information on the frozen ground features and further interpretations of the small mammal sequence at the Wasden Site (Owl Cave), Bonneville County, Idaho. *Tebiwa* 12(1):58–83.
   1972a  The Holocene in the Desert West and its cultural significance. In *Great Basin cultural ecology: A symposium*, edited by D. D. Fowler. *Desert Research Institute Publications in the Social Sciences* 8:5–12.
   1972b  The Holocene or post glacial ecological crisis on the Snake River Plain. *Tebiwa* 15(1):49–63.
Casteel, R. W.
   n.d.    A treatise on the minimum number of individuals index: An analysis of its behavior and a method for its prediction. Unpublished manuscript.
   1978   Faunal assemblages and the "weigemethode" or weight method. *Journal of Field Archaeology*, 5:71–77.
Casteel, R. W., D. P. Adam, and J. D. Sims
   1977   Late-Pleistocene and Holocene remains of *Hysterocarpus traski* (Tule Perch) from Clear Lake, California and inferred Holocene temperature fluctuations. *Quaternary Research* 7:133–143.
Chaplin, R. E.
   1971   *The study of animal bones from archaeological sites*. New York: Seminar Press.

Cope, E. D.
    1878  Pliocene man. *American Naturalist* 12:125–126.
    1889  The Silver Lake of Oregon and its region. *American Naturalist* 23:970–982.
Cressman, L. S.
    1942  Archaeological researches in the Northern Great Basin. *Carnegie Institute of Washington Publication* 538.
Cressman, L. S., H. Williams, and A. D. Krieger
    1940  Early man in Oregon. *University of Oregon Monographs, Studies in Anthropology* 3:1–78.
Dalke, P. D., D. B. Pyrah, D. C. Stanton, J. E. Crawford, and E. Schlatterer
    1960  Seasonal movements and breeding behavior of sage grouse in Idaho. *Transactions of the North American Wildlife Conference* 25:396–407.
    1963  Ecology, productivity, and management of sage grouse in Idaho. *Journal of Wildlife Management* 27:810–841.
Davis, S.
    1977  Size variation of the fox, *Vulpes vulpes* in the palaearctic region today, and in Israel during the late Quaternary *Journal of Zoology* 182:343–351.
Dort, W., Jr.
    1968  Paleoclimatic implications of soil structures at the Wasden Site (Owl Cave). *Tebiwa* 11(1):31–38.
Durrant, S. D.
    1952  Mammals of Utah, taxonomy and distribution. *University of Kansas Publications, Museum of Natural History* 6.
Durrant, S. D., M. R. Lee, and R. M. Hansen
    1955  Additional records and extensions of known ranges of mammals from Utah. *University of Kansas Publications, Museum of Natural History* 9:69–80.
Flinders, J. T., and R. M. Hansen
    1972  Diets and habitats of jackrabbits in northeastern Colorado. *Range Science Department Science Series, Colorado State University* 12.
    1973  Abundance and dispersion of leporids within a shortgrass ecosystem. *Journal of Mammalogy* 54:287–291.
    1975  Spring populations responses of cottontails and jackrabbits to cattle grazing shortgrass prairie. *Journal of Range Management* 28:290–293.
Franklin, J. F., and C. T. Dyrness
    1973  Natural vegetation of Oregon and Washington. *USDA Forest Service General Technical Report* PNW-8.
Graham, R. W.
    1976  Late Wisconsin mammalian faunas and environmental gradients of the eastern United States. *Paleobiology* 2:343–350.
Grayson, D. K.
    1973  On the methodology of faunal analysis. *American Antiquity* 39:432–439.
    1974  The Riverhaven No. 2 vertebrate fauna: Comments on methods in faunal analysis and on aspects of the subsistence potential of prehistoric New York. *Man in the Northeast* 8:23–29.
    1976  The Nightfire Island avifauna and the Altithermal. In *Holocene environmental change in the Great Basin*, edited by R. Elston. *Nevada Archaeological Survey Research Paper* 6:74–102.
    1977a  On the Holocene history of some Northern Great Basin lagomorphs. *Journal of Mammalogy* 58:507–513.
    1977b  Paleoclimatic implications of the Dirty Shame Rockshelter mammalian fauna. *Tebiwa: Miscellaneous Papers of the Idaho State University Museum* 9.
    1977c  A review of the evidence for early Holocene turkeys in the Northern Great Basin. *American Antiquity* 42:110–114.
    1978  Minimum numbers and sample size in vertebrate faunal analysis. *American Antiquity* 43:53–65.

Hall, E. R.
   1946   *Mammals of Nevada*. Berkeley: University of California Press.
Hampton, E. R.
   1964   Geological factors that control the occurrence of ground water in the Fort Rock Basin,
           Lake County, Oregon. *U.S. Geological Survey Professional Papers* 383B:1–29.
Hansen, H. P.
   1942a  The influence of volcano eruptions upon post-Pleistocene forest succession in central
           Oregon. *American Journal of Botany* 29:214–219.
   1942b  Post-Mazama forest succession on the east slope of the central Cascades of Oregon.
           *American Midland Naturalist* 27:523–534.
   1947a  Postglacial forest succession, climate, and chronology in the Pacific Northwest. *American
           Philosophical Society Transactions* 37.
   1947b  Postglacial vegetation of the Northern Great Basin. *American Journal of Botany* 34:164–
           171.
Hansen, R. M., and J. T. Flinders
   1969   Food habits of North American hares. *Range Science Department Science Series, Col-
           orado State University* 1.
Hester, T. R.
   1973   Chronological ordering of Great Basin prehistory. *Contributions of the University of
           California Archaeological Research Facility* 17.
Hoffman, R. S., and D. L. Pattie
   1968   *A guide to Montana mammals: Identification, habitat, distribution, and abundance*.
           Missoula: University of Montana Printing Services.
MacArthur, R. W., and E. O. Wilson
   1967   *The theory of island biogeography*. Princeton: Princeton University Press.
Malde, H. E.
   1964   Ecological significance of some unfamiliar geological processes. In *The reconstruction of
           past environments*, edited by J. J. Hester and J. Schoenwetter. *Fort Burgwin Research
           Paper* 3:7–15.
Mayr, E.
   1939   The sex ratio in wild birds. *American Naturalist* 73:156–179.
Mehringer, P. J., Jr.
   1977   Great Basin late Quaternary environments and chronology. In *Models in Great Basin
           prehistory*, edited by D. Fowler. *Desert Research Institute Publications in the Social
           Sciences* 12:113–167.
Moir, W. H., J. F. Franklin, and C. Maser
   1973   Lost Forest Natural Area. In *Federal research natural areas in Oregon and Washington: A
           guidebook for scientists and educators*, by J. F. Franklin, F. C. Hall, C. T. Dyrness, and
           C. Maser. Portland: USDA Forest Service, Pacific Northwest Forest and Range Experi-
           ment Station.
Orr, R. T.
   1940   The rabbits of California. *Occasional Papers of the California Academy of Sciences* 19.
Patterson, R. L.
   1952   *The sage grouse in Wyoming*. Denver: Sage Books.
Phillips, K. N., and A. S. Van Denburgh
   1971   Hydrology and geochemistry of Abert, Summer, and Goose lakes and other closed-basin
           lakes in south-central Oregon. *U.S. Geological Survey Professional Papers* 502-B.
Rice, D. G.
   1972   The Windust Phase in Lower Snake River region prehistory. *Washington State University
           Laboratory of Anthropology Reports of Investigations* 50.
Sevareid, J. H.
   1950   The pigmy rabbit (*Sylvilagus idahoensis*) in Mono County, California. *Journal of Mam-
           malogy* 31:1–4.

Smith A. T.
  1974a The distribution and dispersal of pikas: Consequences of insular population structure. *Ecology* 55:1112–1119.
  1974b The distribution and dispersal of pikas: Influences of behavior and climate. *Ecology* 55:1368–1376.
Trivers, R. L.
  1972 Parental investment and sexual selection. In *Sexual selection and the descent of man,* edited by B. Campbell. Chicago: Aldine.
Warren, C. N.
  1967 The San Dieguito complex: A review and hypothesis. *American Antiquity* 32:168–185.
Wiley, R. H.
  1973 Territoriality and non-random mating in the sage grouse, *Centrocercus urophasianus.* *Animal Behavior Monographs* 6:68–169.
Williams, H.
  1942 The geology of Crater Lake National Park. *Carnegie Institute of Washington Publication* 540.
  1969 *The ancient volcanoes of Oregon.* Eugene: Oregon State System of Higher Education.
Williams, H., and G. Goles
  1968 Volume of the Mazama ash-fall and the origin of Crater Lake caldera. In Andesite Conference guidebook, edited by H. M. Dole. *Oregon Department of Geology and Mineral Industries Bulletin* 62:37–41.

# 15

# Sunset Crater and the Sinagua: A New Interpretation

PETER J. PILLES, JR.

## INTRODUCTION

One of the major volcanic areas in the United States is the San Francisco Mountain volcanic field, named after one of northern Arizona's most prominent landmarks, the San Francisco Peaks. This field covers over 5200 km² (2000 mi²) and, although most of its 200 cinder cones were formed during the Pleistocene (Colton 1967:4), its most recent period of activity began 900 years ago, when the initial eruption of Sunset Crater brought the dormant field back to life.

During the period of volcanic activity, this area was inhabited by a prehistoric group called the Sinagua, a cultural tradition characterized by paddle-and-anvil-produced brownware ceramics, lack of indigenous decorated ceramics, and extended burials. Most of the knowledge and theories about this group are due to work by Harold S. Colton and his co-workers, John C. McGregor, Lyndon L. Hargrave, and Katharine Bartlett. Colton began the first systematic survey of the Flagstaff area in 1916 and, when pit houses were found buried beneath the Sunset Crater ash in 1930 (Colton 1932:582; McGregor 1936b), investigating the effects of the eruption on local cultural developments became one of his major research pursuits.

### The Traditional Model

In 1932, Colton first postulated his hypothesis that cinder fall from the eruption acted as a moisture-retaining mulch, creating thousands of hectares

459

of new farmlands in areas where it had not been possible to farm before because of a lack of moisture (Colton 1932:589). He believed news of these farmlands spread far and wide, precipitating a prehistoric land rush that caused a tremendous population increase over the next century (Colton 1932:589; 1937:11; 1945b:12). At first Colton did not specify a source for these new people, but after McGregor's work at Winona Village (McGregor 1937, 1941), it was believed that population increase was caused by immigration into the area by people of diverse cultural backgrounds. This flood of immigrants largely submerged the pre-eruptive Sinagua characteristics and, because of cultural mixing, caused a post-eruptive pattern dramatically different from earlier times. New artifact forms, burial patterns, ceramic types, ceramic manufacturing techniques, and architectural forms appear that are traditionally believed to have been the result of these immigrant groups (Colton 1932:589; 1936:4; 1937:11; 1938:65; 1945a:345; 1945b:12; 1946:259; 1949:24; 1962:171).

### The Alternative Interpretation

Colton's interpretation has been accepted by most Southwestern archaeologists over the years and, although several have added to it, none have seriously questioned the validity of the interpretation. In the 40 or so years that have passed since the Colton model was formulated, much new data has been accumulated and methods of interpreting archaeological data have changed considerably. Because of this, it seems appropriate to reexamine the traditional model in the light of new developments.

This chapter will present a new interpretation by demonstrating that (a) much of the evidence for these migrations is based on outdated theoretical concepts and inadequate data; (b) many of the supposed migrant-derived traits actually have pre-eruptive antecedents; (c) the supposed population increase is, for the most part, a change in the proportions of various site types and reflects changes in settlement and subsistence practices through time; and (d) the magnitude of post-eruptive Sinagua change is no greater than that observed elsewhere in the prehistoric Southwest during this time.

## THE PRE-ERUPTIVE CULTURAL PATTERN

(Note: Because of recent developments in the dating of Sunset Crater, which will be discussed presently, it should be noted that the term *pre-eruptive* refers to events before A.D. 1064 and *post-eruptive* to events after this date.)

The Sinagua tradion is first recognizable between A.D. 500 and 700 in the Flagstaff area. Three phases have been proposed to cover the pre-eruptive period—the Cinder Park, Sunset, and Rio de Flag phases, correlating with the Basketmaker III, Pueblo I, and early Pueblo II periods of the Kayenta sequence, respectively. On the whole, these three phases appear to have been fairly uniform relative to ceramics, architectural styles, settlement plans, and subsistence strategies. In the early part of the period, trade was primarily with the Basketmaker III Kayenta, probably those living along the Little Colorado River. Sparse amounts of Santa Cruz Red-on-buff and Gila Plain (Colton 1946:245) show contact with the Hohokam from the very start of the Sinagua sequence. Later, during the Rio de Flag phase, there was considerable trade with the Cohonina to the west and northwest, possibly reflecting their occupation of areas close to the Sinagua, such as Medicine Valley and Deadmans Wash (Figure 15.1).

Considerable variation in pit house forms and construction techniques is evident, especially during the century before the eruption. Circular or sub-square pit houses with a four- or six-post roof support and a lateral entry or antechamber were typical. A peripheral pattern suggesting a tipi-like superstructure is also known from the Cinder Park type site (Colton 1946:243–247, 268–269). Walls were usually of plastered earth, although others were lined with wooden planks or poles (Colton 1946:63–65, 268–269; Breternitz 1959:66). In boggy areas, earth platforms were constructed and above-ground timber structures built on the platforms (Colton 1946:45–46), although these "platform houses" are thought by some to be natural formations (Kelly 1969:114–115). Stone was used in the construction of some Rio de Flag phase houses, usually a row of rocks on the ground surface around the house, and at least two structures were completely lined with stone (Colton 1946:214–215). Some pre-eruptive pit houses in Medicine Valley, along the Sinagua–Cohonina frontier, also had masonry in their construction (Hargrave 1933:54–55; Colton 1946:119–121).

Although a lack of recognizable ceremonial architecture is considered characteristic of the Sinagua, large pre-eruptive structures apparently associated with ceremonial and intercommunity activities have been identified. Breternitz excavated a large circular structure, 8.5 m in diameter, that he believed to be a Cinder Park phase Great Kiva (Breternitz 1959:66). In the Rio de Flag phase, unusually large pit houses have been excavated that are also interpreted as ceremonial or community-related structures (Wilson 1969:133; DeBoer 1976:5–7), and others are inferred from surface indications at other sites. The so-called forts found on the periphery of the Sinagua area, such as Medicine Fort and Deadmans Fort (Hargrave 1933:49–54; Colton 1946:81–84, 134–135) (Figure 15.1), might also have been similar community structures and are additional examples of the pre-eruptive use of masonry in the Flagstaff area. Small masonry surface structures are also known for the Rio de Flag phase. They are found as storage rooms on pit house sites and at a distance from the pit houses, where they probably served as field houses (Colton and Colton 1918:124–125; Colton 1946:38–39; Pilles 1978).

The major centers of Sinagua population shifted through time, apparently in response to changing environmental conditions. During the Cinder Park phase of about A.D. 500–700, the population was mostly concentrated in the area around Turkey Tanks, Cinder Park, and the east base of Anderson Mesa (Colton 1946:247) (Figure 15.1). Lesser numbers of people may have been

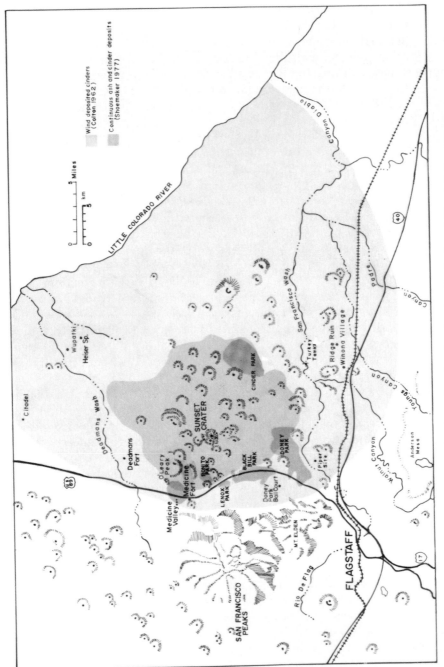

FIGURE 15.1 Map of the Flagstaff region showing the distribution of Sunset Crater ashfall.

scattered along the edges of other parks, since one Cinder Park phase site is known on the edge of Black Bill Park (Breternitz 1959). Today, the area in which most of these sites are found is transitional between ponderosa pine and piñon–juniper forests. Although other explanations are possible, this may indicate that the earliest inhabitants located in areas from which the resources of two environmental zones could be most easily exploited. The climate at that time is presumed to have been more moist than at present and was adequate to allow agriculture in the alluvial parks (Hevly, personal communication; Schoenwetter 1966:22–23).

Beginning around A.D. 900, with the Rio de Flag phase, the climate became warmer and drier (Hevly, personal communication) and the population apparently moved to the parks at higher elevations, along the flanks of Mount Elden and the San Francisco Peaks. Water was available in these areas, since more moisture falls around the Peaks than anywhere else in the region (Colton 1958b:12). In addition, most of the few springs in this region are found along these mountains, a further attraction for human settlement during drier times.

The pre-eruptive settlement pattern seems to have been one of pit house villages scattered along the margins of the large intercone basins, or parks. At least two farming strategies could have been employed within the parks: dry farming in the center, and floodwater farming along the edges where they are entered by numerous washes from the Peaks or surrounding lava flows. Farming away from the parks is indicated by masonry field houses, which are sometimes associated with check dams. In addition to check dams, irrigation ditches were also used in the Rio de Flag phase, possibly even earlier (Breternitz 1957a:30; 1957b:52).

It appears that the pre-eruptive Sinagua settlement pattern reflects utilization of optimal resource zones available to them through time. Although hunting and gathering were undoubtedly an important part of Sinagua subsistence, settlement around the parks might be expected when considering the agricultural aspect of their economy, since the parks contain some of the best arable soils in the region (Robertson and Schinzel, personal communications; Schinzel and Meurisse 1972; Pilles 1978:120) and retain more moisture than adjacent areas (Colton 1967:37).

The pre-eruptive Sinagua are generally thought to have had little social integration beyond the nuclear family (Reed 1956:16–17; Stanislawski 1963b:525). However, the distribution and nature of sites suggests there was some sort of community organization during pre-eruptive times. Many of the pit house villages appear to have been important centers because of their larger size and the presence of a community structure (Coconino National Forest Archeological Survey Files). Since they are located on the edge of the parks, they may have had some control over areas within the parks used as farmland. Pit house sites at a greater distance from the parks are smaller in size, have fewer structures, and lack community structures. Such sites are, however, more numerous than the presumed community centers. Although still unclear, there are indications that natural geographic features such as washes or ridge lines may form boundaries by which these communities of pit house sites may be recognized.

## THE ERUPTIONS OF SUNSET CRATER

In the middle part of the 1060s, the Rio de Flag phase and the pre-eruptive period ended with the initial eruption of Sunset Crater. Many different dates for the eruption have been given, ranging from A.D. 875 (McGregor 1936b:24) to about 1600 (Robinson 1913:90), with most discussion centering between 1064 and 1067 (Colton 1945a:351, 1947:44; Smiley 1958:186; Breternitz 1966:10). Archaeological and geological evidence indicate at least two periods of cineritic activity (McGregor 1936b:16; Smiley 1958:186; Breternitz 1967:73), and Breternitz finally suggested an initial eruption between the growing seasons of 1064 and 1065 and a second eruption between 1066 and 1067 (Breternitz 1967:73). Most recently, Eugene M. Shoemaker and Duane E. Champion, of the U.S. Geological Survey and California Institute of Technology, have been studying the eruption history of Sunset Crater. Although their work is still in progress, paleomagnetic dating of lava flows associated with Sunset Crater indicates that eruptions began between 1064 and 1065 and continued episodically for nearly 200 years (Shoemaker and Kieffer 1974:22; Shoemaker 1977). This new data has been incorporated with previous interpretations (Colton 1929, 1932, 1937, 1945a,b, 1947, 1960, 1967; Smiley 1958; Breed 1976) to present the following history of Sunset Crater. It is based not only on studies of the crater and its vicinity, but also on comparisons with Parícutin, which in some ways is considered a twin of Sunset Crater (Smiley 1958:186).

The eruption cycle of Sunset Crater began about September 1064 (Smiley 1958:190) when a 15-km-long (9.5 mi) fissure opened in a northwest direction from Vent 512 (Shoemaker 1977), accompanied by an 8-km-long (5 mi) lava flow east of the vent (Figure 15.2). Deep accumulations of black cinders were deposited over a $19 \times 24$ km ($12 \times 15$ mi) area around the crater (Colton 1932:586, 1945a:586; Shoemaker 1977) (Figure 15.1).

Within a few decades of activity, the first stage of the cone was formed by reddish black and black cinders (Shoemaker, personal communication). Another major eruption may have occurred between the growing seasons of 1066 and 1067 (Breternitz 1967:73), although this is questioned by some (Shoemaker, personal communication). Periodic cinder production over the next 30 or so years covered the surrounding area and contributed over a half billion tons of cinders to form the cone (Colton 1945b:8; Shoemaker 1977).

During most of this time, prevailing winds were apparently from the southwest, for the rim of the cone is higher on the northeast side, downwind from the eruption (Colton 1929:); 1945b:8), and most of the ash is found north and east of the crater (Figure 15.1). However, a substantial quantity of cinders were also blown to the southeast, and have been identified in Meteor Crater, 55

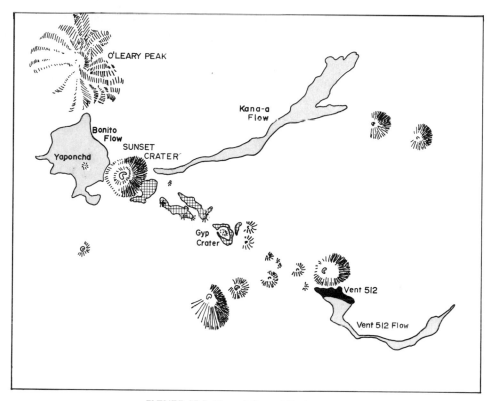

FIGURE 15.2 Map of Sunset Crater area.

km (34 mi) away (Shoemaker, personal communication). Eventually a half billion tons of cinders covered over 2000 km² (800 mi²), ranging from the San Francisco Peaks on the west and the Little Colorado River on the north and east to Interstate 40 on the south (Colton 1945b:8) (Figure 15.1). Windblown ash has been found in archaeological sites near Cameron, 56 km (35 mi) north of the cone (Breternitz 1962:64; Ward 1976:70–73), and may even have gone as far as Kansas (Colton 1945b:8).

In 1150, the Kana-a flow emerged from the base of the cone and moved down Kana-a Wash for 11 km (7 mi), filling it with molten lava (Colton 1945b:8; Shoemaker 1977) (Figure 15.2). Red and black cinders erupted at this time to cover a 4-km-wide (2.5 mi) area on the east side of the cone. Its extent in other directions has not yet been determined (Shoemaker, personal communication). Between 1180 and 1200, a series of fumaroles formed along the fissure between Vent 512 and Sunset Crater, depositing red cinders over the area. Gyp Crater was also formed at this time (Shoemaker 1977) (Figure 15.2).

In 1220, the Bonito Flow broke forth from a vent in the west side of Sunset Crater, filling a park on the south side of O'Leary Peak (Figure 15.2). Localized cineritic activity, covering not more than 1 km (0.5 mi) to the east, ensued to repair the breach formed by the Bonito Flow (Shoemaker, personal communication). An accumulation of cinders on top of the Bonito Flow has been named Yaponcha, after the Hopi wind god (Colton 1932:1). It may be a part of the wall of the breached cone (Shoemaker, personal communication) or a small satellite crater that issued from a fissure across the Bonito Flow (Colton 1945b:10, Smiley 1958:187). Yaponcha was disturbed by later lava movement, and squeeze-ups, or anosma, were produced on the edge of the Bonito Flow as new lava was extruded between the cracks in the older flow (Colton 1929:1, Smiley 1958:186–187). More lava appeared in the center of the Kana-a Flow as a lava tube drained onto its surface (Breed 1976:5).

Finally, in 1250, the volcano gave its last gasp and produced a deposit of red cinders along the rim of the cone (Shoemaker, personal communication). In 1885, this rose-hued cap prompted Major John Wesley Powell, then director of the U.S. Geological Survey, to name the crater "Sunset," for it appeared to be perpetually bathed in the rays of the setting sun (Colton 1945b:1).

### Effects of the Eruptions

It would seem likely that the eruptions of the crater had a profound psychological, emotional, and supernatural effect on the Sinagua people (Breternitz 1967:74), yet the Hopi, whom many researchers believe are descended in part from the Sinagua, have no legends directly attributable to the eruption. Only one is recorded that might refer to the eruption, and this simply refers to smoke and fire being noticed near the San Francisco Peaks (Voth 1905:241; Colton 1932:23). Sunset Crater does, however, figure in Hopi mythology as the home of the Kana-a kachinas. Yaponcha, the wind god, is believed to be imprisoned below the Bonito Flow, and offerings of pahos and pottery have been placed in an ice cave in the Bonito Flow as recently as 1963 (Colton 1945b:12; Schroeder 1977:33). In any case, it seems curious that an event of this supposed magnitude is not more strongly represented in Hopi mythology.

Although volcanic eruptions continued over a 200-year period, the initial activity between 1064 and 1067 can be considered to have had the greatest impact on the Sinagua. Subsequent activity was more localized, but by that time the area around the crater had been abandoned and was never again reoccupied. The effects of the eruptions were doubtless most strongly felt by the Sinagua living in the immediate vicinity of the volcano, in the area around Bonito Park (Figure 15.1).

These inhabitants may have had some warning of impending eruption, such as preliminary earthquakes and smoke issuing from the ground (Colton 1937:10). In studying wood from pit houses, McGregor suggested the people had sufficient time to dismantle their houses and take the main beams away (McGregor 1936a:14). Similarly, Colton (1960:28) observed that little wood was found in pit houses excavated in the area and the floor around postholes was damaged, further suggesting beams were pulled for reuse. He also interpreted a dearth of artifacts in the pit houses as further indication that people had sufficient notice of impending eruption to gather their belongings and move to new areas (Colton 1937:10).

Drainage patterns and associated field areas may have been changed by the damming of Kana-a Wash, and the dense ash deposits that blanketed the slopes could have altered erosion patterns (e.g., Segerstrom 1961:225) and clogged washes. One park on the south side of O'Leary Peak was completely filled by ashes and later by the Bonito Flow, destroying it for any future cultivation. Prior to the major eruption, Bonito Park may have contained an intermittent pond, fed by snowmelt and summer rains (Colton 1937:9). This water source would have been eradicated, although, as at Parícutin, lava flows could have served as an aquifer, catching and storing water, which would have led to the formation of new springs near the edge of the flow (Segerstrom 1960:15). Other springs in the vicinity may have become more productive because of the increased recharge afforded by the lava field (Segerstrom 1960:15). However, such water sources may not have been very important to the Sinagua, since they no longer lived in the area near the crater.

All vegetation within a 3-km (2-mi) radius of the cone was completely destroyed (Eggler 1966:82), largely by being buried with ash, but also by forest fires (McGregor 1936a:4), noxious gases, and acid rains (e.g., Wilcox 1959:452–453). The heavy ash fall likely stripped leaves and needles from plants and broke branches as they became weighed down by the ash. Pine trees, though less vulnerable to ash overloading, are quite susceptible to damage by volcanic gases (Wilcox 1959:451). Vegetation destruction over the rest of the cinder fall area was probably more selective, depending upon plant size, species, wind direction, and other factors. If Parícutin can be used as an example, vegetation was adversely effected to varying degrees for as far as 24 km (15 mi) from the volcano (Eggler 1959:270). The deleterious effects of the eruption on tree growth are evident in the tree-ring record, where extremely curtailed growth is apparent after 1064 and for many years thereafter (Smiley 1958:190; Dean, personal communication).

Most of these effects were of a short-term nature; however, the blanket of ash and cinders from 1064 to 1067 has long been recognized as the most important, long-lasting product of the eruptive cycle. The cinders, being loose and porous, act as a mulch that absorbs rainfall and snowmelt, thus retarding water runoff. Water infiltration is improved by a mulch since it regulates moisture passage into the soil at a steady rate (Stewart and Donnelly 1943:43), preventing puddling and sealing of fine surface soils such as those in the Flagstaff area.

## The Cinder Mulch

The cinders have been found to retain moisture for long periods of time. This promotes tree growth (McGregor 1933:314) and allows ponderosa and juniper to grow 300 m (1000 ft) lower in elevation than their normal range of occurrence (Colton 1932:589–590; 1945a:345; 1945b:13). The effectiveness of a light cover of ash to promote corn growth has also been demonstrated (Maule 1963:30–31; Colton 1965b:78). Experiments have shown that a light cover of cinders promotes germination and increases plant height. The heat-absorbing qualities of the black ash might also promote growth by artificially lengthening the growing season, an important factor in this region where the growing season is rather short, ranging from about 120 days at Flagstaff to 200 days along the Little Colorado River (H. Smith 1956:46; Green and Sellers 1964:177).

Although it was never suggested by Colton, many researchers have assumed the cinder fall also added mineral nutrients to the soil (e.g., Malde 1964:14; Eggler 1966:95; Breternitz 1967:74). However, volcanic ash by itself often lacks the nutrients necessary for plant growth and may be a poor medium for agriculture (Eggler 1959:267–268; Wilcox 1959:457–459). Although the weathering and soil-forming processes of cinders may ultimately lead to rich farming soils typical of many volcanic areas, this process sometimes takes a very long time (Wilcox 1959:461), often too long to be of any immediate benefit. Such decomposition does not appear to have been a significant factor in soil enrichment in the Flagstaff area, since the region was already covered by cinders from earlier periods of volcanic activity. These earlier cinders also retain moisture and, in certain areas, were used for farming by the pre-eruptive Sinagua (Olson 1964:130; Schroeder 1977:21, 33). However, Sunset cinders, being finer-sized and more angular than those exposed for millennia, may have been more effective in holding moisture and providing a looser growing medium.

In any event, it seems likely that the importance of Sunset cinders as the cause of improved growing conditions has been overestimated, because the cinder mulch was not the only factor influencing increased moisture availability. There is evidence for a period of above-average rainfall between 1050 and 1130, accompanied by a warming trend; in fact, this period has the highest moisture values of any time during the 1000–1200 period in the Flagstaff area (Robinson and Dean 1969; Fish 1974:21; Hevly, et al. Chapter 16 in this volume). Hevly et al. (Chapter 16) discusses this in greater detail and concludes that much of this moisture was in the form of summer rains that would have greatly benefited crop growth. Thus, although the cinder cover did act as a mulch, a climate change to very moist conditions was probably even more important for permitting agriculture at lower elevations that could not be farmed before the eruption.

In fact, increased moisture at this time may have forced the Sinagua to move into the ash fall area. Faunal evidence from the post-eruptive Piper Site and Ridge Ruin (Figure 15.1) includes bones of geese, ducks, and cranes, indicating the presence of ponds of standing water (McGregor 1941:258–259; Bliss and Ezell 1956:131). If the Ridge Ruin area 32 km (20 mi) east of the Peaks and 1600 m (1000 ft) lower in elevation was moist enough to support a population of water birds, the area along the flanks of the Peaks would have been even wetter, because of the greater precipitation that falls around the mountains.

Such increased moisture could have made the parks at the base of the mountains too wet for farming, forcing the Sinagua to abandon these pre-eruptive fields and move to areas of less precipitation at lower elevations, which only coincidentally conform in part to the Sunset cinder belt.

According to Colton's interpretation, people would have settled along the periphery of the ash fall, where cinders were shallow enough to permit agriculture (i.e., 30 cm) or less (Colton 1962:171; 1965b:78). Therefore, if his interpretation is correct, one would expect a greater site density on the edge of the ash fall than in areas outside it (Colton 1932:588; Schroeder 1977:34). However, equivalent densities are found outside the ash fall zone in the area south of Interstate 40, on top of Anderson Mesa above Walnut Canyon, and along the base of Anderson Mesa near upper Canyon Padre (Coconino National Forest Archeological Survey Files) (Figure 15.1), indicating that factors other than the cinder cover must be considered. Even though Colton (1932:589; 1962:171) suggested the cinder cover may have blown away in certain areas, the sites along Anderson Mesa are distant enough from known occurrences of cinders to support the observation that high-site densities do exist outside the ash fall area.

The Coconino National Forest has conducted numerous intensive and extensive surveys in the south half of the cinder fall area. These surveys have found that sites are most densely concentrated along washes, particularly those with large deposits of alluvium, and are sparse between drainages. This same correlation of sites with washes occurs in areas outside the cinder fall area, suggesting that washes and the availability of arable soils are more important determinants of site locations than deposits of Sunset cinders.

The cinder mulch could, however, have made a difference in the northern part of the region along the Little Colorado River Valley. This desert area has the lowest elevation, warmest temperatures, and lowest precipitation of any area occupied by the Northern Sinagua. The pollen and tree-ring records also suggest the moist period following the eruption was accompanied by warming temperatures. This would have made the Little Colorado Basin even warmer than before, and it is possible that the cinder mulch helped counteract this effect.

## INFLUX OF MIGRANT POPULATIONS

Migration has often been used to explain culture change and was a common explanatory concept in archaeology when major concepts of Sinagua prehistory were formulated. However, nowhere in the Southwest has this concept been so commonly invoked as in the Sinagua region. By choosing selected traits, various writers have postulated that six different cultural groups migrated to the region following the major eruption: the Hohokam, Kayenta, Cohonina, Prescott, Mogollon, and Chaco.

Colton proposed three contemporaneous focii to explain the 1066–1130

manifestations thought to have resulted from these various groups. He preferred the McKern (1934) classification system rather than Gladwin's (Gladwin
and Gladwin 1934) but considered McKern's "focus" and Gladwin's "phase" to
be more or less synonymous (Colton 1939:7; 1946:14). Over the years the term
*phase* has been generally accepted, but attempts to use Colton's tripartite
division have led to confusion because the distinction between *focus*—a site-
unit intrusion—and *phase*—a stage in the unilineal development of an area—
has been overlooked. The Winona focus represented the Hohokam colonists
(Colton 1939:50; 1946:270), the Angell focus the Mogollon immigrants
(McGregor 1941:281), and the Padre focus the merging of Hohokam, Mogollon, Anasazi, and Sinagua traits (Colton 1939:40). This amalgamation led to
the later Elden (1130–1200), Turkey Hill (1200–1300), and Clear Creek (1300–
1400) focii. Each of the supposed migrations and their contributions to the
post-eruptive Sinagua pattern will now be discussed.

## Hohokam

Of these groups, the strongest case for an actual movement of people can
be made for the Hohokam. Pit houses, ball courts, trash mounds, cremations,
clay human figurines, locally produced red-on-buff ceramics, shell ornaments,
projectile points, molded clay spindle whorls, and other diagnostic Hohokam
items appeared in the area about this time. Despite numerous mentions of
Hohokam occupation, the type site of Winona Village (McGregor 1937, 1941)
(Figure 15.1) is the only place yet found where an actual site-unit intrusion can
be postulated, for only here do all these traits occur together as an assemblage.
All other occurrences of Hohokam-like material are limited to one, or only a
few, items.

It is significant that many of the Hohokam-like objects found at Winona
Village originated in a single Hohokam-like pit house and its adjacent trash
mound (McGregor 1941:92). This localization implies individual, rather than
group, presence, in contrast to the usual interpretation that organized groups,
perhaps totaling as much as 1000 people (Schroeder 1975:35), migrated here
from the Verde Valley.

No sites have yet been excavated where more than a single Hohokam-like
house has been found, unlike colonial Hohokam sites in central Arizona (e.g.,
Breternitz 1960; Morris 1970; Weed and Ward 1970). Even at Winona Village,
there is only one Hohokam-style house, NA2133A, the type structure. Although nine pit houses have been attributed to the Hohokam-derived Winona
phase (McGregor 1941:101, 288; Colton 1946:149–150; W. Smith 1952:20–25),
only three (NA2133A, NA618X, NA1814C) actually resemble Hohokam structures, and two of these are dominated by Cohonina pottery. All occur as
isolated examples within a site and are not associated with any significant
quantity of Hohokam materials.

At least 13 ball courts are known in the area but need not necessarily have
been constructed by Hohokam people. An expanding northern acceptance of a

Mesoamerican ceremonial complex represented by ball courts, human figurines, incense burners, and cremations might also explain such items without suggesting a substantial colonial presence.

There is also evidence that Hohokam influence predates the main eruption. Hohokam ceramics are found in the earliest Sinagua sites (Hargrave 1932:27; Colton 1946:46, 245; Bliss and Ezell 1956:138; Breternitz 1957b:53; DeBoer 1976:24), and most of the human figurines found in the area have coffee-bean eyes typical of the Santa Cruz phase (Figure 15.3) rather than the subsequent Sacaton phase, to which most of the other Hohokam items can be assigned. A Hohokam origin for these figurines can also be questioned, since many are more similar to figurines found in the Prescott culture area (cf. McGregor 1941:77; Scott 1960:17; Stanislawski 1963b:238; Barnett 1974:104) (Figure 15.3). Trash mounds have also been found at pre-eruptive sites devoid of other Hohokam items (Colton 1946:117; Coconino National Forest Archeological Survey Files). Surface sherds from the Doney Park Ball Court Site (NA3205 and NA3206) and the Winona Ball Court (NA2132; McGregor 1941:88) suggest a pre-eruptive occupation, and spindle whorls, shell ornaments, three-quarter grooved axes, vesicular basalt cylinders, and serrated projectile points are all known from pre-eruptive sites (McGregor 1936a:32–35, 45; Breternitz 1957a:24; 1957b:50). Even a possible cremation has been found in a pre-eruptive pit house (Breternitz 1957b:48).

FIGURE 15.3 Human figurines from the Flagstaff area. Note the coffee-bean eyes in a and b, which are typical Santa Cruz phase Hohokam characteristics. C–e more closely resemble Prescott figurines. A and c are from Winona Village (NA3644 and NA2135), b is from the Kahorsho Site (NA10,937), d is from Cinder Hill Village (NA8721), and e is from Elden Pueblo (NA142).

At one time Colton considered Winona Brown to have been manufactured by the Hohokam (Colton 1946:25) and thought the Hohokam were
responsible for the development of pottery in the Flagstaff area (Colton and
Hargrave 1937:173). The latter idea was never developed any further, and
Winona Brown occurs in large numbers in numerous Sinagua sites that completely lack Hohokam items. Additionally, Winona Brown can be traced directly back to earlier Alameda Brown Ware types (Colton 1958a: Ware 14, Type
5).

Although the forms of local red-on-buff ceramics are certainly Hohokam-
inspired (Figure 15.4), they lack the full Sacaton Red-on-buff assemblage. The
repeated small elements, life forms, massive designs, curvilinear scrolls, and
solid red areas so common in Sacaton Red-on-buff are not present in local
types. With few exceptions, the usual design is identical to Flagstaff or Walnut
Black-on-whites. In addition, the general execution of local varieties is unlike
true Hohokam pottery and is more like that found on occasional painted
Sinagua types such as Sunset and Turkey Hill White-on-reds. It should also be
noted that locally made red-on-buff sherds are extremely rare in occurrence
(McGregor 1941:44). Less than 200 sherds were found in the Winona and Ridge
Ruin excavations and less than 12 complete or fragmentary vessels of these
types are known (Museum of Northern Arizona Catalogue Files). It thus seems
likely that these rare red-on-buff vessels are local Sinagua copies, rather than
the products of immigrant Hohokam potters.

Perhaps the Hohokam presence in the Flagstaff area may be better explained as part of the general expansion of the Hohokam out of the Gila–Salt
River Valleys, possibly to facilitate the acquisition and distribution of raw
materials and other trade goods. With this interpretation, the single pit house
at Winona Village might represent a Hohokam trader engaged in a trading
post-like situation, rather than part of a transplanted Hohokam village (e.g.,
Fish 1974:17). Other explanations related to internal Sinagua social growth and
development, rather than Hohokam migration, have also been suggested (Fish
and Pilles in press).

Lastly, it is difficult to picture groups of riverine-oriented Hohokam farmers leaving the fertile bottomlands, permanent water, and mild climate of the
Verde Valley for the poorer soils, marginal farming conditions, and colder
temperatures of the Flagstaff area, particularly when areas in the Verde were
still available for exploitation.

## Kayenta

Evidence for Kayenta "migration" is questionable, since they had been
living in the northern edge of the area along the Little Colorado River and
Deadmans Wash since Basketmaker III times (Colton 1946:101–102, 261–263).
In addition, there is no evidence for Kayenta colonies farther south in the
Sinagua nuclear area.

The Kayenta are thought to have introduced the use of masonry and

**FIGURE 15.4** Locally manufactured red-on-buff jar from Winona Village (NA2133). Photograph retouched to illustrate design.

construction of above-ground pueblos to the Sinagua, although Hargrave (1933:47–49) once suggested this was due to Cohonina rather than Kayenta influence, since the Cohonina also had full-height masonry architecture before the initial eruption (Hargrave 1933:49–52; Colton 1946:84; Cartledge in press). The fact is, however, that the Sinagua had masonry architecture long before the eruptions. Above-ground, masonry field houses are numerous in the Rio de Flag phase (Pilles 1978:129) and masonry was used in pre-eruptive pit houses even as early as the Cinder Park phase (Colton 1946:214–215; Breternitz 1957a:26; Olson 1964:114–116; Wilson 1969:116).

### Cohonina

The Cohonina coexisted with the Sinagua in the Medicine Valley and Bonito Park localities just prior to the eruption and so cannot be said to have migrated into the area. Other than the introduction of masonry architecture (as suggested by Hargrave 1933), their only other contribution is a single projectile point form (McGregor 1941:279). Consequently, a strong case for Cohonina influence on post-eruptive Sinagua cannot be made, particularly when they are no longer recognizable after about 1100 (Colton 1939:25). Their influence appears to be limited to a high frequency of San Francisco Mountain Gray Ware in many sites along Deadmans Wash; however, what this means has yet to be demonstrated.

### Prescott

The sole evidence for a Prescott migration is found at Nalakihu, where Prescott Gray Ware constitutes 35% (925 sherds) of the ceramic count (King 1949:114–115). Very large jars are typical of this pottery and it would not take many to account for this number of sherds. Colton originally made this observation and suggested the sherds represented vessels made in the Prescott area (Colton 1946:54). Nonetheless, he still interpreted them as evidence for a Prescott colony (Colton 1958a: Ware 17, Type 1; 1968:10).

### Mogollon

McGregor believed the architectural and ceramic traits of the Angell phase, such as posts set into house walls, rectangular vestibule entry, and the coil-and-scrape ceramic types of Winona Smudged, Winona Corrugated, and Winona Red, resulted from a migration of Mogollon or Mogollon-derived people from the east (McGregor 1941:121, 281). Colton included Flagstaff Red, an early term for Sunset Red, in Mogollon Brown Ware (Colton 1965a:7), and McGregor also thought that Sunset Red was related to the Mogollon (McGregor 1937:49). In a similar vein, Schroeder views the Padre phase as being introduced by the Alpine Branch of the Mogollon. Diagnostic traits, he believes, consist of deep, rectangular, masonry pit houses with ridge-pole roof and ventilator; a lack of kivas; the coil-and-scrape Winona Series of pottery; and flexed burials (Schroeder 1961:63–65).

Once again, however, most of these can be explained without suggesting a migration of Mogollon people. As mentioned previously, masonry-lined pit houses occur in pre-eruptive sites and are most easily explained as a local development (Colton 1946:270, 311; Wilson 1969:409–413), whereas the vestibule entry can easily be seen as an outgrowth from earlier ramp entries and alcoves. A lack of kivas was often cited as a Sinagua characteristic prior to Schroeder's 1961 article; however, since that time a number of Sinagua kivas have been identified (Lee 1962; Stanislawski 1963b:516–517; Kelly and Skinner 1966), and it is now generally accepted that the Sinagua did use kivas (McGregor 1965:302; Martin and Plog 1973:143). The coil-and-scrape pottery types are trade items. They did not occur in any great abundance at Winona Village or Ridge Ruin (McGregor 1941:49–50), and even Colton considered them intrusive (Colton 1941:28–31). Elden Corrugated is a locally made Mogollon style of pottery; however, it post-dates the Padre phase and is not commonly found until the latter part of the Elden phase of 1130–1200. Even then, it is found only in small numbers.

Flexed burials can more likely be attributed to Anasazi influence, considering their closer proximity and long history of contact with the Sinagua. As with other traits under discussion, flexed burials seem to be found in the pre-eruptive Rio de Flag phase (Colton 1946:166; Breternitz 1957b:45). In post-eruptive contexts, those referred to by Schroeder are mostly from a single trash mound near Ridge Ruin (McGregor 1941:272), and their occurrence does seem unusual. Nonetheless, extended burials are more typical of this time period (Colton 1946:295; Stanislawski 1963a:313).

Although there are definite Mogollon similarities in the Sinagua culture, they are not due to a post-eruptive migration of people from eastern Arizona. Many of the pre- and post-eruptive similarities between the two can be most easily understood by viewing the Sinagua as the western arm of the Mogollon (Colton 1939:33). Pre-eruptive Sinagua pit houses are identical to those found in the nuclear Mogollon area. Although the Sinagua lacked early painted ceramics and did not use the coil-and-scrape manufacturing technique of the Mogollon, there are similarities in the ceramic technology of the two groups, such as plainwares, smudging, polishing, red slip, some vessel forms, and vertical obliterated corrugation. Ceremonial architecture is also similar in the use of community rooms and rectangular kivas with benches. Both areas had flexed burials early in their history, changing to extended inhumations with vertical occipital deformation and occasionally cremations later. Adobe-encased burials or coverings of poles or slabs are also common to both (Fewkes 1927:215; Stanislawski 1963a:310).

Many of these traits characterize the Western Pueblo tradition; Reed (1950:128) and Johnson (1965:57) considered the Sinagua to be part of this development. To be sure, there are many differences between the Flagstaff-area Sinagua and eastern Arizona Mogollon, just as there are between other Mogollon divisions, but nonetheless there are sufficient similarities to consider the Sinagua as part of the Mogollon cultural tradition.

## Chaco

The last group to be considered is the Chaco Anasazi. McGregor initially suggested a Chacoan presence was indicated by the masonry style and wall thickness at Ridge Ruin (McGregor 1941:153, 156). Similarly, Stanislawski viewed the banded masonry at Wupatki and the Citadel (Nalakihu should also be mentioned in this regard) as due to settlement by Chaco people (Stanislawski 1963b:528) (Figure 15.5). Chaco migration was further indicated to Stanislawski by the amphitheater at Wupatki, the local equivalent of a Great Kiva (Stanislawski 1963b:536), and the presence of extended burials (Stanislawski 1963a:313).

However, masonry in these Sinagua sites cannot seriously be considered

FIGURE 15.5 "Chaco"-style masonry at Wupatki (top) and Nalakihu (below).

the handiwork of Chaco masons since it lacks the shaping and precision of Chaco masonry. The "Chaco" masonry at Ridge Ruin is simply chinking and is more similar to Kayenta masonry styles. Wall thickness is largely dependent upon the nature of the building stones available. Chaco burial practices are not well understood and it seems premature to credit them with introducing extended burials to the Sinagua. In any event, it would seem simpler to explain extended burials in the Sinagua as another example of their Mogollon relationship, since extended burials became common in Mogollon sites in eastern Arizona by 1100 (Stanislawski 1963a:310).

## Summary of Cultural Migration Evidence

In sum, it can be seen that the evidence for migrant groups in the area is limited to a few isolated traits, rather than trait assemblages, which would more likely reflect an intrusive cultural pattern. In the 1930s and 1940s, when archaeological cultures were defined primarily by the presence or absence of various traits, such items were adequate to infer migration; however, the evidence from the Flagstaff area does not meet the more recent criteria for migration as formulated by Rouse (1958:64–66) and Haury (1958:1). Only the Hohokam presence at Winona Village may meet these criteria, yet alternate interpretations to migration can also explain these items (Fish and Pilles in press). Most Sinagua researchers have failed to consider the possibility of trait-unit intrusion, rather than site-unit intrusion (Willey et al. 1956:7–8) or the influence of trade and development of social hierarchies, for example, to explain the occurrence of exotic items and changing cultural patterns.

## POPULATION INCREASE

As evidence of the prehistoric land rush to the Flagstaff area, Colton published several articles showing a dramatic population increase (Colton 1936; 1946:261–266; 1949, 1962) as well as an expansion of the area inhabited. Although the magnitude of migrations was a debated point even in the 1940s (McGregor, personal communication), Colton believed about 3000 people migrated into the area (Colton 1949:23). However, his estimates can be questioned, as Colton himself was aware (Colton 1936:1,3; 1949:21–23). The most serious problem with his studies is that the data they were based upon were neither systematic nor complete, and it is questionable whether they are representative of the region. Because Colton was keenly interested in the effects of the Sunset Crater eruption, the work of the Museum of Northern Arizona concentrated on the ash fall zone and neglected the peripheries. Even within the ash fall zone, Colton's coverage varies in completeness, resulting in an erroneous impression of site densities. For example, his data for the west and south sides of Doney Park suggest a density of 5 to 10 sites per square kilometer (Museum of Northern Arizona Archaeological Survey Files; Colton 1962); however, recent intensive surveys of these areas have found 52 to 78 sites per square kilometer (Museum of Northern Arizona Archaeological Survey Files; Coconino National Forest Archeological Survey Files).

At the time Colton's various site-density maps were prepared, too few sites had been recorded to perform population studies with any accuracy. Although Colton intended his figures in "The Sinagua" (Colton 1946:261–266) to show the distribution of cultural groups through time, they imply population growth through territorial expansion and have been interpreted in this manner by some researchers. His original notes used to compile the 1946 figures indicate only about 600 sites were included in the study (Museum of Northern Arizona, Colton Archives).

By the time they are divided into temporal groups and townships, there are too few sites to make meaningful statements. This situation is improved in his 1962 study, where 1032 sites were used. But, once again, the figures shown are based upon unsystematic surveys and do not accurately reflect the actual proportions of pre-eruptive and post-eruptive sites. One example of this inaccuracy can be seen in T22N, R8E, where Colton shows 4 pre-eruptive sites and 25 post-eruptive sites (Colton 1962:172), which, at face value, supports his hypothesis of post-eruptive population increases. In this township, 32.6 km$^2$ (12.6 mi$^2$) have been completely surveyed by the Coconino National Forest, locating 108 pre-eruptive and 54 post-eruptive sites, indicating exactly the reverse of Colton's data.

Besides the issue of representativeness, there are other problems with Colton's estimates. Over the years, concepts of what constitutes a single site have changed. Sometimes groups of structures were assigned a site number, but at other times individual structures were given separate numbers (McGregor 1941:Figure 3; Colton 1946:45, 47, 63–65, 97, 109, 132). Obvious problems result from this inconsistency when changes in the numbers of "sites" are being investigated.

The most serious problem with this approach, however, is that it fails to consider functional differences between various kinds of sites. Many sites in the region can be identified as field houses, functionally specific structures utilized only for short periods of time (Pilles 1978), and a serious bias results if such sites are given the same weight as multiroom pueblos and pit house villages. Colton attempted to compensate for this when preparing his estimates of population change, but in so doing biased his results in favor of post-eruptive increases. In his original study, he recognized such structures as storage units and did not include them in his pre-eruptive site estimates (Colton 1936:3). However, when they occurred in post-eruptive contexts, he considered them residential units and included them in his figures (Colton 1936:3; 1937:12; 1938:65; 1960:12–13).

Recent surveys suggest field houses constitute about 25% of pre-eruptive sites but 60–80% of post-eruptive sites. Thus, it can be argued that post-eruptive site increases are primarily an increase in field houses rather than permanently occupied habitation sites. This change can be attributed to the distribution of arable soils in the region occupied by the Sinagua after the eruption. Soils here are thin and occur only along major washes or in solution-weathered pockets in bedrock. In order to exploit these scattered soil pockets, the population had to disperse itself over the landscape, and a single family would probably use several field houses during a single season. Additionally, the shallow depth of the pockets probably resulted in soil depletion, causing a rapid turnover in the use of the pockets and their associated field houses (Pilles 1978:130). Since most of the region occupied by the Sinagua before the eruption was abandoned, it seems evident that the post-eruptive occupation was not a population expansion but a shift of the indigenous population into a different resource zone.

It is clear that more intensive, systematic surveys are needed before pre-eruptive and post-eruptive population differences can be adequately determined. Present data suggest there is a 19% increase in post-eruptive habitation sites (Pilles 1978). This is in rather sharp contrast to Colton's estimates of an 1100% increase (Colton 1962:171).

One final factor should be considered in assessing Sinagua population growth. As Gumerman has noted (1970:31), a great increase in the number of sites is common over the entire upland area of the Southwest after 1050. Such increases have been noted for Black Mesa (Gumerman 1970:31), the Kaiparowits Plateau (Fowler and Aikens 1963:4), Walhalla Glades (Hall 1942:16), Hopi Buttes (Gumerman and Skinner 1968:190), the Red Rock Plateau (Lipe 1970:112), Vosberg Valley (Cartledge 1977:98), the Hay Hollow Valley (Longacre 1970:12), and many other parts of the Southwest. Thus, it can be seen that the Sinagua region simply reflected a demographic trend experienced by most of the Southwest at this time.

The purpose of this section has been not to deny that there was a population increase after the eruption but to indicate that the magnitude of this increase has been greatly overrated. The population probably did increase after the eruption, but it seems more likely that this increase was due to indigenous population growth, possibly fostered by favorable climatic conditions, rather than to a large influx of migrant groups.

## SUMMARY

For 40 years, the supposed effects of the major eruption of Sunset Crater on the course of cultural developments in the Flagstaff area have stood as a classic study of human responses to volcanism, migration, and population growth. These traditional interpretations have been questioned in this chapter and alternative explanations offered.

Recent geologic work indicates that, rather than being an isolated event, the main eruption in 1064–1065 was one episode of volcanic activity that lasted for nearly 200 years. Although people living in the vicinity of the crater were displaced, the cinder fall has usually been considered to have been the only long-term effect. However, the importance of this ash for improved growing conditions has been overestimated, and may have been of only localized importance. A climatic change to more moist conditions is thought to have been of far greater consequence.

The theory of migration as an explanation of cultural change in the centuries after the main eruption has been countered by demonstrating that there is little evidence in the archaeological record for multicultural migrations. Traits presumably indicating such migrations have been shown to be spurious, local developments or isolated occurrences, rather than trait assemblages that would be expected in instances of true migration. Trade and stimulus diffusion more likely explain these exotic materials in the Flagstaff area.

Evidence for dramatic population increase has been found to be based on inadequate data and can more logically be interpreted as redistributions of the existing population. It has been suggested that changing land-use patterns, rather than an influx of migrant populations, better explains Sinagua responses to new environmental factors.

To be sure, the 150 years following the main eruption of Sunset Crater were marked by more rapid growth and development than the previous 450 years; however, this change was probably more influenced by new environmental, sociological, and climatic conditions than by the eruption itself. Influences from the north, south, and east were accepted into the existing cultural tradition and incorporated into the rapidly changing sociocultural fabric characterizing the post-eruptive Sinagua.

Traditional explanations of post-eruptive change have tended to view the Flagstaff area with blinders. While accepting local changes as due to outside causes, they have ignored the fact that similar changes and probably population growth occurred at this time over the entire Southwest, and not only in the Flagstaff region. Most importantly, this is the time when the Western Pueblo tradition (Reed 1948; Johnson 1965) was developing, and many of the post-eruptive changes in the Sinagua would seem to indicate they were a part of this important cultural event.

The fact that such changes occur over the entire Southwest indicates even more clearly that traditional interpretations of Sinagua prehistory have overemphasized the influence of Sunset Crater. It is apparent that future efforts to understand Sinagua cultural dynamics must view them in a broader geographic and cultural perspective and not as an isolated island of culture in the shadow of a volcano.

## ACKNOWLEDGMENTS

In challenging the time-honored traditions discussed in this paper, the author has benefited from the comments of a number of people, and their assistance is gratefully acknowledged. E.

Charles Adams, Donald K. Grayson, John C. McGregor, Albert H. Schroeder, and others reviewed an earlier draft of the manuscript and provided useful suggestions for its improvement. Grammatical revisions were suggested by William J. Beeson and Thomas R. Cartledge, and illustrations were prepared by Marvin Marcroft and Mark Middleton. Particular thanks are extended to Paul and Suzanne Fish for their help in evaluating the evidence for migration and for their many conversations with the author on Sinagua prehistory. Richard Hevly supplied paleoclimatic information and Eugene M. Shoemaker kindly permitted use of his and Duane E. Champion's unpublished data on their investigations of Sunset Crater. Insights into the development of archaeological concepts for the Flagstaff area were provided by Katharine Bartlett and John C. McGregor. Finally, editors Payson D. Sheets and Donald K. Grayson are thanked for their infinite patience in dealing with a tardy author.

The full story of Sunset Crater and the Sinagua is far from being understood, and the revisionist effort of this chapter would not have been possible without the pioneer work and dedication of Dr. Harold S. Colton. It is hoped that Dr. Colton would approve of the attempt of the present chapter to contribute to this story.

# REFERENCES

Barnett, F.
   1974  Excavation of main pueblo at Fitzmaurice Ruin: Prescott culture in Yavapai County, Arizona. *Museum of Northern Arizona Special Publication*.
Bliss, W. L., and P. H. Ezell
   1956  The Arizona section of the San Juan pipeline. In *Pipeline archaeology*, edited by Fred Wendorf, Nancy Fox, and Orian L. Lewis. Santa Fe and Flagstaff: Laboratory of Anthropology and Museum of Northern Arizona.
Breed, W. J.
   1976  Molten rock and trembling earth: The story of a landscape evolving. *Plateau* 49 (2):2–13.
Breternitz, D. A.
   1957a  1956 Excavations near Flagstaff: Part I. *Plateau* 30(1):22–30.
   1957b  1956 Excavations near Flagstaff: Part II. *Plateau* 30(2):43–54.
   1959  Excavations at two Cinder Park phase sites. *Plateau* 31(3):66–72.
   1960  Excavations at three sites in the Verde Valley, Arizona. *Museum of Northern Arizona Bulletin* 34.
   1962  Excavations at the New Leba 17 site near Cameron, Arizona. *Plateau* 35(2):60–68.
   1966  An appraisal of tree-ring dated pottery in the Southwest. *Anthropological Papers of the University of Arizona* No. 10.
   1967  The eruption(s) of Sunset Crater: Dating and effects. *Plateau* 40(2):72–76.
Cartledge, T. R.
   1977  Human ecology and changing patterns of co-residence in the Vosberg locality, Tonto National Forest, central Arizona. *USDA Forest Service Southwestern Region, Cultural Resources Report* No. 17.
   in press  Cohonina adaptation to the Coconino Plateau: A re-evaluation. *The Kiva* 44(4).
Colton, H. S.
   1929  Sunset Crater and the lava beds. *Museum Notes* 2(4):1–3.
   1932  Sunset Crater: The effect of a volcanic eruption on an ancient Pueblo people. *The Geographical Review* 22(4):582–590.
   1936  The rise and fall of the prehistoric population of northern Arizona. *Science* 84(2181):337–343.
   1937  The eruption of Sunset Crater as an eye witness might have observed it. *Museum Notes* 10(4):9–12.
   1938  The economic geography of the Winona phase. *Southwestern Lore* 3(4):64–66.

1939    Prehistoric culture units and their relationships in northern Arizona. *Museum of Northern Arizona Bulletin* 17.

1941    Winona and Ridge Ruin Part II: Notes on the technology and taxonomy of the pottery. *Museum of Northern Arizona Bulletin* 19.

1945a   A revision of the date of the eruption of Sunset Crater. *Southwestern Journal of Anthropology* 1(3):345–355.

1945b   Sunset Crater. *Plateau* 18(1):7–14.

1946    The Sinagua: A summary of the archaeology of the region of Flagstaff, Arizona. *Museum of Northern Arizona Bulletin* 22.

1947    A revised date for Sunset Crater. *Geographical Review* 37(1):144.

1949    The prehistoric population of the Flagstaff area. *Plateau* 22(2):21–25.

1958a   Pottery types of the Southwest, Wares 14, 15, 16, 17, 18. Revised descriptions. *Museum of Northern Arizona Ceramic Series* 3 D.

1958b   Precipitation about the San Francisco Peaks, Arizona. *Museum of Northern Arizona Technical Series* No. 2.

1960    *Black Sand: Prehistory in northern Arizona.* Albuquerque: University of New Mexico Press.

1962    Archaeology of the Flagstaff area. *Guidebook of the Mogollon Rim Region*, New Mexico Geological Society, 13th Field Conference.

1965a   Check list of Southwestern pottery types. *Museum of Northern Arizona Ceramic Series* No. 2, revised.

1965b   Experiments in raising corn in the Sunset Crater ashfall area east of Flagstaff, Arizona. *Plateau* 37(3):77–79.

1967    *The basaltic cinder cones and lava flows of the San Francisco Mountain volcanic field* (revised ed.) Flagstaff: Museum of Northern Arizona.

1968    Frontiers of the Sinagua. In *Collected papers in honor of Lyndon Lane Hargrave*, edited by Albert H. Schroeder. Santa Fe: Archaeological Society of New Mexico.

Colton, H. S., and L. L. Hargrave

1937    Handbook of northern Arizona pottery wares. *Museum of Northern Arizona Bulletin* 11.

Colton, M. R. F., and H. S. Colton

1918    The little-known small house ruins in the Coconino Forest. *Memoir of the American Anthropological Association* 5(4).

DeBoer, W. R.

1976    Archaeological explorations in northern Arizona. NA10754: A Sinagua settlement of the Rio de Flag phase. *Queens College Publications in Anthropology* No. 1.

Eggler, E. W.

1959    Manner of invasion of volcanic deposits by plants with further evidence from Parícutin and Jorullo. *Ecological Monographs* 29(3):267–284.

1966    Plant succession on the recent volcano, Sunset Crater. *Plateau* 38(4):81–96.

Fewkes, J. W.

1927    Archaeological field work in Arizona: Field season of 1926. *Smithsonian Miscellaneous Collections* Vol. 78.

Fish, P. R.

1974    Prehistoric land use in the Perkinsville Valley. *The Arizona Archaeologist* 8:1–36.

Fish, P. R., and P. J. Pilles, Jr.

In press   Colonies, traders, and traits: The Hohokam in the North. Current Issues in Hohokam Prehistory, *Arizona State University Anthropological Research Papers*, Tempe, Arizona.

Fowler, D. D., and C. M. Aikens

1963    1961 excavations, Kaiparowits Plateau, Utah, *University of Utah Anthropological Papers* No. 76.

Gladwin, W., and H. S. Gladwin

1934    A method for designation of cultures and their variations. *Medallion Papers* XV.

Green, C. R., and W. D. Sellers
  1964  *Arizona climate*. Tucson: The University of Arizona Press.
Gumerman, G. J.
  1970  *Black Mesa: Survey and excavation in northeastern Arizona—1968*. Prescott: Prescott College Press.
Gumerman, G. J., and S. A. Skinner
  1968  A synthesis of the prehistory of the central Little Colorado Valley, Arizona. *American Antiquity* 33(2):185–199.
Hall, E. T., Jr.
  1942  Archaeological survey of Walhalla Glades. *Museum of Northern Arizona Bulletin* 20.
Hargrave, L. L.
  1932  The Museum of Northern Arizona archaeological expedition, 1932. *Museum Notes* 5(5):25–28.
  1933  Pueblo II houses of the San Francisco Mountains, Arizona. *Museum of Northern Arizona Bulletin* 4.
Haury, E. W.
  1958  Evidence at Point of Pines for a prehistoric migration from northern Arizona. In *Migrations in New World culture history*, edited by R. H. Thompson. *University of Arizona Social Science Bulletin* 27.
Johnson, A. E.
  1965  *The development of the Western Pueblo culture*. Doctoral dissertation, Dept. of Anthropology, University of Arizona.
Kelly, R. E.
  1969  Salvage excavations at six Sinagua sites. *Plateau* 41(3).112–132.
Kelly, R. E., and S. A. Skinner
  1966  Ceremonialism and kivas in Sinagua prehistory. Paper presented at the Thirty-first Annual Meeting of the Society for American Archaeology, Reno.
King, D. S.
  1949  Nalakihu: Excavations at a Pueblo III site on Wupatki National Monument, Arizona. *Museum of Northern Arizona Bulletin* 23.
Lee, T. A., Jr.
  1962  The Beale's Saddle Site: A nonconformity? *Plateau* 34(4):113–128.
Lipe, W. D.
  1970  Anasazi communities in the Red Rock Plateau, southeastern Utah. In *Reconstructing prehistoric Pueblo societies*, edited by W. A. Longacre. Albuquerque: University of New Mexico Press.
Longacre, W. A.
  1970  Archaeology as anthropology: A case study. *Anthropological Papers of the University of Arizona* No. 17.
Malde, H. E.
  1964  The ecologic significance of some unfamiliar geologic processes. In *The reconstruction of past environments. Proceedings of the Fort Burgwin Conference on Paleoecology: 1962*, edited by J. J. Hester and J. Schoenwetter. Fort Burgwin Research Center.
Martin, P. S., and F. Plog
  1973  *The Archaeology of Arizona: A study of the Southwest region*. New York: Doubleday/Natural History Press.
Maule, S. H.
  1963  Corn growing at Wupatki. *Plateau* 36(1):29–32.
Morris, D.
  1970  Walnut Creek Village: A ninth century Hohokam–Anasazi settlement in the mountains of central Arizona. *American Antiquity* 35(1):49–61.
McGregor, J. C.
  1933  Volcanic cinder fall and tree ring growth. *The Pan-American Geologist* 60(4):313–314.

1936a Culture of sites which were occupied shortly before the eruption of Sunset Crater. *Museum of Northern Arizona Bulletin* 9.

1936b Dating the eruption of Sunset Crater. *American Antiquity* 2(1):15–26.

1937 Winona Village: A XIIth century settlement with a ball court near Flagstaff, Arizona. *Museum of Northern Arizona Bulletin* 12.

1941 Winona and Ridge Ruin Part I: Architecture and material culture. *Museum of Northern Arizona Bulletin* 18.

1965 *Southwestern archaeology* (2nd ed.). Urbana: University of Illinois Press.

McKern, W. C.
1934 The Midwestern taxonomic method as an aid to archaeological culture study. *American Antiquity* 4(4):301–313.

Olson, A. P.
1964 The 1959–1960 Transwestern Pipeline survey, Window Rock to Flagstaff: A report on archaeological survey and excavation. *Archives of Archaeology* No. 25.

Pilles, P. J., Jr.
1978 The field house and Sinagua demography. In *Limited activity and occupation sites: A collection of conference papers. Contributions to Anthropological Studies* No. 1., edited by Albert E. Ward. Albuquerque: Center for Anthropological Studies.

Reed, E. K.
1948 The Western Pueblo archaeological complex. *El Palacio* 55(1):9–15.

1950 Eastern-central Arizona archaeology in relation to the Western Pueblos. *Southwestern Journal of Anthropology* 6(2):120–138.

1956 Types of village plan layout in the Southwest. In *Prehistoric settlement patterns in the New World. Viking Fund Publications in Anthropology* No. 23, edited by Gordon R. Willey.

Robinson, H. H.
1913 The San Franciscan volcanic field, Arizona. U.S. Geological Survey Professional Paper 76.

Robinson, W. J., and J. S. Dean
1969 Tree-ring evidence for climatic changes in the prehistoric Southwest from A.D. 1000 to 1200. 1967–1968 Report to the National Park Service from the Laboratory of Tree-Ring Research.

Rouse, I.
1958 The inference of migrations from anthropological evidence. In *Migrations in New World culture history. University of Arizona Social Science Bulletin* 27, edited by R. H. Thompson.

Schinzel, R. H., and R. T. Meurisse
1972 *Soil resource inventory, Elden Ranger District, Coconino National Forest.* Albuquerque: U.S. Forest Service Southwest Region.

Schoenwetter, J.
1966 A re-evaluation of the Navajo Reservoir pollen chronology. *El Palacio* 73(1):19–26.

Schroeder, A. H.
1961 The pre-eruptive and post-eruptive Sinagua patterns. *Plateau* 34(2):60–66.

1975 The Hohokam, Sinagua, and the Hakataya. *Imperial Valley College Occasional Paper* No. 3. El Centro: Imperial Valley College Museum Society.

1977 *Of men and volcanoes: The Sinagua of northern Arizona.* Globe, Ariz.: Southwest Parks and Monuments Association.

Scott, S. D.
1960 Pottery figurines from central Arizona. *The Kiva* 26(2):11–26.

Shoemaker, E. M.
1977 Eruption history of Sunset Crater, Arizona. Investigator's annual report. MS on file at Wupatki-Sunset Crater National Monument.

Shoemaker, E. M., and S. W. Kieffer
1974 *Guidebook to the geology of Meteor Crater, Arizona.* Thirty-seventh Annual Meeting of

the Meteoritical Society, Publication No. 17. Tempe: Center for Meteorite Studies, Arizona State University.

Segerstrom, K.
1960   Erosion and related phenomena at Parícutin in 1957: Geologic investigations in Mexico. *Geological Survey Bulletin* 1104. Washington: U.S. Government Printing Office.
1961   Deceleration of erosion at Parícutin, Mexico. Short Papers in the Geologic and Hydrologic Sciences, Articles 293–435. *Geological Survey Professional Paper* 424-D. Washington: U.S. Government Printing Office.

Smiley, T. L.
1958   The geology and dating of Sunset Crater, Flagstaff, Arizona. In *Guidebook of the Black Mesa Basin, Northeastern Arizona*, New Mexico Geological Society, Ninth Field Conference.

Smith, H. V.
1956   The climate of Arizona. *Agricultural Experiment Station Bulletin* 279. Tucson: The University of Arizona Press.

Smith, W.
1952   Excavations in Big Hawk Valley, Wupatki National Monument, Arizona. *Museum of Northern Arizona Bulletin* 24.

Stanislawski, M. B.
1963a  Extended burials in the prehistoric Southwest. *American Antiquity* 28(3):308–319.
1963b  *Wupatki Pueblo: A study in cultural fusion and change in Sinagua and Hopi prehistory.* Doctoral dissertation, Dept. of Anthropology, University of Arizona.

Stewart, G. R., and M. Donnelly
1943   Soil and water conservation in the Pueblo Southwest. *The Scientific Monthly* 56:31–44, 134–144.

Voth, H. R.
1905   The traditions of the Hopi. *Field Columbian Museum Publication* 96. Anthropological Series Vol. 8.

Ward, A. E.
1976   Ariz. I:3:1. In *Papers on the archaeology of Black Mesa, Arizona*, edited by G. Gumerman and R. C. Euler. Carbondale: Southern Illinois University Press.

Weed, C., and A. E. Ward
1970   The Henderson Site. *The Kiva* 36(2):1–12.

Wilcox, R. E.
1959   Some effects of recent volcanic ash falls with especial reference to Alaska. *Geological Survey Bulletin* 1028-N:450–474.

Willey, G. R., C. C. DiPeso, W. A. Ritchie, I. Rouse, J. H. Rowe, and D. W. Lathrap
1956   An archaeological classification of culture contact situations. In *Seminars in archaeology: 1955. Memoirs of the Society for American Archaeology* No. 11, edited by R. Wauchope.

Wilson, J. P.
1969   The Sinagua and their neighbors. Doctoral dissertation, Dept. of Anthropology, Harvard University.

# 16

**Comparative Effects of Climatic Change, Cultural Impact, and Volcanism in the Paleoecology of Flagstaff, Arizona, A.D. 900–1300**

RICHARD H. HEVLY
ROGER E. KELLY
GLENN A. ANDERSON
STANLEY J. OLSEN

For decades Southwestern archaeologists have been intrigued by the problem of prehistoric shifts of settlement patterns and changing density of local populations. Pioneering biological studies by Douglass (1929), Miller (1932), and Hargrave (1939) and geomorphological studies by Gregory (1917), Bryan (1941), and Hack (1942) brought attention to environmental changes as a potential cause of demographic changes. As archaeological data became more complete and more precise, attention was extended to other influences to explain more completely the problem of changing prehistoric population densities and settlement patterns (Reed 1944; Colton 1962; Ellis 1964; Jett 1964, 1965 a,b; Davis 1965).

Among the suggested influences were such factors as volcanism, soil nutrient depletion, disease, and biotic changes resulting from human disturbance and exploitation. Several sites located near Flagstaff, Arizona (Figure 16.1) provide the opportunity of assessing the relative significance of these suggested influences, since the sites lie in an area that was extensively occupied and presumably exploited as well as subjected to periodic volcanic eruption (e.g., Sunset Crater, 1066). Comparative analysis of tree-ring, pollen, and paleontological data from these and other sites will also afford excellent opportunities for supplementing and substantiating paleoecological interpretations based primarily on pollen studies from archaeological sites at lower elevations on the Colorado Plateau (Hill and Hevly, 1968; Euler *et al.* 1979; Schoenwetter and Dittert, 1968).

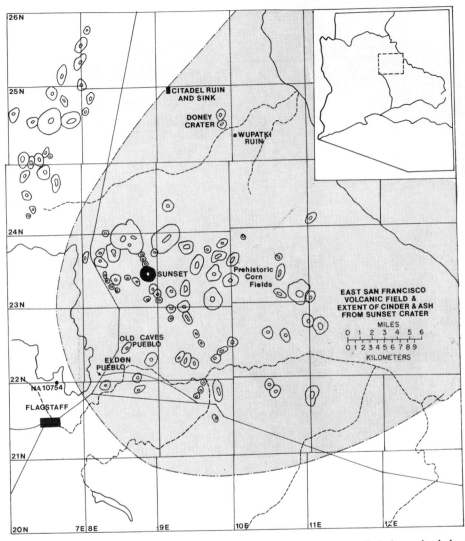

**FIGURE 16.1** East San Francisco volcanic field showing the extent of cinder and ash from Sunset Crater (Colton 1962). Also shown are the locations of NA 10754, Elden Pueblo (NA 142), Old Caves Pueblo (NA 72), Citadel Sink, and prehistoric corn fields whose paleontological data have been used to reconstruct past environments.

## MATERIALS AND METHODS

Pollen data from this study are derived primarily from three sites; NA 10754, Elden Pueblo, and Old Caves Pueblo, all of which are located in mixed ponderosa and piñon pine, Utah juniper, and gambel's oak at an approximate elevation of 2121 m within or immediately outside the ashfall area of Sunset Crater (Figure 16.1). These data obtained from prehistoric structures have

been supplimented by data derived from "more natural," open contemporaneous sites (Citadel Sink and a prehistoric cornfield) at lower elevations in
Juniper-Savana but still in the Sunset Crater ashfall area.

Thirty pollen samples spanning the period A.D. 900–1300 were obtained
from floors, fills, and alluvium of Citadel Sink, the Sunset Crater cornfield, and
pit houses and pueblos noted above. Dating of pollen samples was based on
stratigraphic correlations, associated ceramics, and other artifacts as well as
tree-ring dates. Pit-house samples assigned to the first half of the Rio de Flag
phase on the basis of ceramics (Pilas, Hudgens and Fish, personal communication) were obtained from NA 10754. Elden Pueblo (actually Room 37) overlies
an Angell-Winona–Padre phase pit house dated ceramically between 1070 and
1120 (Kelly 1971). Unconfirmed dates of 1159–1162 based on a single tree-ring
sample from Elden Pueblo have been published (Robinson *et al.* 1975) and are
in agreement with ceramic dating of this Elden-phase Pueblo. Pottery and
other data suggested that Rooms 29, 35, and 39 were relatively similar and
earlier than Room 36, which was in turn similar to or slightly earlier than Room
37 (Kelly, personal communication). The Sunset Crater cornfield appears
ceramically equivalent to Elden Pueblo (Berlin *et al.* 1977). Old Caves Pueblo,
on the other hand, ceramically and on the basis of tree rings dates from the
middle of the thirteenth century (Colton 1946).

Standard extraction and analysis procedures were followed (Kummel and
Raup 1965) but were supplemented by determination of arboreal (AP)–
nonarboreal (NAP) pine–juniper, pine, and economic pollen ratios and the
relative abundance of broken grains, fungal spores, and parasite eggs. Pollen
identification was based on a small reference collection and keys to pollen types
such as those by Faegri and Iverson (1964). At Citadel Sink sediment composition was also analyzed to determine depositional history, including the deposition of ash from the eruption of Sunset Crater. Data for comparative purposes
were obtained from previous studies of vegetation, climate, tree rings, and
modern pollen rain in northern Arizona and other climatic indexes from the
Northern Hemisphere.

## RESULTS AND DISCUSSION

Salient features of sediment composition, pollen density, preservation,
and composition are summarized for Citadel Sink in Figure 16.2. This natural
sink apparently has preserved a pollen record of vegetation conditions prior to,
during, and following the eruption of Sunset Crater. Pollen preservation is
generally excellent and pollen density generally uniform (except in levels with
most cinder). However, AP proportions are generally fewer than found in
surface or near-surface levels. Likewise, pine in pine–juniper ratios, though
fluctuating, still generally remains below the levels observed in surface and
near-surface levels. Though the AP–NAP ratios are generally consistent from

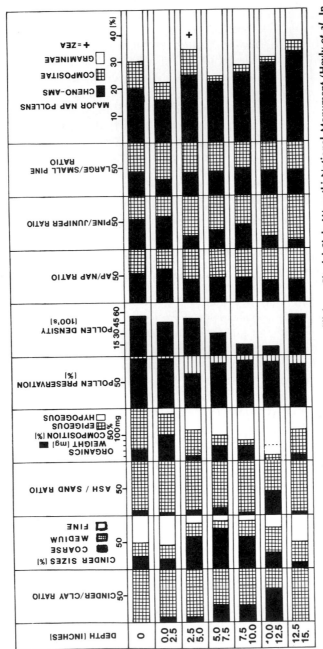

FIGURE 16.2 Sediments and pollen from the upper 37.5 cm fill from Citadel Sink, Wupatki National Monument (Hevly *et al.* In preparation). Cinder from Sunset Crater is concentrated between 12.5 and 31.25 cm below the surface. The cinder contains very little organic material originating above ground (*epigeous*) but does contain abundant organic materials originating below ground (*hypogeous*). The cinder contains a reduced density of pollen compared with the over- and underlying layers of clay. These long-term trends probably reflect changes of floristic composition initiated by the deposit of cinder and essentially continued to the present. The only interruption of this trend is shown in the sample at the 6.25- to 12.5-cm depth, from which corn pollen was recovered, indicating prehistoric cultivation of this horizon. Pollen of Compositae, which were previously a minor component of the pollen spectra, become and remain a more major component from this level to the present, suggesting that disturbances originating with cultivation have also had a long-lasting effect.

level to level, the proportion of Gramineae and Compositae increases at the expense of Cheno-Ams in the NAP fraction, probably reflecting altered edaphic conditions resulting from the deposition of cinder and possible use of the sink bottom for corn cultivation. The trends noted in this open, essentially "natural" deposition environment were also detected in part at a nearby prehistoric contemporaneous cornfield, where again AP proportions were below those of modern samples (probably reflecting prehistoric disturbance), but pine–juniper ratios approximated modern levels (Berlin *et al*. 1977).

Salient features of the pollen, spore, and egg data from the higher-elevation, better-dated prehistoric structures are summarized in Figures 16.3–

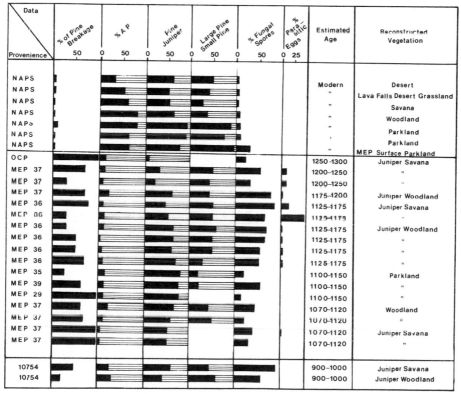

FIGURE 16.3  Pollen, spore, and parasite egg data from Elden Pueblo (MEP) and Old Caves Hill Pueblo (OCP). Modern pollen samples (NAPS) are shown for comparative purposes. (Adapted from Hevly 1968a, 1970.) Pollen preservation, as indicated by percentage of pine breakage, is excellent in modern samples but is much poorer in fossil samples. High proportions of arboreal pollen (AP) are found only in the savanna, woodland, and forest or parkland. Highest proportions of juniper and piñon pine (small pine) pollen can be obtained only in the piñon–juniper woodland or in close proximity to trees within the juniper savanna. Note that the fossil samples generally contain higher proportions of juniper and piñon pine than do surface samples of the study site (surface parkland). Comparison of the pollen abundance, the proportion of arboreal pollen in total pollen and the proportion of pine in pine–juniper ratios with percentage pine breakage and relative abundance of fungal spores fails to show an evident correlation.

16.9. Preservation of fossil pollen in this study is poorer (percentage of pine breakage greater) than that obtained from modern soil samples, but it was still possible to identify 85% or more of the pollen observed. Nonarboreal pollen is conspicuously more abundant than arboreal pollen throughout the fossil record, and variations of the AP–NAP ratio are statistically nonsignificant when 99% confidence limits are applied to the data (Figure 16.3).

Comparison of the AP–NAP ratio with those of modern pollen samples would suggest generally treeless conditions in the Mount Elden area from the tenth to the thirteenth century; however, several factors may reduce the probability of this interpretation. Such treeless conditions could have resulted from a catastrophic fire, but such fires were apparently rare in the past (Dunkley 1973). Furthermore, pine and juniper regeneration should be palynologically detectable within the several centuries spanned by the occupation of the sites sampled unless unfavorable climate or aboriginal lumbering impeded the succession process. Furthermore, the treeless interpretation of the AP–NAP ratios is not supported by the pine–juniper ratios (Figure 16.3) which have been shown to parallel the AP–NAP ratios of modern plant communities (Hevly 1968a). The fossil pine–juniper ratios may therefore indicate severe overrepresentation of NAP. Such overrepresentation could be due to any of several factors, including (*a*) altered climatic conditions favoring NAP production or reducing AP production; (*b*) increased local abundance of NAP producers as a result of disturbance; (*c*) differential deposition; or (*d*) differential transport of NAP-bearing plant parts into the rooms by prehistoric people (Leopold *et al.* 1963; Hevly 1964; Martin and Byers 1965; Potter 1967). Overrepresentation of NAP types (other than cultigens) has also been detected by Bohrer (1972), suggesting the inherent danger in utilizing AP–NAP ratios for vegetative reconstruction and necessitating emphasis on analysis of the AP record in this instance.

## Pine–Juniper Pollen Ratios and Changes of Relative Effective Moisture

The AP record is composed primarily of three pollen types: pine (mostly *Pinus ponderosa* and *P. edulis*, juniper (*Juniperus* spp.), and oak (*Quercus gambellii*). Previous studies have shown that pine–juniper ratios are good indicators of vegetation types (Hevly 1968a). Fossil pine–juniper ratios from Elden Pueblo are lower than similar ratios from soil surfaccs of modern pine parkland in all but the first half of the twelfth century (Figure 16.2 and Figure 16.3a). Direct comparisons with modern ratios suggests a more open, juniper-dominated vegetation type than is now found about the sites. Substantiation of the woodland above its present elevation in the study area is afforded by evidence of increased use of piñon pine and juniper as construction timbers after A.D. 900–1050 (Figure 16.3a) and by previous pollen studies from nearby areas (Schoenwetter 1965; Bennett 1967; Wilson 1969). However, several questions regarding the factors influencing arboreal pollen proportions must be explored—for example, the relative significance of differential pollen transport and preservation versus the effects of various biotic and physical factors on the environment.

Pollen is known to be differentially preserved (Faegri and Iversen 1964). Since the percentage of broken pine grains is much greater in the fossil spectra from Elden Pueblo than in modern surface samples (Figure 16.2), the hypothesis of differential destruction of arboreal pollen must be examined. Bradfield (1973) and Potter (1967) suggest that juniper pollen is differentially destroyed in Southwestern soils. Hence, one would anticipate pine–juniper ratios dominated by pine rather than juniper in fossil spectra. No differential preservation of these were detected in studies of the modern pollen rain of northern Arizona (Hevly 1968a) and Havinga (1971) has recently shown that pine and juniper are about equally well preserved in experimental studies. Furthermore, differential destruction of either pine or juniper should be detectable in the pine–juniper ratios by persistent and consistent departures of the proportions of these two types from modern values if the pollen of one type were to be more readily destroyed. Since it is not (Figure 16.2), it may be concluded that no differential destruction of pine and juniper occurred.

Several studies have also recently suggested that various wind-transported arboreal pollen types are not equally transported, and hence do not appear in the same proportions outside and inside depositional environments with restricted openings, such as lakes surrounded by forest, rooms of archaeological sites and caves or rock fissures (Hevly 1968b, 1970; Tauber, 1967; Currier and Kapp, 1974). The failure of pine to be better represented in the fossil record is particularly noteworthy in view of the excellent transport capabilities and general overrepresentation of pine pollen in most depositional environments, suggesting that the low pine proportions might be due to limited entry portals during the time of pollen release. Since the records of roofed structures (Figure 16.2, Rooms 29, 35, and 36) and partially walled, unroofed areas (Figure 16.2, Rooms 37 and 39) are similar, it may be concluded that differential transport resulting from wind does not seem to account for the low pine pollen percentages. Though human transport of pollen on plant parts is feasible and demonstrable, it results in augmentation of the arboreal pollen record by only a few percent (Adam et al. 1967; Bohrer 1972; Hevly in preparation).

If the fluctuating but essentially continuously depressed proportions of pine in the pine–juniper ratios cannot be attributed to differential preservation and transport of pollen, one is forced to conclude that these ratios reflect an environment in which the composition of vegetation was different from that of today (or at least the production of pollen differed) as the result of changes of selective impacts of the biotic and physical factors of the environment. Ponderosa pine is relatively disease resistant except under severe drought stress or damage from fire and hence extensive prolonged damage or death would be infrequent (Pearson 1950). Forests can also be modified by the cultural activities of man (Simmons 1969). If the lower proportion of pine is due to prehistoric lumbering of a mixed stand of juniper and pine, one would anticipate reduced pine in a pine–juniper ratio, since pine was the preferred species for construction purposes during the period of this study (Figure 16.3). However, the study area is in fact immediately bordered by a large pine forest to the south (Mogollon Rim) and to the west (San Francisco Peaks). Hence, removal of local juniper and pine in close proximity to the sites would have resulted in lowered local pollen production and overrepresentation of long-distance transported, in modern pollen studies elsewhere in Arizona (Solomon 1976). Therefore, the increase in pine pollen proportions during the period of maximal population growth, followed by decline as human population density diminished could be interpreted as a reflection of relative lumbering effect on the pollen record (Figures 16.4 b,c). Fortunately, the pine–juniper ratios of the preceding and following periods indicate that fluctuations of human populations and related construction activities were not parallel (Figures 16.4 b,c). Therefore, although prehistoric lumbering activities could be the cause of generally diminished overall pine pollen proportions, changes in proportions of pine and juniper must be due to other factors.

Physical environmental factors that might account for reduced proportions of pine include different relative tolerance of pine and juniper for altered climatic conditions, fire, or special physical events such as the eruption of Sunset Crater only 12.8 km from Elden Pueblo. Undoubtedly, the ash and cinder fall as well as gaseous emissions affected nearby vegetation, as for example, they did at Parícutin, Mexico, a volcano of comparable size and type, where moderate or severe damage was noted in an 8–16-km radius (Eggler 1948).

The depth of residual cinder surrounding Sunset Crater indicates that the prevailing wind direction during that phase of the eruption was away from the Elden Pueblo. Though the cinder was probably deeper in the past, only a trace of it remains in the Elden Pueblo area today. The concentration of gaseous emissions is of course unknown, but was probably minimal because of distance from the source and likelihood of similar prevailing wind directions. However, the tolerance of pine to certain toxic gases, including those typically produced in volcanic eruptions, is less than that of juniper (Treshow 1972) and could have resulted in local differential leaf damage, defoliation, and tree death, as noted in other volcanic events elsewhere (Eggler 1948; Malde 1964). Leaf damage, defoliation, and abnormal growth responses as indicated in studies elsewhere by Eggler (1967), Lovelius (1970), and Oswalt (1957) should manifest themselves for only 1 or 2 decades; however, if tree death had occurred, many decades would have been necessary for regeneration of the forest and could

---

FIGURE 16.4 Comparison of (a) the wood types utilized for construction purposes and the habitat locations of archaeological sites of the Flagstaff area (adapted from Stein 1964) with (b) the demographic records of the Flagstaff, (1) Jeddito Valley (2) and Hay Hollow Valley (3) areas of northern Arizona (Hack 1942; Colton 1962; Zubrow 1971) and (c) the fluctuating pine pollen proportions of Hay Hollow Valley (Hevly In preparation) with that of the Elden Pueblo study area. Note that yellow pine utilization is always the most predominant, but use of piñon pine and juniper increases as the prehistoric agriculturalists expand into marginal lands of the savanna and desert

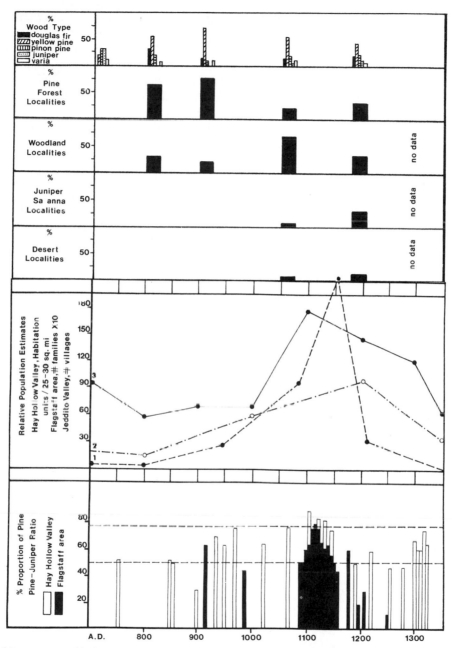

habitats, presumably in response to changing edaphic and climatic factors. Dashed lines in bottom diagram equal mean proportion of pine in pine–juniper ratios of Flagstaff (upper) and Hay Hollow Valley (flower). Note parallelism of pollen trends in these areas separated by a distance of over 240 km, but also the generally higher proportions of pine pollen in the Hay Hollow Valley area (piñon–juniper woodland–savana) than found in the Elden Pueblo study area (piñon–juniper woodland–yellow pine parkland).

account for the overall reduction of pine pollen in the pine–juniper ratios (McKelvey 1954; Nicholls 1959; Eggler 1966).

At Parícutin, an orderly succession of plant associations was reestablishing the original vegetation in less than 20 years except on lava flows and deeper cinder deposits (Eggler 1959; Beaman 1961). Succession proceeds much more slowly in the semi-arid environment of the Flagstaff area, and the progressive changes of pollen proportions might be interpreted as suggestive of secondary success following the eruption of Sunset Crater were they not already in progress before the eruption and also reversed within a few decades (Figures 16.2, 16.3, and 16.4). Succession has therefore been largely restricted to the lava flows and deeper cinder deposits. The depressed pine pollen proportions, if attributable to volcanic activity, are probably related to modified edaphic conditions resulting in cinder accumulations that favor dry-habitat plants (Lindsey 1951; Eggler 1966; Kargalitadze 1972). A similar modification of phytogeography has been noted also in the White Mountain volcanic field in east-central Arizona (Hevly 1968a).

Tree species also differ in their ability to survive various types of fires and in their ability to regenerate following a fire (Stewart 1956). Fires could have been started by lightning, or accidentally by prehistoric man, or even by the eruption of Sunset Crater. Such fires could have locally altered the composition of the vegetation surrounding Elden Pueblo as a result of differential regeneration following a fire (juniper will crown-sprout, whereas pine will not) or failure of pine seedlings to become established because of unfavorable climatic or edaphic conditions. Catastrophic fires of large extent apparently were rare in the San Francisco Peaks–Elden area (Dunkley 1973). For their palynological consequences to have persisted for several centuries, their effects on floristic composition would have to have been perpetuated by repetition and by cultural and climatic factors.

The final factor to be evaluated as to influence on pine–juniper ratios is climate. Recent studies have shown that both temperature and precipitation influence growth and cone (seed and pollen) production, temperature assuming greater importance at the upper elevational limits of a species and precipitation having greater importance at the lower elevational limits (Roeser 1941; Daubenmire 1960; Lester 1967; Fritts 1965; Shoulders 1967). It is not possible to tell whether the decreased proportions of fossil pine pollen in the pine–juniper ratios from Elden Pueblo are due to decreased local abundance or decreased pollen production of pine or to increased local abundance or increased pollen production of juniper. Nevertheless, juniper increases or pine decreases would most likely be due to less mesic (i.e, warmer and/or drier) conditions in the late tenth or early thirteenth centuries (except during the periods noted when pine pollen proportions increased) than are now found in the Elden Pueblo area.

Comparison of the above pollen data with tree-ring data from the same area and time period (Figure 16.5) permits testing of the above interpretation and at the same time resolution of the problem of changed floristic composition versus relative pollen production. Stratigraphic and chronological arrangement of the departures of pine pollen proportions from the modern value found in the study area reveals fluctuating proportions of pine culminating in the highest values found in this study in Rooms 29–35, and 39, dated 1100 and 1125 or 1135. Spectra of Room 36 reveal a reversal of this trend, climaxing in extreme negative departures. This pattern of variation in pine pollen proportions may be seen (Figure 16.4) to parallel (keeping in mind the relative precision of dating of pollen and tree-ring data) the growth records of trees in the Flagstaff area for the same period of time obtained from previous studies by Fritts (1965) and Robinson and Dean (1969).

Similar parallelisms of pine pollen proportions and tree-ring growth records have been noted elsewhere on the Colorado Plateau (Hill and Hevly 1968; Schoenwetter 1970). The change in pollen proportions proceeds too quickly (50 years) to be a reflection of changed floristic composition, since personal observations indicate that 65–85 years are required for pine seedling establishment and growth of saplings to sufficient age to produce abundant pollen. Since the pine pollen proportions and tree-growth records parallel one another, the pollen record may be interpreted to reflect the same factors regulating growth as shown by studies by Fritts (1965)—for example, effective moisture residual from the preceding winter and current growing season temperature.

In summary, therefore, the arboreal pollen record from prehistoric structures of the Elden area is depressed greatly below the levels characteristic for the modern habitat. The phenomenon of fluctuating but generally reduced AP has been observed in many archaeological sites of various ages elsewhere on the Colorado Plateau (Hevly 1964; Dickey 1971; Bohrer 1972). This depression is in large part the result of NAP overrepresentation, but examination of pine–juniper ratios permits recognition of ancient floristic composition. The general depression of pine in the pine–juniper ratios of Elden Pueblo appears to reflect an altered floristic composition such that juniper was locally more abundant relative to pine than it is today, suggesting a floristic composition more typical of plant communities of less mesic environments.

Previous pollen studies from prehistoric structures at lower elevations near Flagstaff also yielded similar interpretations using somewhat different criteria (Schoenwetter 1965; Bennett 1967; Wilson 1969). The most probable contributing factors accounting for the notable reduction of pine in the fossil pine–juniper ratios include selective prehistoric lumbering, differing relative tolerance to gaseous emissions from the eruption of Sunset Crater, differential survival or regeneration of juniper following fire, and generally less mesic climatic conditions.

Pollen studies elsewhere on the Colorado Plateau but in similar vegtation type do not exhibit a general pattern of pine being greatly reduced below modern levels in pine–juniper ratios but do nonetheless reveal a pattern of fluctuating proportions of these two types (Hevly 1964; Schoenwetter and Eddy 1964; Schoenwetter and Dittert 1968; Dickey 1971). These fluctuating pro-

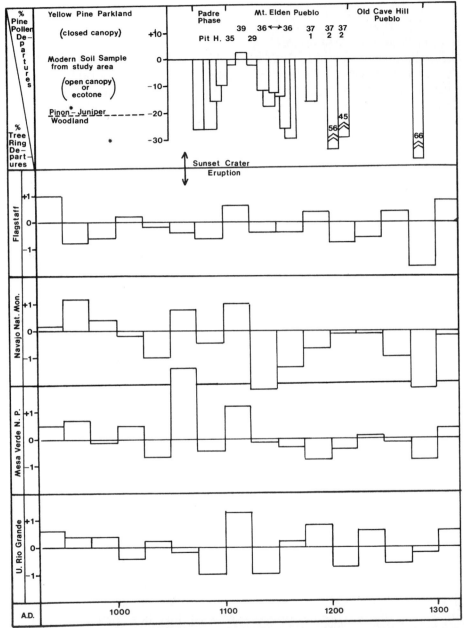

FIGURE 16.5 Comparison of departures of fossil pine proportions from modern levels with departures of tree-ring widths from standardized growth curves of the Flagstaff area (25-year means from Fritts 1965). Note the continual depression of pine pollen proportions below modern levels, suggesting some sort of environmental effect extending over at least 1 century (e.g., climate, prehistoric lumbering, eruption of Sunset Crater, fire, or some combination of these). Note that samples from the 900–1000 A.D. interval (*) also manifest depressed pine pollen departures; hence such depression is unrelated to the eruption of Sunset Crater but does not exclude the possibility of

portions of pine in the pine–juniper ratios generally parallel the local tree-ring records (Hill and Hevly 1968; Schoenwetter 1970). The inferences drawn from pollen and tree-ring data are in essential agreement with those derived by Smiley (1961) and Woodbury (1961) from archaeological and paleontological data elsewhere in northern Arizona.

The fluctuating proportions of pine and juniper seem explicable only in terms of relative pollen production (as regulated by effective moisture from the preceding winter and current growing season temperature), since duration and rapidity of change are too brief to permit seedling establishment and sapling growth to reproductive age. Periods of increased proportions of juniper pollen (1075–1100, 1150–1175, 1200–1225 and 1275–1300) should, therefore, reflect less mesic climatic trends, whereas periods characterized by increased proportions of pine pollen (1100–1135 and 1175–1200) should reflect more mesic climatic trends. Though there exist widespread regionality and synchroneity of pollen and tree-ring data, there are also real minor local variations (Fritts 1965; Schoenwetter 1970). Nevertheless, the palynological changes noted have now been shown to be a phenomenon not only of the savanna and woodland, but also of the pine parkland region. Comparison between pollen data for these communities would suggest change of comparable magnitude (Figures 16.4c and 16.5a).

## Pine Pollen Composition and Changes of Temperature and/or Seasonal Distribution of Precipitation

Previous studies have shown that proportions of piñon pine relative to ponderosa pine exceed 50% only when piñon pine is the dominant pine species present or nearby (Hevly 1968). Furthermore, piñon pine usually grows at lower, drier, and warmer elevations than ponderosa pine and under a climate characterized by significant summer rain and winter precipitation less pronounced than typical for higher-elevation conifer forests (Fritts 1965; Hevly 1968; Jameson 1969). Analysis of fossil pine pollen composition at Elden Pueblo reveals a period of increased local density of piñon pine from the 1000s to late 1100s or early 1200s relative to present conditions (Figure 16.6). Although pine pollen composition could not be determined for the 975–1050 and 1250–1300 periods, adjoining spectra suggest trends of increasing abundance of ponderosa pine relative to piñon. If these trends are valid, then very good correspondence can be obtained with studies of fossil pine pollen composition in eastern and northern Arizona by Dickey (1971), Hevly (1964), and Sears (1961), and in

---

effects from earlier eruptions. Note that although many parallelisms may be noted between the tree-ring records of the various areas, periods exist in which the growth trends are not parallel. The pine pollen departures as dated within the limits of tree rings, pottery typology, and stratigraphy of a series of samples from Elden Pueblo more or less parallel the tree-ring growth records from Flagstaff. Should relative proportions of pine pollen always parallel local tree-growth records, regional differences of the pollen records should be anticipated as the precision and, it is hoped, the accuracy of dating of pollen samples increases.

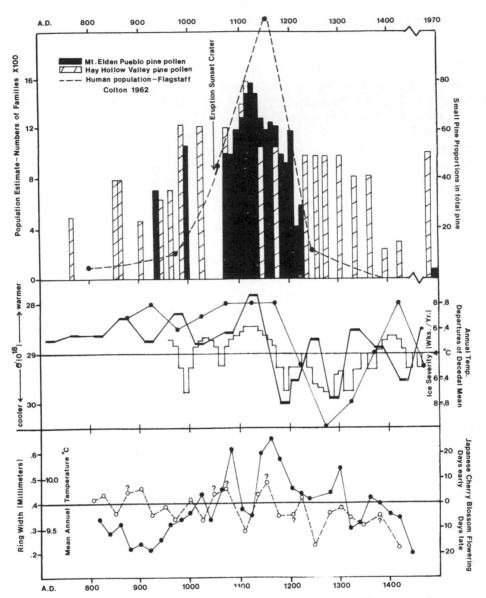

**FIGURE 16.6** Comparison of (a) fossil small pine proportions (a probable measure of temperature and/or seasonal distribution of precipitation) and the demographic record of the Flagstaff area with Northern Hemisphere temperature indicators; (b) the Greenland oxygen isotope (Dansgaard *et al.* 1969; heavy bars and lines) with the Iceland ice severity index (Bergthorsen 1969; lines or circles and lines); and (c) the White Mountain bristle-cone pine growth records (LaMarch 1974; solid circles) with the Japanese cherry blossoming records (Arakawa 1966; open circles). Note the apparent agreement between pollen records from east-central Arizona (Hevly In preparation) and the Flagstaff area and the apparent correlation of this climatic event with the demographic record (actually climatic trends may slightly precede demographic events as presently understood). Also note that these trends are paralleled at least in part by the ice core and tree-ring growth records. Failure of ice core and tree-ring temperature trends to parallel one another exactly probably reflects geographic separation of the study areas and discrepancy of dating between ice cores (based on ice accumulation rates determined by [14]C dating) and tree rings (based on annual increments of growth). Nevertheless, all records indicate warm conditions during the twelfth century.

northwestern New Mexico by Martin and Schoenwetter (n.d.). Reasons for such change in the proportion of pine pollen types include actual prehistoric change in the floristic composition of the surrounding plant community and changed relative production of pollen as the result of climatic change. Pine pollen production is known to be affected by both temperature and moisture (Leiberg *et al.* 1904; Daubenmire 1960). However, the long persistence of pine dominance in the fossil record is strongly suggestive of actual floristic change rather than variation in relative pollen production from year to year. Change of floristic composition in the Flagstaff area is also strongly indicated by changes in location of archaeological sites and the composition of woods utilized for construction purposes (Figure 16.4a) and by other paleontological data.

Pine pollen proportions thus substantiate the earlier interpretation of less mesic woodland conditions in the Flagstaff area from the late tenth to the fourteenth century, but at the same time they permit inferences regarding changes of temperature and/or seasonal distribution of precipitation. The increasing proportions of piñon pine pollen composition in the late 1000 and early 1100 period (or as early as A.D. 925–975 in other nearby areas) suggest warming temperatures and/or diminished annual effective moisture particularly during the winter season, if altered floristic composition is indicated by the above data. The decreasing proportion of piñon pine pollen in the latter part of the 1150–1250 period suggests cooling temperatures and/or increased effective annual moisture with increased precipitation during the winter months. The preceding analysis of the pine–juniper ratios and tree-ring data are incompatible with these inferences regarding effective annual moisture but not necessarily with the inferences regarding the significance of winter moisture or of the inferred temperature trends.

The temperature trends suggested by the proportion of pine types are also found (Figure 16.6) in oxygen isotope data from ice cores in Greenland (Dansgaard *et al.* 1969) and ice-severity index from Iceland (Bryson 1975; Schell 1967), as well as in the growth records of high-elevation pine in western North America (LaMarch 1974) and historical records of cherry blossoming in Japan (Arakawa 1966). Analysis of a variety of archaeological, pedological, and paleontological data from the midcontinental area of North America as well as historical information from western Europe has resulted in similar interpretations of warmer weather during the 1000–1200 period, followed by cooling weather (Lamb 1966; Bryson and Wendland, 1967) beginning in 1200. It may therefore be suggested that the piñon pine proportions probably reflect changing temperature trends of that period similar to those observed elsewhere in the Northern Hemisphere.

## Nonarboreal Pollen, Paleoecology, and Agricultural History

Analysis of archaeobotanical and palynological data received from the study sites and other contemporaneous sites in the Flagstaff area indicates that corn, beans, pumpkins, squash, and maybe even cotton were grown during the

A.D. 900–1300 period; however, little is known about the use of native plants (Jones 1941; Conner 1943; King 1949; Rixey and Voll 1962; Stanislawski 1963; and Wilson 1969). Analysis of the NAP record from the study sites suggests that native plants played an important role in the economy of the prehistoric occupants, since the pollen of such plants is extraordinarily abundant (Figure 16.7). Furthermore, the proportions of native pollen types and of pollen from cultivated species parallel the demographic trends of the Flagstaff areas as well as some of the paleoclimatic trends noted above.

Entomophilous native pollen types and pollen from cultivated species parallel the demographic record, generally increasing to about 1150 or 1200 and thereafter diminishing (Figure 16.7a). Entomophilous pollen types, and the large anemophilous pollen of corn, are limited in their ability to be transported by wind more than a few meters from their point of origin (Jones and Newell 1946; Faegri and Iversen 1964). Therefore, the occurrence of these pollen types in archaeological sediments indicates their transport by humans into such proveniences on pollen-bearing plant parts. Assuming that the records of cultivated plant pollen and the pollen of native entomophilous species reflect at least generally the relative utilization of these plants, one becomes curious about causal relationships for the observed trends of pollen proportions. It would appear that corn agriculture was only sporadically successful, the highest proportions of corn pollen being found in the early twelfth and thirteenth centuries. Native entomophilous pollen types were several times more abundant than the pollen of cultivated species and more consistently recovered but in diminishing proportions after 1200.

The native entomophilous pollen category is composed of taxa that flower in either spring or early summer and are often, but not always, perennial pioneer species in disturbed ground. Thus, the trends of native entomophilous pollen proportions could reflect relative disturbance by the prehistoric people whose regional and local population trends parallel the record of this pollen catagory. It is also possible that the record of this pollen category is unrelated to disturbance by changing human population densities but merely reflects the relative utilization of plants belonging to this category as necessitated by the relative success of domestic crops in meeting the dietary needs of the population. Furthermore, these changing pollen proportions of native entomophilous taxa may simply reflect opportunistic exploitation of plants whose population densities varied as did those of domestic crops in response to changing temperature, annual effective moisture, and the seasonal distribution of precipitation.

A possible explanation of the relative significance of the various alternatives posed above is provided by analysis of additional independent NAP pollen data (Figure 16.7b). Three major nonarboreal pollen categories contribute to the NAP overrepresentation at the study sites: Cheno-Ams, Compositae, and Gramineae. The Compositae can be divided into three categories: *Artemesia*, low-spine Compositae (*Ambrosia* type), and high-spine Compositae (*Helianthus* type). Since Gramineae, Cheno-Ams, and low-spine Compositae are abundant and wind-pollinated, changes of their pollen proportions, though

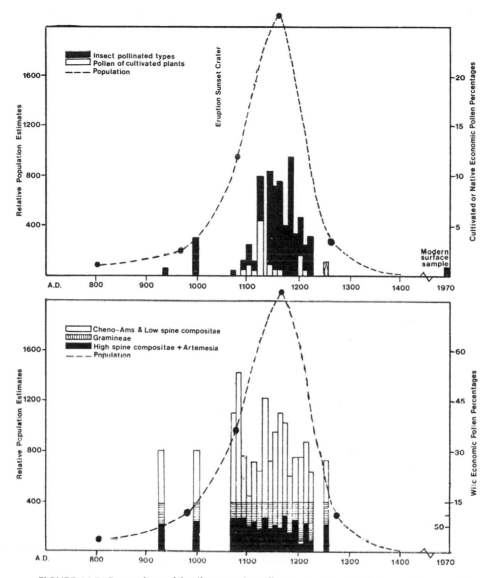

FIGURE 16.7 Comparison of fossil economic pollen proportions (including cultigens such as corn) and the demographic record from the Flagstaff area. Note the parallelism of the two records suggesting a probable relationship between food and population. The economic pollen record also parallels that of climate (Figures 16.3 and 16.4), suggesting a relationship between climate and population via the impact of climate on food resources. Analysis of wild economic pollen percentages in excess of modern background levels (lower diagram) may reflect relative pollen production of pioneer species in soils whose degree of disturbance is changing with variation of human population density or opportunistic utilization of noncultivated food sources as necessitated by relative success of agricultural attempts. In either situation the relative abundance and pollen production of these plants would have been influenced by seasonal availability of moisture: Grasses begin to flower in late spring and early summer, and the Cheno-Ams and Compositae flower from midsummer to early autumn (Hevly and Renner 1979). Diminished proportions of the latter types (lower diagram), if reflecting diminished summer rain, could explain the observed diminished proportions of corn pollen (upper diagram), a species dependent on summer moisture.

likely to be exaggerated by human transport, are not dependent on this activity for their occurrence in the fossil record, and may therefore provide environmental data less biased by human behavior.

Studies of the modern pollen rain in the Flagstaff area indicate that the above pollen types are produced in a seasonal sequence: Gramineae in late spring and early summer; Cheno-Ams and low-spine Compositae in midsummer; high spine Compositae and *Artemesia* in late summer (Hevly and Renner 1979). It is probable that the population densities (and thereby the pollen production) of these frequently annual pioneer taxa would be directly influenced by local weather conditions, particularly the availability of adequate moisture for seed germination, seedling establishment, and flowering. Though the fluctuating but declining overrepresentation of Cheno-Ams and low-spine Compositae pollen could reflect reduced disturbance or reduced gathering by a diminishing human population (Figure 16.7b), analysis of the independent relative proportions of Gramineae and other Compositae suggest otherwise.

The proportions of pollen of the Gramineae and other Compositae in the fossil samples, though variable, nonetheless remain within or very close to the range of values found in modern samples. Their relative proportions to each other, however, change during the occupancy of the study sites. If their growth and flowering are related to seasonal distribution of moisture as suggested above, then the high proportions of Cheno-Ams, Compositae, and reduced proportions of Gramineae pollen early in the fossil record should be indicative of abundant moisture from mid- to late summer but reduced moisture in the spring. The diminished proportions of these types and increased abundance of Gramineae later in the fossil record should be indicative of increased spring and early summer moisture with reduced mid- and late-summer moisture.

The above interpretation accords well with the previous interpretation of arboreal pollen, which likewise suggested a shift in seasonal distribution of moisture as well as changed temperature. The independent AP and NAP analyses suggest the following paleoclimatic interpretations: Temperatures were generally warmer than at present after the eruption of Sunset Crater and to about 1150 or 1200 but cooled thereafter. Annual effective moisture varied throughout the occupancy of these sites but was generally less than found in the area today except for a brief episode in the early twelfth century immediately preceding or coincident with the increase of human population of the Flagstaff area. Moisture was predominantly distributed in the mid- and late summer from the time of the eruption of Sunset Crater to about 1150 or 1200, after which it was predominantly winter, spring, or early summer in seasonal distribution. Presumably such spring moisture should have promoted piñon pine growth and flowering (Fritts 1965), but apparently cold temperatures inhibited this.

Such climatic trends were apparently responsible for the observed sporadic and declining proportions of corn pollen and presumably of agricultural success. That this should be so is not surprising, since corn, beans, and squash, the commonly cultivated triad of economically important plants, are all summer rain annuals favored by warm temperatures. Corn, for example, requires a frost-free period of at least 90–120 days, mean summer temperatures and mean night temperatures above 19° and 13° C, respectively, and at least 20 cm of effective moisture during the growing season (Jenkins 1941). Comparison of these requirements with climatic conditions prevailing near study sites indicates that annual effective moisture should not normally be a limiting factor for corn, beans, and squash; however, temperature and seasonal distribution of moisture during the growing season of these plants can be. Therefore, it is unlikely that such crops could have been a major reliable source of sustenance at the study sites unless environmental conditions were more favorable in the past. The magnitude of environmental change necessitated could have been quite minimal, Colton (1965) has demonstrated the feasibility of dry-land farming of corn using cinder as a mulch near Sunset Crater only 91–121 m below the elevation of Elden Pueblo. Cinder for mulch was available following the eruption of Sunset Crater and would have preserved residual winter moisture for germination of crops in those seasons when spring and early summer precipitation was inadequate. That such techniques were practiced is amply demonstrated by prehistoric fields known in the vicinity of Sunset Crater (Schaber and Gummerman 1969; Berlin et al. 1977). Conceivably cinder and ash might also have functioned as a source of soil nutrition; however, such fertilization appears to proceed slowly in arid environments as compared with more mesic ones (Eicher 1957; Shields 1957; Berlin et al. 1977).

Therefore, if climatic conditions were warmer and with moisture predominantly summer in distribution, conditions would have been more favorable for cultivation of corn, beans, and squash, and at least corn pollen, which is produced in large quantities, would have become abundant in floor- and trash-fill spectra as the result of human transport of pollen-bearing plant parts. Such was the case (Figure 16.7a) at Mount Elden Pueblo as well as at other sites in northern Arizona that also exhibit increased relative abundance of economic pollen during this same period (Hevly 1964; Dickey 1971; Ward 1975). Unfortunately, these climatic conditions seem not to have persisted, and in many areas the finer cinders may even have been blown away, thus eliminating the special combinations of geological and climatic conditions favoring prehistoric agriculture and occupation of the Flagstaff area. It is also known that fertility of soils was depleted by prehistoric agriculture and 700 years has not been sufficient for pedogenic processes to restore fertility (Schaber and Gummerman 1969; Berlin et al. 1977).

## Fungal Spores, Human Parasite Eggs, and Disease

In the course of analysis of pollen from sediments of the study sites, numerous spores of fungi were recovered (Figure 16.2). The proportion of such spores in the samples of Elden Pueblo greatly exceeds the spore percentages in modern soil samples and many other previously examined archaeological samples. Though many of the spores could not be identified, some could be tentatively assigned to genera that are noted for decomposition of wood and other plant materials, as noted early by Dickey (1971). Studies by Hevly et al.

(1978) have shown a general trend of increasing spore density paralleling increasing organic content of samples. Such spores may ultimately prove useful, but much additional work needs to be done regarding techniques of identification.

Though spores may ultimately prove useful for paleoenvironmental reconstruction, it should be noted that some human parasite eggs sufficiently resemble fungal spores that they can be confused with them (Figure 16.8). Careful study of some organic objects originally identified as "possible fungal spores" showed them to be the eggs of well-known human parasites not infrequently recovered in Southwestern archaeological sites from human coprolites (Samules 1965; Fry and Moore 1969; Moore *et al*. 1969). Eggs of pinworm, tapeworm, and whipworm are essentially absent in early samples, but their relative abundance and incidence increase manyfold in the trash fill of rooms used as latrines during the final stage of occupancy of the pueblo (Figure 16.2). Though these parasites are not sufficiently detrimental to have caused death of the entire population or even to have caused serious disease in most instances (whipworm is sometimes associated with appendicitis), their increased incidence is nevertheless indicative of the general health of the population. Since parasitism generally increases under conditions of malnutrition and declining vigor of the population, one may conclude that at Elden Pueblo malnutrition and low population vigor were present in the final stages of occupancy. It would appear that disease may have accompanied abandonment of some Southwestern pueblos, as was suggested many years ago by Colton (1936).

## Climatic Changes and Sinagua Demography

The relationship of pueblo peoples to their environment has been studied by Hack (1942), and it is apparent that the pueblo people were and are dependent on the vagaries of nature for their survival. The crops that form the basis of their subsistence are hereditarily limited to those areas that provide the requisite conditions of effective moisture and temperature; thus the living areas of the peoples cultivating these crops are restricted. When climate changed so as to be unsuitable for their current mode of subsistence, the prehistoric aborigines were faced with three alternatives: (*a*) to change cultural practices to a type suitable for the new climate; (*b*) to continue current cultural practices with disastrous decline or extinction of local population; or (*c*) to move to a new area where current cultural practices could be continued. Examination of the cultural record reveals that extensive migration of pueblo peoples occurred, and comparison of the chronology of this movement with the chronology of environment change revealed by pollen and other studies (summarized in Table 16.1) suggests that a relationship may exist between the two.

Population growth and expansion of pueblo peoples into higher-elevation areas was not an event unique to the Flagstaff area (Woodbury 1961; Zubrow 1971). Therefore, the population explosion experienced by the Sinagua in the later 1000s and early 1100s was not necessarily the sole result of the deposition

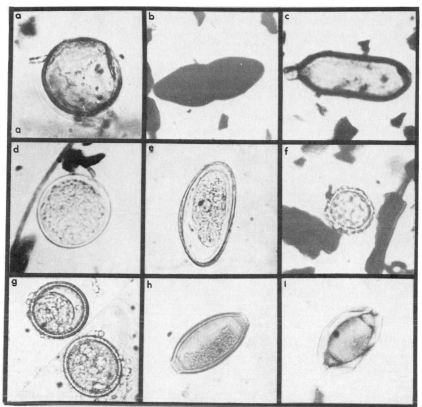

FIGURE 16.8 Parasite eggs and fungal spores recovered from sediments of Elden Pueblo following pollen extraction. Note the absence of protoplasmic content in the fungal spores and its persistence in the worm eggs. Fungal spores (a–c, I). (a) Probable sporangium or conidiophore, × 500. (b)–(c) spores of Ascomycetes or Basidiomycetes, × 2000. (i) spore of unidentified fungus, × 750. Parasitic worm eggs (d–h). d. Egg of Hymenolepid tape worm, probably *Hymenolepis* sp. × 720. Adult worms inhabit the small intestines of mice, rats, and humans, which become infected by ingesting infected grain beetles. (e) Egg of the pin worm *Enterobius vermicularis* × 720. This worm is a relatively common inhabitant of the large intestine of humans, who become infected by swallowing the egg. (f) Egg of *Ascaris lumbricoides* × 360. The adult worms occupy the small intestine of humans. Infection is by ingestion of the eggs. (g) Eggs of a Taeniid tapeworm, × 720. Taeniid tapeworms are parasites of the small intestine of carnivores such as dogs and omnivores such as humans which become infected by eating infected rabbits, deer, and other herbivores. h. Egg of the whipworm *Trichuris* sp., × 720. This roundworm lives in the caecum and large intestine of several mammals, including humans, which are infected by ingesting the eggs. The concentration and frequency of parasite eggs increases late in the occupancy of Elden Pueblo suggesting a deterioration of the general health of the human population. Frequency of occurence and concentration of fungal spores, which are most frequently derived from plant parasite or saprophytes, is greatest in the sediments with highest organic residues.

of cinder from the eruption of Sunset Crater, making formerly nonarable land suitable for cultivation—as has been hypothesized by Colton (1946)—but was probably the result of a fortuitous combination of favorable climatic and geologic events. Abandonment of the Flagstaff area in the latter half of the twelfth and early part of the thirteenth centuries likewise is not unique to the

**TABLE 16.1**
**Correlation of Climatic, Floristic, Geologic, and Demographic Events in the Flagstaff Area**[a]

| Chronology (A.D.) | Probable local floristic changes (pine–juniper and pine comp.) | Probable temperature trends (pine comp. ratio) | Probable seasonal moisture distribution | | | Probable geological trends | Probable demographic trends |
|---|---|---|---|---|---|---|---|
| | | | Autumn and Winter (pine–juniper) | Spring and Summer (economic NAP) | Summer and Autumn (economic NAP) | | |
| 1300 | Parkland | Cold | +[b] | ±[c] | ± | Stable or aggradation | Total abandonment of the Flagstaff area |
| 1275 | Savanna–woodland (decreasing agricultural success) | Cold | −[d] | + | − | Aeolian and alluvial degradation (or stable) | |
| 1250 | | Slightly warmer | ± | + | − | | |
| 1225 | | Cold | − | + | − | | |
| 1200 | | Cold | − | + | − | | |
| 1175 | | Warm | ± | + | − | | |
| 1150 | | Very warm | − | ± | ± | | |

| Year | Vegetation | Climate | | | Stable or aggradation | | Internal population growth |
|---|---|---|---|---|---|---|---|
| 1125 | Woodland–parkland | Like present | + | + | | + | |
| 1100 | | | | | | | |
| 1075 | Savanna–woodland (Increasing agricultural success) | Very warm | − | ± | Aeolian and alluvial degradation | + | |
| 1050 | | Very warm | − | ± | S.S. Crater eruption | + | |
| 1025 | | Like present | − | ± | Alluvial aggradation (or stable) | + | |
| 1000 | | Warm | ± | ± | | ± | |
| 975 | Savanna–woodland | Warm | + | ± | Alluvial aggradation (or stable) | ± | |
| 950 | | Cool but warming | − | ± | | ± | |
| 925 | Savanna–woodland | Cold | + | ± | Alluvial aggradation (or stable) | ± | |
| 900 | | Cold | + | ± | | ± | |

[a]Pollen indices utilized for probable floristic changes and inferred climatic trends are enclosed by parentheses. In addition to this study paleoecological data are derived from the following sources: tree-rings (Fritts 1965); pollen (Schoenwetter 1965; Bennett 1967; Wilson 1969); demography (Colton 1961; Pilles 1977); geology (Brady 1936; Cooley 1962; Karlstrom et al. 1974).

[b]A + sign indicates that precipitation of that season was greater than at present.

[c]A ± sign indicates that precipitation of that season was about the same as at present.

[d]A − sign indicates that precipitation of that season was less than at present.

Flagstaff area (Smiley 1961; Woodbury 1961) and thus need not be explained as the result of aeolian deflation of fine cinder (Colton 1961).

The combination of changing relative effective moisture, temperature, and/or seasonal distribution of precipitation undoubtedly had varying effects in different localities, depending on microhabitat variation and aboriginal utilization of local microhabitats. In some areas the effects of drought may have been more pronounced as the result of headward erosion of drainage systems, which lowered the water tables (cf. Bryan 1941), and the inability of the prehistoric people to obviate such microhabitat alterations by modification of water-control measures. Certainly abandonment need not have been synchronous, since abandonment for reasons of insufficient moisture could have occurred at any time, depending on local erosion and evapotranspiration rates, and could have been the initial stimulus for migration into higher-elevation areas. Abandonment of high-elevation sites was probably stimulated in the late twelfth century with the onset of a cooling trend that continued into and intensified during the fourteenth century. The xeric condition of the thirteenth century need not have been important to the prehistoric people because it could have been partially offset by lowering temperatures of that period (particularly for summer rain plants). However, if such drought were augmented by the effects of lowered water tables resulting from erosion now documented to have occurred at many localities particularly in the twelfth and thirteenth centuries (Cooley 1962; Karlstrom et al. 1974), the conditions were probably important.

It may be concluded that Jett's observation that "regional climatic difference cannot explain population shifts within the relatively homogeneous plateau region [1964, 1965b]" is not supported by a variety of data. The Colorado Plateau is not edaphically, climatically, or biotically homogenous today, and there is no reason to suggest that it was so in the past, since both pollen and tree-ring data do exhibit distinctive local variations through time. Furthermore, climatic variations occurred that affected local plant and animal communities quite profoundly and certainly appear to be correlated with aboriginal movements in the Flagstaff area. It might even be suggested that such influences at least indirectly may have provided a stimulus for cultural evolution (Schwartz 1957).

## Geoclimatic Stimuli, Biotic Migration, and Evolution

The geoclimatic events of the A.D. 900–1300 period are reflected not only by altered proportions of species in local plant communities but also by biotic migration (in addition to that of man) and evolution. One such recently recognized example is the flowering plant species *Penstemon clutii*. Since this species is restricted to the ash of Sunset Crater, its evolution must postdate the eruption of that volcano. The morphologically most similar, and therefore most likely ancestral, species occurs today at lower elevations to the south. It is probable that ancestral plants established themselves from wind-transported seeds on the black ash in the warm episode of the 1050–1200 period. Subsequent selection during the shift to cooler climates of the thirteenth century probably initiated shifts in the gene pool of the local population, resulting in its current ecological and reproductive isolation.

Studies of vertebrate remains recovered from archaeological sites in northern Arizona have primarily yielded evidence of taxa identical to those found at present in the study area (Table 16.2); however, some taxa were also found that do not occur locally at present or that are different anatomically from local populations presently occupying the area. Studies of bird remains from abandoned Indian dwellings in the Flagstaff area and elsewhere on the Colorado Plateau by Miller (1932), Hargrave (1939, 1970), and Rea (1973) have revealed not only a bird fauna similar to that of today, but also the occurrence of species no longer native to northern Arizona. Most of the exotic forms, such as the scaled quail reported by Rea (1973), presently occur at lower elevations south of the Mogollon Rim in Arizona. Their prehistoric occurrence in northern Arizona is concentrated in the 1050–1300 period (some occurrences may be traced to the tenth century). Their prehistoric occurrence in the Flagstaff area tends to substantiate the existence of savanna and woodland conditions where pine parkland now occurs as well as the warmer temperatures inferred from the pollen data of Elden Pueblo.

Studies of mammal remains, as with the birds, have revealed a fauna primarily composed of species native to the study area. Of the 18 taxa recovered from Elden Pueblo, 3 are no longer found in the area today: Sonoran mud turtle, humpbacker sucker , and bighorn sheep. Bighorn sheep were known historically in the Flagstaff area, where they have probably occurred since prehistoric times only to be eventually exterminated by man. The Sonoran mud turtle and the humpback sucker presently occur in the valleys of the Verde and Little Colorado Rivers, 606–909 m lower than Elden Pueblo and 88 km south and north of it, respectively. The small number of bones of the turtle and fish suggest that the species were either not used if locally present in the past or were actually present at considerable distance from Elden Pueblo and brought to it. Since the items were commonly used by the prehistoric occupants of these river valleys, the latter hypothesis is reasonable. However, it should be noted that the occurrence of whipworm eggs suggests that the prehistoric inhabitants of Mount Elden Pueblo probably visited such warm, moist environments, which may have been located as close as Oak Creek Canyon only 40 km south of Mount Elden, during the 1150–1300 warm period.

Prehistoric occurrences of vertebrates are not, however, always best explained as the result of human transport or historic local extinction. For example, Stein (1963), Lawrence (1951), and Diggs (1979) have documented the prehistoric occurrence of bison, a grassland species not known historically in the state, at several localities in east-central Arizona. Likewise, Ward (1975) has documented the prehistoric occurrence at Hay Hollow Valley (east-central Arizona) of Mohave rattlesnake, a species currently occurring only at lower elevations west of the Flagstaff area. The occurrence of these species is contemporaneous with the warmer and occasionally more mesic interval noted

**TABLE 16.2**
**Vertebrate Remains from Elden Pueblo**[a]

| Name of animal | Room 35 Fill | Room 36 Fill | Room 37 Fill | Room 38 Fill | Room 39 Fill |
|---|---|---|---|---|---|
| *Fish*: | | | | | |
| Gila sucker | + | | | | |
| (*Catstomus insignis*)* | | | | | |
| *Reptiles*: | | | | | |
| Sonoran mud turtle | | | | + | |
| (*Kinosternon hemionus*)* | | | | | |
| *Mammals*: | | | | | |
| 1. Mule deer | + | + | + | + | + |
| (*Odocoileus hemionus*) | | | | | |
| 2. Bighorn Sheep* | | + | + | | |
| (*Ovis canadensis*) | | | | | |
| 3. Pronghorn antelope | | | | | + |
| (*Antilocapra american*) | | | | | |
| 4. Coyote | + | | | | |
| (*Canis latrans*) | | | | | |
| 5. Bobcat | + | | | | |
| (*Lynx rufus*) | | | | | |
| 6. Black bear | + | + | | | |
| (*Euarcto americanus*) | | | | | |
| 7. Jack rabbit | + | + | + | + | + |
| (*Lepus californicus*) | | | | | |
| 8. Cottontail rabbit | + | + | + | + | + |
| (*Syvilagus auduboni*) | | | | | |
| 9. Squirrel | + | + | + | | |
| (*Sciurus sp.*) | | | | | |
| 10. Rock squirrel | + | | + | | |
| (*Citellus sp.*) | | | | | |
| 11. Pocket gopher | + | | | + | |
| (*Thomomy's bottii*) | | | | | |
| *Birds*: | | | | | |
| 1. Hawk | + | | | | |
| (*Buteo sp.*) | | | | | |
| 2. American kestrel | + | | | | |
| (*Falco sparverium*) | | | | | |
| 3. Raven | + | | + | | |
| (*Corvus corax*) | | | | | |
| 4. Crow | + | | + | | |
| (*Corvus brachrhynchus*) | | | | | |
| 5. Lewis woodpecker | + | | | | |
| (*Asyndemus lewis*) | | | | | |

[a]Identifications by S. Olsen and A. Rea (University of Arizona) and C. Minkley (Northern Arizona University). Adapted from file reports to the Museum of Nothern Arizona. Species indicated by asterisk are not presently native to the Flagstaff area.

above for the range expansion of other animals. The number and types of bones argue against long-distance transport. Hence, the prehistoric occurrence of bison and Mohave rattlesnake in these areas is best viewed as reflecting altered environmental conditions that permitted the expansion of the range of these species far beyond their present occurrence. Prehistoric regional extinction resulting from cultural or climatic factors ultimately followed.

Just as the pollen data indicate variation of past climatic conditions, so also do the vertebrate data. One bird record, that of the magpie (Hargrave 1939), a species presently known only in the northeastern portion of Arizona, may be indicative of the more mesic conditions noted in the pollen record of about 1125. Similarly, statistical comparison of cranium size and dentition patterns of several small modern and fossil rodents at Wupatki National Monument (48 km north and 667 m lower than Elden Pueblo) indicate possible cooler temperatures and more mesic conditions between 1073 and 1230 (Lincoln 1962). Cooling temperatures are indicated in the fossil pollen record after 1150 but less mesic conditions than present were prevalent during this entire period except about 1125, as noted above.

Small rodent remains not only may confirm the pollen data, but at the same time may provide possible data on rates of evolutionary adaptation to changing environments during the period of occupation of Wupatki Pueblo. If it could be demonstrated that changes of cranium size and dentition pattern of the small rodents noted by Lincoln (1962) actually occurred during occupancy (and not just during the past 800 years), it would be possible to suggest that evolutionary adaptation occurred within a relatively brief period perhaps as the result of isolation by cinder and ashfall from Sunset Crater of small, rapidly reproducing populations. (The larger body size, indicated by a larger cranium than that of the present, probably was not environmentally adaptive to these rodents during the interval of increased temperatures prior to 1150, if in fact they occupied the Wupatki area during that period.) Similarly, the dentition pattern of the prehistoric rodents, which differs from that of their contemporaneous desert relatives, may reflect prehistoric adaptation to food sources differing from those of the present as the result of modified edaphic conditions following deposition of cinder from Sunset Crater or of altered climatic conditions. It may be suggested that environmental changes of greater magnitude, duration, and complexity than previously recognized from tree-ring and geomorphological studies have occurred in the past millennium in the Flagstaff area, and that, although of geologically short duration, they were significant in biotic migration and possibly even evolution.

## SUMMARY AND CONCLUSIONS

The volcanic field east of the San Francisco Peaks near Flagstaff, Arizona, is extensive and contains several recently active volcanoes, of which the eruption of one, Sunset Crater, has been tree-ring-dated at 1066. This same volcanic field was intensively occupied by man for several millennia, and survey data clearly indicate some increase in the density of the human population of the Flagstaff area beginning in the eleventh century, culminating about 1150, and abruptly declining in the thirteenth century (Pilles, Chapter 15 in this volume). It has been speculated that population increase by migration and internal growth was stimulated by favorable edaphic conditions resulting from the deposition of cinders from Sunset Crater. Abandonment, on the other hand, was suggested to have been necessitated by wind deflation of fine, water-retaining cinder and depletion of soil nutrition through heavy cropping (Colton 1961).

These population trends were thought to have been peculiar to the Flagstaff region or at least exaggerated to some extent beyond those found elsewhere on the Colorado Plateau. However, recent surveys indicate that the demographic trends described above were present at other localities where coincident volcanic events are absent (Hack 1942; Longacre 1964). Furthermore, the trend of increasing population found in the Flagstaff region is not exceptional in magnitude, except in areas of relatively shallow Sunset Crater ash deposit, and was in fact initiated prior to the eruption of Sunset Crater (Pilles, Chapter 15 in this volume). These findings suggest that population movements to higher elevations and internal growth of populations were initiated in response to factors other than modified edaphic conditions resulting from deposition of cinder from Sunset Crater. Furthermore, it must be concluded that the population growth of the twelfth century resulted not solely from cinder mulch creating favorable agricultural conditions over a larger area than formerly existed. Such observations do not negate the effects of volcanism, for certain local impacts would have been inevitable. Lava flows and deep cinder deposits would have made some formerly nonarable land arable so long as the proper thickness and texture of cinder and ash mulch persisted. While not negating the beneficial effects of cinder mulch, which are well documented and were also known by the prehistoric agriculturalists, the above observations do indicate the necessity of evaluating other factors when attempting to explain the observed demographic phenomena.

Other such factors include man's impact on the local environment as well as the effects of climatic change. Elucidation of the paleoenvironments of the prehistoric people inhabiting the Flagstaff area prior to and following the eruption of Sunset Crater was attempted utilizing a variety of paleontological data obtained from archaeological sites located in the east San Francisco volcanic field.

The principle types of data recovered were pollen and macroscopic plant and animal remains. These were supplemented by studies of wood types and tree-growth records, parasite eggs, soil chemistry of aboriginal fields, and studies of the effects of modern volcanism on vegetation.

Macroscopic animal and plant remains are not abundant in the open sites of the Flagstaff area and those of plants occur almost exclusively in a charred condition. Animal remains are those of taxa native to the Flagstaff area today except for the occurrence of bighorn sheep, Mexican macaw, scaled quail, Sonoran mud turtle, and humpback sucker, all of which occur today or historically at lower, warmer elevations. Macroscopic plant remains fall into three categories: cultivated (e.g., beans, corn, cotton, and squash), nonwoody wild (e.g., prickly pear, yucca, pigweed, and lamb's quarters), and woody types (e.g., juniper, pine, fir, and aspen). All species except cotton may be found in the Flagstaff area today; however, the use of piñon pine and juniper increases after the eruption of Sunset Crater. Examination of the tree-growth record reveals a generally long period of deficient growth interrupted on only a few occasions by average or above-average growth, suggesting generally warmer, drier climates from the tenth through the thirteenth century than now prevail in the Flagstaff area.

Palynological data was used to reconstruct past vegetation and patterns of pollen production to elucidate probable trends of temperature, as well as the amount and seasonal distribution of effective moisture. Reconstruction of past vegetation and effective moisture trends was attempted utilizing AP–NAP and pine–juniper ratios. All AP–NAP ratios were indicative of desert environments when compared with modern pollen spectra. Since the AP–NAP and pine–juniper ratios generally yield parallel inferences, it is apparent that NAP is generally overrepresented in the archaeological pollen record of the Flagstaff area. It is apparent, nevertheless, that the local vegetation was generally that more typically found at slightly lower, warmer elevations (e.g., woodland–savanna) rather than the present ecotone of forest and woodland. Thus, the pollen record, vertebrate bone record, and tree-ring records reflect parallel fluctuations from the tenth through the thirteenth century. These may be interpreted as long-term trends of effective moisture influencing pollen production and tree growth, since they occur within a time interval less than that necessary for establishment of pollen-producing trees. The most mesic of these brief intervals occurred in the first half of the twelfth century, whereas the most xeric occurred in the last half of the thirteenth century.

The great overrepresentation of nonarboreal pollen in the fossil pollen spectra from the archaeological sites suggests plants gathered and utilized by man. These plants have been shown by phenological observations to flower in a seasonal sequence apparently dependent upon late-spring to early-autumn rains. If so, the change of proportion of these types in the fossil pollen spectra could well reflect a shift from predominantly mid- and late-summer moisture to a late-spring and early-summer moisture regimen about the middle of the A.D. 900–1300 period. Moisture may therefore have been inadequate for maturation of summer plants having a long growing season, such as corn and squash. It is therefore not surprising to find corn pollen well represented only in the first half of the fossil pollen record.

Examination of the arboreal pollen record also indicates that yellow pine pollen was being produced less abundantly than piñon pine in the late tenth through the twelfth century but that this trend was reversed during the thirteenth century. Both temperature and moisture are known to affect pollen production and tree growth. The effects of moisture on pine pollen production and tree growth have been addressed earlier, and the comparison with various Northern Hemisphere temperature records indicates that pine composition phenomena are coincident with numerous documents of elevated temperatures during the eleventh and twelfth centuries. Hot, dry, late-spring winds are known to cause abortion of pine pollen. If this occurred during the pollen production season of yellow pine and not of piñon pine, the peculiar pine pollen proportion data would be explicated and also yield useful seasonal temperature data. If correct, such data would suggest a trend of earlier or greater warming in the late spring (a lengthening of the growing season) beginning about A.D. 950–1000 and a reversal of this trend (a shortening of the growing season) beginning about A.D. 1150–1200. For economic plants such as corn and squash that require warm temperatures and long growing seasons, shortening of the growing season could have been disastrous, and it is therefore not surprising to find the proportions of corn pollen diminishing in the thirteenth century.

It would appear that prior to A.D. 1000 there was already a sizable population in the Flagstaff area residing in a woodland- and savanna-like environment. The occurrence of this woodland- and savanna-like environment appears unique to the Flagstaff area at that time and elevation on the Colorado Plateau and probably reflects the edaphic conditions resulting from several millennia of volcanic activity in the east San Francisco volcanic field. Episodic drought, as well as warming temperatures beginning about this time and culminating in 1150 may well have stimulated movement of people to higher elevations. This demographic trend was probably further stimulated by erosion and lowering of water tables, which also was common in the 1000–1300 period. Deposition of cinder mulch resulting from the eruption of Sunset Crater was simply a coincidental event that was significant in determining local settlement patterns. Internal growth of the population was probably occasioned by favorable agricultural conditions resulting from the brief coincidence (1120–1140) of warmer springs (longer growing seasons), increased effective moisture residual from the previous year, and adequate mid- and late-summer moisture. The effect of population growth in the Flagstaff area would have been largely detrimental as trees were cut for construction and firewood and nutritional levels of arable soils diminished. A reversal of the above favorable climatic trends resulted in reduced carrying capacity and movement of people from the Flagstaff area. Exhaustion of soil nutrition and wind deflation of cinder mulch from marginal soils accelerated the abandonment of the area, which was complete by 1300. At least at Elden Pueblo the incidence of parasitic disease increased during the final stages of occupation and is an indication of generally declining health of the population coincident with deterioration of environmental conditions essential for the prehistoric agricultural communities.

Therefore, at least in the Flagstaff area, the archaeologically demonstrated variation of density, settlement pattern, and movement of prehistoric populations appears coincident with a number of geological, paleontological, and paleobotanical indications of environmental changes. Such changes include episodic alluviation and erosion, volcanism, and altered floral and faunal distributions. Tree-ring and pollen studies when integrated with the above data strongly suggest that climatic change must be regarded as the ultimate cause of the given demographic trends. More proximate causes would be (a) altered local edaphic conditions modifying availability of moisture (e.g., elevation, slope and exposure, volcanism, erosion); (b) altered soil nutrition; and (c) fluctuating temperature, all of which affect the carrying capacity of a locality. Human beings could have been in part responsible for erosion through removal of local vegetation and other cultural activities, but their effect was probably only an acceleration or exaggeration of climatically induced trends necessitating shifting of field locations or ultimate abandonment of an area. Humans probably also were responsible for mineral depletions of some soils through improper cropping techniques. Volcanism probably modified on a very local scale the availability of moisture, leading to specific settlement densities and patterns. The ultimate and major stimulating factor for demographic movement, however, was the carrying capacity of the locality as influenced by temperature and effective moisture. These same factors were also responsible for population growth and decline. Because temperature and available moisture vary from locality to locality, there is no reason to assume all populations will respond in exactly the same way or at the same trend. When they do, however, the overriding significance of climatic change over local conditions is even more strongly implicated. Volcanism without doubt exerted strong local impacts that, because of the persistence of basalt and cinder, prevailed for many centuries in the semi-arid environments of the Southwest (Figure 16.8).

## REFERENCES

Adam, D. P., C. W. Fergusson, and V. C. Lamarch, Jr.
   1967   Enclosed bark as a pollen trap. *Science* 157:1067–1068.
Arakawa, H.
   1966   Addenda to climatic change as revealed by the data from the Far East. In *Pleistocene and post pleistocene climatic variations in the Pacific area*, edited by D. I. Blumenstock. *Honolulu: Museum Press*.
Beaman, J. H.
   1961   Vascular plants on the cinder cone of Parícutin Volcano in 1960. *Rhodora* 63(756):340–344.
Bennett, P. S.
   1967   Pollen and soil analysis at I:14:30. File Report to Roger Kelley. Museum of Northern Arizona.
Bergthorsen, E.
   1969   In R. A. Bryson, A perspective on climatic change. *Science* 184:753–760.
Berlin, G. L., J. R. Ambler, R. H. Hevly, and G. G. Schaber
   1977   Identification of a Sinagua agricultural field by aerial thermography, soil chemistry, pollen plant analysis, and archaeology. *American Antiquity* 42(3):588–600.

Bohrer, V. L.
    1972   Paleoecology of the Hay Hollow site, east of Snowflake, Arizona. *Fieldiana*
           (Anthropology) 63(1):1–30.
Bradfield, M.
    1973   Rock cut cisterns and pollen "rain" in the vicinity of Old Oraibi, Arizona. *Plateau*
           46(2):69–71.
Brady, L. F.
    1936   The arroyo of the Rio De Flag. *Museum Notes* (Museum of Northern Arizona) 9(6):33–
           37.
Bryan, Kirk
    1941   Pre-Columbian agriculture in the Southwest, as conditioned by periods of alluviation.
           *Annual Association of American Geologists* 31(4):219–242.
Bryson, R. A.
    1975   The lessons of climatic history. *Environmental Conservation* 2(3):223–230.
Bryson, R. A., and W. M. Wendland
    1967   Tentative climatic patterns for some late glacial and post-glacial episodes in Central
           America. In Life, land and water, edited by W. J. Mayer-Oakes. Winnipeg: University of
           Manitoba Press.
Colton, H. S.
    1936   Rise and fall of the prehistoric population of Northern Arizona. *Science* 84(2181):337–343.
    1946   The Sinagua. *Museum of Northern Arizona Bulletin* (22).
    1961   *Black sand*. Albuquerque: University of New Mexico Press.
    1962   Archaeology of the Flagstaff area. In *New Mexico Geological Society 13th Field Confer-
           ence guide book of the Mogollon Rim region, east-central Arizona*, edited by R. H. Weber
           and H. W. Peiree.
    1965   Experiments in raising corn in the Sunset Crater ashfall area east of Flagstaff, Arizona.
           *Plateau* 37(3):77–79.
Conner, S.
    1943   Excavations at Kinnikinnick, Arizona. *American Antiquity* 8:376–379.
Cooley, M. E.
    1962   Late Pleistocene and recent erosion and alluviation in parts of the Colorado River system,
           Arizona and Utah. U.S.G.S. Proffessional Paper 450B:48–50.
Currier, P. J., and R. O. Kapp
    1974   Local and regional pollen rain components at Dairs Lake, Montcalm County, Michigan.
           *Michigan Academician* 7(2):211–225.
Dansgaard, W., S. J. Johnsen, and J. Møller
    1969   One thousand centuries of climatic record from Camp Century on the Greenland Ice
           Sheet. *Science* 116:377–381.
Daubenmire, R. F.
    1960   A seven-year study of cone production as related to xylem layers and temperature in
           *Pinus ponderosa*. *American Midland Naturalist* 64:189–193.
Davis, E. M.
    1965   Small pressures and cultural drift as explanations for abandonment of the San Juan area,
           New Mexico and Arizona. *American Antiquity* 30:353–355.
Dickey, A. M.
    1971   Palynology of Hay Hollow Valley. Master's thesis, Dept. of Biological Sciences, Northern
           Arizona University.
Diggs, R.
    1979   Prehistoric exploitation of a microenvironment in east-central Arizona, Master's thesis,
           Dept. of Anthropology, Northern Arizona University.
Douglass, A. E.
    1929   The secret of the Southwest solved by talkative tree-rings. *The National Geographic
           Magazine* (December).
Dunkley, John
    1973   Report to Wilson Foundation. File report at Museum of Northern Arizona, Flagstaff.

Eggler, W. A.
1948  Plant communities in the vicinity of the volcano El Parícutin, Mexico, after two and a half years of eruption. *Ecology* 29(4):415–436.
1959  Manner of invasion of volcanic deposits by plants, with further evidence from Paricutín and Jorullo. *Ecological Monographs* 29(3):268–284.
1966  Plant succession on the recent volcano Sunset Crater. *Plateau* 38(4):81–96.
1967  Influence of volcanic eruptions on xylem growth patterns. *Ecology* 48(4):644–647.
Eicher, G. J. Jr., and G. A. Rounselfell
1957  Effects of lake fertilization by volcanic activity on abundance of salmon. *Limology and Oceanography* 1(2):70–76.
Ellis, F. H.
1964  Comment on Jett's "Pueblo Indian migrations." *American Antiquity* 30(2):213–215.
Euler, R. C., G. J. Gummerman, T. N. V. Karlstrom, J. S. Dean, and R. H. Hevly
1979  Cultural dynamics and paleoenvironmental correlates on the Colorado Plateaus. *Science* in press.
Faegri, K., and J. Iversen
1964  *Textbook of modern pollen analysis.* Copenhagen: Munksgaard.
Fry, G. F., and J. G. Moore
1969  *Enterobius vermicularis*: 10,000 year old human infection. *Science* 166:1620.
Fritts, Fritts, H. C.
1965  Tree-ring evidence for climatic changes in western North America. *Monthly Weather Review* 93(7):421–443.
Gregory, H. B.
1917  Geology of the Navajo country. U.S.G.S. Professional Paper No. 93.
Hack, J. T.
1942  The changing physical environment of the Hopi Indians of Arizona. *Papers Peabody Museum, Harvard University* 35(1):1–87.
Hargrave, L. L.
1939  Bird bones from abandoned Indian dwellings in Arizona. *Condor* 41:206–210.
1970  Mexican macaws: Comparative osteology and survey of remains from the Southwest. *University of Arizona Anthropology Paper Number* 20:1–20.
Havinga, A. J.
1971  An experimental investigation into the decay of pollen and spores in various soil types. In *Sporopollenin*, edited by J. Brooks *et al.* New York: Academic Press. 446–478.
Hevly, R. H.
1964  Pollen analysis of lacustrine and archaeological sediments from the Colorado Plateau. Doctoral dissertation, Dept. of Botany, University of Arizona.
1968a  Studies of the modern pollen rain in northern Arizona. *Journal of Arizona Academy of Science* 5(2):116–127.
1968b  Sand dune cave pollen studies. In A. Lindsay *et al.*, Survey and excavation north and east of Navajo Mountain, Utah, 1959–1962. *Museum of Northern Arizona Bulletin* 45:393–399.
1970  Botanical studies of sealed storage jar cached near Grand Falls, Arizona. *Plateau* 42(4):150–156.
In preparation  Paleoecological studies of archaeological sites: East central Arizona, A.D. 500–1500.
Hevly, R., R. Berry, and R. Schley
In preparation  Sediment and pollen studies of Sunset Crater cinder, Citadel Sink, Wupatki National Monument, Arizona.
Hevly, R. H., M. L. Heuett-Graff, and S. J. Olsen
1978  Paleoecological reconstruction from an upland Patayan rock shelter, Arizona. *Journal of Arizona Academy of Science* 13:67–78.
Hevly, R. H., and L. E. Renner
1979  Atmospheric pollen and spores in Flagstaff, Arizona. *Research Reports, Museum of Northern Arizona.* #18.

Hill, J. N. and R. H. Hevly
  1968   Pollen at Broken K Pueblo: Some new interpretations. *American Antiquity* 33(2):200–210.
Jamesen, D. A.
  1969   Rainfall patterns on vegetation zones in Northern Arizona. *Plateau* 41(3):105–111.
Jenkins, M. T.
  1941   Influence of climate and weather on growth of corn. *Yearbook of agriculture.*:308–320.
Jett, S. C.
  1964   Pueblo Indian migrations: An evaluation of the possible physical and cultural determinates. *American Antiquity* 39(3):281–300.
  1965a  Comment on Davis's hypothesis of Pueblo Indian migrations. *American Antiquity* 31(2):376–377.
  1965b  Reply to Ellis's "Comment 'on' Pueblo Indian migrations." *American Antiquity* 31(1):116–118.
Jones, M. D., and L. C. Newell
  1946   Pollination cycles and pollen dispersal in relation to grass improvement. *University of Nebraska College of Agriculture Experiment Station Research Bulletin* 148:1–43.
Jones, V. H.
  1941   The plant materials from Winona and Ridge Ruin. *Museum of Northern Arizona Bulletin* No. 18:295–300.
Kargalitadze, N. A.
  1971/1972   The history of forests of the northwestern part of Trialeti Range in the Holocene according to pollen analysis. *Journal of Palynology* 7:69–75.
Karlstrom, T. N. V., G. Gummerman, and Robert Euler
  1974   Paleoenvironmental and cultural changes, Black Mesa region, Arizona. In *Geology of northern Arizona*, edited by T. N. V. Karlstrom *et al*. Flagstaff: Geological Society of America (Rocky Mountain Section Meeting).
Kelly, R. E.
  1971   Diminishing returns twelfth and thirteenth century Sinagua environmental adaptation in north central Arizona. Doctoral dissertation, Dept. of Anthropology University of Arizona.
King, D. S.
  1949   Nalakihu: Excavations at a Pueblo III site on Wupatki National Monument, Arizona. *Museum of Northern Arizona Bulletin* No. 23.
Kummel, B., and D. Raup
  1965   *Handbook of paleontological techniques*. San Francisco: Freeman.
LaMarch, V. C., Jr.
  1974   Paleoclimatic inferences from long tree-ring records. *Science* 183:1043–1048.
Lamb, H. H.
  1966   *The changing climate*. London: Methuen.
Lawrence, B.
  1951   Mammals found at Awatove Site. *Papers of the Peabody Museum, Harvard University*. 35:1–44.
Leiberg, J. B., T. F. Rixon, and A. Dodwell
  1904   Forest conditions of the San Francisco Mountains Forest Reserve, Arizona. U.S.G.S. Professional Paper No. 22.
Leopold, L. B., E. B. Leopold, and F. Wendorf
  1963   Some climatic indicators in the period A.D. 1200–1400 in New Mexico. In *UNESCO and world meteorological organization symposium on changes of climate with special reference to the arid zone, Rome, 2–7 October, 1961*, Arid Zones Research Series No. 20. Paris: UNESCO.
Lester, D. T.
  1967   Variation in cone production of red pine in relation to weather. *Canadian Journal, Botany* 45:1683–1691.
Lincoln, E. P.
  1962   Mammal remains from Wupatki ruin. *Plateau* 34(4):129–134.

Lindsey, A. D.
  1951  Vegetation and habitats in a Southwestern volcanic area. *Ecology Monograph* 21:227–253.
Longacre, W. A.
  1964  A synthesis of Upper Little Colorado prehistory, eastern Arizona. In Martin *et al.*,
         Chapters in the prehistory of eastern Arizona, II. *Fieldiana* (Anthropology) 55:1–262.
Lovelius, N. V.
  1970  The effect of volcanic activity on the vegetation of Kamchatka (in Russian). *Bot ZH*
         55(11):1630–1633.
Malde, H.
  1964  The ecological significance of some unfamilar geological processes. In *The reconstruction
         of past environments*, edited by J. Hestor and J. Schoenwetter. #3. Taos. N. M.: Ft.
         Burgwin Research Center Publication No. 3.
Martin, P. S., and W. Byers
  1965  Pollen and archaeology at Wetherill Mesa. *American Antiquity* 31(2, Part 2):122–135.
Martin, P. S., and J. Schoenwetter
  n.d.  Pollen stratigraphy of a great kiva from Chaco Canyon.
McKelvey, P. J.
  1954  Forest colonization after recent volcanic activity at West Taupo. *New Zealand Journal of
         Forestry* 6(5):435–448.
Miller, Alden H.
  1932  Bird remains from Indian dwellings in Arizona. *Condor* 34:138–139.
Moore, J. B., G. F. Fry, and E. Englert Jr.
  1969  Thorny-headed worm infection in North American prehistoric man. *Science* 163:1324–
         1325.
Nicholls, J. L.
  1959  The volcanic eruption of Mt. Tarawera and Lake Rotomahana and effects on surround-
         ing forests. *New Zealand Journal of Forestry* 8(1):133–142.
Oswalt, W. M.
  1957  Volcanic activity and Alaskan spruce growth in A.D. 1783. *Science* 126:928–929.
Pearson, G.
  1950  Management of ponderosa pine in the Southwest. U.S.D.A. Managr. #6.
Pilles, P. J.
  1977  The field house and Sinagua demography. In *Limited activity and occupation sites: A
         collection of papers, edited by A. E. Ward. Albuquerque: Cont. to Anthr. St. 1. Center for
         Anthropological Studies*.
Potter, L
  1967  Differential pollen accumulation in water tanks and adjacent soils. *Ecology* 48(6):1041–
         1013.
Rea, A. M.
  1973  The scaled quail (*Callipepula squamata*) of the Southwest: Systematic and historical
         consideration. *Condor* 75(3):322–329 ,
Reed, E.
  1944  The abandonment of the San Juan region. *El Palacio* 51(4):61–74.
Rixey, R. and C. B. Voll
  1962  Archaeological materials from Walnut Canyon cliff dwellings. *Plateau* 34:85–96.
Robinson, W. J., and J. S. Dean
  1969  Tree-ring evidence for climatic changes in the prehistoric Southwest from A.D. 1000 to
         1200. 1967–1968 Report to the National Park Service, Dept. of Int., Wash. D. C. from the
         Lab. Tree-Ring Res., Univ. of Arizona, Tucson.
Robinson, W. J., B. G. Harrill, and R. L. Warren
  1975  *Tree-ring dates from Arizona I-Flagstaff area*. Tucson: Tree-ring Laboratory, University of
         Arizona.
Roeser, J. Jr.
  1941  Some aspects of flower and cone production in ponderosa pine. *Journal of Forestry*
         39:534–536.

Samules, R.
    1965   Parasitological study of long dried fecal samples. *American Antiquity* 31(2, Part 2):175–179.
Schaber, G. G., and G. J. Gumerman
    1969   Infrared scanning images—An archaeological application. *Science* 164:712–713.
Schell, I. J.
    1967   Sea ice—Climatic change. In *Encyclopedia of Atmospheric Sciences and Astrogeology*, edited by R. W. Fairbridge. New York: Reinhold.
Schoenwetter, J.
    1965   Pollen analysis at Arizona 1:15:18. File Report to R. Kelley. Museum of Northern Arizona.
    1970   Archaeological pollen studies of the Colorado Plateau. *American Antiquity* 35(1):35–48.
Schoenwetter, J., and A. E. Dittert, Jr.
    1968   An ecological interpretation of Anasazi settlement patterns. *Anthropological Archaeology in the Americas*. Washington D.C.: The Anthropological Society.
Schoenwetter, J., and F. Eddy
    1964   Alluvial and palynological reconstruction of environments, Navajo reservoir district. *Navajo Project Studies* XI: 1–155.
Schwartz, D. W.
    1957   Climatic change and cultural history in the Grand Canyon region. *American Antiquity* 22(4):372–377.
Sears, P. S.
    1961   Palynology and the climatic record of the Southwest. In Solar variations, climatic change and related geophysical problems, edited by R. W. Fairbridge. *Annals of the New York Academy of Science* 95(1):632–641.
Shields, L. M.
    1957   Algal and lichen floras in relation to nitrogen content of certain volcanic and arid range soils. *Ecology* 38(4):661–663.
Shoulders, E.
    1967   Fertilizer application, inherent fruitfulness, and rainfall affect flowering of longleaf pine. *Forest Science* 13:376–383.
Simmons, I. G.
    1969   Evidence for vegetation changes associated with mesolithic man in Britain. In *Domestication and exploitation of plants and animals*, edited by P. J. Ucko and G. W. Dimbleby. Chicago: Aldine.
Smiley, T. L.
    1961   Evidences of climatic fluctuations in Southwestern prehistory. In Solar variations, climatic change and related geophysical problems, edited by R. W. Fairbridge. *Annals of New York Academy of Science* 95(1):697–704.
Solomon, A. M.
    1976   Pollen analysis of alluvial sediments: Implications of experimental evidence from sedimentary and atmospheric pollen samples. *American Quaternary Association Abstracts*:159.
Stanislawski, M. B.
    1963   Wupatki Pueblo: A study in cultural fusion and change in Sinagua and Hopi prehistory. Doctoral dissertation, Dept. of Anthropology, University of Arizona.
Stein, T.
    1963   Mammal remains from archaeological sites in the Point of Pines region, Arizona. *American Antiquity* 29(2):213–220.
    1964   Comparison and analysis of modern and prehistoric tree species in the Flagstaff area, Arizona. *Tree-ring Bulletin* 36(1–4):6–12.
Stewart, O. C.
    1956   Fire as the first great force employed by man. In *Man's role in changing the face of the Earth*, edited by Thomas, W. L. Chicago: University of Chicago Press.

Tauber, Henrik
1967    Investigations of the mode of pollen transfer in forest areas. *Rev. Palaebot. Palynol.*
        3:277–286.
Treshow, M.
1972    *Environment and plant response.* New York: McGraw-Hill.
Ward, J.
1975    Intra-site subsistence at the joint site as a function of paleoclimate. Master's thesis, Dept.
        of Biological Science, Northern Arizona University.
Wilson, J. P.
1969    The Sinagua and their neighbors. Doctoral Dissertation, Dept. of Anthropology, Harvard
        University.
Woodbury, R. B.
1961    Climatic changes and prehistoric agriculture in the southwestern United States. In Solar
        variations, climatic change, and related geophysical problems, edited by R. W. Fair-
        bridge. *Annals of New York Academy of Science.*
Zubrow, E.
1971    Carrying capacity and dynamic equilibrium in the prehistoric Southwest. *American
        Antiquity* 36(2):127–138.

# 17

# Environmental and Cultural Effects of the Ilopango Eruption in Central America

PAYSON D. SHEETS

## INTRODUCTION

Intermittent geological and archaeological research conducted during the past 60 years in El Salvador, Central America, has indicated that a natural disaster in the form of a large volcanic eruption occurred some 2000 years ago (summarized in Sheets 1976). The evidence consists of a number of sites where artifacts of the Late Preclassic Maya are buried by volcanic ash. Until recently it was not known whether these sites were buried by ash from separate eruptions or from a single eruption, and guesses as to when the eruption or eruptions occurred ranged from 1400 B.C. to well within the Christian era.

During 1975–1976, a pilot project sponsored by the National Science Foundation demonstrated that the ash layers burying highland sites are not spatially and temporally isolated phenomena, but are part of a single, complex eruption. The source, according to the German Geological Mission to El Salvador, is Volcán Ilopango, and our data are in agreement with their findings (Steen-McIntyre 1976, 1978).

Not much is known about the time elapsed between the components of the eruption, the ash flows, and the airfall units other than that it was brief. Basic data from fieldwork and labwork, including determination of the petrographic and chemical "fingerprint" of this *tierra blanca* tephra, are needed in order to trace the volcanic impact over the Maya highlands and adjacent areas. Based on the observed repercussions of a number of historic eruptions (cf. Sheets 1976), there were three zones of environmental–social impact: (*a*) a

VOLCANIC ACTIVITY AND HUMAN ECOLOGY

zone of lethal damage where biotic survival was not possible, such as the ash-flow areas; (b) a zone of ecologic impact too severe for contemporary technology to cope with the suddenly changed circumstances, but where people could have physically survived the tephra fall; and (c) a zone of significant environmental impact by the airfall tephra, but where adaptive adjustments allowed for continued sedentary, agriculturally based occupation. The predominant source area for migrants would have been the second zone, and a conservative estimate of the area of this zone based on fragmentary evidence, is 8000 km². A minimal population density figure would have been 40 people/km², yielding 320,000 as a very conservative estimate of potential emigrants.

Airfall ash deposits almost a meter thick were encountered as far as 77 km from Ilopango, and ash-flow (or base-surge) deposits as far as 43 km indicate a geological event of considerable magnitude. It might rank with the major historically known explosive eruptions, such as Thera (see Renfrew, Chapter 18 in this volume), Karakatoa, Mount Pelée, and Cosegüina (Bullard 1976). Because the area affected apparently was densely inhabited and subsistence had been agriculturally based for more than a millennium, this extreme geological event must have been a major disaster for the highland Maya.

Sufficient cases of natural disasters have been recorded for Mileti *et al.* (1975:61) to compile a set of factors that isolate major disasters from those of lesser impact. The most disruptive events are characterized by (a) suddenness of occurrence, (b) high uncertainty, (c) prolonged duration, (d) broad scope of damage (physical destruction, death, and injury), (e) occurrence at night, and (f) survivors' exposure to the dead and injured. Certainly the first four of these characteristics apply to the Ilopango eruption. As McLuckie (in Mileti *et al.* 1975:97) generalizes, the degree of stress on a social system varies inversely with the predictability of the disaster agent and directly with the magnitude of the disaster agent. Viewed from both perspectives, the Ilopango eruption must have been a massive natural disaster.

And because there are indications of at least partial abandonment of areas in the Southeast Maya Highlands, in what is now El Salvador (Sharer 1974, 1978), migrations must have occurred out of the devastated area as well as back into the area when vegetative and soil recovery was sufficient to support reoccupation. Recovery from the disaster, involving soils, flora, fauna, and human recovery, is the focus of current research in El Salvador. The recovery process may have required 200 years for significant human reoccupation, and that would have followed the at least partial recovery of soils, vegetation, and animal populations.

## CLIMATE AND FLORA OF EL SALVADOR

El Salvador, between 13 and 14.5° north latitude, lies within the latitudinal tropics (Figure 17.1). The high heat and humidity common in the tropics are

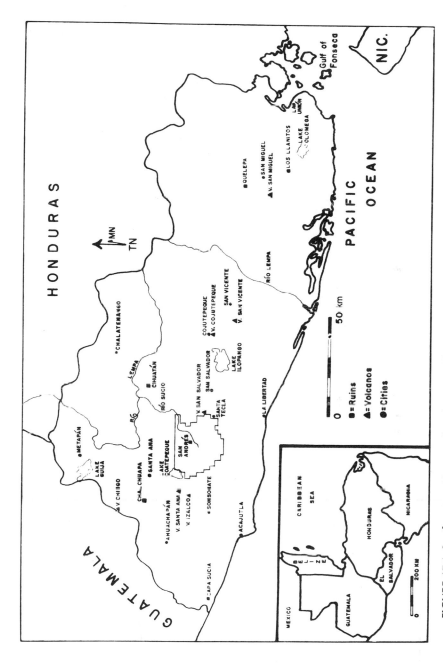

FIGURE 17.1 Southeastern Mesoamerica. Ilopango's third century eruption severely affected human settlements, flora, and fauna in central and western El Salvador. Details of the 1978 research area, the irregular boxed area around San Andrés, are shown in Fig. 17.2. Map by Kevin Black.

ameliorated by elevation and the fact that Salvador lies entirely in the Pacific watershed. The Pacific side of Central America receives rain on a seasonal basis, whereas the Caribbean side receives convectional rain throughout the year. The single best description of the Salvadoran climate, topography, geology, flora, and fauna is by Daugherty (1969). Although frequently hampered by inadequate data, he was able to trace in general terms the changes in land use, population, fauna, and vegetation from prehistoric times through the historic period up to the present.

With the exceptions of the higher peaks, El Salvador lies within the zone of seasonal wet-dry tropical climates (Daugherty 1969:23). Approximately 90% of the annual average precipitation falls in the 6-month rainy season from May through October. In some cases this seasonal dichotomy can be even more extreme; on the basis of 20 years of records I have calculated that fully 94% of the precipitation at Chalchuapa falls in the rainy season. The average annual precipitation in El Salvador ranges from 1600 to 2200 mm. The rains primarily are convectional and orographic, caused by the northward movement of the intertropical convergence zone and prevailing onshore winds. The dry season is created by the southern position of the subtropical high-pressure cell and the resulting northerly winds, leaving the western slope of Central America in a rain shadow (Daugherty 1969:32).

Daugherty (1969:95, 225) notes the rich variety of lake, river, and marine habitats, along with high soil fertility, that were exploited by Pre-Columbian populations. The high productivity of inhabited environments allowed for population growth and intensive human exploitation without significant soil degradation. Applying David Harris's (1972) argument to El Salvador helps explain the edaphic and climatic factors that underlay the vitality of Southeast Maya Preclassic society.

Harris (1972) defines an "intermediate tropical zone" between the humid tropics and the more arid temperate areas of the world. He notes that seed-based agriculture has come to characterize temperate areas, whereas vegeculture, focusing primarily on root crops such as yams or manioc, often dominates tropical rainforest areas. However, he argues that many of the root crops derived originally from the intermediate tropical zone, where a rainy–dry season climatic regime prevails. El Salvador, with its seasonal climate and volcanically derived soils, is an area of high agricultural productivity as predicted by Harris.

Daugherty (1969:43–53) was able to reconstruct the climax vegetation of El Salvador in terms of five lowland formations and three upland formations. The 1000-m line is used to divide upland from lowland formations. Because Pre-Columbian habitation was rare above this line, the upland formations will not be considered here. The lowland formations most pertinent to prehistoric habitation are the tropical evergreen forest, the deciduous forest, and the galeria forest.

The galeria forest is found along streams and is predominantly evergreen because of the constancy of water supply. Some galería forests still remain, but most have been removed or extensively modified during the past 2 centuries.

The deciduous forest is the primary climax vegetation formation of El Salvador, covering some .9 of the countryside. It ranged from dense stands of trees in the central and eastern areas of the country to more open woodland in the east.

The tropical evergreen forests of El Salvador were located along the coastal piedmont, in inland basins, and on flood plains where year-round moisture was available. The upper story was composed of 30-m-high wide-crowned trees, with the understory a combination of saplings, shrubs, and smaller trees.

The present topography, dominated by the chain of some 20 major volcanoes, is predominantly a product of Pliocene and Pleistocene volcanism. Sedimentary and metamorphic formations exist only in the northern edge of the country. Detailed descriptions of topography, hydrography, and geology are presented by Daugherty (1969), by Sayre and Taylor (1951), and by Williams and Meyer-Abich (1955).

Volcanologic and tectonic stability in the Preclassic period allowed these geologic and climatologic factors to combine forces, generating rich soils (Olson 1978) and allowing for the dense populations and cultural exuberance of the time (Sharer 1974, 1978). The stable agroeconomic base of Preclassic peoples may be contrasted with instability during the past 2 millennia. The nature of that agroeconomic base is the topic of the next section.

## PRE-COLUMBIAN SUBSISTENCE AND EXPORT AGRICULTURE

A variety of sources, particularly ethnohistory, archaeology, and contemporary ethnography, can be consulted to reconstruct at least the general outlines of Pre-Columbian societies' agroeconomic base in El Salvador. From these sources a picture of agricultural diversity emerges, for Pre-Columbian peoples apparently were cultivating seed crops, root crops, and tree crops, using both dry-land and irrigation techniques.

Autochthonous subsistence agriculture recorded in the sixteenth century involved maize, beans, chiles, avocados, cucurbits (squash), and a wide variety of other crops (Browning 1971:5). On the basis of palynological evidence Tsukada and Deevey (1967) have traced maize agriculture as far back as the Classic Period (sixth century A.D.) in western El Salvador. Contemporary with Chalchuapan maize agriculture is the agricultural field at Cerén buried by the sixth-century Laguna Caldera eruption (see pp. 539–547). The field was a monocultural maize field buried by tephra early in the growing season. Such evidence of maize cultivation in the Classic period may be extended back in time. Excavations at Chalchuapa in 1969 encountered corncobs in sealed

Late Preclassic structural fill (Sharer 1978). Pre-Columbian subsistence agriculture probably involved root crops as well; at least three varieties of manioc
are cultivated in El Salvador today. And export-related production, or "cash
cropping," of cacao, cotton, balsam, and indigo was known to varying degrees
in Pre-Columbian and early historic El Salvador.

Several scholars have documented the importance of cacao (cocoa, or
chocolate) production and export in Postclassic El Salvador (Daugherty 1969;
Browning 1971; Dahlin 1976). According to Browning (1971:54) and L.
Feldman (personal communication 1976), the Sonsonate area of southwestern
El Salvador was the single most important producer of cacao in Mesoamerica.
It was the prime producer in what Dahlin (1976:182) calls the "cacao belt," a
20–60-km-wide zone along the Pacific piedmont of Guatemala and El Sal-
vador. It is likely that the Postclassic distribution and intensity of cacao produc-
tion is indicative of earlier production, but this remains to be demonstrated.
Other areas of cacao production in El Salvador, according to Browning, were
the Santa Ana-to-Ahuachapan area, the Rio Jiboa Valley, and near San
Miguel.

The sap of the Balsamo hardwood tree (*Myroxylon pereirae*) was collected,
processed, and traded widely in the fifteenth and sixteenth centuries (Browning
1971:61–62). How far back into the Pre-Columbian past its use goes is unknown. It was in demand for its medicinal and aromatic properties. The area
around Sonsonate became known as the "Costa del Balsamo" because it was
the leading balsam producer of the Americas. Other likely Pre-Columbian
export crops include indigo (Browning 1971:66–68) and cotton.

Specific data on Formative (Preclassic) land use are rare. Earnest (1975)
recently discovered an intensive agricultural system along the Rio Lempa west
of the town of Chalatenango. It apparently was an irrigated field, perhaps used
for cash-cropping cotton. The immediate superposition of the Ilopango tephra
both preserved the agricultural field in pristine condition and provided the date
of construction and use as the end of the Preclassic.

Slash-and-burn (swidden) agriculture has been indicated but not
confirmed by the limnologic work of Tsukada and Deevey (1967) in the Chalchuapa area during the Classic and Postclassic. More intensive land-use types,
involving annual cropping and in some cases irrigation, are also likely but yet
undemonstrated for aboriginal Chalchuapa.

The overall, if hazy, picture of Pre-Columbian agriculture in the Salvadorean area of Mesoamerica is a mixture of swidden and irrigation agriculture,
largely for subsistence but at least partly for export.

Daugherty (1969:121ff) provides a generalized assessment of the Pre-
Columbian impact on the natural environment. Focusing particularly on the
Postclassic, he suggests there was a significant impact on vegetation by swidden
cultivation in flat-lying areas below 1000 m, resulting in savannas and the
expansion of pyrophytic species. At the time of the conquest hilly areas were
densely forested, indicating little or no human modification (with the obvious
exception of hunting, some collecting, and the extraction of balsam).

Daugherty believes the Pre-Columbian depletion of fauna was slight, for the species that were most exploited, such as deer and rabbits, can sustain heavy exploitation and often increase in population with swidden cultivation. Geographically, the central and southern portions of the country supported the heaviest populations, and these areas experienced greater degrees of ecosystem modification.

## THE ILOPANGO ERUPTION AND THE SOUTHEAST MAYA HIGHLANDS

The above generalized descriptions of the Salvadorean climate and Pre-Columbian adaptations serve as a framework to examine pre- and post-eruption conditions in El Salvador. The portions of the Prehistoric era that are the focus of this chapter are the Late Preclassic, from 500 B.C. to A.D. 300 (with the Protoclassic being the last century of that period), and the Classic period, from A.D. 300 to 900 (dates from Sharer 1978). These two periods are separated by the Ilopango eruption, dated to the latter part of the third century A.D. (Table 17.1).

It should be noted from the onset that the discovery of prehistoric artifacts under volcanic ash in El Salvador is not new. In fact, intermittent research on artifacts under volcanic ash has been conducted during the past 60 years. From these various finds a sketchy view of Late Preclassic habitation may be reconstructed, particularly for central and western El Salvador.

### Early Research

Jorge Lardé (1926) was the first scholar to note volcanic ash burying ancient artifacts, and Lothrop (1927) expanded Lardé's findings with an extensive description and interpretation of the Cerro Zapote site. The site, located in the southeastern sector of San Salvador, the capital of El Salvador, is 17 km from Ilopango (Figures 17.1 and 17.2). It contained Preclassic ceramics, figurine fragments, and lithic artifacts. It was a sedentary agricultural settlement, but its original size is unknown. Lothrop recognized the Lowland Classic Maya character of the artifacts **above** the ash layer, suggesting human reoccupation of the devastated area may have come from the north.

During the past 40 years Stanley Boggs has excavated numerous sites buried by volcanic ash. Boggs (in Longyear 1944:53–54) encountered sherds towards the bottom of a white volcanic ash layer at Tula (Figure 17.1), approximately 30 km southeast of Ilopango. Both architecture and artifacts were found by Boggs (personal communication 1970) below the ash at Tazumal. Tazumal is in the southern portion of the Chalchuapa site-zone, some 77km west-northwest of Ilopango (Figure 17.1). Three Preclassic burial urns were found in a humic horizon buried by volcanic ash at Loma del Tacuazín near San Salvador (Boggs 1966). Three other localities in the San Salvador area with

**TABLE 17.1**
Chronological Summary of the Zapotitán Valley

| Periods[a] | Volcanic events | No. of survey sites (pre-historic) | Cultural events |
|---|---|---|---|
| Republic A.D. 1821–present | San Salvador lava flow 1917 Izalco 1770–1965 | | |
| Colonial A.D. 1524–1821 | San Marcelino lava flow 1722 El Playón eruption and lava flow 1658–1671 | | |
| Postclassic A.D. 900–1524 | | 11 (Early Postclassic) | Population decline in basin Cambio occupied (after San Andrés Formation) |
| Late Classic A.D. 600–900 | San Salvador eruption (San Andrés Formation) | | |
| | Laguna Caldera eruption (Cerén Formation) A.D. 590 ± 90 | 42 | Cerén occupied; Cambio occupied; major occupation of San Andrés site |
| Early Classic A.D. 300–600 (Protoclassic A.D. 200–300) | Ilopango eruption A.D. 260 ± 114 | 0 | Basin apparently abandoned |
| Late Preclassic 500 B.C.–A.D. 300 | | 5 | Cambio occupied; light occupation of Cerén |
| Middle Preclassic 900–500 B.C. | | 1? | |
| Early Preclassic 1200–900 B.C. | | 1? | |

[a]Dates from Sharer 1978.

comparable deposits of ash and artifacts are Hospital Cardiovascular, Modelo Bridge (Boggs 1966), and Barranco Tovar (Porter 1955). Longyear (1944:33; 1966) found volcanic ash underlying Classic Period construction at Los Llanitos in eastern El Salvador.

## Chalchuapa Archaeological Project

Numerous instances of Late Preclassic artifacts and architecture having been buried by a white volcanic ash were encountered during Chalchuapa Archaeological Project excavations in 1967–1970 (Sharer 1968:301; 1974, 1978). Primary deposits of air-fall ash at Chalchuapa, 77 km from Ilopango, measured as much as 60 cm in depth, indicating the eruption's magnitude (Figures 17.3 and 17.4). Stratigraphic excavations at Lake Cuzcachapa documented the

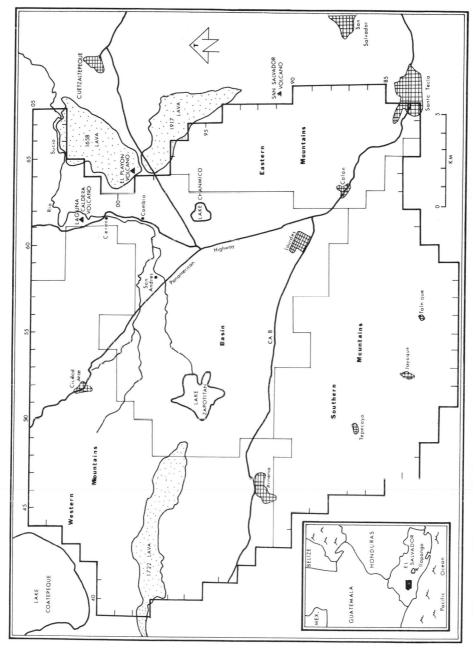

FIGURE 17.2 The Zapotitán Valley, with the four subregions defined for 1978 fieldwork. Fifteen percent of each subregion was surveyed, and excavations were conducted at Cerén and Cambio archaeological sites.

FIGURE 17.3 1967 excavations at Chalchuapa: Monument 3 sealed under Ilopango "tierra blanca" at the base of Structure E3-1 (see Sharer 1978). Note that the ash is mottled, not a pristine white layer of ash, because of proximity to the present ground surface, root action, and weathering. (Photograph courtesy of Robert J. Sharer.)

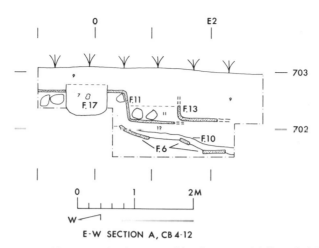

**E-W SECTION A, CB 4-12**

FIGURE 17.4 Pyramid construction interrupted by Ilopango ashfall at Chalchuapa. Feature 6 is the final surfacing layer, which was not completed when the Ilopango ash (Feature 10) fell. Later, in the Classic period, the valley was reoccupied and the pyramid was enlarged. Feature 11 is the edge of the refurbished pyramid. It later was enlarged still further (Feature 13) and then abandoned and used for deposition of trash (Feature 17) just prior to the arrival of the Spanish. (From Sheets in Sharer 1978.)

development of the "Protoclassic" constellation of characteristics (as defined by Willey and Gifford 1961; Willey et al. 1965; Willey et al. 1967:298) capped by the volcanic ash layer.

Just prior to the Ilopango eruption Chalchuapa was a major residential, economic, and ritual center in the southeast Maya highlands. The central ritual zone alone was 2 km in length, composed of numerous 15-m-high pyramids (e.g., Figure 17.4) separated by formal plazas and surrounded by habitation areas of the elite, artisans, and agriculturalists. Obsidian was traded in large quantities from Ixtepeque and manufactured by craft specialists into numerous kinds of tools. Monuments, some with elaborate sculpture and calendric dates, were carved and erected in the ritual zones. Elaborately decorated ceramics were manufactured, commonly using Usulután decoration combined with the mammiform tetrapod vessel form. Ceramic decoration during the Preclassic sometimes involved polychrome painting, evidenced by the occasional polychrome sherd in Preclassic horizons at Lake Cuzcachapa.

## The Protoclassic Project

Until 1975 these various finds of volcanic ash burying Preclassic artifacts scattered across central and western El Salvador could be interpreted in two different ways. They could represent a series of eruptions from numerous vents along the volcanic axis of Central America, each of which affected a limited area and number of people and was separated from the other eruptions by decades or centuries. Or these finds could be indicative of a massive eruption that simultaneously affected a large area with a dense population.

Geologists working in El Salvador after World War II encountered the ash layer burying Preclassic artifacts, and they christened it the *tierra blanca*. Considerable speculation has revolved around its source or sources. Research by the German Geological Mission within the past 10 years, most of it yet unpublished, has resolved this issue. The source is Volcán Ilopango (H.S. Weber personal communication; Schmidt-Thomé 1975).

The 1975–1976 research of the Protoclassic Project (Sheets 1976; Steen-McIntyre 1976) worked on the same problem by judiciously sampling artifacts and tephra (Figures 17.5 and 17.6) along a northwest–southeast transect through central and western El Salvador, and then subjecting the tephra samples to petrographic and granulometric analysis. Results confirm the Ilopango area to have been the source of the *tierra blanca* tephra.

According to data collected in 1975, the eruption consisted of three closely spaced components (Steen-McIntyre 1976, 68–78). The first two were ash flows, perhaps involving a base surge, and these were followed shortly by the deposition of a blanket of airfall ash over the southeast Maya Highlands. An ash flow (*nuée ardente*, or glowing cloud) is an incandescent cloud of ash and pumice that travels downhill at high speed (Bullard 1976). Ash-flow temperatures are often greater than 200°C, above the kindling temperature of wood; the *nuée ardente* that killed 30,000 people in St. Pierre (on the island of Martinique in the Caribbean) in 1902 had a temperature over 700°C. Velocities range from 58 to 480 km/hr, and are usually accompanied by great turbulence.

The two ash flows apparently headed north and west, devastating low-lying agricultural and habitation areas in the Rio Acelhuate–Sucio–Lempa drainages. The maximum known distance covered by these ash flows is 45km, longer than that of most historically known ash flows. It is unlikely that any vegetation, fauna, or people would have survived an ash flow. But, from a regional perspective, a relatively small fraction of the land rendered uncultivable by the eruption was hit by a *nuée ardente*.

Most of the damage was done, from the regional view, by the ashfall deposits. The ashfall deposits are stratigraphically superior to the ash-flow layers, and they extend much farther—how much farther depends on analyses of deep-sea cores. The ashfall in El Salvador was at least .5–1 m in thickness. Human physical survival under such conditions is likely, but according to the observations made during comparable eruptions (cf. Segerstrom 1950; Malde 1964; Bullard 1976; and summary in Sheets 1976) an ash depth of 10–25 cm suppresses nonindustrialized agriculture at least for a few years. The time intervening between impact and agricultural recovery depends on a number of factors, primarily climate, weathering rate, the mafic-to-sialic character and particle size of the tephra, the depth and stability of the tephra, plant recolonization, and specific nutriments needed by cultivars.

## 1978 Protoclassic Project Research: Preliminary Results

Staff of the Protoclassic Project spent the first half of 1978 in El Salvador conducting archaeological surveys and excavations and sampling soils, pollen,

FIGURE 17.5 Cartografía Site in the northeastern corner of San Salvador, being recorded by Robert Koll and Mike Foster in 1975. Visible are (A) Preclassic soil with artifacts; (B) basal coarse pumiceous layer from Ilopango; (C) thick beds of ash-flow and air-fall material from Ilopango; and (D) the weakly developed contemporary soil horizon. Site is 15km from Ilopango. (After Sheets 1975:Figure 37.)

FIGURE 17.6 La Cuchilla site midway between Ciudad Arce and San Andrés (see Figure 17.2) being recorded in 1975 by M. Foster and V. M. Mejia. Preclassic soil contains dense Late Preclassic artifacts. Central part of site, consisting of a pyramid–plaza complex, is 100m to right. Ilopango, 45km distant, deposited a thin basal coarse horizon followed by air-fall and ash-flow material. (After Sheets 1975:Figure 32.)

and tephra. Although the data are far from completely analyzed, a preliminary report is now available (Sheets 1978). Contributions of the project staff, although tentative and suject to revision as the data are analyzed in detail, are summarized below.

The geographic focus of 1978 research was the Zapotitán Valley in west-central El Salvador (Figure 17.2). It was chosen over other areas because it best met the demands underlying the research objectives. The principal goal of the 1978 season was to enlarge understanding of the processes underlying recovery from the Ilopango eruption. Human recovery from the disaster is predicted on soils and vegetation recovery, yet virtually nothing was known about these recovery processes.

That Salvador is in an extremely active volcanic area is evidenced by Hart's documenting (1978) three air-fall eruptions more recent than the Ilopango eruption (Table 17.1). Each eruption deposited an airfall tephra layer in at least part of the Zapotitán Valley. But each varied in its nature, thicknesses, and distribution, so they would have varied considerably in their effects on inhabitants in the area.

The Cerén tephra was the first to follow the Ilopango eruption. It was erupted from the Laguna Calera Volcano (Figure 17.2) a few centuries after Ilopango. It is radiocarbon dated to A.D. 590 ± 90. Approximately 4m of tephra were deposited at the Cerén site, 1.4km to the south. The eruption occurred in 15 recognizable phases, each varying in particle size, temperature of emplacement, and force of the eruption. Although deposits of the Cerén Formation are very thick within a 2-km radius of Laguna Caldera, the deposits thin rapidly, and only a few-cm-thick blanket reached 5km to the south. Hence it intensely affected a very restricted area, quite different from Ilopango.

The San Andrés tuff (Figure 17.7) was deposited all over the Zapotitán Valley from a flank eruption of Volcán San Salvador apparently sometime between A.D. 800 and 1000. It varies in thickness from only 7 cm on the western side of the valley to 6m on the rim of Volcán San Salvador. It is a dense, finely laminated layer, in exposures often containing plant casts of trees and grasses

FIGURE 17.7 Stratigraphic sequence of air-fall eruptions and cultural horizons for the Zapotitán Valley is exposed in the Cambio site road cut. (A) Preclassic soil with artifacts, including vessel exposed; (B) Basal pumice layer from Ilopango 37km distant; (C) lower (T2) component of the Ilopango tephra; (D) upper (T1) component of the Ilopango tephra; (E) soil developed out of the Ilopango tephra, with Late Classic artifacts; (F) thin layer of the Laguna Caldera tephra (which buried the Cerén site to the north); (G) San Andrés tuff; (H) Late Classic artifacts and floor construction; and (I) tephra from the Playón eruption of 1658–1771. (See Chandler 1978.) (Courtsey of C. Zier.)

growing in areas where it was deposited. A unique opportunity to study vegetation patterns associated with human settlement in the Late Classic period is
provided by such preservation.

The most recent airfall unit to affect the valley is the El Playón tephra
(Figure 17.7) from El Playón Volcano. Followed by earthquakes, El Playón
began erupting in 1658, depositing a thick layer of dark pumice in the eastern
portion of the valley. Deposits between 10 and 60 cm in thickness are common
in the area. Most of the volume erupted was not airfall but was involved with
an andesitic lava flow.

To Hart's sequence I would add two more recent lava flows that have
buried substantial areas of the valley: (a) the San Marcelino lava flow of 1722 in
the western mountains, and (b) the 1917 flow from a side vent of San Salvador
Volcano along the eastern margin of the valley (see Figure 17.2). Neither of
these recent eruptions was accompanied by significant air-fall tephra units.

Recent research by Steen-McIntyre (1976, 1978) has added to the understanding of the internal components of the eruption that deposited the *tierra
blanca* across the Salvadorean countryside. To the best of our knowledge at
present, the eruption occurred in three stages, each of which resulted in
visually and petrographically distinguishable deposits. The first part of the
eruption sequence, with its source vent located near the northwest shore of
present Lake Ilopango, blasted a relatively coarse pumice into the air. It rapidly
settled on the countryside, extending some 45km to the west and north. The
deposit ranges in thickness from 1–2 cm at its most distant exposures to 40cm
near the lake shore.

The second major component of the Ilopango *tierra blanca* is much
thicker than the basal coarse member. This "T2" unit is composed of a series of
fine to coarse vitric ash subunits ranging in thickness from about 20 cm at
Laguna Seca (Chalchuapa area, 77 km from the source) to 6 m at a site within a
few km of the source. Some of the T2 was air-fall deposited, but other subunits
may have been deposited by a low-temperature ash flow. It is possible that the
extraordinary fineness of the T2 tephra may be explained by the tephra's
having been erupted through an ancestral Lake Ilopango. The chilling effect of
the water would have caused fracturing and fragmentation of the hot pumice
pieces, resulting in a finely divided tephra.

The topmost unit of the *tierra blanca*, the "T1," is more tan in color and is
composed, at least in large part, of hot ash-flow material. The heat is indicated
by charred logs from forests that were uprooted and then destructively distilled
by the hot tephra matrix. The deposit is 25 cm thick at Laguna Seca and it
increases to over 9 m in thickness close to the source. The lack of weathering or
evidence of soil formation between these three layers indicates that they were
all deposited within a very short span of time.

Analysis by Olson (1978) of the soils that existed in the basin during the
past 2 millennia indicates the most fertile soil was the Preclassic soil prior to the
eruption of Ilopango (the emplacement of the *tierra blanca* tephra). It was well
weathered, creating a soil with good structure, good water-retention capability,

almost neutral pH, and a suitable clay content. Soils that have developed on top of the various layers of volcanic ejecta during the past 2000 years (i.e., on the Cerén Formation, the San Andrés tuff, and the Playón tephra) are less suitable for cultivation. From that we can conclude that soils have yet to recover completely from the tephra damage sustained during the past 2000 years.

The archeological survey of the Zapotitán Valley, under the field direction of Black (1978), sampled 15% of each of the four subregions indicated in Figure 17.2. Out of the 546-km² study area a total of 82 km² was surveyed, in which 54 sites were located. Population was dense in the basin and the western mountains, and sparse to nonexistent in the southern and eastern mountains.

Preliminary ceramic dating of the sites indicated the following numbers of sites occupied per period: 1 Early or Middle Preclassic, 5 Late Preclassic, 0 Early Classic, 42 Late Classic, 11 Early Postclassic, and 0 Late Postclassic (Beaudry 1978; Black 1978). More detailed ceramic analysis presently in progress will necessitate some adjustments in the above figures, but the general pattern should not change. The surprisingly low figure of only 6 Preclassic sites may be explained in two alternative ways: (a) The figure is representative, and Preclassic occupation was in fact very slight; or (b) Preclassic occupation was more substantial but ashfalls and ash flows have buried the Preclassic settlements deeper than more recent settlements, resulting in their underrepresentation in our data. Judging from known densities of Preclassic occupation in surrounding areas (e.g., Chalchuapa to the west and Quelepa to the east), I feel the latter explanation is more likely.

The lack of Early Classic occupation is not surprising, given the intensity of the Ilopango natural disaster in the basin (Steen-McIntyre 1976, 1978). The basin area appears to have been abandoned, at least for a few decades, and perhaps as long as 2 or 3 centuries. But human recovery from the Ilopango disaster was effected during the Late Classic, as evidenced by the 42 sites occupied in our survey sample. Soil recovery, though far from complete, was sufficient to sustain the agroeconomic base of Maya society.

The reason for the population decline in the Postclassic is unknown, but it might be related to the emplacement of the San Andrés tuff from Volcán San Salvador throughout the basin. Alternatively, the decline may be related to the collapse of Classic Maya society to the north, or to internal difficulties unrelated to either of these factors.

One of the most significant discoveries of the 1978 season is the buried structures at Cerén (Sheets 1979). Laguna Caldera erupted, probably during the sixth century A.D., and deposited some 4m of hot ash, cinders, and pumice on structures and fields (Figures 17.8–17.12).

At Cerén the preservation is extraordinary. Portions of a farmhouse and a work platform were excavated from underneath the Laguna Caldera tephra. The multiroom farmhouse (Figure 17.9–17.11) was constructed of fired clay floors and columns, wattle-and-daub exterior and interior walls, with a palm thatch roof. Numerous artifacts were found on the floor where they had been

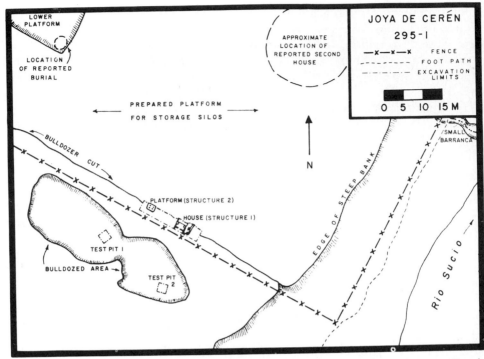

**FIGURE 17.8** Location of features within the Cerén site. Both the farmhouse (Structure 1) and the platform, or ramada (Structure 2), were buried by a layer of hot tephra 4m thick from the Laguna Caldera 1.5km to the north. The agricultural fields exposed in Test Pits 1 and 2 were likewise buried and preserved. Local informants described another house, completely destroyed by bulldozing, located approximately 50m northeast of the two structures, indicating a dispersed settlement pattern of farmhouses surrounded by their agricultural lands. (Courtsey of C. Zier)

used. The sudden tephra fall preserved the functioning household far better than most structures are preserved archaeologically.

The eruption apparently struck the area with little warning. No major earthquake presaged the eruption, for the walls and columns are intact. And the nature of the tephra bedding with no unconformities indicated rapid deposition. Both the structures and the agricultural fields were struck simultaneously by large hot lava bombs and a fall of ash and pumice. According to local informants who witnessed the bulldozing of the hillside that led to the first discovery of the house in 1976, bodies were found on the floor of the house. Unfortunately, that area of the house is now completely destroyed, but it does indicate that people were trapped and killed by the eruption. Death could have come from asphyxiation by volcanic gases, burning, suffocation from tephra, or a combination of these factors. Clasts within the tephra were over 575°C when they landed (R. Hoblitt personal communication 1979).

Local informants also reported the bulldozing and complete destruction of another house, similar to the one we excavated, located perhaps 50 m to the northeast of the Cerén house (Figure 17.8). If accurate, these amounts do give a general idea of the pattern of settlement: isolated farmhouses with the

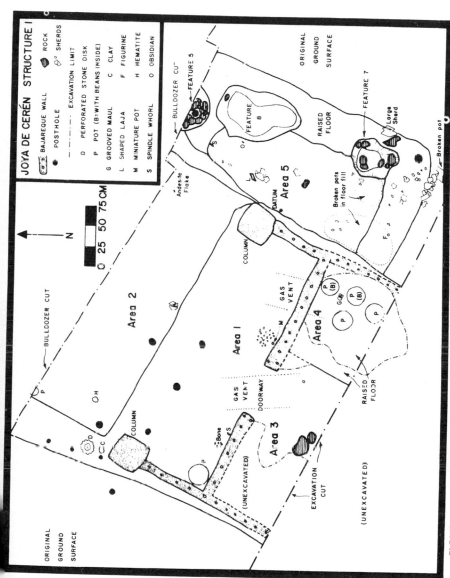

FIGURE 17.9  The Cerén farmhouse (Structure 1). The western part of Area 1 was used for artifact storage and perhaps for weaving of cotton garments, and the eastern part may have been a small girl's room. Area 2 apparently was used for pottery manufacture. Area 4 was the pantry; it is likely that the kitchen of the house lies in the yet unexcavated area to the south. Two of the four vessels in the pantry stored beans. Area 5 was a storage and work area, with vessels suspended from the roof. (Courtesy of C. Zier)

FIGURE 17.10 The Cerén farmhouse, looking southwest. Note the Preclassic soil horizon, with an occasional artifact, buried by the white volcanic ash from Ilopango. House and fields indicate at least partial recovery of soils and human population from the Ilopango disaster. Floor and columns were made of adobe, wet-laid and then fired, and wingwalls were made of wattle and daub. The bottom of the doorway is visible in the center. The hole through the left wingwall may have been caused by the hot tephra degassing itself after it fell, creating a gas vent. The gas vent exposes the floor of the pantry (Area 4). The air-fall layers of tephra from the Laguna Caldera eruption are 4 m deep above the floor. (Courtsey of C. Zier)

outbuildings of one house separated from those of another by some 50–100 m of agricultural fields.

Area 4 of Structure 1, within the Cerén farmhouse (Figure 17.9), is the pantry. Two large ceramic vessels full of beans, two nearby empty vessels (they may have contained a liquid such as water or chicha, a maize beer, or they may have been empty), and a grooved stone maul were all found on the elevated

FIGURE 17.11 The Cerén farmhouse (middle ground) and platform-ramada (foreground) buried by the Laguna Caldera tephra and later eruptions. The palm-thatch roof of the platform was supported by posts along its perimeter. Most of the chipped stone at the Cerén site was found around the platform, perhaps indicating it was a male work area. A walkway of tuff and lava blocks connected the two structures. An extended burial was intruded through the Ilopango tephra below the corner of the platform while it was being constructed. (Courtsey of C. Zier)

floor. I suspect the kitchen is adjacent, to the south, in the yet unexcavated portion of the house.

Area 3 and most of Area 1 were clear of floor artifacts. This is not surprising in that the doorway provided access from room to room along the long axis of the house. But the western corner of the Area 1 room was used for artifact storage, for a spindle whorl and a ceramic vessel were found there. The spindle whorl was mounted on the spindle at the time of the eruption. Along the south wall of the room were found a crude miniature clay vessel, upside down, and 20 potsherds. Because of the ethnographically and ethnohistorically known tendency for Maya potters to be female, and the Maya numerical system to be vigesimel (base-20), these discoveries have given rise to the speculation that this part of the Area 1 room may have been a little girl's room. If so, she would have been old enough to be able to make a crude miniature pot as well as learn to count, indicating a possible age of 5 or 6 years.

Area 2 is better known functionally. The prepared ball of potter's clay and the ground hematite encountered on the western side indicate it was a pottery-making area. If tests on the clay and the hematite indicate polychrome pottery was being manufactured at the house, some rethinking of archaeological assumptions will be called for. It has been supposed that commoners living in a rural farmhouse in the middle part of the Classic period would not have been using large numbers of polychrome vessels, let alone making them. Lithic analysis (Sheets 1978) indicates the inhabitants of the house were having some difficulties obtaining obsidian—that is, that they were "poorer" than most of their contemporaries in the valley with respect to the chipped-stone industry. Yet not only could they "afford" polychrome pottery, but they also owned numbers of vessels, and these exhibited a great variety of style and technique. Beaudry (1978) tabulates the following polychrome types from Cerén: Copador, Arambala, Gualpopa, Campana, and Nicoya. At least in this case, the presence of polychromes, even in abundance and great variety, cannot be taken as an indicator of elite status.

Area 5 is a storage and work area appendaged onto the house. It had an irregular clay floor with storage pits, no walls, and a palm thatch roof supported by vertical posts. Floor artifacts included a spindle whorl (also mounted on the spindle), an obsidian flake, a figurine head, some broken pots, and a grass floor mat. The pots apparently were suspended from roof beams, probably in net bags, instead of resting on the floor, as did the other vessels found inside the house (the evidence being the tephra found beneath the broken vessels—some tephra had settled on the floor before the roof collapsed and burned). The floor mat covered such an irregular surface that it probably was not a sleeping mat; more likely it was used to make sitting on the clay floor more comfortable while working. Both the spindle whorl, indicating the weaving of cotton garments, and the pottery vessels probably imply a female work area. In fact, all five excavated areas of the farmhouse seem to have been associated with female activities. It is likely that the more northerly portions of the house, destroyed by the bulldozer, were associated with sleeping areas for all the family as well as with common areas.

About 7 m from the house is a platform that may have served as a male work area. It is a solid clay platform with posts supporting a palm thatch roof. The evidence for walls is equivocal; I interpret the evidence as indicating that the platform (Structure 2 in Figure 17.8 and Figure 17.11) had walls but that they had fallen into disrepair and were removed before the eruption. Most of the chipped stone from the Cerén site excavations was found around the platform. Because chipped stone tool manufacture is ethnohistorically known to be a male activity in Mesoamerica, we interpret the Structure 2 ramada to be a male work area. A walkway of irregular slabs of laja and volcanic tuff connected the farmhouse with the ramada.

Two test pits excavated some 20 m to the southwest of the farmhouse and ramada yielded evidence of an agricultural field (Figures 17.8, 17.12). Plant casts, of what has been tentatively field-identified as corn, were found on top of low ridges. Plants were spaced about .5 m apart, and each was about 5 10 cm high. If the field was not irrigated, as the data indicate, then the height of the corn in the highly seasonal Salvadorean climate indicates that the Laguna Caldera eruption probably occurred in May.

## Summary of the Effects of the Ilopango Eruption in El Salvador

A massive natural disaster, in the form of the air-fall and ash-flow deposits from Ilopango, struck the southeast Maya highlands in the third century A.D. Most people living within a few kilometers of Ilopango, or living in the path of the ash flows, would have been killed. Many others, numbering probably in the tens of thousands, could have survived, but would have had to migrate because of tephra damage to agricultural fields.

In the years following the eruption, weathering of the top of the ash blanket formed a soil sufficient for plant recolonization and eventual human recolonization of the devastated area. Although soils had not recovered all of their high pre-eruption fertility by the Late Classic period (A.D. 600–900), they had weathered sufficiently to support a vigorous human reoccupation. That human recovery apparently was well underway as early as the sixth century A.D. At least some of the agroeconomic basis of that recolonization was maize and bean agriculture in cleared, intensively prepared fields.

## POSSIBLE REGIONAL REPERCUSSIONS OF THE ILOPANGO ERUPTION

Beyond the areas of southern Mesoamerica with direct environmental damage by substantial ashfall, it does appear that indirect economic, ecologic, social, and demographic effects may have been felt. But the data on these possible effects are equivocal, and this section therefore is quite speculative. Most speculations here are testable, given careful sampling in stratigraphic excavations, with petrographic or other specific analyses of suitable samples. Flooding is one of the possible long-distance effects of the Ilopango eruption. Floods resulting in the deposition of extensive sterile deposits of mud on riverine settlements apparently occurred toward the end of the Preclassic in northern Belize (Bruce Dahlin personal communication, 1973), in west-central Belize (Willey et al. 1965:565), and in northwestern Honduras (Stone 1972:57–62; Sheets 1977). Such flooding may have been caused by ash damage to

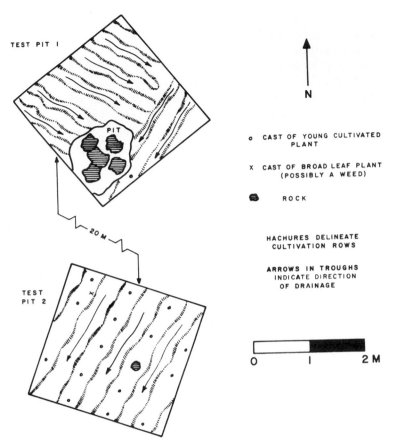

TEST PIT I

PIT

20 M

TEST
PIT 2

N

o    CAST OF YOUNG CULTIVATED
            PLANT

X    CAST OF BROAD LEAF PLANT
        (POSSIBLY A WEED)

⬤       ROCK

HACHURES DELINEATE
CULTIVATION ROWS

ARROWS IN TROUGHS
INDICATE DIRECTION
OF DRAINAGE

0              I              2 M

FIGURE 17.12 Test Pits 1 and 2 excavated to the south of the Cerén structures. Agricultural field ridging apparently was for retention of rainfall moisture early in the rainy season. Small corn plants preserved as casts by the Laguna Caldera tephra indicate the eruption probably occurred in May. The pit in Test Pit 1 was created by a large, hot scoria block falling on the cornfield early in the eruption. (Courtsey of C. Zier)

vegetation in the drainage basin of the lowland rivers, resulting in increased runoff and a heavy particulate load in floodwaters (Sheets 1975:133).

## Migration Theory and the "Protoclassic Problem"

The initial formulations of Burton *et al.* (in press) toward a general theory of human behavioral adjustments to natural disasters are used here as a context within which to view migration. They categorize responses into four modes of increasing severity: loss absorption, loss acceptance, loss reduction, and radical change. These modes are separated by three different threshold levels: awareness, action, and intolerance.

Loss absorption involves incidental adjustments with no conscious program of change. Loss acceptance involves conscious awareness (the first threshold), and generally the losses of the victims are borne by a larger group of people. Loss reduction incorporates direct action, the second threshold, by the victims to reduce their losses. However, as the scale of the disaster increases, the third threshold is crossed, that of toleration, and radical action must be undertaken. Such radical action involves in situ fundamental adaptive changes, or, in extreme cases where the environmental changes are beyond the human technological capacity to cope, migration occurs.

But the assumption cannot be made that all societies at the same point of sociotechnological complexity will react to the same stress by the same adjustments. The condition of the society and the success of its land and resource use must be taken into account; a society experiencing ecologic difficulties likely would react to the same stress by crossing a higher threshold than a comparable society with a more successful adaptation.

Migration may be defined simply as the relatively permanent movement of people over space (Peterson 1968:268, du Toit 1975:1). Migrations vary in the degree of permanence, the number of people, the distance traveled, in whether they are voluntary or coercive, and in their underlying causes. According to du Toit (p. 6) people tend to migrate to places where they are known and about which they have general knowledge or where they already have established kin ties. This general rule may be applicable to Protoclassic migrations to the lowlands immediately following the Ilopango eruption, but it is of limited utility in investigating the reoccupation of the devastated area.

Peterson (1968:289–290) identifies types of migration by the underlying reason for the migrants' move. Innovating migrants are those who move to achieve something new, whereas conservative migrants are attempting to retain the old way of life by moving to a new locality. Either of these can be a forced migration, depending on circumstances, or it can be a free migration in that the migrants largely retain the power of decision.

As Safa (in Safa and du Toit 1975:1) and Peterson (1968) have pointed out, the limitations of many earlier migration studies have derived from the scale of phenomena examined. Many of these studies assumed or concluded that migrants left because of the limited employment opportunities in their areas of origin and the hopes of better economic situations elsewhere. Much of migration theory is economically based, and it derives from the large nineteenth-century migrations of Europeans into North America. Safa suggests that a wider context than the individual's or the family's making "economic" decisions is needed, since national and international political and economic factors may be deeply involved.

It seems that the emplacement of the Ilopango *tierra blanca* exceeded the toleration level of communities in central and western El Salvador, requiring radical action in the form of forced migrations. If migrations were necessary, do the analyses of artifacts in southern Mesoamerica divulge any areas likely to have received immigrants? There are some intriguing similarities between sub-ash artifacts in Salvador and some intrusive artifacts to the north. The archaeological literature of the Maya lowlands contains indications of an intrusion of foreign artifacts into a domestic continuum. Explaining that intrusion is what is called the "Protoclassic problem."

The Protoclassic problem was first recognized from data deriving from the 1910–1911 excavations at Holmul, Guatemala, directed by R. E. Merwin under the sponsorship of the Peabody Museum, Harvard University. Merwin's excavations, to my knowledge the first stratigraphic excavations in the Maya area, divulged a series of five periods. It is the first, Holmul I, that we can now identify as Protoclassic, based on the presence of mammiform tetrapod vessels (Figure 17.13), Usulután-style decoration, pot-stands, and spouted vessels. Merwin and Vaillant (1932:62, 65) note that the Holmul I materials bear strong stylistic resemblances to sub-ash artifacts encountered by Lothrop (1927) in central El Salvador. Explanation of the resemblance was not attempted by Vaillant.

Numerous other sites in the Maya lowlands have been found to have experienced a similar site-unit intrusion sometime near the time of Christ. The artifacts have been known by a number of phase names, including Holmul I, Cimi, Matzanel, Salinas, Cantutse, and Floral Park. The name *Protoclassic* is used here to subsume these similar materials. Still the best summary of Protoclassic artifacts' distribution in the Maya area, particularly in the lowlands, is that by Willey and Gifford (1961). They note the full occurrence of the Protoclassic at such sites as Barton Ramie, Holmul, Poptun, Mountain Cow, Douglas, Nohmul, Santa Rita, and Pomona. It has also been found at Altar de Sacrificios (Willey *et al.* 1967:298, Willey 1973:34–39).

Sites with a few scattered Protoclassic artifacts, or lacking the evidence of a major, sudden infusion of the complete "Protoclassic package," are Uaxactún, Tikal, Finca Arevalo, Seibal, Chiapa de Corzo, Monte Albán, Kaminaljuyú,

FIGURE 17.13 Ceramic vessels from the intrusive Floral Park Complex at Barton Ramie, Belize. These highly distinctive mammiform tetrapod vessels are characteristic of the Protoclassic complex, here interpreted as a manifestation of the migrations from the Ilopango disaster northward into the Maya lowlands. (A) Aguacate Orange vessel with section drawing; (B) Guacamallo Red-on-orange vessel with section drawing. Comparable vessels are found under the Ilopango ash in El Salvador. (After Willey *et al.* 1965:Figure 204, with permission.)

and sites in southern Campeche. Willey and Gifford (1961:167) view the Protoclassic at Barton Ramie as an intrusion of a constellation of foreign characteristics into an indigenous continuum. The intrusion does not break that continuum, and the intrusive styles are eventually absorbed into the general Early Classic culture. Barton Ramie is examined in more detail in the next section of this chapter.

Although the Protoclassic spread of artifacts could be explained by changes in a trade network, Willey and Gifford (1961:168, 170) claimed that a migration was the most likely explanation. They were not certain of the source, suggesting northern Honduras or the Guatemalan highlands.

Tikal Project excavations beginning in 1956 have demonstrated that the roots of the Maya Classic fluorescence in the Petén are deeply embedded in the Petén Preclassic. Therefore, I do not argue that lowland Maya civilization mechanistically was derived from the highlands. The lowlands were on their way toward civilization and would have achieved it whether or not there was a disaster in the highlands. However, it is likely that the sudden arrival of large numbers of people on the peripheries of the central Petén "core area" necessitated an intensification of social and political control mechanisms, therefore accelerating the rate of cultural development. The Protoclassic, in my opinion, acted more as a catalyst increasing the rate of cultural change than as a critical reagent in the emergence of Maya civilization.

The controversy over the Protoclassic continues among Mayanists. Norman Hammond (personal communication 1975) feels the Protoclassic-style artifacts in Belize may be the result of local development, and thus they do not represent a migration. Willey (1973) and Adams (1971) disagree on the explanation for the large amounts of Protoclassic materials encountered during excavations at Altar de Sacrificios (Salinas phase, A.D. 150–450). They agree the Salinas phase was a time of rapid change in ceramics, trade, architecture, artifacts, and iconography. Adams explains this by an immigration into Altar accompanied by some violence, with the invaders establishing themselves as rulers over the earlier resident population. Willey had earlier favored the invasion explanation, but more recently he (1973:38–39) expressed doubts by stating the Protoclassic phenomenon was "culture change benefiting by stimulation from foreign contacts."

Willey (1973:36) notes that the indigenous continuum of Altar utility wares was unbroken from the preceding Plancha phase through the Salinas phase. Significantly, the first "luxury ware" at Altar is the intrusive Protoclassic. If there was a Protoclassic immigration, then these two ethnic traditions and their socioeconomic concomitants may be used to trace the dynamics of ethnic interactions. Immigrant groups usually enter pre-existing societies on either the lowest or the highest socioeconomic level; rarely are they accepted for long by the bulk of the local residents at the latter's own level. It therefore would appear, based on artifactual analysis, that the highland immigrants at Altar and Barton Ramie were socially (and probably politically and economically) dominant in the first few generations after the migration.

However, as Schwartz (1970:178) notes in discussing the developmental phases of postmigration cultures, the community may eventually develop along lines quite different from those of either of the two separate groups or of the original postmigration community. Even though at Altar and Barton Ramie the intrusive sociotechnic artifacts are initially dominant, the overall Maya lowland developmental continuum from the Late Formative into the Classic overwhelmed the Protoclassic cultural intrusion within a few generations. All the sites that may have received a significant site-unit intrusion were, by the fifth century, fully assimilated within the ongoing Classic Maya civilization.

The late facet of the Cantutse (terminal Preclassic) phase at Seibal contains numerous elements of the Protoclassic complex, such as mammiform tetrapod vessels (thick, flaring-sided bowls with mammiform supports), orange-slipped ceramics, and imitation "Usulután" types (Sabloff 1975:11, 231–232). The Protoclassic influence at Seibal is similar to that at Tikal; in both cases there is no published evidence of a full site-unit intrusion and sudden population increase, but rather Protoclassic elements evidently appeared by trade or imitation. Even though the earliest ritual construction at Seibal dates to the late facet of the Cantutse phase, Seibal experienced a population and "cultural" decline (Sabloff 1975:11–12), with "much of the site abandoned to the jungle."

### Barton Ramie: Ilopango-induced Flooding and Immigration?

At Barton Ramie during the Protoclassic Floral Park phase a number of cultural and natural events occurred at approximately the same time and may have been interconnected. The dating of these events is not precisely known. They occurred sometime between 100 B.C. and A.D. 300 (Willey et al. 1965:26–27, 565), placing them within the time frame of the Ilopango eruption. These changes include a more than doubling of population, as evidenced by a more than twofold sudden increase in "house occupations," and new ceramic characteristics (Figure 17.13) interjected into an indigenous continuum—barkbeaters, perforated potsherd discs (probably spindle whorls for cotton processing), and probably other artifacts. Among the ceramic changes a new ceramic type appears at Barton Ramie—Aguacate Orange, which is so similar to sub-ash ceramics in El Salvador as to be indistinguishable to the ceramicists working at Chalchuapa and Barton Ramie (Sharer and Gifford 1970). These cultural changes were also accompanied by the virtual disappearance of freshwater mussels and univalves from Barton Ramie. The high sensitivity of aquatic species (animals and plants) to damage by tephra is noted by Malde (1964).

If the tephra fall radically diminished shellfish, fish, and related aquatic species in Belize, the same phenomenon should have occurred elsewhere. I suggest it may be more than coincidence that the Ocós area of Pacific Guatemala-Chiapas was abandoned in the Early Classic (Shook 1965:185–186), and then thoroughly reoccupied a few centuries later (Coe and Flannery 1967(84–91). The Formative occupation at Ocós was heavily dependent on aquatic protein resources (Coe and Flannery 1967). At Bilbao, more inland on the Guatemalan Pacific coastal plain, Parsons (1967:24) encountered a marked diminution of cultural materials after the Preclassic that may be indicative of a population decline in, but not an abandonment of, the area. Because the Bilbao inhabitants apparently were not heavily relying on aquatic protein resources, I would expect their agricultural subsistence base to have been less sensitive to a tephra-induced perturbation than that of their Ocós neighbors.

The evidence of substantial flooding of the Belize River was encountered at Barton Ramie in the form of a sterile brown clay deposit stratigraphically separating the Preclassic from later occupations and humic horizons. "We noted that mounds whose construction began in Preclassic Period times had their bases directly on or slightly into the [buried] black soil stratum, while mounds whose construction began at a later date had brown clay intervening between the mound base and the black soil [Willey *et al.* 1965:31]."

This flooding occurred at about the same time as the decrease in mollusks and the Protoclassic intrusion. Flooding is common during explosive volcanic eruptions, resulting from tephra damage to vegetation in the drainage basin of rivers combined with the greatly increased rainfall.

## Possible Effects of the Ilopango Eruption on Western Honduras

In a paper (Sheets 1977) on the possible effects of the Ilopango eruption on western Honduras, I reviewed the geological and hydrological literature on the long-distance effects of comparable eruptions. Using that comparative literature as a basis, I was able to make general predictions of possible effects. The archaeological literature was examined with those predictions in mind, to see whether there was any possible evidence of the predicted effects. A surprisingly high number of sites did show evidence that could be interpreted as deriving from the Ilopango eruption. But it should be pointed out that weaknesses in Honduran data, particularly the lack of chronometric dating and the lack of petrographic and geochemical data, leave that causality unconfirmed. Until these predictions are tested, this section must be considered speculative.

If the Ilopango eruption did cause flooding at Barton Ramie or in western Honduras, then there should be Ilopango tephra in sediments. The archaeologist excavating a possible Ilopango-induced flood deposit should not expect to see a pure, white layer of ash such as exists close to the source as air-fall deposits. Rather, the layer will be darker in color as the ash is incorporated with eroded soils and other sediments. The Ilopango ash may be present as a very small percentage of the bulk of flood deposits, requiring careful petrography and electron microprobe analysis to discover and reliably identify it.

The predicted effects for western Honduras are summarized later, followed by an overview of the archaeological literature in the area. With large explosive eruptions, air-fall ash decreases in particle size with increasing distance from the source. The fine-grained ash that falls farther away from the source is less permeable, resulting in increased runoff and erosion. As Segerstrom (1960) has shown at Parícutin, volcanic ash is unstable on slopes. Many slopes greater than 15° with ash less than 25cm in thickness were re-exposed down to the pre-eruption humic soil horizon in 10 or 15 years from the end of the eruptions.

The rapid erosion of ash increased stream particulate load with its specific gravity. That resulted in both accelerated channel cutting where stream gradients were high and flooding with rapid sedimentation downstream where gradients lessened. Some samples of tephra-laden water contained as much ash as water, and one sample was 79% particulate load by volume. The increase in specific gravity, up to 2.08, resulted in transport of large boulders that had been immobile with the same amount of clear-water stream flow. As the Rio Itzicuaro changed gradient from 2.2 to 1.2%, where it reached its flood plain, it began dropping sediment load. Coarser materials are deposited earlier than finer sediments, so that in the lower flood plain only fine-grained sediments compose the alluvial deposit.

Not all tephra on slopes end up in stream deposits. Lahars are volcanic mudflows or landslides composed of water-saturated volcanic material (Bullard 1976:541), and they commonly accompany explosive volcanic eruptions. Herculaneum, for example, was buried by a lahar from Mount Vesuvius a short time after Pompeii was struck by airfall tephra during the A.D. 79 eruption (see Jashemski, Chapter 19 in this volume).

Before discussing possible evidence of lahars and floods at archaeological sites in Honduras, it should be noted that nothing is known yet about the distribution of Ilopango tephra outside El Salvador. Given the approximate 1-m original depth at Chalchuapa, 77 km west-northwest of Ilopango, and the findings of considerable deposits northward along the Rio Lempa (Earnest 1975; Sheets 1976), significant air-fall deposits should have reached western Honduras.

The straight-line distance from Ilopango to Copán is 134km, so it is likely that the Rio Amarillo drainage received measurable amounts of ash. Since the basin is not large, major flooding is not likely, but minor flooding in mid-valley is to be expected. Lahars might have occurred along the valley margins, but it is unlikely that they were large. Archaeologically, flooding has been documented, but not well dated, for the valley. It did occur sometime prior to the Classic period, for Longyear (1948, 1952) found Classic construction and occupational debris overlying a 40-cm-thick sterile river clay deposit that represents the flooding. Animal bones, charcoal, and lithic artifacts were encountered under the flood deposit.

The Ulúa River drainage is considerably larger than the Amarillo, and the closest portion is only 80km from Ilopango. The southern part of the drainage is dissected, with high stream gradients, but streams coalesce into a low-gradient broad flood plain north of Lake Yojoa. Here the prediction is of lahars and stream entrenchment in the southern part of the drainage basin, with fewer lahars but major flooding north of Lake Yojoa. The most marked aggradation in Honduras is to be anticipated in this flood plain.

Archaeological evidence from the Ulúa flood plain indicates this actually may have occurred. At Playa de los Muertos (Popenoe 1934; Strong et al. 1938) a very thick flood deposit has been encountered that separates Formative from Classic occupation (Figure 17.14). In the Formative occupation layer were living floors, charcoal (undated), sherds, prismatic blades, grinding stones, hammerstones, and evidence of wattle-and-daub housing. The Classic layer is characterized by the presence of polychrome ceramics, extended burials, bifacial obsidian implement manufacture, and groundstone artifacts. These deposits are separated by as much as 5m of flood-lain silt.

The Santa Rita site, 10km downstream from Playa de los Muertos, has a similar stratigraphy: (a) Preclassic materials in the lower layer, (b) a flood deposit, followed by (c) a Classic occupation (Strong et al. 1938:46ff). The presence of Usulután ceramics and orange ware (Aguacate Orange?) help identify the lower layer as Middle or Late Preclassic. That tightens the apparent time span of the flood to Late Preclassic or Early Classic, which does not weaken the possibility that Ilopango was the cause.

Los Naranjos has a similar stratigraphic sequence (Figure 17.15), but it is not located along a river. It is on the north shore of Lake Yojoa. The lower layer (Strong et al.:102ff) contains abundant Preclassic artifacts. It is capped by a 60–80-cm-thick layer of mixed sterile yellow clay and gravel. It is possible that the deposit could be a lahar deriving from the very high mountains (up to 2835 m) to the west of the site. Classic occupation was effected on top of this deposit.

In summary, the Honduran archaeological data are intriguingly close to the predictions based on what is now known about the scale of the Ilopango eruption and the effects of comparable, historically documented eruptions. Sampling of archaeological sites and of lake and stream alluvial strata, followed by chronometric dating and petrography plus geochemistry, will generate data to confirm or deny the hypothesized Ilopango eruptive effects on western Honduras.

## Long-range Social and Economic Effects

Since the 1960s scholars have been studying natural hazards and disasters from a social science perspective. These studies, conducted primarily by geographers, sociologists, and economists, in a number of ways have broadened our understanding of human behavior under the threat or impact of extreme physical events. Their deliberate avoidance of the old pitfall of studying only the most extremely affected has resulted in insights into the differential impact of hazards or disasters as related to social class, ethnicity, location, or economic sector of a society. In fact, some sectors of a society benefit from

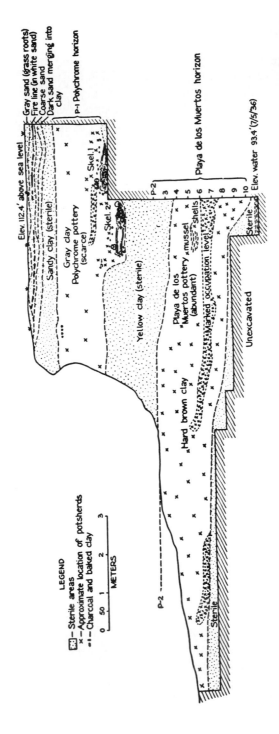

FIGURE 17.14 Playa de los Muertos, Honduras: North wall of Excavation 1. The lower Playa de los Muertos horizon apparently is Preclassic, and it is capped by a sterile yellow clay deposit that might have been caused by Ilopango ashfall and increased rainfall in the headwaters of the Ulúa River. The flooding was followed by Classic period occupation, as indicated by the polychrome horizon. (After Strong *et al.* 1938:64.)

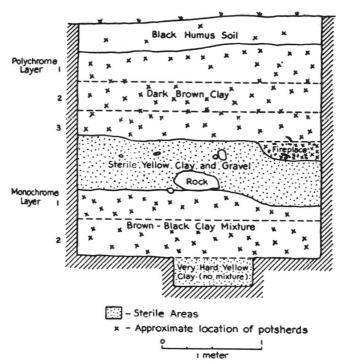

FIGURE 17.15 Los Naranjos, Honduras: Excavation A near Site I. The lower, apparently Preclassic, horizon is separated from the upper, apparently Classic, horizon by a sterile layer of clay and gravel. The sterile deposit may be a lahar (volcanic mudflow) resulting from the Ilopango eruption. (After Strong et al. 1938:113.)

disasters, in the short run or the long run. And social scientists are just beginning to understand some of the long-term effects of disasters.

Much of the earlier literature on natural disasters concentrated on the immediacy of the extreme event. These studies are generally humanistic, rather than social science, in their focusing on a single unique event rather than using a comparative framework, and in their tendency to examine only the extreme cases of hardship within a short time perspective.

Natural hazard and disaster research is a subarea of the study of human ecology, the investigation of the dynamic interrelationships between people and their environment. And, of course, no environment is entirely constant; mean annual rainfall or temperature figures are statistical abstractions. In response, all human societies have built flexibility into their adaptive strategies to deal with secular environmental fluctuations. Hazard and disaster research investigates the extreme of this continuum, the mentalistic and behavior adjustments that people make to greater-than-normal environmental perturbations.

One of the oft-noted limitations of the present state of hazard and disaster research is the time frame. Social science researchers consider a time frame of a few years to as much as a few decades a long-term perspective. True, compared with the earlier studies, which often ended their accounts only a few hours or days after the disaster struck, this is an improvement. But disaster researchers are anxious to extend their temporal coverage, and archaeologists can contribute valuable information. Although contemporary techniques do not allow for as fine-grained a resolution of people and events in the past as would be desirable, archaeology is capable of examining the effects of natural disasters on political, economic, and social systems over decades or centuries. Renfrew (Chapter 18 in this volume) notes the possible effects of the Thera eruption on Minoan and Mycenaean civilizations, Workman (Chapter 11) argues for major effects on inland Yukon populations from intermittent ashfalls, and below I propose that the economic disruptions of the southern Maya were beneficial to the Maya at Tikal, at least for 2 centuries. And it now is clear that neither the soils nor the Maya human populations in El Salvador completely recovered from the Ilopango eruption.

Not all components of a society struck by a natural disaster are affected equally. Researchers have found significant variation by class and by residence in the years following a disaster. "Disasters by their very nature were inequitous [Burton et al. in press I-16]." Some sectors of the society bear the negative effects for long periods, yet "there are gainers as well, those who by work or location are well placed to profit from a hazard [Kates 1970:16]."

An example is provided by Bowden and Gelman (1975:II-7, 8), who note the differential impact of the 1972 Managua earthquake on various components of Nicaraguan society. Some elite and upper-class members rapidly took advantage of new business opportunities made available by the disrupted social and economic systems. Many poor people suffered greatly during the months of indecision regarding construction. Similarly, after the 1906 San Francisco earthquake, the recovery by class and by ethnic group varied considerably.

A survey of cases wherein social, economic, and/or political processes were accelerated by disasters was compiled by Mileti et al. (1975:137–139). Occasionally new institutions emerge following the disaster, but more common is the strengthening and interorganizational reformulation of some institutions and the weakening of others. Other cases where acceleration of pre-existing processes occurred because of a disaster are the 1953 Holland floods, the 1972 Managua earthquake, the 1972 Rapid City flood, the 1964 Alaskan earthquake, and the 1906 San Francisco earthquake (Kates 1975; Mileti et al. 1975:137–139). And the long-range beneficial effects of a disaster can vary considerably in different zones when viewed from a regional perspective, as Rogers (1970) found in studying the economic impact of the 1964 Alaskan earthquake.

Such hazard and disaster research findings may be applicable to Mayan prehistory in helping to explain the economic growth of the lowland Classic Maya by a shift in the major Mesoamerican trade route as a result of the Ilopango eruption. The major trade route linking southeastern and northwestern Mesoamerica was southern during the Preclassic (Sabloff and Rathje 1975;

Sharer 1975; Dahlin 1976), with El Salvador as a key segment at the southeastern end. Preclassic trade included large quantities of obsidian, salt, jade, cotton, cacao, and probably other items, such as balsam.

It long has been noted that the lowland Maya crossed the threshold into full civilization by A.D. 300; in fact, the recognition of A.D. 300 as the beginning of the Classic period reflects this understanding. But not all areas of the lowlands were undergoing culture change at the same rate. As Haviland (1978) notes, Tikal "took an early lead in the development of lowland Maya civilization [p. 762]". If the Ilopango disaster did have an accelerating effect on lowland cultural development, the effects were not uniformly distributed. Some sites, because of fortuitous geographic location (and perhaps other factors difficult to reify archaeologically, such as aggressive leadership, early information on the disaster, or pre-established social and economic contacts) may have benefited more than others from the misfortunes of the southeast Maya highlands.

During the Classic period, I suggest at least partially resulting from the Ilopango disaster, the dominant trade route shifted northward to the Petén, benefiting sites such as Tikal (Dahlin 1976:182–184). Thus, the social, ecologic, and economic disruptions caused by the eruption may have been exploited during the Early Classic by the lowland Maya and particularly Tikal, as the major Mesoamerican trade route became trans-Petén. Delegations from Teotihuacán, the major power at the other end of the trade route, established themselves at Tikal during the Early Classic, probably to maintain control or at least a strong influence over the exchange system.

But, on a longer-term basis, the success and growth of the lowland Maya at Tikal declined by approximately A.D. 500 (Dahlin 1976) as Tikal entered the "Dark Ages" of the Middle Classic. Elite-directed activities of monumental construction, stela erection, caching, and manufacture of elaborate polychrome ceramics declined. Paradoxically the importation of obsidian **increased** by late in the sixth century. This apparent paradox might be explained by the success of the Chortí southward expansion at about this time to control both Ixtepeque and Media Cuesta obsidian sources.

Ixtepeque, and probably Media Cuesta, would have been severely affected by the Ilopango ashfall, and abandonment of their environs would have been likely. It is possible that the reoccupation of the southeast Mayan highlands was a direct colonization scheme directed by the lowland Maya as mediated through Quiriguá and Copán for the purpose of "owning" their own obsidian sources. That 86% of a sample of 28 obsidian artifacts from Quirguá that were analyzed for trace elements derived originally from the Ixtepeque source (Stross personal communication 1978) adds substantiation for this argument.

In overview, the trend acceleration phenomenon of natural disasters may have operated on the Maya in two ways. The sudden arrival of large numbers of immigrants in the well-settled lowlands could have required an intensification of agriculture. Such a migration should have lead to greater political control mechanisms, an enlarged local system of production and distribution of needed commodities, and more accentuated social class differentiations. Here demographic change can be viewed as an independent variable affecting others in such a way as to accelerate tendencies that were underway prior to the eruption.

The second major effect on the lowland Maya was initially economic. With the dominant southern trade route of the Preclassic in shambles, the benefit to the group obtaining control of an alternative route would have been great indeed. I suspect it is not by accident that Tikal prospered so markedly in the Early Classic, for it is located midway along the portage between the Rio Hondo and Belize River to the east and the San Pedro Martir River to the west.


## SUMMARY

The region of the southeast Maya highlands that is now called El Salvador was first settled by agricultural populations at approximately 1200 B.C. (Sharer 1974, 1978). Population growth in the resource-rich intermediate tropics led to a complex society with an intensive agroeconomic adaptation by the time of Christ. The region was dotted with agricultural villages surrounding large economic, political, and ritual centers. The vitality of southeast highland societies was ended abruptly during the third century by a violent Peléan eruption of Volcán Ilopango. Some low-lying areas near the volcano were devastated by hot ash flows. Survival of plants and animals, including people, would not have been possible in these areas. However, human survival in the larger area receiving the ashfall would have been possible, but migration would have been necessary because of damage to agricultural land.

Some Maya lowland sites apparently received the highland immigrants, and Barton Ramie is an example. The sudden arrival of people in lowland communities may have acted as a catalyst to intensify agricultural adaptations as well as to accelerate trends toward greater political centralization and class differentiation. And some lowland sites that apparently did not receive disaster victims directly, such as Tikal, may have profited greatly by seizing control of the major northwest–southeast Mesoamerican trade route and successfully rerouting it through the Petén.

Recovery from the damage done to soils, vegetation, and human populations was slow, in part because of the sialic (acid) nature of the tephra and its thickness even at large distances from Ilopango. Soils had at least partially recovered 2 centuries later to support a moderately dense human reoccupation. A major component of the agroeconomic adaptation of these Classic-period Maya was corn and bean farming. Subsequent volcanic eruptions in western El Salvador affected soils, vegetation, and people in various ways, but not to the same degree as the Ilopango eruption. Soils analysis indicates the soils of the past two millennia have yet to recover the fertility of the Preclassic soils that existed prior to the eruption of Ilopango.

## ACKNOWLEDGMENTS

A number of institutions have supported research, which has, directly and indirectly, contributed information about the Ilopango eruption. They are the National Science Foundation (grants SOC75-04209 and BNS77-13441), the Ford Foundation, the University Museum of the University of Pennsylvania, and the Council on Research and Creative Work of the University of Colorado. Their assistance is most gratefully acknowledged.

Salvadoreans have, over the years, lived up to their national reputation for hospitality in facilitating our research. In particular I would like to mention Stanley Boggs, Carlos deSola, Roberto Huezo, Victor Manuel Mejía Murcia, Roberto Galicia, John Sartain, Coronel Tejada, Edwin Nuila, Daysi Contreras de Ríos, and many others.

Numerous specialists have aided in the detailed analyses of artifacts, soils, charcoal, and tephra. Without their contributions little would be known about the Ilopango disaster. They include Virginia Steen-McIntyre, Robert Sharer, Sam Valastro, Gerald Olson, Howard Earnest, William Fowler, Richard Crane, Fred Stross, Marilyn Beaudry, Susan Short, Peter Mehringer, Irving Friedman, Bill Rose, Fred Trembour, Rick Hoblitt, Steven Self, Branson Reynolds, and Diana Kamilli.

This manuscript benefited from critical readings by Don Dumond, Don Grayson, and Marilyn Beaudry.

## REFERENCES

Adams, R. E. W.
   1971   The ceramics of Altar de Sacrificios. *Papers, Peabody Museum, Harvard University* 63:1.
Beaudry, M.
   1978   Preliminary analysis of ceramics from the 1978 Protoclassic Project, Zapotitán Basin, El Salvador. In Research of the Protoclassic Project in the Zapotitán Basin, El Salvador, edited by P. D. Sheets. Unpublished ms., Dept. of Anthropology, University of Colorado.
Black, K.
   1978   Preliminary report on the survey of the Zapotitán Basin area. In Research of the Protoclassic Project in the Zapotitán Basin, El Salvador: A preliminary report of the 1978 season, edited by P. D. Sheets. Unpublished ms., Dept. of Anthropology, University of Colorado.
Boggs, S. H.
   1966   Pottery jars from the Loma del Tacuazín, El Salvador. *Middle American Research Records* 3:5. Tulane University.
Bowden, M., and K. Gelman
   1977   Models of the reconstruction process: Evolutionary model of reconstruction following disaster. In *Research on reconstruction following disaster*, edited by J. Haas, R. Kates, and M. Bowden. Cambridge: MIT Press.
Browning, D.
   1971   *El Salvador: Landscape and society*. Oxford: Clarendon Press.
Bullard, F. M.
   1976   *Volcanoes of the earth*. Austin: University of Texas Press.
Burton, I., R. Kates, and G. White
   In press   *The environment as hazard*. New York: Oxford University Press.
Chandler, S. M.
   1978   Preliminary report of excavations at the Cambio Site (336–1). In Research of the Protoclassic Project in the Zapotitán Basin, El Salvador: A preliminary report of the 1978 season, edited by P. D. Sheets. Unpublished ms., Dept. of Anthropology, University of Colorado.

Coe, M. D., and K. V. Flannery
    1967    Early cultures and human ecology in South Coastal Guatemala. *Smithsonian Contributions to Anthropology* 3.
Dahlin, B. H.
    1976    An anthropologist looks at the Pyramids: A Late Classic revitalization movement at Tikal, Guatemala. Doctoral dissertation, Dept. of Anthropology, Temple University. University Microfilms, Ann Arbor.
Daugherty, H. E.
    1969    Man-induced ecologic change in El Salvador. Doctoral dissertation, Dept. of Geography, UCLA. University Microfilms, Ann Arbor.
duToit, B.
    1975    Introduction: Migration and population mobility. In *Migration and urbanization*, edited by B. duToit and H. Safa. The Hague: Mouton.
Earnest, Howard H., Jr.
    1975    Second preliminary report on archaeological investigations on the Hacienda Sta. Barbara, Chalatenengo, El Salvador. Unpublished ms.
Harris, D. R.
    1972    The origins of agriculture in the tropics. *American Scientist* 60:180–193.
Hart, William J., Jr.
    1978    Recent volcanism in the Zapotitán Basin. In Research of the Protoclassic Project in the Zapotitán Basin, El Salvador: A preliminary report of the 1978 season, edited by P. D. Sheets. Unpublished ms., Dept. of Anthropology, University of Colorado.
Haviland, W. A.
    1978    The rise of the Maya. *Science* 199:761–762.
Kates, R.
    1970    Natural hazard in human ecological perspectives: Hypotheses and models. Natural Hazard Research Working Paper 14, Institute of Behavioral Science. Boulder: University of Colorado.
    1977    Major insights and central issues. In *Research on reconstruction following natural disaster*, edited by J. Haas, R. Kates, and M. Bowden. Cambridge: MIT Press.
Lardé, Jorge
    1926    Arqueología Cuzcatleca. Vestigos de una población pre-Mayica en el Valle de San Salvador, C. A., sepultados bajo una potente capa de productos volcanicos. *Revista de Etnologia, Arqueologia, y Linguistica* 1:3 and 4.
Longyear, J. M. III
    1944    Archaeological investigations in El Salvador. *Memoirs of the Peabody Museum, Harvard University* IX:2.
    1948    A sub-pottery deposit at Copán, Honduras. *American Antiquity* 13:248–249.
    1952    Copán Ceramics. Carnegie Inst. Washington, Pub. 597.
    1966    Archaeological survey of El Salvador. In *Handbook of Middle American Indians*, edited by R. Wauchope 4:132–156. Austin: Univ. of Texas Press.
Lothrop, S. K.
    1927    Pottery types and their sequence in El Salvador. *Indian Notes and Monographs* 1:4:165–220. Museum of the American Indian, Heye Foundation, New York.
Malde, H.
    1964    Ecologic significance of some unfamiliar geologic processes. In *The reconstruction of past environments*, edited by J. J. Hester and J. Schoenwetter. Fort Burgwin Research Center.
Merwin, R. E., and G. C. Vaillant
    1932    The ruins of Holmul, Guatemala. *Memoirs of the Peabody Museum, Harvard University* III:2.
Mileti, D., T. Drabek, and G. Haas
    1975    Human systems in extreme environments: A sociological perspective. Monograph 21, Program on Technology, Environment and Man, Institute of Behavioral Science. Boulder: University of Colorado.

Olson, G. W.
  1978  Field report on sampling of buried Maya soils around San Salvador and in the Zapotitán
        Basin, El Salvador, Central America: An evaluation of soil properties and potentials in
        different volcanic deposits. In Research of the Protoclassic Project in the Zapotitán Basin,
        El Salvador: A preliminary report of the 1978 season. Unpublished ms., Dept. of An-
        thropology, University of Colorado.
Parsons, L. A.
  1967  Bilbao, Guatemala. *Milwaukee Public Museum, Publications in Anthropology*, 1(11).
Peterson, W.
  1968  Migration: Social aspects. *International Encylopedia of the Social Sciences* 10:286–292.
Popenoe, D. H.
  1934  Some excavations at Playa de los Muertos, Ulúa River, Honduras. *Maya Research*
        1(2):61–85.
Porter, M. N.
  1955  Material Préclasico de San Salvador. *Communicaciones del Instituto Tropical de Inves-
        tigaciones Científicas* 3/4:105–12.
Rogers, G. W.
  1970  Economic effects of the earthquake. In *The great Alaska earthquake of 1964*, edited by the
        National Research Council. Washington: National Academy of Sciences.
Sabloff, J. A.
  1975  Excavations at Seibal, Department of Petén, Guatemala: Ceramics. *Memoirs, Peabody
        Museum, Harvard University* 13:2.
Sabloff, J. A., and W. L. Rathje
  1975  The rise of a Maya merchant class. *Scientific American* 233:4:72–82.
Safa, H., and B. duToit (Eds.)
  1975  *Migration and Development*. The Hague: Mouton.
Sayre, A. N., and G. C. Taylor, Jr.
  1951  Ground-water resources of the Republic of El Salvador, Central America. USGA, Water
        Supply Paper 1079-D. Washington, D.C.: U.S. Government Printing Office.
Schmidt-Thomé, Michael
  1975  The geology in the San Salvador area, a basis for city development and planning. *Geol.
        Jb. B.* 13:207–228.
Schwartz, D. W.
  1970  The postmigration culture. A base for archaeological inference. In *Reconstructing prehis-
        toric pueblo societies*, edited by W. Longacre. Albuquerque: University of New Mexico
        Press.
Segerstrom, K.
  1950  Erosion studies at Parícutin. U.S. Geological Survey Bulletin 965A. Washington.
  1960  Erosion and related phenomena at Parícutin in 1957. U.S. Geological Survey Bulletin
        1104-A. Washington.
Sharer, R. J.
  1968  Preclassic archaeological investigations at Chalchuapa, El Salvador: The El Trapiche
        Mound Group. Doctoral dissertation, Dept. of Anthropology, University of Pennsylvania.
        University Microfilms, Ann Arbor.
  1974  The prehistory of the southeastern Maya periphery. *Current Anthropology* 15(2):165–187.
  1975  The southeast periphery of the Maya area: A prehistoric perspective. Paper presented at
        the Seventy-fourth Annual Meeting, American Anthropological Association, San Fran-
        cisco.
Sharer, R. J. (General Ed.)
  1978  *The prehistory of Chalchuapa, El Salvador* (3 vols.) Philadelphia: University of Pennsyl-
        vania Press.
Sharer, R. J., and J. C. Gifford
  1970  Preclassic ceramics from Chalchuapa, El Salvador, and their relationships with the Maya
        lowlands. *American Antiquity* 35:441–462.

Sheets, P. D.
  1975 Review of D. Stone, Precolumbian man finds Central America. *American Anthropologist* 77:132–134.
  1976 Ilopango Volcano and the Maya Protoclassic. University Museum Studies #9. Carbondale: Southern Illinois University
  1977 Possible repercussions in western Honduras of the third century eruption of Ilopango Volcano. Paper presented to the Conference on the Southeast Maya Area, UCLA Latin American Center, Los Angeles.
  1979 Maya recovery from volcanic disasters: Ilopango and Cerén. *Archaeology* 32(3):32–42.
Sheets, P. D. (Ed.)
  1978 Research of the Protoclassic Project in the Zapotitán Basin, El Salvador: A preliminary report of the 1978 season. Unpublished ms., Dept. of Anthropology, University of Colorado.
Shook, E. M.
  1965 Archaeological survey of the Pacific Coast of Guatemala. In *Handbook of Middle American Indians*, edited by R. Wauchope 2:180–94. Austin: University of Texas.
Steen-McIntyre, Virginia
  1976 Petrography and particle size analysis of selected tephra samples from western El Salvador: A preliminary report. Appendix 1 in *Ilopango Volcano and the Maya Protoclassic* by P. D. Sheets. University Museum Studies #9. Carbondale: Southern Illinois Museum
  1978 Progress report, Protoclassic Project: Section on geology. In Research of the Protoclassic Project in the Zapotitán Basin, El Salvador: A preliminary report of the 1978 season. Unpublished ms., Dept. of Anthropology, University of Colorado.
Stone, D.
  1972 *Precolumbian man finds Central America*. Cambridge, Mass.: Peabody Museum Press.
Strong, W. D., A. V. Kidder II, and A. J. D. Paul, Jr.
  1938 Preliminary report on the Smithsonian Institution—Harvard University archaeological expedition to northwestern Honduras, 1936. *Smithsonian Miscellaneous Collections* 97(1):1–129.
Tsukada, M., and E. Deevey
  1967 Pollen analyses from four lakes in the southern Maya area of Guatemala and El Salvador. In *Quaternary paleoecology*, edited by E. J. Cushing and H. E. Wright, Jr. New Haven: Yale University Press.
Willey, G. R.
  1973 The Altar de Sacrificios excavations: General summary and conclusions. *Papers, Peabody Museum, Harvard University* 64:3.
Willey, G. R., W. Bullard, Jr., J. Glass, and J. Gifford
  1965 Prehistoric Maya settlements in the Belize Valley. *Papers, Peabody Museum, Harvard University* 54.
Willey, G. R., T. P. Culbert, and R. E. W. Adams (Eds.)
  1967 Maya lowland ceramics: A report from the 1965 Guatemala City Conference. *American Antiquity* 32:289–315.
Willey, G. R., and J. C. Gifford
  1961 Pottery of the Holmul I style from Barton Ramie, British Honduras. In *Essays in Precolumbian art and archaeology*, edited by S. K. Lothrop. Cambridge, Mass.: Harvard University Press.
Williams, H., and H. Meyer-Abich
  1955 Volcanism in the southern part of El Salvador. *University of California Publications in Geological Sciences* 32:1–64.

# 18

# The Eruption of Thera and Minoan Crete

COLIN RENFREW

For more than a century it has been realized that the volcanic island of Thera in the Aegean Sea contained prehistoric remains buried in an eruption of some magnitude (Fouqué 1879). Pottery and buildings, buried deep in volcanic ash, gave ample evidence of a prehistoric cataclysm that at first could not be dated with any precision. Excavations were renewed on the island in 1967 by the late Professor Spyridon Marinatos, at the site of Akrotiri (Marinatos 1968–1976). The results have been exceedingly spectacular, revealing extensive areas of a very well preserved urban settlement, with buildings standing in places up to a height of more than 6 m. In some cases the basement rooms, the ground floor, and the first floor are still in position. The wide range of finds, notably the wall paintings, offer what is in many ways the most comprehensive body of material from this period in the Aegean, datable to around 1500 B.C. They underline once again that though volcanism is, from the standpoint of the inhabitant, a very destructive force, from that of the archaeologist it can be one of the best preservative agencies known, in favorable cases blanketing the remains in their pristine state under several meters of sterile ash.

It has been postulated that the effects of the Thera eruption were felt at a much greater distance, and that the widespread destruction of the palaces and settlements of Minoan Crete, datable to approximately the same period, may have been the direct result of it, whether by blast, tsunami, ashfall, or earthquake (Marinatos 1939; Page 1970; Hiller 1975). The "Minoan" eruption of Thera would then have been responsible not simply for the destruction (and preservation) of settlements on Thera itself, but for what has been regarded by

565

some as the end of a civilization. In such a case the consequences for Aegean history would be very considerable, and it has indeed been suggested that the direct effects of the eruption were experienced also in Cyprus and even Egypt.

But were the effects in Crete as dramatic as has been suggested? Was the widespread Cretan destruction in fact simultaneous with the great eruption of Thera? Or if, as now seems likely, the Cretan sites were destroyed some time after the destruction of the site of Akrotiri on Thera, could this have been in the later phase of a long eruptive sequence? And if the Thera eruption was not responsible for the disasters in Crete, what was? It was Marinatos who realized clearly that these were vulcanological questions just as much as archaeological, and in collaboration with Peter Nomikos, he organized in 1969 an international conference on Thera, attended by vulcanologists as well as archaeologists, to try to resolve some of these problems. The second such conference, organized by Nomikos, was held in the summer of 1978. The archaeological interest has given great impetus to the work of vulcanologists, and some of the questions are now nearer solution. In what follows only an outline of the complicated issues can be given: More details are found in the conference publications (Marinatos and Ninkovich 1971; Doumas 1978; Doumas, in press). And though it now appears likely that the effects of the eruption may not have been so dramatic or so destructive, outside Thera, as once thought, the whole program of research over the past decade has produced results that will be of great interest to all students of the historical effects of volcanic activity. The effects of blast, gaseous emission, seismic shock, ashfall, ash flow, mudflow, and tsunami have all been separately considered, and the studies in question have a relevance far beyond Thera.

## THERA

The island group of Thera, also known as Santorin or Santorini, constitutes the southernmost of the Cycladic Islands, lying in the Aegean Sea between mainland Greece and Crete (Figure 18.1). Today it has at its center (Figure 18.2) a deep caldera, which provides a well-sheltered harbor, although one too deep for ships to lie at anchor.

Geological research, summarized in the papers for the two conferences (Marinatos and Ninkovich 1971; Doumas 1978) makes abundantly clear that 4000 years ago Thera was a single island, probably rising to a peak in the center, rather than the present ring of land surrounding the deep caldera. The present configuration is mainly the result of a single eruption, dated by archaeological means (but see pages 574–577) to about 1500 B.C. The volcano remains active; indeed, the small island of Kameni, lying in the middle of the caldera, first appeared in classical Greek times and has continued to grow since then.

Geological and geochemical studies allow the major eruption in question to be classed as of Plinian type (Vitaliano and Vitaliano 1971)—the term being

FIGURE 18.1 Map of area discussed, showing the island of Thera and other areas mentioned.

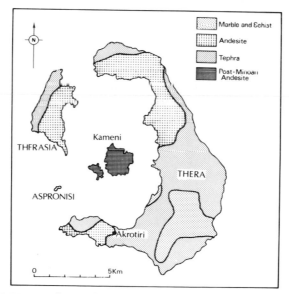

FIGURE 18.2 Thera today, showing outline geology of the island.

taken from the cataclysmic eruption of the volcano of Vesuvius in A.D. 79, described by the contemporary historian Pliny the Younger. Such eruptions occur suddenly with explosive force, and with the rapid ejection up to a height of several kilometers of great quantities of molten matter. Blocks of hot solid material ("bombs") accompanied by pumice and ash rain down over a radius of several kilometers. The dust cloud formed will, if a wind is blowing, carry tephra tens and even hundreds of kilometers away, often with serious consequences for vegetation and human settlement. Ash flows and mudflows, traveling horizontally and with great rapidity, can overwhelm all settlement within several kilometers of the active center. It was the ashfalls associated with the eruption of Vesuvius that buried Roman Pompeii (Jashemski, Chapter 19 in this volume); other settlements near Vesuvius, including Herculaneum, were buried, either in ash or in mud resulting from the eruption. The volcanism associated with Plinian eruptions is often cyclical in character, and it is clear that the major eruption of Thera, often termed the "Minoan" eruption on account of its possible effects on Minoan Crete, was not its first. Nor need it have been the last: In a few millennia, Thera may once again be a single unified mass, on the brink of a further cataclysm.

The dramatic nature of the Minoan eruption is suggested by the cliffs at the middle of the island group, overlooking the caldera. They fall sheer more than 200 m to the sea and continue in places as far as 300 m below sea level: The caldera to the north of Kameni is even today more than 400 m deep. There is no doubt that great quantities of ash and other material were ejected in the course of an eruption of some violence, and that the roof of the emptied volcanic chamber thus formed collapsed or subsided, allowing the sea to enter and hence to form the present deep and water-filled caldera. The sudden subsidence of the newly emptied magma chamber in this way can form great tsunamis ("tidal waves"), although they can also be formed by underwater earthquakes that need not be associated with volcanic eruptions.

The evidence suggests the validity of comparing the Minoan eruption with more recent eruptions of Plinian type elsewhere (Neumann van Padang 1971). The eruption of Krakatoa, east of Java, on 27 August 1883, for instance, was well described by a number of observers. A tsunami 30 m high, traveling at 550 km/hr inundated the Sunda Straits, and the death toll of the eruption has been estimated at over 30,000. It is indeed worth noting that in Plinian eruptions most deaths are usually due to secondary effects of this kind. At Thera, as at Pompeii and Herculaneum, nearly all the occupants escaped from the settlement itself, thus avoiding burial in the ashfall, although at Pompeii a number of refugees were overcome by fumes before they could make good their escape.

The underlying geology of such eruptions is now well understood, and the tectonic setting of the Mediterranean volcanoes has been well described (Ninkovich and Hays 1971). It should be mentioned, however, that the caldera collapse is not always a sudden one, and it need not be assumed that the Minoan eruption of Thera was accompanied by great tsunamis on the Krakatoan scale, although the eruption itself was undoubtedly as great in terms both of energy and of volume of material ejected (Hedervári 1971).

There are several problems involved in judging the scale of the wider effects of the great Minoan eruption on Thera, and in determining their precise date. Before turning to these it will be useful to say a little of their undoubted consequences on Thera itself, and of the arguments that have been put forward about the destruction of Minoan Crete.

## AKROTIRI

Small-scale excavations have been carried out at a number of locations on Thera, and chance finds of prehistoric material have been made during quarrying excavations. The remarkable archaeological potential of the island was first revealed, however, by the late Professor Spyridon Marinatos. He began work in 1967 on the south side of the island, in a dry stream bed, where winter rainfall had removed more than 6 m of volcanic ash. Elsewhere on the island the depth of the ash layer is generally greater than 6 m, and in places attains 60 m in depth.

Well-preserved building remains were revealed, and work in succeeding seasons (Marinatos 1968–1976; Doumas 1975; Schachermeyr 1976) has unearthed a district of a prehistoric town. The excavated area now exceeds 12,000 m². The buildings are of dressed stone, and in favorable cases more than one floor is preserved, so that there are basement as well as ground-floor rooms in much of the settlement.

Human remains, as already indicated, are not found, and it is clear that the population had ample time to escape before the settlement was buried by ash. Indeed Doumas, who has continued work at Akrotiri after the death of Marinatos, has presented the clear evidence for a brief reoccupation of the site (Doumas 1974, 1978). It seems likely that the settlement was first damaged by earthquakes and that some repair work was carried out there before the more dramatic events that caused abandonment by the inhabitants, and eventual burial.

Much of the pottery is locally made, but there are also numerous imports from Crete. It is these which offer the best evidence for the date of the eruption, since the ceramic sequence has been well studied in Crete since the time of Sir Arthur Evans. The imported pottery found on Thera is predominantly in the Late Minoan Ia style. The date of this material on Crete is established on the basis of links between Minoan Crete and ancient Egypt, where the archaeological sequence can be correlated with the historical chronology. The links between Crete and Egypt at this time are documented both by Egyptian objects imported in Crete and by Minoan finds in Egypt. On this basis, and assuming that pottery of the Late Minoan Ia style was in use in Crete and in Thera at the same time, the destruction of Akrotiri can be dated to about 1500 B.C. The chronological question is discussed further in what follows.

The major finds at Akrotiri so far fall into three main classes: the architecture, the pottery, and the wall paintings. For each of these the condition and the quantity of material far surpass anything else found in the prehistoric Aegean. But it is clear that the inhabitants had the opportunity to take many movable objects with them before the most cataclysmic phase of the eruption; thus there are none of the finds of precious metal encountered, for instance, in the rich burials of the Minoan and Mycenaean world, and, indeed, few of the beautifully worked vases in exotic stone found, for instance, in the palace of Kato Zakro in East Crete (Zakro is one of the major sites in Crete for which the eruption of Thera has been suggested as the destructive agency). A fourth class of find is eagerly awaited: the Akrotiri archive. For it is well documented that the palaces and major centers of Crete at this time have records on clay, inscribed in the Minoan Linear A script, and recent excavations on the island of Melos, to the northwest of Thera (Renfrew 1978a) indicate that such an archive must also have been maintained there. It is likely, in view of the very strong Minoan links reflected on Thera in both the pottery and the wall paintings, that the discovery of a Linear A archive is to be awaited. Presumably it will be complete, and there is the hope that it will be well preserved. By analogy with the Cretan finds, the records are not likely to be narrative historical documents, but simply administrative accounts (which is why the term *prehistoric* still seems an appropriate one applied to the early Aegean, despite a measure of literacy). But the records might nonetheless clarify important social questions and give evidence of the administration of the settlement, just as the later Linear B documents found at Knossos in Crete and at several Mycenaean sites on the mainland have clarified Aegean administration at a period some 1½ centuries later.

The excavations at Akrotiri have concentrated upon the excavation of a single area, revealing a whole series of houses, and one of the streets of the town, now referred to as Telchines Street. The town does not have the same regular and rectangular layout as the Cretan palaces, although ashlar masonry is just as widely used to face buildings (Shaw 1978). It should be noted too that the buildings are in general made entirely of stone, and Cretan scholars are already reconsidering their previous conclusion that the upper stories of the Minoan towns and palaces were often made of mud brick.

At the north end of the excavation is a building, the first to be excavated, containing large *pithoi* (storage jars); loom weights were found in the same room. Fifteen meters to the south lies the "House of the Ladies," so called from the vivid fresco found within it (see Figure 18.3). Twenty meters to the south again is the "West House," in which the Ship Fresco, the most interesting of the wall paintings so far recovered, was unearthed (Figure 18.4). The "West House" is at the northwestern end of a large excavated urban area, some 90 m north to south by 60 m east to west, where the visitor has the vivid impression

**FIGURE 18.3 Fresco from the "House of the Ladies." The lady of the west wall.**

of walking along a street, with buildings rising on each side of him, just as at Pompeii. But in this case the buildings are some 1500 years earlier.

This is not the place to describe the finds in any detail: They are well illustrated in the preliminary reports by Marinatos. But it is appropriate to say that Akrotiri has now yielded more complete vases than have ever been recovered from a single Aegean site in the past. The great importance of the assemblage is that the entire settlement was destroyed and abandoned simultaneously—as in any volcanic destruction deposit. One therefore obtains a clear and quantitatively valid view of the entire repertoire of pottery used on the site. This has its value in functional terms—and the plant remains preserved inside the storage jars document the vegetable component of the diet, as the bones do the animal contribution. Moreover, it is the best available indicator of the commerce of the time. Typological studies clearly show that pottery was reaching Thera from Crete and from the Greek mainland, and these have

FIGURE 18.4 The fresco of the "Naval Expedition." The "Flagship" of the small fleet (detail).

been backed up by petrological study of thin sections (Williams 1978) and by trace-element analysis (Jones 1978) to give well-documented data on exchange.

The great revelation of Akrotiri, however—and in this case *revelation* is not too strong a term—has been the wall paintings. The mural art of the time was already well known from Minoan Crete, where compositions have been carefully restored from numerous small fragments. At Akrotiri complete compositions have been found, in one case actually in situ on the wall. As well as increasing the repertoire of Aegean Late Bronze Age art, they offer, in their completeness, an unparalleled opportunity to understand the aesthetic qualities of the composition—the disposition of motifs, the balance, the use of space—in a way simply not possible with the fragmentary examples hitherto available.

Among the most remarkable of the wall paintings (which may not have been painted in the *al fresco* technique [Asimenos 1978]—thus the term *fresco* is best avoided) are those found in the West House. In Room 5 were found two large-scale depictions of fishermen bringing back the day's catch (Figure 18.5). And in the same room, on three walls, was found the "Miniature Fresco," depicting a sea battle, a splendid and imaginary landscape, and a remarkable composition showing a fleet of ships leaving one town and arriving at another amid scenes of public welcome (Page 1976). These miniature scenes constitute perhaps the most important historical document yet recovered for the Aegean Bronze Age. The ships offer a wealth of new information about the seafaring of the time, the town views give a whole new insight into the urban landscape,

**FIGURE 18.5 The fisherman fresco from the West House.**

and the historical subject depicted has given rise to much speculation. Marinatos himself judged that part of the painting refers to Libya (Marinatos 1974; Page 1976), but this view has been questioned by others, some of whom see the whole scene sequence as referring to Thera, or perhaps to Crete also (Morgan Brown 1978).

Archaeologically the wall paintings supplement the already rich documentation from the other material finds at Akrotiri. Artistically they offer some of the finest and best-preserved examples of one of the world's major early art styles.

## CRETE

Though the effects of the great eruption on Thera cannot be doubted, the magnitude of their impact on neighboring areas, notably Crete, is at present hotly disputed. The arguments are fairly complicated, depending inevitably upon the chronological sequence for Crete, elucidated by the study of the changing ceramic styles.

In the early years of this century, as the chronology of Minoan Crete was established, it became clear that the Minoan civilization as a whole had suffered a considerable setback during what could, in ceramic terms, be designated the Late Minoan Ib period. Most of the major settlements, including the palaces of Zakro and Mallia, were destroyed at this time, some of them never to be reoccupied. The only major site to survive and flourish was the main center of Knossos itself. There are clear suggestions that in the succeeding Late Minoan II period Knossos had much stronger links with the Mycenaean societies of mainland Greece. The pottery of Late Minoan Ib type found in the various destruction deposits, and Egyptian cross-dating, suggests that they should be set around 1450 B.C.

Sir Arthur Evans, the excavator of the great palace at Knossos, and in his time the foremost authority on the Minoan civilization, attributed the disaster that overcame Crete at this time to a severe earthquake (Evans 1921–1935, IV: 942). His assistant and successor, Pendlebury, offered two alternative explanations: a national revolt of Cretans against foreign rulers (Pendlebury 1939: 229), or the sack of Crete, including Knossos, by invading Mycenaeans (p. 230). Here he could cite in support the ancient legend of Theseus, the Athenian prince who successfully attacked and slew the bull-monster, the Minotaur, in his labyrinth, with the aid of the princess Ariadne. He painted the picture in graphic terms (Pendlebury 1939):

> And in the last decade of the fifteenth century on a spring day when a strong South wind was blowing which carried the flames of the burning beams almost horizontally northwards, Knossos fell.
> The final scene takes place in the most dramatic room ever excavated—the Throne Room. It was found in a state of complete confusion. A great oil jar lay overturned in one corner, ritual vessels were in the act of being used when the disaster came. It looks as if the king had been hurried here to undergo too late some last ceremony in the hope of saving the people. Theseus and the Minotaur. Dare we believe he wore the mask of a bull? [231]

In fact, subsequent research has made clearer that Knossos survived the Late Minoan Ib destructions seen in much of the rest of Crete. The final disaster at Knossos has now been placed in the Late Minoan IIIA1 period, around 1380 B.C. (although Professor L. R. Palmer (1965) would place it even later). In a sense, though, this chronological revision supports Pendlebury's second theory, that of Mycenaean invasion. For in the Late Minoan II period, which is now documented principally by finds at Knossos, there are Mycenaean features that were not present before. The Linear A script is replaced by Linear B, which since the decipherment by Michael Ventris (Ventris and Chadwick 1973) is widely accepted as a Mycenaean adaptation of Linear A, and was used to inscribe records in the Greek language. There are burials near Knossos at this time, the so-called Warrior Graves, which have grave goods of predominantly Mycenaean character, that can be used to support Pendlebury's second theory. Indeed, a version of it, taking account of all the more recent discoveries and refinements in chronology, is argued at present by Sinclair Hood (1978).

In the year 1939, Marinatos made the new and arresting suggestion that the eruption of Thera might instead have been responsible for the marked decline in Cretan civilization, and in particular for the destruction of the major centers (Marinatos 1939). This explanation could account in a direct and straightforward way for the widespread destructions in Crete at the end of the Late Minoan Ib period, as earthquakes or ashfall or tsunamis destroyed most of the Minoan settlements. Only Knossos, it seems, survived, and Knossos is well inland and might have escaped some of these destructive effects.

Marinatos was in fact led to this suggestion by his discovery of pumice in the debris of the ruined site of Amnissos, on the north coast of Crete, which was destroyed at precisely this time. And it was in search of further evidence bearing on this great issue that he later began his excavation at Akrotiri with such dramatic results.

This hypothesis of the destruction of Minoan Crete as a direct result of the Thera eruption won many adherents (e.g., Page 1970, 1978). It was at once claimed that other, comparable effects of the great eruption could be recognized elsewhere in the east Mediterranean—ash and pumice in Cyprus and the Levant (Åstrom 1978), and even in the Seven Plagues of Egypt (van Bemmelen 1971). But confronting some of these views there is a substantial chronological problem.

## THE CHRONOLOGICAL PROBLEM

At the time Marinatos wrote his brilliant and influential article (Marinatos 1939) the chronological distinction between the Late Minoan Ia and Ib ceramic styles was not universally accepted. Indeed, as we have seen, Pendlebury could still consider the Late Minoan Ib destructions contemporary with the final destruction of the palace at Knossos, now agreed by virtually all scholars to have been half a century later.

More recent work in Crete, including excavations by Sinclair Hood at Knossos itself, show that the most characteristic pottery of the Late Minoan Ib period—termed the "Marine Style" on account of its lively use of octopus, argonaut, and other maritime motifs—does indeed represent a distinct phase of manufacture and use. The Late Minoan Ib period is set by many writers between about 1500 and 1450 B.C. Pottery of this kind is found buried in many of the destruction deposits at various sites in Crete.

But here is the problem. The pottery buried at Akrotiri is of the Late Minoan Ia style, and there is not a single Marine Style sherd from the site that has been acknowledged as such by competent specialists. The question then emerges, Were the two events—the eruption of Thera and the destructions in

Crete—simultaneous? A controversy has developed in which there are two polar views. The first (e.g., Hood 1971) argues that the Cretan destructions took place some 50 years after the eruption of Thera, and that if direct effects of that eruption are to be found in Crete, they may perhaps be recognized in several less striking indications of damage at the end of the Late Minoan Ia period. This would imply that the major Late Minoan Ib destructions in Crete had an altogether different cause. The second view argues that a different interpretation of the pottery on Thera, and a consideration of the mechanics of import, can lead to the conclusion that the two events were simultaneous (Luce 1976). It has been argued that there are a few sherds from Akrotiri that have a Marine Style character, and that there may have been delays before Thera obtained the latest products then current in Crete (Luce 1976; Thorpe-Scholes 1978). On this view, therefore, Marinatos's original theory would be upheld.

An intermediate view is that the eruption may have had several phases of activity lasting over several decades. The first of these would have destroyed Akrotiri in an earthquake, and the last perhaps 50 years later would have covered eastern Crete in ash and culminated in the sudden collapse of the volcano chamber, forming the Thera caldera and causing destructive tsunamis. Certainly it is the case that the settlement at Akrotiri was first damaged by earthquakes, and then was reoccupied for a while, during which time some efforts were made to repair the damage (Doumas 1974, 1978). Then came the great eruption with its ashfalls, burying the town completely. This compromise view would thus allow the first destructive earthquake in Thera to have taken place around 1500 B.C., near the end of the Late Minoan Ia period. Akrotiri would then have been reoccupied for a while before the final eruption itself, which might, on this view, have proceeded in several phases of activity, perhaps over several decades. It would have been the last phase of the eruption, around 1450 B.C., that had such disastrous effects upon Crete, with the destruction of the Late Minoan Ib palaces.

Over the past 10 years these questions have been much discussed. In the view of most archaeologists, the pottery at Akrotiri is of undoubted Late Minoan Ia character, without any traces of the Late Minoan Ib style. Thus, if there was a single and short eruptive sequence, it happened in the Late Minoan Ia period, and its effects in Crete were evidently not decisive.

But, on the other hand, if the eruptive sequence had been a long one, the chronological problem would disappear. Volcanologists on the whole incline to the view that the earthquakes heralding a major volcanic eruption do not in general precede it by more than 2 or 3 years (Hedervári et al. in press). Attempts to detect weathering deposits among the debris at Akrotiri resulting from the earthquake (Money 1973)—which would imply that the period was a longer one—have been effectively rebutted (Doumas 1974; Davidson 1978).

Moreover, most volcanologists are now agreed that the period from the initial eruption to the final collapse of the caldera could be at most little more than a year or 2 (Bond and Sparks 1976; Pichler in press). One worker suggests a period of 19 hours (Wilson 1978).

It should be noted that radiocarbon dating has not so far been of help in resolving this problem. Of course it is doubtful whether the method could in any case reliably confirm or disprove a difference of only 50 years, even if abundant samples were available from the Cretan Late Minoan Ib destructions, as well as from Akrotiri. But in fact there has been so wide a variation in the determinations from Akrotiri (Michael 1978), all supposedly documenting a single destruction episode, that special distorting effects, perhaps carbon dioxide of volcanic origin, have had to be postulated to explain them (Biddle and Ralph in press).

It may be, then, that a consensus is developing among volcanologists that there is likely to have been no great duration between the first earthquake, the onset of the eruption, and its last paroxysm. If that were so, the theory of the volcanic destruction of Crete would seem untenable. Certainly this view seems sufficiently authoritative to rebut the suggestion of Pomerance (1978) that the final eruption did not take place until the thirteenth century B.C. But the specialists are far from unanimous: Sparks *et al*. (1978) argue that a long eruptive sequence remains possible. So it is once again to Crete that we must turn, to see what firm evidence is there for volcanic effects at the time of the Late Minoan Ib destructions, or indeed earlier. Evidence from elsewhere in the Aegean can also be relevant.

## THE EFFECTS IN CRETE AND BEYOND

It might well be thought that decisive evidence as to the nature of the destruction of the Late Minoan Ib sites should come from these sites themselves. That this has not so far been the case is due in part to the limited nature of the relevant publications, since many of the excavations took place early in this century, and stratigraphic detail, with good sections, is not available. Secondly, there has been much uncertainty among archaeologists as to the precise nature of the evidence they are seeking. In most cases there is undoubted evidence of the collapse of buildings, often accompanied by fire and the burial of valuable objects.

### Earthquakes

Earthquakes are common occurrences in the Aegean, and the theory of earthquake destruction was one of the first to be put forward. Seismologists have pointed out that consistent patterns in the direction of collapse of walls might well be anticipated in such a case, but these have not yet been systematically sought. Tectonic earthquakes have indeed been associated with volcanic activity in the Aegean (Vitaliano and Vitaliano 1971), and Galanopoulos

(1971) has argued the case for an association with the Thera eruption. But at a conference Hedervári and his colleagues (in press) showed that the interval between earthquake and eruption, for those earthquakes actually accompanying an eruption, is not generally more than 2 years. Of course the destruction of Crete as a result of a tectonic earthquake around 1450 B.C. is perfectly possible, but in such a case there would seem no good reason to regard it as causally linked with an eruption of Thera about 1500 B.C. They could in such a case be regarded as independent events.

## Tsunamis

Tsunamis occur relatively commonly in parts of the Pacific, and the destructive effects of those associated with the Krakatoa eruption were described by a number of writers. It is considered likely by many workers (e.g., Marinos and Melidonis 1971; Yokoyama 1978) that substantial tsunamis were associated with the Thera eruption, although Pichler (in press) has stressed that the caldera collapse may have occurred at several stages, so that the tsunamis so formed may not have been large. No direct effects of tsunamis from the Thera eruption have been recognized in the Aegean—the ash deposits on the island of Anaphi have been shown not to relate to this eruption (Keller in press). It has been pointed out that tsunamis do not actually transport floating pumice or other debris across the sea: Their destructive action occurs when they gain in height as they near the shore.

My own personal view is that, notwithstanding the brief nature of the available excavation reports, it is almost inconceivable that the sites in question could have been destroyed by a massive inrush of sea water, with all the accompanying debris, and then its equally violent backwash, without leaving very clear indications in the stratigraphy. One would anticipate quite deep deposits of waterborne material, as well as large chunks of debris immediately recognizable as intrusive. Whereas damage by earthquake may in some cases readily be confused with destruction by human agency (sometimes with the use of fire), I cannot imagine this to be true of damage by tsunami. This is perhaps a case where excavation of sites known to have been destroyed by tsunamis—such as those suffering in the Krakatoa eruption of 1883—would be of value for comparative purposes. Certainly I know of no cases from the archaeology of the Mediterranean where destruction by tsunami has been plausibly claimed by the excavator on the basis of his own documented observations.

## Ashfall

The most interesting work on the effects of the Thera volcano has been carried out in relation to the ashfall that accompanied it. Clearly a heavy fall of ash could have ruined agricultural production, with disastrous consequences for the human population. Deep-sea cores have now elucidated a complicated sequence of ashfalls from different eruptions in the Mediterranean region over the past several hundred thousand years (Ninkovich and Heezen 1965; Cita and Ryan 1978; Keller et al. 1978; Watkins et al. 1978). Refractive index studies on tephra sherds, accompanied by trace-element work, allow the recognition of the source of the material in most cases. At least three of the ashfalls derive from the Thera volcano. The most recent of them can be correlated with the "Minoan" eruption of about 1500 B.C. (Figure 18.6). The earlier eruptions of Thera are of course well documented from the geology of the island (Pichler in press), and the plants buried in them on Thera itself have been dated by means of radiocarbon determinations (Friedrich 1978).

Interpretation of the depth of ash recovered from different cores indicates that the eastern half of Crete, which lies 120 km southeast of Thera, was subjected to a substantial fall of ash. Depths of up to 4 cm are found in the cores, but it is not clear precisely what would be the equivalent depth on land.

Pichler and Schiering (1977) have argued against significant effects in Crete arising from an ashfall of no more than 5 cm, and Blong (in press) considers in detail the effects on buildings, humans, plants, and animals of a fall of this depth. He concludes that the direct effects on humans and on buildings would be negligible. The effect on plants would depend on other factors, such as rainfall, but might in his view be beneficial rather than deleterious. The effects on animals would probably be small, unless they were exposed to fluorine poisoning. Similar views have been reached by Thorarinsson (1971; 1978).

Ash deposits on Crete itself have been sought. Although none visible to

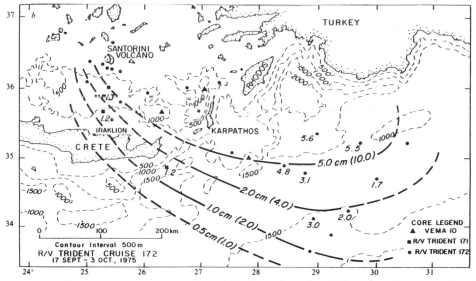

FIGURE 18.6 Isopachs (contours of equal thickness) for tephra fall from the Minoan eruption of Thera (Santorini), as determined from deep-sea cores. The figures in brackets estimate the corresponding depth, prior to compaction, of tephra falling on land. [From Watkins et al. 1978.]

the naked eye has been found, microscopic examination (and refractive index determination) of samples from archaeological contexts has given positive indications of Thera tephra (Cadogan *et al*. 1972; Vitaliano 1974). On the island of Melos, 100 km to the northwest of Thera, tephra have also been found stratified in late Bronze Age deposits (Renfrew 1978a). These finds are not conclusive: They confirm the existence of ashfalls in Crete but do not suggest any great depth of deposit.

Direct effects of the Thera eruption elsewhere in the Mediterranean have been claimed. But they are inherently difficult to substantiate, since without better chronological evidence than is at present available, they could as well be attributed to quite different events. Certainly Thera pumice could be identified, where it occurs, by chemical and refractive index criteria, and this has been done for Crete. But large quantities of pumice float around the Mediterranean after many volcanic eruptions and do not in themselves document destruction or disaster. So far the evidence beyond the Aegean is unconvincing.

In the absence of clear evidence of volcanic destruction, it is permissible to ask what other causes the Cretan destructions might have had. That question goes rather beyond the scope of the present chapter, but the views of Pendlebury have been summarized above. Hood (1978) has been a consistent advocate of a destruction of Late Minoan Ib Crete by Mycenaean military activity. My own view is that neither volcanic action nor the arrival of Mycenaean invaders is necessary to account for the sudden collapse of an early state society. Comparable collapses have been observed in many civilizations—indeed, the Mycenaean civilization itself underwent a perhaps analogous decline some 3 centuries later. It has often been the assumption that a sudden collapse must have a sudden cause, but the application of catastrophe theory to archaeology has shown that this need not be the case (Renfrew 1978b). It may be suggested that the decline of Minoan Crete around 1450 B.C. is another example of the phenomenon of systems collapse (Renfrew 1979). As in so many cases, it would be followed by, not caused by, the arrival of newcomers from outside the boundaries of the original territory. In the Minoan case these newcomers appear to have been Mycenaeans from the mainland, who have plausibly been documented at Late Minoan II Knossos. However, it must be confessed that these suggestions are as speculative as those which precede them, and will remain so until further data have been collected and adequately published.

## LOST ATLANTIS?

The Greek philosopher Plato is our only important source for the story of Atlantis. In his *Timaeus* he recounts the alleged narrative of the Greek sage Solon, who visited Egypt about 590 B.C., heard the Atlantis story from an Egyptian priest, and related it on his return to Athens to his friend Dropides, the greatgrandfather of Critias, from whom Plato claimed to have heard it.

The story tells of the island of Atlantis, home of a great and wonderful empire that at the early date in question was at war with Athens. "But afterwards there occurred violent earthquakes and floods; and in a single day and night of misfortune all your warlike men in a body sank into the earth, and the island of Atlantis in like manner disappeared in the depths of the sea."

The narrative involves much circumstantial detail, and Plato locates Atlantis beyond "the Pillars of Hercules"—the present Straits of Gibraltar—and thus in the Atlantic Ocean, at a time some 8000 or 9000 years before Solon.

Generations of scholars have speculated on the location of this legendary island, and early in this century, shortly after the discovery of the Minoan civilization of Crete, the Irish scholar K. T. Frost (1909) suggested that Crete itself might be the starting point for the legend. He was able to point to a number of striking similarities in detail between Plato's account and the new findings of Minoan archaeology, although of course he had to reduce the distance of Atlantis from Egypt and the time span from the date of Solon, both by a factor of about 10.

The suggestion by Marinatos in 1939 that the Thera eruption caused the end of the Minoan palace civilization added an interesting new ingredient to the story. For if Crete was indeed destroyed in the Thera cataclysm, the dramatic end of Atlantis recounted by Plato might have an altogether respectable basis in fact. An outlandish legend might become well-documented prehistory.

Several accounts have now been written, linking together the various elements of the story, some of them rather fanciful, going so far as to bring the Plagues of Egypt into the picture (e.g., van Bemmelen 1971). The case is most carefully set out by Luce (1969), but of course the theory is only plausible if it can indeed be shown that the Thera eruption did destroy Minoan Crete. Yet even if Crete was little affected by the Thera eruption, there is nothing to prevent the more modest conclusion that Thera itself was the origin for the legend of the great lost continent, the empire that disappeared "in a single day and night of misfortune."

There are nonetheless many discrepancies to be overcome between Plato's account and what we know of Crete or Thera—not least the error by a factor of 10 in the date and location of Atlantis. Any critical evaluation of the arguments must at its most generous lead to a verdict of not proven. We have good precedent in suspecting that Plato made up the whole story himself: The Greek philosopher Aristotle dismissed it with the comment, "The man who dreamed it up made it vanish."

## CONCLUSION

There is no doubt that the exceptional preservation of the prehistoric settlement at Akrotiri on Thera makes it one of the most important and promising archaeological sites in the Old World. The extent of wider effects of the Thera eruption, however, now seems very questionable. Interdisciplinary researches are leading to a much more precise picture of the depth of ashfall in Crete and other Aegean islands, and comparisons with well-understood cases in Iceland, Indonesia, and elsewhere allow a more balanced evaluation of its effects.

There is no archaeological evidence whatever for damage by tsunami in Crete. Earthquake damage is in any case very common in the Aegean without any need of reference to cataclysmic volcanic eruptions of the Plinian type.

In reaching a balanced assessment of the effects of a natural cataclysm, it is necessary to recognize and to discount the common tendency among archaeologists and historians to assume a causal link between the distant and often widely separated events of which they may have knowledge. An eruption here, a destruction there, a plague somewhere else—all are too easily linked in a hasty surmise by a process of fallacious reasoning that I have termed (Renfrew 1971) the "method of suppositious correlation." Disparate events simply should not be cross-correlated unless their simultaneity can independently be documented. Indeed there is the risk that the volcanic eruption may on occasion replace the migratory horde as an easy explanation to be assumed on convenient occasions by those who will not critically evaluate their evidence.

The Thera eruption thus suggests a useful archaeological moral. It stresses the opportunity that exceptional preservation offers for important discoveries on Thera itself, and the value of careful interdisciplinary study by archaeologists and geologists of the effects of volcanic eruptions. These opportunities, both for energetic excavation and for an interdisciplinary approach, were brilliantly seized by the late Professor Marinatos. It offers no comfort to those who, taking a cataclysmic view of history, wish to answer all their questions with a single, easy solution.

## REFERENCES

Åstrom, P.
    1978   Traces of the Thera eruption in Cyprus? In *Thera and the Aegean World* I, edited by C. Doumas. London.
Asimenos, K.
    1978   Technological observations on the Thera wall paintings. In *Thera and the Aegean World* I, edited by C. Doumas. London.
Biddle, M., and E. K. Ralph.
    In press   Radiocarbon dates from Akrotiri—Problems and a strategy. In *Thera and the Aegean World* II, edited by C. Doumas. London.
Blong, R.
    In press   The possible effects of Santorini tephra fall on Minoan Crete. In *Thera and the Aegean World*, II, edited by C. Doumas. London.
Bond, A., and R. S. J. Sparks
    1976   The Minoan eruption of Santorini, Greece. *Journal of the Geological Society of London* 132:1–16.
Cadogan, G., R. K. Harrison, and G. E. Strong.
    1972   Volcanic shards in Late Minoan I Crete. *Antiquity* 46:310–313.

Cita, M. B., and W. B. F. Ryan
  1978   The deep-sea record of the Eastern Mediterranean in the last 150,000 years. In *Thera and the Aegean World* I, edited by C. Doumas. London.
Davidson, D. A.
  1978   Aegean soils during the second millennium B.C. with reference to Thera. In *Thera and the Aegean World*, I, edited by C. Doumas. London.
Doumas, C.
  1974   The Minoan eruption of the Santorini Volcano. *Antiquity* 48:110–115.
  1975   Anaskaphi Theras. *Praktika tis Archaiologikis Etaireias* 1975:212–229.
  1978   The stratigraphy of Akrotiri. In *Thera and the Aegean World*, I, edited by C. Doumas. London.
Doumas, C., (Editor)
  1978   *Thera and the Aegean World* I (papers presented at the Second International Scientific Congress, Santorini, Greece, August 1978). London.
  In press   *Thera and the Aegean World* II (Proceedings of the Second International Scientific Congress, Santorini, Greece, August 1978). London.
Evans, A. J.
  1921–1935   *The Palace of Minos at Knossos* (Vols. I–IV). London: Macmillan and Co.
Fouqué, F.
  1879   *Santorin et ses éruptions*. Paris: Masson.
Friedrich, W. I.
  1978   Plants from Weichselian palaeosols, Santorini. In *Thera and the Aegean World I*, edited by C. Doumas. London.
Frost, K. T.
  1909   The Critias and Minoan Crete. *Journal of Hellenic Studies* 33:189–206.
Galanopoulos, A. G.
  1971   The Eastern Mediterranean trilogy in the Bronze Age. In *Acta of the First International Scientific Congress on the Volcano of Thera, 1969*, edited by S. Marinatos and D. Ninkovich. Athens: Archaeological Services of Greece.
Hedervari, P.
  1971   Energetical calculations concerning the Minoan eruption of Santorini. In *Acta of the First International Scientific Congress on the Volcano of Thera, 1969*, edited by S. Marinatos and D. Ninkovich. Athens: Archaeological Services of Greece.
Hedervári, P., G. Komlos, and S. Meszaros
  In press   Tectonic earthquakes related to the activity of Santorini. In *Thera and the Aegean World* II, edited by C. Doumas. London.
Hiller, S.
  1975   Die Explosion des Vulkans von Thera. *Gymnasium* 82:32–72.
Jones, R. E.
  1978   Composition and provenance studies of Cycladic pottery with particular reference to Thera. In *Thera and the Aegean World*, I, edited by C. Doumas. London.
Hood, M. S. F.
  1971   Late Bronze Age destructions at Knossos. In *Acta of the First International Scientific Congress on the Volcano of Thera, 1969*, edited by S. Marinatos and D. Ninkovich. Athens: Archaeological Services of Greece.
  1978   Traces of the eruption outside Thera. In *Thera and the Aegean World*, I, edited by C. Doumas. London.
Keller, J.
  In press   Prehistoric pumice tephra on Aegean islands. In *Thera and the Aegean World* II, edited by C. Doumas. London.
Keller, J., W. B. F. Ryan, D. Ninkovich, and R. Altherr.
  1978   Explosive volcanic activity in the Mediterranean over the past 200,000 yrs. as recorded in deep-sea sediments. *Geological Society of America Bulletin* 89:591–604.
Luce, J. V.
  1969   *The end of Atlantis*. London: Thames and Hudson.

1976    Thera and the devastation of Minoan Crete: A new interpretation of the evidence.
        *American Journal of Archaeology* 80:9–18.
Marinatos, S.
    1939    The volcanic destruction of Minoan Crete. *Antiquity* 13:425–439.
    1968–1976  *Excavations at Thera* (Vols I–VII). Athens: Archaiologike Etaireia.
Marinatos, S., and D. Ninkovich (Editors)
    1971    *Acta of the First International Scientific Congress on the Volcano of Thera*, 1969, Athens:
        Archaeological Services of Greece.
Marinos, G., and N. Melidonis
    1971    On the strength of seaquakes (tsunamis) during the prehistoric eruptions of Santorini. In
        *Acta of the First International Scientific Congress on the Volcano of Thera*, 1969, edited by
        S. Marinatos and D. Ninkovich. Athens: Archaeological Services of Greece.
Michael, H. N.
    1978    Radiocarbon dates from the site of Akrotiri, Thera, 1967–77. In *Thera and the Aegean
        World*, I, edited by C. Doumas. London.
Money, J.
    1973    The destruction of Akrotiri. *Antiquity* 47:50–53.
Morgan Brown, L.
    1978    The ship procession in the miniature fresco. In *Thera and the Aegean World*, I, edited by
        C. Doumas. London.
Neumann van Padang, M.
    1971    Two catastrophic eruptions in Indonesia. In *Acta of the First International Scientific
        Congress on the Volcano of Thera*, 1969, edited by S. Marinatos and D. Ninkovich.
        Athens: Archaeological Services of Greece.
Ninkovich, D., and J. Hays
    1971    Tectonic setting of Mediterranean volcanoes. In *Acta of the First International Scientific
        Congress on the Volcano of Thera*, 1969, edited by S. Marinatos and D. Ninkovich.
        Athens: Archaeological Services of Greece.
Ninkovich, D., and B. C. Heezen
    1965    Santorini tephra. In *Colston Research Papers* 17 (Submarine Geology and Geophysics,
        Proceedings of the Seventeenth Symposium of the Colston Research Society, Bristol);
        413–435.
Page, D.
    1970    *The Santorini Volcano and the desolation of Minoan Crete*. London: Society for the
        Promotion of Hellenic Studies (Supplementary Paper no. 12).
    1976    The miniature frescoes from Akrotiri, Thera. *Praktika tis Akademeias Athinon* 51:
        135–152.
    1978    On the relation between the Thera eruption and the desolation of eastern Crete c. 1450
        B.C. In *Thera and the Aegean World*, I, edited by C. Doumas. London.
Palmer, L. R.
    1965    *Mycenaeans and Minoans*, London: Faber.
Pendlebury, J. D.
    1939    *The archaeology of Crete*. London: Methuen.
Pichler, H.
    In press    The Minoan eruption of Santorini. In *Thera and the Aegean World*, II, edited by C.
        Doumas. London.
Pichler, H., and W. Schiering
    1977    The Thera eruption and Late Minoan Ib destruction on Crete. *Nature* 267:819–822.
Pomerance, L.
    1978    Improbability of a Theran collapse during the New Kingdom, 1503–1447 B.C. In *Thera
        and the Aegean World*, I, edited by C. Doumas. London.
Renfrew, C.
    1971    Comments on the paper by van Bemmelen. In *Acta of the First International Scientific
        Congress on the Volcano of Thera*, 1969, edited by S. Marinatos and D. Ninkovich.
        Athens: Archaeological Services of Greece.

1978a  Phylakopi and the Late Bronze I period in the Cyclades. In *Thera and the Aegean World*, I, edited by C. Doumas. London.

1978b  Trajectory discontinuity and morphogenesis. The implications of catastrophe theory for archaeology. *American Antiquity* 43:202–222.

1979  Systems collapse as social transformation: Catastrophe and anastrophe in early state societies. In *Transformations, mathematical approaches to culture change*, edited by C. Renfrew and K. L. Cooke. New York: Academic Press.

Schachermeyr, F.

1976  *Die Mykenische Zeit und die Gesittung von Thera* (Die Ägäische Frühzeit, Vol. 2). Vienna: Österreichische Akademie der Wissenschaften.

Shaw, J. W.

1978  Consideration of the site of Akrotiri as a Minoan settlement. In *Thera and the Aegean World*, I, edited by C. Doumas. London.

Sparks, R. S. J., H. Sigurdsson, and N. D. Watkins

1978  The Thera eruption and Late Minoan Ib destruction on Crete. *Nature* 271:91.

Thorarinsson, S.

1971  Damage caused by tephra fall in some big Icelandic eruptions. In *Acta of the First International Scientific Congress on the Volcano of Thera, 1969*, edited by S. Marinatos and D. Ninkovich. Athens: Archaeological Services of Greece.

1978  Some comments on the Minoan eruption of Santorini. In *Thera and the Aegean World*, I, edited by C. Doumas. London.

Thorpe-Scholes, K.

1978  Akrotiri: Genesis, life and death. In *Thera and the Aegean World*, I, edited by C. Doumas. London.

van Bemmelen, R.

1971  Four volcanic outbursts that influenced human history. In *Acta of the First International Scientific Congress on the Volcano of Thera, 1969*, edited by S. Marinatos and D. Ninkovich. Athens: Archaeological Services of Greece.

Ventris, M., and J. Chadwick

1973  *Documents in Mycenaean Greek*, 2nd ed. Cambridge: University Press.

Vitaliano, C. J., and D. B. Vitaliano

1974  Volcanic tephra on Crete. *American Journal of Archaeology* 78:19–24.

Vitaliano, D. B., and C. J. Vitaliano

1971  Plinian eruptions, earthquakes and Santorin, a review. In Marinatos and Ninkovich 1971, 88–108.

Watkins, N. D., R. S. J. Sparks, H. Sigurdsson, T. C. Huang, A. Federman, S. Carey, and D. Ninkovich

1978  Volume and extent of the Minoan tephra from Santorini Volcano: New evidence from deep-sea sediment cores. *Nature* 271:122–126.

Williams, D. F.

1978  A petrological examination of pottery from Thera. In *Thera and the Aegean World*, I, edited by C. Doumas. London.

Wilson, L.

1978  Energetics of the Minoan eruption. In *Thera and the Aegean World*, I, edited by C. Doumas. London.

Yokoyama, I.

1978  The tsunami caused by the prehistoric eruption of Thera. In *Thera and the Aegean World*, I, edited by C. Doumas. London.

# 19

# Pompeii and Mount Vesuvius, A.D. 79

WILHELMINA F. JASHEMSKI

## POMPEII BEFORE A.D. 79

No volcanic eruption has so captured the popular imagination or been the object of as much discussion as the eruption of Vesuvius in A.D. 79, which destroyed the flourishing cities of Pompeii and Herculaneum, as well as countless villas in the surrounding countryside (Figure 19.1). The vivid eyewitness account of this catastrophe in two letters written by Pliny the Younger (*Letters* 6. 16,20) is justly famous. Nor has the keen public interest in the excavation of the sites buried by this eruption ever flagged since these excavations were begun over 200 years ago. The term *Plinian eruption* used by volcanologists to describe all eruptions similar to that of Vesuvius in A.D. 79 is derived from the name of Pliny the Younger, nephew of the famous natural scientist Pliny the Elder, who lost his life in this eruption while attempting to rescue friends. In fact, some of the basic concepts of volcanism are founded on information derived from records of the activity of Vesuvius (Bullard 1976:223).

Pompeii owed its origin as well as its destruction to Vesuvius. It was built on a volcanic ledge formed by a stream of lava flowing out toward the sea in prehistoric times (Figure 19.2). The abrupt cessation of the lava flow gave the site a natural protection on the sea side. The original lava flow, however, became overlaid with various layers of volcanic debris from subsequent eruptions prior to the building of a town on the site. The origins of the city are unknown and must await further subsoil excavations. The earliest remains found thus far are those of the Doric temple in the Triangular Forum, which

*VOLCANIC ACTIVITY AND HUMAN ECOLOGY*

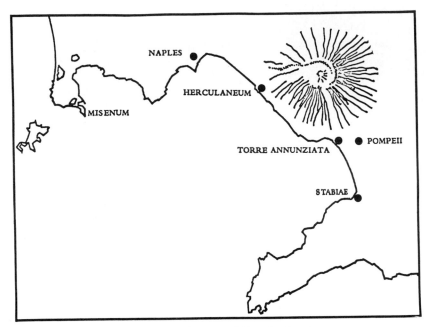

**FIGURE 19.1 The area of Vesuvius (map).**

date from the sixth century B.C. The final walls of the city followed the shape of
the lava ledge on the west, south, and east, thus giving the city an irregular oval
outline (Figure 19.3).

Although this chapter is primarily concerned with conditions at Pompeii at
the time of the eruption of A.D. 79 and the effect of this eruption on subsequent
human habitation in the area, it might be well to examine briefly what is known
of the earlier history of the volcano and of its possible physical appearance at
the time of the A.D. 79 eruption. The morphological evolution of the volcano
from prehistoric times to the present is shown in Figure 19.4. There was a
Plinian eruption in the twelfth century B.C., less violent than that of A.D. 79,
which left the volcano with a wide summit crater. Persistent activity then built
up within this crater a cone that eventually became about 3000 m high.
Another Plinian eruption took place in the eighth century B.C., this time a
powerful one. "Subsequently the crater was filled level with collapsed material"
so that the mountain then had a "single summit with a crater plateau
[Rittmann 1962:129]." This was how the volcano appeared before the eruption
of A.D. 79.

It is interesting to compare this description arrived at by volcanologists
with accounts by the ancient writers that give much the same picture. The
geographer Strabo, who lived at the turn of the Christian era, gives a good
picture of the volcano at this stage:

> Mt. Vesuvius, . . . save for its summit, has dwellings all around, on farm-
> lands that are absolutely beautiful. As for the summit, a considerable part of it

**FIGURE 19.2 Volcanic ledge on south.**

is flat, but all of it is unfruitful, and looks ash-colored, and it shows pore-like cavities in masses of rock that are soot-coloured on the surface, these masses of rock looking as though they had been eaten out by fire; and hence one might infer that in earlier time this district was on fire and had craters of fire and then, because the fuel gave out, was quenched [Strabo 5.4.8].

This was the condition of the volcano some years earlier in 73 B.C., when the gladiator Spartacus used the crater as a fort for his band of insurgent slaves and gladiators, when they were besieged by the praetor C. Claudius Glaber (Broughton 1952:109). Around the flat plateau that was the filled-in crater were high rocks over which were wild vines; from these the besieged men fashioned ropelike ladders to effect their escape (Plutarch, *Crassus* 9.2).

The Roman poet Martial (4.44), who lived at the time of the eruption, gives an attractive picture of Vesuvius before the catastrophe:

This is Vesuvius, green yesterday with viny shades; here had the noble grape loaded the dripping vats; these ridges Bacchus loved more than the hills of Nysa; on this mount of late the Satyrs set afoot their dances: this was the

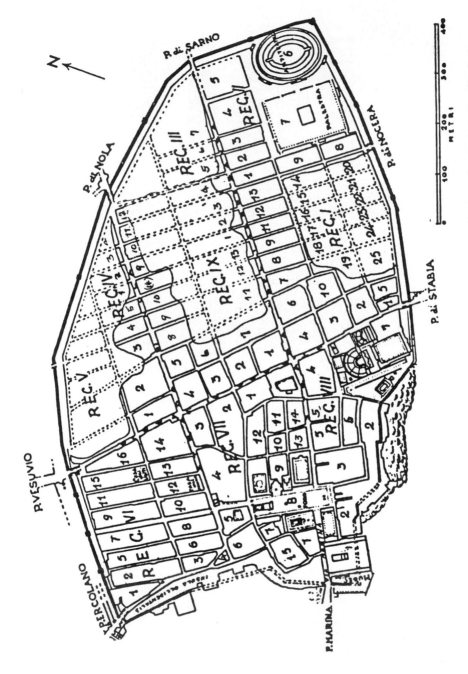

FIGURE 19.3  Pompeii (plan): Sites in Pompeii are easily located by the system introduced by the archaeologist G. Fiorelli, who became director of the excavations in 1860. He divided the city into nine regions; each region was subdivided into numbered *insulae* (blocks), and each entrance in each *insula* was assigned a number. Thus each door has an address of three numbers. (Adapted from A. Maiuri, *Pompeii*.)

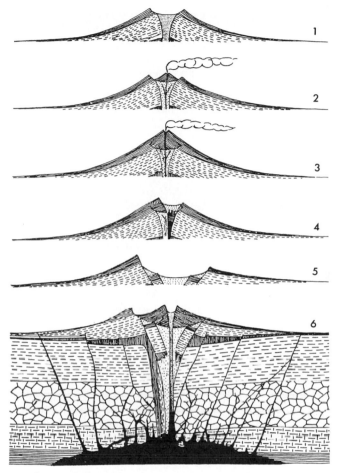

**FIGURE 19.4 The morphological evolution of the volcano from prehistoric times to today: (1) Crater after last prehistoric eruption; (?) central cone subsequently built up in crater; (3) cone 3000 m high in eighth century B.C.; (4) volcano before A.D. 79 eruption; (5) after A.D. 79 eruption; (6) today. (From Rittmann 1950, Figure 6.)**

> haunt of Venus, more pleasant to her than Lacedaemon; this spot was made glorious by the name of Hercules.

But writing as he did, after the eruption, Martial goes on to comment on the present state of the area: "All lies drowned in fire and melancholy ash; even the High Gods could have wished this had not been permitted them."

The painting found on a household shrine in the "House of the Centenary" (IX. 8.3,6) at Pompeii (and now in the National Archaeological Museum in Naples) (Figure 19.5) shows grapes growing on the lower slope of a mountain, which scholars believe suggests the shape of the cone of Vesuvius before the A.D. 79 eruption (Cocchia 1899:47–56; Rittmann 1950:474). Bacchus, god of the vine, an extremely popular deity in ancient Pompeii, stands nearby clothed in grapes.

**FIGURE 19.5  Lararium painting of Vesuvius before A.D. 79.**

Vesuvius rises in the Campanian plain, which owes its famed fertility to the volcano. The ancient writers speak extravagantly of the beauty and fertility of this plain. Strabo (5.4.3) calls it "the most blest of all plains and round about it lie fruitful hills." Strabo goes on to say that the area was so fertile that it bore as many as four crops a year: "Some areas were seeded twice with spelt, a third time with millet, and others still the fourth time with vegetables."

Pliny the Elder, who as admiral of the Roman fleet, was stationed at Misenum, across the Bay of Naples from Pompeii, describes in much the same way the plain of Campania, which he says "surpasses all the lands of the world." In his *Natural History* (18.111) he tells us that the land was kept in crop the whole year through, being planted once with millet, twice with spelt:

> and yet in spring, the fields having had an interval of rest produce a rose with a far sweeter scent than the garden rose, so far is the earth never tired of giving birth; hence there is a common saying that the Campanians produce more scent than other people do oil.

In another passage Pliny (*IIN* 3.40) rhapsodizes over the "blissful and heavenly loveliness" of the Campanian plain. The epitomizer Florus (*Epitome* 1.16.3) calls it "the fairest of all regions, not only in Italy, but in the whole world. . . . Nowhere is the soil more fertile." Strabo (5.4.8) suspected what scientists now know, namely that Campania owed its fertility to Vesuvius, just as at Catania, the volcanic ash from Mount Aetna made the land ideal for growing vines.

It was the fertile soil of Campania that was the source of the economic prosperity of the region. At the time of the eruption, the Campanian countryside was dotted with an extraordinarily large number of villas. Scattered literary references tell us of their existence, but only excavation informs us as to their actual appearance. Unfortunately, of those discovered thus far few are still visible today, and some of these have been uncovered only in small part (Jashemski in press). Villas destroyed by Vesuvius were first discovered by the Bourbon rulers, who haphazardly explored them by digging large subterranean tunnels. Between the years 1749 and 1782 Bourbon engineers tunneled through numerous villas in the vicinity of Stabiae and Gragnano. At Herculaneum they found the spectacular "Villa of the Papyri" in 1750, and by means of tunnels, which have now become filled, removed quantities of impressive sculpture, today displayed in the National Archaeological Museum in Naples.

Many of the villas that have been discovered were on private property and only partially excavated, for owners were anxious to fill in the temporary excavations and return the valuable land to cultivation. At no site have any separate farm buildings or farmlands attached to a villa been excavated.

The villas excavated thus far differ greatly in size, luxuriousness, and plan. Most appear to be real agricultural estates (*villae rusticae*). Many were obviously run by resident managers but contained quarters of varying degrees of luxury for the use of the owner during his visits; others were the residences of well-to-do farmers. A third type, with bare living quarters, was run by slaves for an absentee owner (Rostovtzeff 1957:564). The chief crops on these villas were grapes first and olives second, but there is evidence that other crops, including grain, vegetables, and fruits, were also raised (Day 1932:173–176).

But it is at Pompeii that we get our most complete picture of life in this area before the eruption, for here we have a complete city preserved and in good part excavated (Figure 19.3). At Herculaneum only four city blocks along with parts of four or five others have been excavated. At Herculaneum the volcanic materials were carried along by torrents of rainwater that created a mud lava that penetrated into every crevice. This hardened into a tufalike fill that is extremely difficult to remove.

Gardens have a prominent place in the layout of both Pompeii and Herculaneum. They are an integral part of the house, and they are also found in connection with many sacred and public buildings. Through the years as I have studied these gardens, I have become aware of the many large open areas in Pompeii. It was commonly believed that these were used for some industrial or business purpose. I suspected that they had been cultivated. At Pompeii, because of the way in which it was destroyed, it is possible to make subsoil excavations and determine whether the land was planted. The careful excavation of a planted area gives the only definite evidence for its actual appearance and use in antiquity. My excavations have discovered large commercial vineyards and orchards, as well as a commercial flower garden and vegetable gardens within the city walls (Jashemski in press).

The identification of all the planted areas large and small has led me to make a comprehensive study of land use at Pompeii, for it was only as I have slowly identified and studied each individual garden that I have come to an appreciation of the layout and character of the city as a whole. Pompeii is of unique importance in this respect, for scholars can do little more than speculate as to the probable uses of land in other ancient cities.

The first step in the excavation of a planted area is the removal of the lapilli until the level of the soil in A.D. 79 is reached. At this point the cavities that were left by the decay of the ancient roots, and into which lapilli had slowly trickled, are clearly visible. With special long-handled tools with tiny spoons at the end, or with long-handled tweezers, we carefully remove the lapilli from the root cavities and reinforce the large ones with heavy wire. (Our cavities ranged in their largest dimension at ground level from less than 1 cm to 230 cm; they were up to 160 cm in depth.) We fill the cavities with cement, which is allowed to harden for 3 or more days. When the soil around the cast is removed, the shape of the ancient root is revealed (Figure 19.6).

The next step is the identification of the casts. Both the shape and size, together with the way in which they are arranged in a planting pattern, are

**FIGURE 19.6  Tree-root cast.**

significant. The contour of the soil and the methods of watering also give valuable clues regarding the identity of the plants. A minute examination of the surface of the soil sometimes yields carbonized nuts, fruits, and vegetables. Carbonized fruits and nuts had previously been found in shops and cupboards but those we have excavated are of special importance, for they show the actual conditions of cultivation. The tools found add further information. The discovery of ancient pollen and the study of the fragments of carbonized twigs and roots give valuable information. The cooperation of many specialists has resulted in a multidiscipline approach that makes possible the recovery of the gardens uniquely preserved by Vesuvius.

## A Large Vineyard

For over 200 years the large city block to the north of the amphitheater at Pompeii (II.5 on plan, Figure 19.3) was known as the *Foro Boario*, or Cattle Market, a name given to it by the early excavators, who had uncovered only a small portion of it. Only subsoil excavation could determine whether this valuable piece of property had been planted. This I began in the summer of 1966. I discovered that excavators, who had uncovered a good part of the *insula* during the 1950s, had removed the volcanic debris down to the original level in most places and later covered the area with backfill of varying depths. But I have found that even in areas that have been previously excavated, it is possible to find root cavities if modern roots have not been too destructive. We found the backfill deep and compacted, excavation exceedingly difficult, and the results discouraging. But we persisted and we eventually found cavities.

The area had been too disturbed, however, to indicate for sure whether there had originally been two cavities at each location, and whether the second cavity found at many locations was that of a stake or another root. The local growers were sure that the vines had been staked in antiquity just as they are staked today. Fruit experts at the University of Naples, on the other hand, believed that the vines had been pruned low and left unstaked, and that the second cavity was also that of a vine root.

The importance of this vineyard made it desirable to excavate a part of the unexcavated portion still covered by the hill of original volcanic fill and to examine cavities in an undamaged area. It was with great excitement that we began to empty the cavities and found that one was always that of a vine root, easily identified by its shape and small lateral roots. The second cavity was always that of a stake (Figure 19.7). The local growers were not surprised! By the end of our 1970 season we had found a total of 2014 vine-root cavities and 58 tree-root cavities (Jashemski 1968, 1970, 1973a).

We also discovered paths dividing the vineyard just as recommended by the ancient agricultural writers (Figures 19.8 and 19.9). The cavities on each side of the path were perfectly preserved. The large cavities were those of posts. The smaller cavities were those left by vine roots (Figure 19.10). The posts had supported an arbored passageway similar to those found in vineyards in the

**FIGURE 19.7 Vine-root and stake cavity; vine-root and stake cast.**

vicinity of Pompeii today. The posts were probably chestnut, recommended in antiquity because of its "obstinate durability" and still preferred today. The ancient writers also recommend planting both the willow and the poplar to furnish withes for tying vines. Today the Pompeian still ties his vines with willow and poplar.

The many bones that we found in the vineyard were identified by Dr. Henry Setzer, mammalogist in charge of the African Section at the Smithsonian Institution (Jashemski 1973b:826). Cleaver marks indicate that they had been split for marrow and suggest that they were debris from vineyard meals that had been served at the two triclinia we found in the vineyard (Figure 19.11). This finally explains why the first excavators mistook this area for the cattle market.

## The Garden of the "House of the Ship *Europa*"

This once noble house (I.15 on plan) (named from the large graffito of a ship labeled *Europa*, found on the north wall of the peristyle when it was excavated in 1957) had obviously been converted to some kind of commercial use by the time of the eruption. The large, open split-level area at the rear had been badly damaged by trucks at the time of original excavation, but even so when we excavated this area in 1972 we soon found two plots with distinct furrows in the lower garden, and many single cavities fairly evenly spaced, with distinct depressions for holding water on each side, similar to those found in the large vineyard.

We found a total of 416 root cavities in this garden (Figure 19.12) (Jashemski 1974). The two furrowed plots appeared to be vegetable plots, but the root cavities of the vegetables were too small to be preserved. The 31 irregularly spaced root cavities in the two small plots were undoubtedly those of small fruit trees, such as are still found in Pompeian vegetable plots today. The regularly spaced small roots in the lower garden appeared to be vine roots, 4½·

FIGURE 19.8 Plan of large vineyard at end of 1970 season: (a, b) unexcavated areas; (c) south entrance; (d, e) masonry triclinia; (f) room with wine press; (g) shed with 10 dolia embedded in the ground; (6) Via dell'Abbondanza; (i) Sarno Gate; (j) unexcavated backfill of 1955 excavations; (k) path along the north wall; (m) wineshop and portico; (x) the crossing of the north-south and east-west paths. Dots indicate vine roots; small circles indicate tree roots 10 cm or less in diameter at ground level; large circles indicate tree roots 11–29 cm in diameter at ground level; solid circles indicate tree roots 30 cm or more in diameter at ground level.

Roman feet apart. The other cavities in the garden are either those of the roots of trees or very old grapevine roots. Perhaps there are some of both. The random way in which the roots were planted and their various sizes are duplicated in orchards in the area today, and indicate that our cavities are probably those of tree roots. The trees in modern orchards are usually small and planted close together. There were only five trees in this garden with their longest dimension over 30 cm.

**FIGURE 19.9 Pathway in large vineyard (toward Vesuvius).**

Most surprising was our discovery of 28 pots embedded in the soil at intervals along the four walls. Fourteen of these were found at the bottom of root cavities ranging in depth from 17 to 42 cm. With one exception, the pots have one hole in the bottom and three on the sides like the pots in Figure 19.13. This is the first time that a large number of pots have been found in a garden at Pompeii or to my knowledge at any Italian site. How had these pots been used? References to pots in the ancient writers are extremely rare. Cato and Pliny the Elder give us helpful information about ancient gardening. They speak briefly of pots being used to make air layers. Pliny also reports that "various countries tried to acclimatize the citron, importing it in earthenware pots provided with breathing holes for the roots." This is a good description of our Pompeian pots. At the time Pompeii was destroyed there was a great interest in introducing new fruits into Italy. Is it possible that in this garden exotic trees, such as the citron, were started in pots?

We have other important evidence to help us identify what was grown in this garden. A considerable number of carbonized fruits, nuts, and vegetables

FIGURE 19.10 Post and vine-root cavities at intersection of pathways.

were found preserved where they were grown. These were identified by Dr. Frederick G. Meyer, Research Botanist in charge of the U.S. National Arboretum Herbarium. There were pieces of almond shells, a piece of carbonized fig, pieces of filberts, perfectly preserved grapes, and even grape seeds. The large number of broad beans, or horse beans, strongly suggests intercultivation, which is common in the area today (Figure 19.14). From a hole in one of

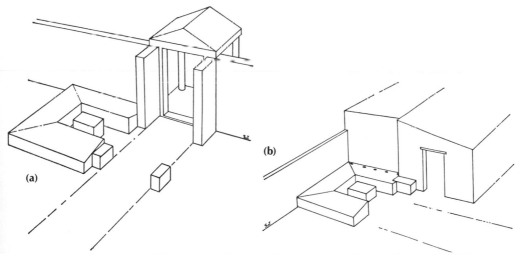

FIGURE 19.11 (a) Triclinium across from amphitheater; (b) triclinium adjacent to building in which wine was made.

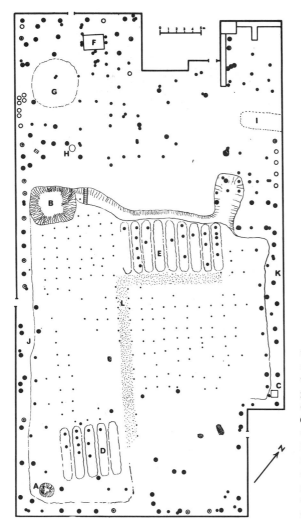

FIGURE 19.12 Plan of garden of the "House of the Ship *Europa*" (I.xv): (D) SW vegetable garden; (E) N vegetable garden. Dots indicate grapevine roots; small solid circles indicate roots 10 cm or less in longest diameter at ground level; large solid circles indicate roots 11–29 cm in longest diameter at ground level; irregular shapes indicate tree roots 30 cm or more in longest diameter at ground level. Large empty circles indicate pots; circles with dots indicate cavities containing pots.

the ancient broad beans entomologists at the Smithsonian Institution, with tiny tweezers, extracted the hind leg of a bruchid, or strawberry weevil. From another they extracted a large part of a bruchid.

Scholars have suggested that there must have been many small market-gardens near ancient Pompeii. Now for the first time such a garden has been found, surprisingly, within the city walls. Modern gardens adjacent to the city continue the age-old planting pattern. Furrowed vegetable plots are found near fruit and nut trees, and often, a small vineyard, frequently for table grapes, is nearby. Small trees are planted underneath larger trees, ready to serve as replacements when the older trees cease to be productive.

FIGURE 19.13 Planting pots found in "Garden of Hercules."

FIGURE 19.14 Broad beans, or horse beans (*Vicia faba* L. var. minor [Peterm. and Harz] Beck). Top: (7–10 mm long × 6–7 mm wide) from U.S. Department of Agriculture Seed Collection, Beltsville, Md. Bottom: (6–9 mm long × 5–7 mm wide) found in garden of "House of the Ship *Europa*." Hind leg of bruchid found in bean on right. (Photo by U.S. Department of Agriculture.)

## "The Garden of Hercules"

We found still another type of garden, in a large area attached to a humble house (II.8.6) to the west of the Great Palaestra (Jashemski in press). We named it the "Garden of Hercules" after the statue of Hercules found near the large shrine and altar in this garden. Since its original excavation in 1953, vegetation had completely filled the garden and it looked as if all evidence had been destroyed. Closer examination showed, however, that in some areas the original lapilli had been left to a sufficient height to preserve the ancient soil contours and cavities.

In the root cavities near the walls we found pots similar to those that we had found in the garden of the "House of the Ship *Europa*" (Figure 19.13). We left one pot in the soil, however, for we did not want to destroy a root cavity in taking out the pot. When we excavated the cast the next year, it was most impressive to see a large root growing out of a small pot; surprisingly, there was a second pot nearby. The young tree that had been started in this pot had not grown, so the ancient owner had planted another in a pot close by. Dr. Fideghelli immediately noted the similarity of this root to that of a citron or lemon tree root—perhaps another suggestion that the ancient Pompeians were painting on their walls the likenesses of the fruit that they grew in their gardens. The carbonized cherries that we found near the shrine probably identify one of the trees found in this garden.

It was in this garden that we made our original discovery of pollen at Pompeii. To the west of the cistern we found a huge tree-root cavity, which had the appearance of a very old olive root; this perhaps explains the very high concentration of olive pollen (a rarity in present-day Pompeii) (Figure 19.15). Our soil samples were analyzed for pollen by Professor G. W. Dimbleby of the Institute of Archaeology at the University of London.

The most striking feature of this garden was the complicated soil contour, which divided the garden into many beds. After a rain, in the most perfectly

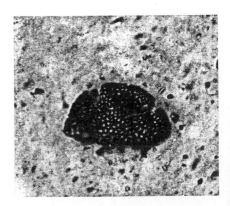

**FIGURE 19.15  Olive pollen found in "Garden of Hercules." (Photo by Institute of Archaeology, University of London.)**

FIGURE 19.16 "Garden of Hercules" (aerial view by Whittlesey Foundation).

preserved beds, we discovered formations that indicated a very small plant in the center with circular depressions for water around it. Careful provision had been made for watering whatever was grown here. Some water was furnished by collecting rainwater from the roof of the house. Additional water had to be carried in because the aqueduct had not reached this zone.

During the summer of 1974 the aerial photos made by the Julian Whittlesey Foundation were an important aspect of our work. For the photographs root and stake cavities were marked with white discs, and the large roots painted white. The photos show clearly the pattern of root cavities and stakes, and in a remarkable way reveal even badly damaged soil contours not clearly seen from the ground (Figure 19.16). The wide beds were separated by water channels in which most of the root cavities were found. The beds were not all on the same level; those on the north of the garden were higher than those on the south. To ensure the flow of water, because the variation in height was very small, water channels in the north part of the garden were above the beds, those on the south below the beds.

The question remains as to what was grown in this garden. There were relatively few tree-roots and quite a few stakes. Most of the garden was laid out in beds, for which water was provided. This makes it certain that one of two things were grown, flowers or vegetables, perhaps some of both. There are indications, however, that this garden might have been a large commercial flower garden. Pompeii was the center of an important flower industry in antiquity, as it still is today. Within a 5–8-km radius of Pompeii, large foreign seed companies have hundreds of hectares of flowers under contract. Perhaps the stakes in the modern gardens throw some light on those that we have in this garden. They too may have formed shelters for the young plants in the beds. Trees in the water channels at the edges of the beds may have furnished shade, as fruit trees do today. One of the largest flower growers in the Pompeii area, after studying this garden, told me that the contours definitely indicated flower culture. He grows flowers in the same type of bed today.

Flowers grown at Pompeii in antiquity were used for two purposes—for making garlands and for making perfume. Perhaps it is significant that a large number of perfume bottles were found in the partially excavated house of this garden. We found more bottles such as these in the garden, as well as fragments of terra-cotta perfume containers. We know that olive oil was an important base in making perfume, which may explain the large amount of olive pollen found in this garden.

## The Peristyle Garden in the House of Polybius

The noble House of Polybius (IX.13.1–3) on the Via dell' Abbondanza was built by the wealthy Samnite inhabitants long before their city was conquered by the Romans. The peristyle garden in this house, which I excavated in 1973, is of special importance because it is the first undisturbed garden to be carefully studied (Jashemski 1975:76–81). It is also important because our knowledge is far from complete about what was grown in peristyle gardens at Pompeii.

We first uncovered the soil next to the west wall. What we found looked like a large sombrero. The soil had been carefully shaped in a high mound, surrounded by a channel for water. In the high mound was a small root cavity. We continued to find these formations along the west wall. It was obvious that something special had been planted here: The plants were still young, and provision had been made for them to get considerable water. When the lapilli were removed from the root cavities, we found in two of them terra-cotta fragments that I recognized as parts of pots with "breathing" holes such as we had previously found. It seems likely that the small roots in the sombrero formations were those of small trees, perhaps the exotic citron or the lemon tree.

Much to our surprise, since we have become accustomed to low, formal plantings in restored peristyle gardens, we found a total of five large tree-root cavities (Figure 19.17). In addition there were various small cavities, several clearly those of stakes, and on two edges of the garden there were rows of smaller cavities that appeared to be those of ornamental shrubs.

Two strange marks in the soil in the center of the garden baffled us. But by the time the complete marks were exposed it became clear that they were the outline of a ladder. It was an exceptionally long and narrow one, 8 m long, 0.5 m wide at the bottom, and 0.3 m wide at the top. It is exactly the same size as those used to pick cherries and pears in Italy today. The ladder was shaped to fit into fruit trees that were tall and full of dense branches. The many carbonized figs we found may help to identify one of the trees.

The five large trees would have almost completely shaded the peristyle garden, but they were evidently pruned enough to allow planting underneath, at least along the edges. Three of the large trees had their branches supported by stakes, which would indicate that they were laden with fruit or nuts. The extraordinarily dense planting in this garden immediately suggests that the soil must have been very fertile for such a small plot of land to have supported five large trees, as well as a number of smaller trees and shrubs.

Professor John Foss of the Department of Agronomy at the University of Maryland studied the morphology and fertility of the soils at Pompeii and Torre Annunziata during our 1978 season. He reports that

> The soils in the garden of the House of Polybius have a succession of buried A horizons (high organic layers) in the upper meter. These soils are the result of additions of volcanic materials prior to A.D. 79 and modifications resulting

**FIGURE 19.17 Garden in House of Polybius (northwest part).**

from man's activities. Stability of landscapes is indicated by organic matter contents exceeding one precent for most horizons in the upper meter. The soils are generally coarse textured, with loamy sands and sandy loams predominating. Although coarse textured, these volcanic soils, because of the porous characteristics of the pumice, contain large quantities of moisture available to plants. These soils show high levels of extractable phosphorus, potassium, calcium, and magnesium. However, some of the calcium in the soils may have resulted from leaching of the overlying pumice and ash. The soils are alkaline (weakly calcareous) with pH values of 7.0 or above; leaching from the overlying pumice and ash has recharged the soils with bases. The well-drained condition, high nutrient status, and high water capacity of the soils made them suitable for a variety of crops. Soils sampled at other gardens at Pompeii and Torre Annunziata were similar to those in the garden of Polybius.

## The Villa at Torre Annunziata

In 1974 we began excavating the gardens in the luxurious villa at Torre Annunziata, on the Bay of Naples, recently identified as the site of the lost Oplontis, also destroyed by Vesuvius. Strabo (5.4.8) tells us that the entire coastline of the Bay of Naples was characterized by an unbroken succession of houses and vegetation that gave it the appearance of a single city. Most of these villas, however, are known only from the ancient literary sources or from inscriptions (D'Arms 1977). For this reason the villa at Torre Annunziata is of special importance. We continued our work at this villa during the summers of 1975–1978. The villa so far has 12 spectacular gardens. A large and complex portico stretches across the back of the villa looking out on a parklike garden whose dimensions, design, and plantings we are only beginning to discover (Figure 19.18). So far we have found five huge tree cavities in this area, as well as many smaller root cavities that are part of the larger landscape design composed of many beds and passageways (Jashemski in press).

## THE ERUPTION OF A.D. 79

Vesuvius had been quiescent for many centuries before the great eruption of A.D 79. The volcano was considered extinct (Diodorus 4.21.5) and its slopes were green with vineyards. An earthquake in A.D. 62 was the first sign of a reawakening of Vesuvius. This earthquake was so severe that a flock of 600 sheep was reported killed, and many people were said to have lost their reason (Seneca NQ 6.1.1–3). Scarcely a building in Pompeii escaped damage. Seventeen years later came the final destruction of the city.

The two letters of Pliny the Younger mentioned earlier give a masterful account of this eruption. These valuable historical documents (which volcanologists consider the beginning of modern descriptive volcanology), together with a scientific study of the stratification of the erupted material

**FIGURE 19.18 Excavating villa at Oplontis under modern Torre Annunziata.**

encountered in excavations, make it possible to reconstruct in detail the character of the eruption.

First Pliny's account. As a lad of seventeen at the time of the eruption he was with his mother at Misenum, where her brother, Pliny the Elder, was in command of the Roman fleet. Years later, replying to inquiries from the historian Tacitus, the younger Pliny described the events leading to his famous uncle's death (6.16). Early in the afternoon on 24 August his mother drew her brother's attention to a cloud of unusual size and appearance rising from a mountain, later known to be Vesuvius (almost 30 km away). It was shaped like an umbrella pine, rising to a great height on a sort of trunk and then splitting off into branches. At times it was white but sometimes black and gray, depending upon the amount of ashes and soil that it contained. Pliny the Elder immediately gave instructions for a ship to be made ready so that he could inspect the phenomenon closer at hand. But as he was leaving the house, a plea for help came from Rectina, who lived at the foot of the mountain, and for whom escape was possible only by boat. Pliny changed his plans and went to rescue those living on the coast. By the time he reached the coast, however, he found that ashes were falling hot and thick, followed by pumice and blackened stones, and that the shore was blocked by debris from the mountain. He decided to go to Pomponianus, who lived at Stabiae. When he arrived he tried to calm his terrified friend by showing his own composure. But during the night when the courtyard that gave access to his room began to fill with ashes and pumice, he came out and joined the rest of the household. After some discussion, since the buildings were shaking with violent shocks, they decided to take their chances in the open, tying pillows on their heads for protection from the falling pumice. Although it was by this time morning elsewhere, at Stabiae it was darker than any ordinary night. Pliny went down to the shore to explore the possibilities of escape by sea, but he found that the waves were too turbulent for that to be possible. While still there he was asphyxiated by the dense fumes and died.

In a second letter Pliny replied to Tacitus's queries about his own experiences (6.20). There had been earth tremors for several days, a not unusual occurrence in Campania, but during the night of 24 August the shocks became so violent that Pliny and his mother got up and spent the rest of the night in the forecourt of the house. Frightened by the danger of the tottering buildings, in the morning they decided to leave town. Extraordinary experiences alarmed them. Carriages on level ground began to run in different directions; the sea sucked back from the shore left quantities of sea creatures stranded on the dry sand; a fearful black cloud parted to reveal great tongues of fire. Then total darkness came and the shrieks of those fleeing on the road could be heard. The ashes began to fall in heavy showers, but at last a genuine, but yellowish, daylight came, and they were terrified to see everything buried deep in ashes like snowdrifts. Even though the earthquakes continued, Pliny and his mother returned to Misenum, to await news of Pliny the Elder.

This vivid account, written some years after the event, was, as Pliny says, based on incidents that he witnessed himself or heard about immediately after the event when reports were most likely to be accurate. The only other description of the eruption is a brief one, in the history of Dio Cassius (66.23), who wrote about 150 years later. He emphasizes the crashing noise connected with the eruption and the fact that two entire cities, Pompeii and Herculaneum, were buried. The volcanic dust, he says, reached as far as Africa, Syria, and Egypt and filed the air above Rome.

When the detailed description of the eruption found in Pliny's letters was compared by the volcanologist Rittmann with a reconstruction of the event based only on the nature and stratigraphy of the erupted materials and on the known physiochemical laws of magmalogy, he found no conflict in the two accounts (Rittmann 1950:467–470; 1962:87–88). Rittmann lists the evidence (Rittmann 1950:459–474; 1962:86–87) from which he reconstructs the nature of the eruption, beginning with the deposits on the Grand Palaestra at Pompeii, which show the following stratigraphical sequence beginning at the bottom (Figure 19.19):

1. Layers of pumice                                                                    260 cm
   a. White, rounded, highly porous
   b. Light gray, slightly rounded, porous
   c. Greenish gray, fairly angular, moderately vesicular
2. Volcanic sand (xenolithic and glassy)                                       5 cm
3. a. Lapilli of allogenic origin                                                    3 cm
   b. Ash (predominantly glassy) with pisoliths                     64 cm
   c. Lapilli                                                                             3 cm

    d.  Ash                                                                  2 cm

    e.  Lapilli                                                 3 cm

    f.  Ash with pisoliths                             30 cm

    Herculaneum, on the other hand, was buried in a very different way, beneath three mudflows; the first contains all the types of pumice enumerated above and also darker pumice somewhat resembling the volcanic sand in (2). "At the northern foot of the volcano, near the present-day village of Somma, a thick lava flow poured out along a broad front as far as Castello di Cisterna" (Rittmann 1962:86–87). Rittmann correlates this flow, on petrological and stratigraphical grounds, with the A.D. 79 eruption.

    Pliny's account, which makes it possible to date events more accurately, contains obvious omissions, for he treated only those phenomena that were connected with his uncle's death. When Pliny first saw the umbrella-shaped cloud, Rittmann says that the eruption had already entered its third stage; this is obvious not only from the description of the cloud but from the fact that people across the bay in the vicinity of Herculaneum earlier had become alarmed and there had been time for a messenger to come to Misenum before Pliny the Elder left. Rittmann (1950:467–468) sees a hint of the first phase of the eruption, namely the expulsion of the vent plug, in the account of Dio Cassius (66.23), who wrote that a great crash was heard and huge stones of extraordinary mass were thrown up as high as the top of the mountain and that afterwards the ashes were thrown out. In the evening, when Pliny the Elder

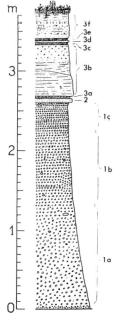

FIGURE 19.19 Stratification of erupted material in the Grand Palaestra at Pompeii. (From Rittmann 1950: Figure 59.)

approached the shore and found it blocked by volcanic debris, Pompeii would already have been buried, but neither Pompeii nor Herculaneum is mentioned by Pliny, for they were not connected with his uncle's last days.

The eruption reached its climax on the morning of 25 August accompanied by violent earthquakes powerful enough to cause destruction within a 20–30 km radius. It was on this day that Pliny the Elder died. A change in the wind that day brought clouds filled with ash that turned day into night at Misenum. These ashes were snow white, a characteristic of ashes coming from the deep magma of the basin. They were thrown out toward the end of the third phase and are not found at Pompeii. But they are found in the last flow of mud at Herculaneum.

Nor does Pliny mention the thick lava flow that poured out at the northern foot of the volcano. Rittmann suggests that this flow probably came on 26 August and continued through the following night, after all the inhabitants had already fled or perished. But it is unlikely that this would have been included in the historical accounts even if it had been observed, for the lava caused no damage in this marshy zone and it in no way compared with the drama caused by the eruption elsewhere (Rittmann 1950:470).

A recent study of the coarse pumice deposit thrown out in the A.D. 79 eruption documents the exceptionally wide dispersal of the pumice (Lirer, Pescatore, Booth, and Walker 1973). A bed 26 cm deep has been found as far south as Agropoli, 72 km from Vesuvius. The authors suggest that this wide dispersal points to an eruption of great vigor, and that the "relative homogeneity of the pumice beds" indicates that the eruption was "short and of continuous gas-blast type." Although the authors are not concerned in this article with the ash deposits overlying the pumice, they note that the ashes include beds that they interpret as pyroclastic flows. This means that the large lithic blocks found in the ashes "were transported by these laterally moving density flows rather than thrown from the vent," as has been previously believed. Earlier volcanologists considered all the volcanic debris to be air-fall that rained down after being blasted out of the volcano and did not realize that in the latter part of the eruption avalanches of hot ashes and blocks slid down the side of the mountain. Dr. Richard Fiske, volcanologist at the Smithsonian Institution, Washington D. C., tells me that "glowing avalanche deposits apparently blanketed a fairly large area in and around Pompeii and would have produced a very hot blanket with an initial temperature 500° to 700°C. Despite the widespread nature of the blanket of heat, the downward propagation of this heat was probably not uniform over the whole area; in some places, where continuous gas-blast type." Although the authors are not concerned in this lying pumice deposit than it did elsewhere [conversation May 23, 1979]."

## AFTER THE ERUPTION OF A.D. 79

When the eruption finally subsided Pompeii was covered to a depth of approximately 4–6 m (Figures 19.20 and 19.21), and Herculaneum had com-

FIGURE 19.20 Stratification of erupted material at Pompeii.

pletely disappeared under a blanket of mud–tufa that in places was up to 20 m deep. The nature and depth of the fill in the surrounding countryside depended upon the distance and the direction of the site from Vesuvius. At Oplontis (modern Torre Annunziata) there are four layers of lapilli, separated by thin layers of ash—altogether a total of only 1.8 m. Above these layers were 5 m of mud (De Franciscis 1975:5) (Figure 19.22).

After the eruption the Emperor Titus sent a commission to Campania to make plans to rebuild the destroyed cities, with instructions to use the property of those who had died without heirs for this purpose (Suetonius, *Titus* 8). There is no mention in the ancient writers, however, of any work being done by the commission. The destruction was obviously considered too great for any rebuilding to be attempted.

We do not know how many inhabitants perished in the eruption. It is believed that most of the population was able to escape from Herculaneum. From the number of skeletons found at Pompeii it has been estimated that more than 2000 persons stayed behind and lost their lives (Russell 1977). It is impossible to say how many fled from the city on the morning of 24 August,

**FIGURE 19.21 Partially excavated house at Pompeii.**

when the eruption began, and found safety, or how many decided too late to flee and lost their lives on the crowded roads leading from the city. Nor do we know where those who escaped eventually settled. Evidence in the excavations, however, shows that many survivors returned to Pompeii after the eruption and removed valuables (Della Corte 1933). The upper parts of houses and the roofs that had not fallen in and become covered afforded landmarks. The nature of the volcanic fill made it relatively easy to dig down from the surface and tunnel from room to room. Scarcely a house has been found that was not entered. Valuable furnishings were for the most part removed, along with much building material, especially from the Forum area. The depth and nature of the fill at Herculaneum, on the other hand, made such salvage impossible and for this reason the finds are much more complete.

The possibility of the rebirth of an active life in the Vesuvian zone was much discussed by scholars at the end of the last century and at the beginning of our century, but never resolved (Sogliano 1915:483–514). The question has recently been reopened by Dr. Giuseppina Cerulli Irelli, Director of the Excavations at Pompeii (Cerulli Irelli 1975:291–298). In her discussion she lists three types of archaeological evidence for the reoccupation of the Vesuvian area after the eruption of A.D. 79: (a) evidence left by the survivors of their efforts to recover valuables from the buried city of Pompeii; (b) traces of habitations in the Vesuvian area; and (c) the late tombs. Some of this evidence has been long known; some has been recently uncovered.

The evidence of scavengers is everywhere at Pompeii. There are the familar holes in walls and ceilings and examples of the removal and shuffling of

FIGURE 19.22 Excavating a large tree-root cavity at Oplontis (note stratification of erupted material).

materials, as for example fragments of wall decorations from a lower story found in the upper level. There are also lamps of a later period found in the ruins, as well as graffiti written on the walls by those entering.

The traces of habitations built after the eruption are, as Dr. Cerulli points out, all outside the ancient city walls and they often are found on the site of a pre-eruption *villa rustica*. She lists the following posteruption remains: a *villa rustica* at Scafati; an *officina* equipped with vats, to the southeast of the amphitheater at Pompeii; an unidentifiable building southwest of Pompeii; another unidentifiable building about 100 m north of the city wall at Pompeii; a bath building with mosaics near Boscoreale; an edifice of unknown use near Boscotrecase; as well as finds at Torre Annunziata, and a *torcularium* at S. Sebastiano, which she excavated in 1964.

The tombs are more numerous, their furnishings poor: a few lamps and coins. Apart from a few scattered tombs near Herculaneum, the tombs are

clustered in two places at Pompeii. There are a few near the *officina* beyond
the amphitheater, but most are found in the area north of the wall at Pompeii.
This Dr. Cerulli believes is a genuine necropolis. It was this area that the
scholar Sogliano early in this century believed to be the site of the Second
Pompeii.

  During the clearance of a site for a public housing project in 1976, some
tombs of poor people were found to the east of ancient Herculaneum on top of
the mudflow of A.D. 79. The tombs did not constitute a city necropolis, but
they are valuable evidence for posteruption occupation of the area. The bodies
had been placed in amphoras for burial; the pottery dates the burials to the
third or the fourth century (Maggi 1976:249). During the summer of 1978,
while going through the recently established photography files at the Direction
at Pompeii, I found two photos of skeletons in amphoras; these burials were
discovered in 1951 during the excavation of the top of the building in the
southeast part of Region II, *insula* 3, across from the Grand Palaestra at
Pompeii (Figures 19.23 and 19.24). These posteruption burials made use of the
same types of amphoras as those found at Herculaneum. One additional bit of
evidence for posteruption occupation in the Pompeii area is cited by Helbig
(1865:234–235), who reports the discovery in a modern vineyard near the
amphitheater of two sepulchral columns (*cippi*) carved in high relief so that
they appeared almost as statues; their style dates the *cippi* at least to the third
century. This evidence from Pompeii and Herculaneum indicates that these
areas were occupied in the third century.

  The evidence of the Peutinger Table, a map for travelers of the third or

**FIGURE 19.23 Post-
eruption burials in am-
phoras at Pompeii. (Photo
by Soprintendenza alle Antichità delle Province di
Napoli e Caserta.)**

FIGURE 19.24 Pos-
teruption burial in amphora
at Pompeii. (Photo by Soprintendenza alle Antichità
delle Province di Napoli e
Caserta.)

fourth century, may also be helpful in our search for evidence of posteruption occupation in the Vesuvian area. Dr. Cerulli points out that on this table Pompeii is indicated by the same symbol that is used for other cities in the area that continued in existence, whereas for Herculaneum, which remained buried, there is only a name (Figure 19.25). She suggests that the name of Pompeii was probably given to a modest agricultural center, to which the posteruption cemetery north of Pompeii is related.

It is difficult, however, to determine how soon the farmland covered by Vesuvius in A.D. 79 was returned to cultivation. After the eruption of 1944, which threw out eruptive material similar to that of A.D. 79 but in much less amount, the humus was built up again after a few years. Cultivation was resumed between 1945 and 1946 at Boscoreale and Terzigno by shoveling off the lapilli. Unfortunately, despite the importance and great renown of Vesuvius, little attention has been paid to the question of the return of soil to cultivation after an eruption. Bottini (1932:289–291) discusses briefly the return of vegetation on lava. Agostini (1975) in more detail records the sequence of the pioneer plants that appeared on the newly formed volcanic soils of Vesuvius after the 1944 eruption. He follows from their first stages the appearance of plants on two fundamental types of substratum. After 10–12 years, the lichen *Stereocaulon vesuvianum* Pers. began to appear on scoriaceous lavas, eventually giving the lava a characteristic gray green, almost opalescent color. Between 1954 and 1957 on pyroclastic masses formed by cinders, lapilli, and pumice, the first plants to appear were *Rumex acetosella* L., *Silene armeria* L., *Artemisia campestris* L.v. *variabilis* (Ten.), *Rumex bucephalophorus* L.,

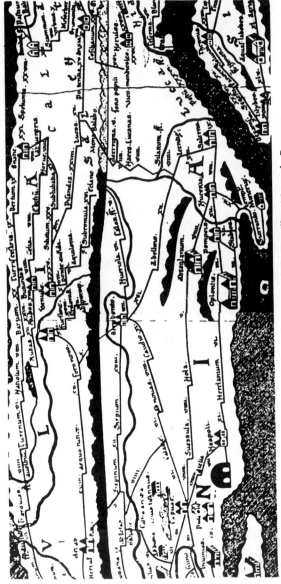

FIGURE 19.25 Peutinger Table (detail) from Miller, Itineraria Romana.

616

*Scrophularia canina* L. subsp. *bicolor* (S. and S.) W. Greuter. These plants were growing in a narrow strip between 700 and 900 m above sea level in the old crater between the present cone of Vesuvius and Mount Somma. The succession of species appearing on both substrata is recorded until 1970.

This report describes conditions in only a very limited part of the area covered by the 1944 eruption, an area in which no effort was made to hasten the return of vegetation. But the information in regard to the length of time elapsing before the return of vegetation on pyroclastic masses is significant for an understanding of conditions after the A.D. 79 eruption. The Pompeii area was covered by lapilli and volcanic ash, Herculaneum, by a mudflow. As we have noted, there was little lava in A.D. 79. The crisis situation after the A.D. 79 eruption and the acute need for food on the part of homeless survivors would suggest that heroic measures would have been taken to make the destroyed area cultivable. A diligent application of nitrogen by the survivors would have hastened the return of the destroyed area to cultivation, but we have no information as to how soon this took place. Dio (66.21), about a century and a half after the A.D. 79 eruption, described the top of Vesuvius as clothed with trees and vines, but said that the circular cavity was abandoned to fire and almost every year cast up stones and ashes.

There have been numerous eruptions since A.D. 79, but the scientific study of the volcano began only with the building in 1845 of the Vesuvian Observatory, one of the few volcano observatories in the world. Confirmed eruptions took place in A.D. 203, 472, 512, 787, 968, 991, 999, 1007, and 1036 (Imbo 1965:9). The volcano was then quiescent until the seventeenth century, when a series of earthquakes during the last half of the year 1631 culminated in a major eruption on 16 December 1631, which destroyed many villages and took the lives of about 3000 people.

A change took place in the eruptive character of the volcano after 1631. Cycles consisting of two stages could thereafter be observed, a quiescent stage when the volcano's mouth was obstructed, and an eruptive one during which it would be almost continuously open (Imbo 1965:9). Investigators give different details, however, as to the number and the length of the various cycles. Some cycles have been as short as 5 years (1850–1855), others as long as 34 (1872–1906) or 38 years (1906–1944) (Bullard 1976:215–218, 345). After the 1944 eruption Vesuvius entered a quiescent stage, but Imbro reports that the first sure signs of a future renewal of activity occurred on 11 May 1964 (1974:19,99).

Today the dominant feature of the volcano is the cone of Vesuvius (1280 m high), which has been building up in the crater formed in the A.D. 79 eruption. To the north, the edge of the A.D. 79 crater (Figure 19.26), known as Mount Somma (1132 m high), is still prominent (Figure 19.27); on the south it has been buried under lava flows from Vesuvius (Figure 19.28). The crater at the top of the cone of Vesuvius formed in the 1944 eruption is over 300 m deep and over 600 m across.

Today vineyards and orchards cover the lower slopes of Vesuvius. Higher up there are oak and chestnut groves. The western and southern slopes above

FIGURE 19.26 Mount Somma and the A.D. 79 crater (detail).

the trees in the early summer are golden when the broom is in blossom; still higher, the cone as well as the inner slope of Mount Somma is almost barren, but during quiescent periods meadow plants appear.

Vesuvius continues to enrich the land. The fertile soil continues to yield at least three crops a year, as it did in antiquity. Grapes are still the most important crop; then come olives, fruits, cereals, vegetables, and flowers. Intercultivation is common. Vegetables grow under vines, and tall trees grow among the vines. Cauliflowers are ready for market by the end of January or February; the same plot may then be planted in potatoes, which mature in 2½ months. Finally flowers are planted and these mature in 3 or 4 months. Frequently the land is sublet in smaller areas for the portion of the year when flowers are raised. Although many flower plots are 2–4 ha in size, many are as small as .1 ha. Flowers are also planted in orchards under trees, as they were in antiquity. Today, however, flowers are not grown for garlands and perfume. Some are grown for cut flowers. But the raising of flower seed is a major

FIGURE 19.27  Mount Vesuvius and Mount Somma today from ancient Pompeii.

industry in the area. As we have noted, large seed companies with headquarters in France, England, Holland, Germany, and Denmark have hundreds of hectares under contract in the Pompeii area to produce seed that is sold throughout the world. Vegetables are also marketed outside the country. Germany and Switzerland import cucumbers, lettuce, and cauliflower in quantity.

Vesuvius has undergone an inverted development in respect to its ejecta.

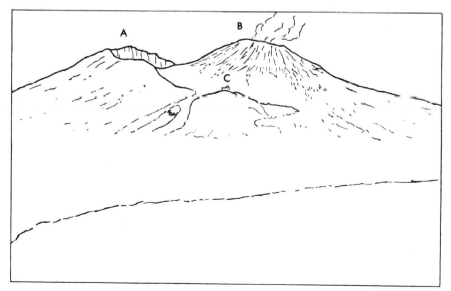

FIGURE 19.28  Mount Vesuvius (B) and Mount Somma (A) from the Bay of Naples. Volcano Observatory (C).

Its magma, which was earlier fairly acid, has now become basic. Its lavas are very rich in potash and contain much leucite.

Professor Foss (1978) found:

> The soils occurring on the present landscape at Pompeii are similar to the soils of A.D. 79 in that a number of buried A horizons occur in the upper meter. Because of limited modification of the present agricultural soils, the layers are probably the result of volcanic deposition after A.D. 79 and subsequent stabilization and organic matter additions. The textures of present-day soils are sandy loam or loamy sand and they show good internal drainage. They also show high levels of extractable calcium, phosphorus, potassium, and magnesium, thus indicating an ample supply of these plant nutrients. Soil pH analysis indicated slightly acid surface layers for the soils near Pompeii, but additional sampling is needed to evaluate soil pH in surrounding areas.

The Vesuvian area today is the home of over 2 million people. The cities along the bay have become industrialized, but the towns on the northern slopes of Vesuvius are still small agricultural centers.

## ACKNOWLEDGMENTS

Funds for the University of Maryland Excavations at Pompeii and Torre Annunziata were in large part provided by the National Endowment for the Humanities, and in part by the General Research Board of the University of Maryland. Our work has been greatly facilitated by the generous cooperation and gracious hospitality of Professor Alfonso de Franciscis, Superintendent of Antiquities for the Provinces of Naples and Caserta (1961–1976), of Dr. Fausto Zevi, Superintendent since 1977; of Dr. Giuseppina Cerulli Irelli, Director of the Excavations at Pompeii, and of Dr. Giuseppe Maggi, Director of the Excavations at Torre Annunziata, who generously made available the facilities at these sites and also furnished trucks, additional workmen, and other valuable assistance. Domenico Pelli, Chief Assistant at Pompeii, and Ferdinando Balzano, in charge of the deposito at Torre Annunziata, gave much valuable help and advice. Nicola Sicigiano was our able and ever helpful foreman.

Gratitude is also expressed to Dr. Frederick G. Meyer, research botanist in charge of the U.S. National Arboretum Herbarium, for his generous help in identifying carbonized plant specimens and comparing them with contemporary specimens; to Dr. John Foss, Professor of Agronomy, University of Maryland, for his invaluable work on the soils at Pompeii; to Professor G. W. Dimbleby, of the Institute of Archaeology, University of London, who so generously made our pollen analyses; to Dr. Henry Setzer, Mammalogist in charge of the African Section of the Smithsonian Institution, who identified the bones found in our excavations; to Dr. Carlo Fideghelli of the Instituto Sperimentale per la Frutticoltura in Rome, who is studying the roots of living trees and bushes in the area today, and gave valuable help in identifying our root casts; to Dr. John Kingsolver, U.S. Department of Agriculture Research Entomologist at the Smithsonian Institution, who identified the carbonized insects found in our excavations; and to Dr. Richard S. Fiske, Volcanologist at the Smithsonian Institution, who also gave valuable help.

All photos and plans unless otherwise stated are by Stanley A. Jashemski.

## ANCIENT AUTHORS

Dio Cassius, *Roman History*, translated by E. W. Cary, 9 vols. Cambridge: Harvard University Press, 1914–1927.

Diodorus, *Library of History*, translated by C. H. Oldfather and others, 12 vols., Cambridge: Harvard University Press, 1933–1967.

Florus, *Epitome*, translated by E. S. Forster, Cambridge: Harvard University Press, 1929.

Martial, *Epigrams*, translated by W. C. A. Ker, 2 vols. Cambridge: Harvard University Press, rev. ed. 1968.

Pliny the Elder, *Natural History*, translated by H. Rackham and others, 10 vols., Cambridge: Harvard University Press, 1938–1963.

Pliny the Younger, *Letters*, translated by Betty Radice, 2 vols. Cambridge: Harvard University Press, 1969.

Plutarch, *The Parallel Lives*, translated by B. Perrin, 11 vols. Cambridge: Harvard University Press, 1914–1926.

Seneca, *Natural Questions*, translated by T. H. Corcoran, 2 vols. Cambridge: Harvard University Press, 1971–1972.

Strabo, *Geography*, translated by H. L. Jones, 8 vols. Cambridge: Harvard University Press, 1917–1933.

Suetonius, *Lives of the Caesars*, translated by J. C. Rolfe, 2 vols. Cambridge: Harvard University Press, 1914.

(The editions listed above are in the Loeb Classical Library which contains both the original text and translation. The translations of passages from Strabo, Martial, and Pliny the Elder in this article are reprinted from the Loeb Classical Library by permission of Harvard University Press.)

## REFERENCES

Agostini, R.
  1975  Vegetazione pioneira del Monte Vesuvio: Aspetti fitosociologici ed evolutivi. *Archivio botanico e biogeografico italiano* 51:11–34, Pl. 1–4.
Bottini, O.
  1932  La regione vesuviana: Studio chimico-geoagrologico. *Annali del Regio Istituto Superiore Agrario di Portici*, 5(Series 3):269–317.
Broughton, T. R. S.
  1952  *The Magistrates of the Roman Republic*. 2 New York: The American Philological Association.
Bullard, F. M.
  1976  *Volcanoes of the earth*. Austin: University of Texas Press.
Cerulli Irelli, G.
  1975  Intorno al problema della rinascita di Pompei. In *Neue Forschungen in Pompeji*, edited by B. Andreae and Helmut Kyrieleis. Recklinghausen: Verlag Aurel Bongers.
Cocchia, E.
  1899  La forma del Vesuvio nelle pitture e descrizioni antiche. *Atti dell'Accademia di Archeologia, Lettere e Belli Arti* 21: 1–66.
D'Arms, J. H.
  1977  Proprietari e ville nel Golfo di Napoli. *Atti del Convegno Internazionale dell'Accademia Nazionale dei Lincei*, Roma 1976. Rome: Accademia Nazionale dei Lincei.
Day, J.
  1932  Agriculture in the life of Pomepii. *Yale Classical Studies* 3.167–208.
De Franciscis, A.
  1975  *The Pompeian wall paintings in the Roman villa of Oplontis*. Recklinghausen: Verlag Aurel Bongers.
Della Corte, M.
  1933  Esplorazioni di Pompeii immediatemente successiva alla catastrofe dell' anno 79. In *Memoria lui Vasile Pârvan*. Bucharest: Associatia Academică "Vasile Pârvan."
Foss, J.
  1978  The soils of Pompeii and Torre Annunziata. Preliminary typescript report.

Helbig, W.
   1865  *Bulletino dell'Instituto di Correspondenza Archeologica*: 234–235.
Imbo, G.
   1965  Vesuvius. In *Catalogue of the active volcanoes of the world including solfatara field*, Pt.
          18, Italy. Rome: International Association of Volconology.
   1974  Encyclopaedia Britannica, 15th ed., s.v. Vesuvius.
Jashemski, W. F.
   1968  Excavations in the "*Foro Boario*" at Pompeii: A preliminary report. *American Journal of
          Archaeology* 72:69–73, Plates. 33–34.
   1970  University of Maryland excavations at Pompeii, 1968. *American Journal of Archaeology*
          74:63–70, Plates 17–18.
   1973a The discovery of a large vineyard at Pompeii: University of Maryland excavations, 1970.
          *American Journal of Archaeology* 77:27–41, Plates 3–4.
   1973b Large vineyard discovered in ancient Pompeii. *Science* 180:821–830.
   1974  The discovery of a market-garden orchard at Pompeii: The garden of the "House of the
          ship *Europa*." *American Journal of Archaeology* 78:39–404, Plates 80–82.
   1975  The gardens of Pompeii: An interim report. *Cronache Pompeiane* 1:48–81.
   In press  The gardens of Pompeii, Herculaneum and the villas destroyed by Vesuvius. New
              Rochelle: Caratzas Brothers.
Lirer, L. and T. Pescatore, B. Booth, G. Walker
   1973  Two Plinian pumice-fall deposits from Somma-Vesuvius, Italy. *Geological Society of
          America Bulletin*. 84:759–772.
Maggi, G.
   1976  Notiziario. *Cronache Pompeiane* 2:249.
Maiuri, Amedeo.
   1970  *Pompeii*. Rome: Istituto Poligrafico dello Stato.
Miller, K.
   1916,  repr. 1962 *Itineraria Romana*. Rome: L'Erma di Bretschneider.
Onorato, G. O.
   1949  La data del terremoto di Pompei, 5 febbraio 62 D.C.. *Atti dell'Accademia nazionale dei
          Lincei. Rendiconti*. ser. 8, vol. 4:644–661.
Parascandola, A.
   1938  L'attività e la forma del Vesuvio nell'antichità e l'origine del suo nome. *Gli Abissi* 1.
Rittmann, A.
   1950  L'Eruzione Vesuviana del 79: Studio magmalogico e vulcanologico. In *Pompeiana. Rac-
          colta di studi per il secondo centenario degli scavi di Pompei*. Naples: Gaetano Mac-
          chiaroli Editore.
   1962  *Volcanoes and their activity*, translated by E. A. Vincent. New York, London: Intersci-
          ence Publishers.
Rostovtzeff, M.
   1957  *The social and economic history of the Roman Empire*. 2nd ed. (2 vols.). Oxford: The
          Clarendon Press.
Russell, Josiah C.
   1977  The population and mortality at Pompeii. International Committee on Urgent An-
          thropological and Ethnological Research, Institut für Volkerkunde, Bulletin No. 19.
Sogliano, A.
   1915  La rinascita di Pompei. *Atti dell'Accademia nazionale dei Lincei. Rendiconti*. ser. 5, vol.
          24:483–514.

# 20

# Volcanic Disasters and the Archaeological Record

DONALD K. GRAYSON
PAYSON D. SHEETS

In this volume we have attempted to provide an overview of volcanism and of the human responses to both past and present volcanic hazards and disasters. As archaeologists interested in volcanism in general and in the interrelationships between volcanoes and human societies in particular, we have incorporated a set of contributions that we hope displays the potential which paleoenvironmental analyses, combined with archaeological analyses, have to add to our general understanding of the effects of volcanic events on human populations. In this concluding chapter we discuss this potential in greater detail.

## HUMAN ECOLOGY, HAZARDS, AND DISASTERS

Human ecology is a dynamic, processual study that focuses on the interactions between human populations and their environments. All human societies have means to cope with variation in environmental parameters, and much has been learned during the past few decades concerning how people adapt to their environments and how they deal with predictable environmental fluctuations. Anthropologists, who have studied the widest range of the adaptations made by human societies to their environments, have tended to focus on how a society adapted to environmental conditions during their year of fieldwork or on how a society adapts to normal conditions. But all societies are faced as well with the potential of surviving in the face of such geophysical extremes as drought,

623

frost, wind, or earthquake, extremes that may exceed situations met by, or in the memory of, any member of that society. Natural hazard and disaster researchers study human adaptability under the threat of, at, and beyond these extremes.

The term *hazard research* encompasses the study of hazards as potential disasters as well as the disasters themselves. In most cases, in order for a hazard to be perceived by a society, an analogous disaster must have occurred, and have been known as such, in the same or similar locality in the past. Under such conditions, societies frequently develop, either by folk or scientific means, some understanding of the probability of an extreme event occurring at a given time. Here, hazard research is used to include hazardous geophysical events, their societal repercussions, and human perceptions of them as potential threats.

The key distinctions between *natural hazards* and *natural disasters* relate to timing and impact. A natural hazard is a problem in the wings, a potential disaster that has yet to occur but which has nonetheless been perceived as a threat to a society by someone either within or outside of that society. Hazards considered in this volume are geophysical in origin, and specifically volcanic, but hazards can also arise from social, biological, psychological, economic, and political sources. Biological hazards may stem from human diseases, plant pests, and even pirañhas in Florida; psychological hazards involve a variety of mental health syndromes and the factors leading to their development. Few people living in western society today are unaware of the social and economic hazards posed by urban overcrowding, underemployment and unemployment, inflation, and civil disturbances; fewer still are unaware of the fact that the possibility of warfare poses a threat of considerable proportions. The threat of a volcanic event is likewise a severe hazard, although one considerably rarer, and affecting far fewer people, than the social and political hazards that seem to threaten western society today.

The very term *disaster* conveys the sense of an extremely deleterious event; whereas a natural hazard is a potential disaster that has yet to occur, a natural disaster is the extreme geophysical event itself that is in progress or has occurred, with harmful social consequences. Coverage of natural disasters by the broadcast and print media reinforce this sense by emphasizing the most severely affected as the disaster is at its most extreme point. Only rarely do the media follow the course of recovery, and almost never do they consider long-term effects. In addition, they consistently neglect the broader picture of those people only minimally affected; only slightly more often, and even then because of political implications (as, for instance, in Nicaragua in 1972), do they consider those who may have actually benefitted by the event. Social scientists conducting hazard research, on the other hand, have made some progress in understanding preparedness and warning, disaster impact, and the recovery process (e.g., White 1974; Mileti, Drabek and Haas 1975; Haas, Kates and Bowden 1977). However, while hazard researchers have begun to discern similarities in human responses to hazard and disaster conditions and have begun to formulate theories of human responses to such conditions, it is also true that the complexity of these situations and the newness of disaster research provides a situation in which a sound model of human behavior under disaster conditions lies far in the future.

Large numbers of natural and cultural components are involved in the impact of a volcanic event, specifically, on a society and its long-range ecological and social repercussions. The natural factors of import in analyzing the impact of such an event include the nature of the local climate, soils, flora and fauna, and the precise nature of the event itself. Volcanic events vary in a number of ways that affect their environmental and social impact: the magnitude of the event, the frequency of similar events in the history of the impacted society, the temperature of ejecta, lava flow versus airfall emplacement, warning and speed of onset, duration, the size of the devastated area relative to sociopolitical and economic boundaries, and many other factors combine to determine the effects of a given volcanic event on the society involved. Components more within the human sphere that are pertinent to short- and long-term effects of volcanic disasters are equally varied. The nature of social organization (the band, tribe, chiefdom, and state typology is a useful starting point), hazard perception and preparedness, land use and resource exploitation (from mobile hunter-gatherers using dispersed wild food resources to intensive agriculturalists and urban populations), settlement pattern, and the extent of social, economic, and political spheres of control are among the numerous complex social and cultural variables that play a role in mediating the interrelationships between volcanic events and human societies. Given the multitude of both natural and sociocultural variables that interact to produce the results of the impact of a given volcanic event on a given society, it should be clear that, although human and natural components in such a situation can be isolated and studied separately, hazard research is above all an interdisciplinary pursuit, one in which researchers from such widely disparate disciplines as geology and psychology must participate if the impacts of a single volcanic episode on a modern human society are to be understood.

## RESULTS OF HAZARD RESEARCH

Natural disasters in general have taken a heavy toll in lives lost and property damaged for thousands of years. Their toll has gradually increased in the past few centuries as population has increased and nucleated and as cultural modifications of the landscape have intensified. Currently, about 30 major natural disasters occur somewhere in the world each year, about half of which affect cities (Haas *et al.* 1977). Reconstruction following a disaster may be relatively rapid, or may take a decade or more. It may seem surprising that people only rarely relocate their cities in safer surroundings: Haas *et al.* (1977) argue that the most recent instance of such relocation is provided by the movement of Antiqua, Guatemala, after the 1773 earthquake. There is, however, a more recent example: After San Salvador, the capital of El Salvador, was destroyed by an earthquake in 1854, it was moved to Santa Tecla and renamed Nueva San Salvador. Clearly, it must be asked why people continue to live in known hazardous locations. The answer, in part, is that the occupants typically value the short-term social and economic advantages more highly than the risks from a potential disastrous event: The short-term advantages are easy to assess; the long-term risks can just as easily be dismissed or ignored.

The benefits of living in volcanically active areas often contribute to the intensity of the disaster when the infrequent eruption occurs. The rich, weathered volcanic soils can support a dense and varied flora and fauna that are attractive to human use by peoples ranging from mobile hunter-gatherers to sedentary agriculturalists. Volcanic soils, as Ugolini and Zazoski (Chapter 3) note, can often attract and maintain intensive agriculture and dense human populations, thus concentrating people and their subsistence activities in areas that have the potential of receiving the heaviest volcanic impact.

Based on moderately extensive comparative cases for all kinds of natural disasters, hazard researchers have been able to formulate some preliminary conclusions about human behavior during a disaster. Quarantelli and Dynes (1972), in an overview of such generalizations, note that the most common myth about disaster behavior is that people panic. Panic descriptions make good wire service copy, but, in fact, researchers have found panic to be very rare. Unfortunately, local officials generally believe the myth of inevitable panic, and thus make pre-disaster announcements that are intended to calm the population rather than prepare them for the impacts of the event. Such a step can, in many instances, compound the impact of the disaster.

Quarantelli and Dynes (1972) also note that it is commonly assumed that disaster victims are in shock, and that they are unable to perform rationally after a disaster. Again, there is a grain of truth to this assumption, for shock does occasionally occur, particularly following intense disasters that occurred with little warning. But shock and irrational behavior are very rare. Generally, people turn first to family and friends for assistance. If these are unavailable or inadequate, people commonly turn next to local agencies. Failing that, the last place people traditionally seek help are such international disaster relief agencies as the Red Cross or United Nations. Irrational responses play little role in such searches for assistance.

Likewise, the fear of chaos and looting after a disaster is largely—though not entirely—unfounded. In case after case, researchers have found both officials and disaster victims expecting crime to increase after the disaster occurred, when, in fact, crime decreased after the event (Quarantelli and Dynes 1972).

Haas et al. (1977) have assessed current knowledge concerning recovery processes following disasters, focused largely on reconstruction efforts. They find that disasters are commonly egalitarian in their initial impact in that they affect people regardless of wealth, class, or position (although some disasters, such as drought or pestilence, may have differential impacts). On the other hand, the recovery process is profoundly nonegalitarian. Some segments of society suffer more intensely for longer periods, whereas others may benefit greatly during reconstruction. The profiteers may benefit because of their knowledge, social or economic position, access to local or distant resources, transportation, and other factors. Being associated with the ruling elite conferred its benefits in Nicaragua in the years following the "Christmas earthquake" of 1972. The net effect with many societies is to heighten social, political, and economic differences that existed prior to the disaster. And, the processes leading to social differentiation are often accelerated by disaster and the recovery from it.

In their critical assessment of the current state of hazard research, White and Haas (1975) noted that there has been an overemphasis on technological solutions to hazards and disasters. Political, social, and economic factors, they suggest, are far too often neglected in contemporary research. These authors summarize current weaknesses in knowledge about volcanic disasters and hazards as follows:

> Although the United States has active or potentially active volcanoes in Hawaii, Alaska, and the Pacific Northwest, knowledge about the physical characteristics of the volcano hazard is insufficient to define many continental hazard areas clearly. Nor has enough attention been paid to the social implications of great eruptions so that public policies can be designed with confidence. Special studies on human response to volcanic action . . . should be expanded in an integrated program by building upon previous geological studies [White and Haas 1975: 25].

They further recommend "postaudits" of the effects of natural disasters, combined with longitudinal studies: "never have there been long-term studies on how communities and families . . . recover from natural disasters [White and Haas 1975: 12]."

It is here, we suggest, that the use of the archaeological record could make a major contribution to hazard research. That record can reveal the nature and frequency of volcanic disasters in the past, the nature of human responses to those disasters, and, in a way that cannot be done with any other source of information on volcanic disasters, the long-term repercussions of interactions between human societies and volcanoes in the disaster setting.

## PAST VOLCANISM AND THE ARCHAEOLOGICAL RECORD

It does not take detailed reading of the literature on modern volcanoes and their impact on the environment to realize that these studies have gathered a wealth of precise information on the effects of volcanic activity on the local flora, fauna, human populations, and virtually all other aspects of the impacted environment. Thorarinsson's contribution to this volume (Chapter 5) provides one such example, while Hodge, Sharp, and Marts (Chapter 8) and

Bullard (Chapter 2) provide others. Given the detailed information that can be gathered by observing modern volcanoes and modern volcanic disasters, why spend the time and money needed to examine such episodes prehistorically? Direct observation and recording of volcanic eruptions provide detailed accounts of the beneficial and detrimental effects of volcanism on human societies and environments and can do this, it might seem, less inferentially than can be done with the archaeological record. What, then, can the archaeological, geological, and paleobiological records (which we shall refer to collectively as the *paleoenvironmental record* of prehistoric volcanism) offer that modern records cannot provide at much less expense?

First, only the paleoenvironmental record can provide a true long-term assessment of the impacts of volcanic events on an entire landscape. While the historic record might provide case studies for detailed assessments extending back a few decades, only the paleoenvironmental record will allow impact assessments extending across centuries. Indeed, since social science disaster research represents a new approach to understanding human behavior, the data needed to use disaster research techniques is generally not available, because it was not gathered, for volcanic disasters more than a few decades in the past. The paleoenvironmental record holds the potential of providing long-term assessments of volcanic impacts on all aspects of an environment, assessments that are available in no other way, and assessments that are likely to produce results and insights very different from those gained from the perspective of a few years or a few decades. Grayson (Chapter 14), for instance, suggests that long-term impacts of the eruption of Mt. Mazama on the fauna of the Fort Rock Basin may have been minimal; two decades after the eruption, however, a very different prediction about the area's faunal history probably would have been offered by the region's inhabitants. How long will the impacts of a volcanic event of a given type on human settlements persist in an area? Clearly, the effects of the eruption of Parícutin on the people of central Mexico may look very different in a few centuries from how they look today; only long-term studies of the sort possible with the archaeological record can provide answers to such questions. In short, the paleoenvironmental approach to the effects of volcanism on human populations can provide answers to questions of a very different scale from those that can be answered using the historical record alone. Both long-term and short-term responses to volcanism need to be known if we are to have full understanding of the interrelationship between volcanism and human societies. While the historical record and direct analyses of contemporaneous volcanic events can approach the issue of short-term responses, only the archaeological record can, at least without an unacceptable wait, allow researchers to address the issue of long-term responses.

Second, and equally important, the use of the archaeological record opens up a far wider range of human societies to analysis than can be studied using the historical record alone. This is an advantage of tremendous scope. No volcanic event in historic times, for instance, has affected a group of hunter-gatherers in a cool desert, yet the eruption of Mt. Mazama did precisely that some 7000 years ago. Indeed, current information on far southeastern Oregon suggests that these people may have been affected by pumice and ashfalls of varying magnitude six times in the last 7000 years, beginning with two falls of Mazama tephra (Blinman, Mehringer, and Sheppard, Chapter 13). Dumond (Chapter 12) describes a similar situation of repetitive volcanism in southeastern Alaska. An opportunity to study the effects of such repetitive volcanic activity on human societies could never be gained from the historical record. Likewise, no recent communities directly comparable to Herculaneum and Pompeii have been destroyed by volcanic activity; again, only the archaeological record allows access to such events. In addition, volcanic disasters are rare occurrences of which the historic record provides only a handful of examples; to restrict analysis to this handful would come close to restricting study to the unique, and would greatly hinder the search for predictable responses to these kinds of disasters across a wide variety of settings. The archaeological record has the potential of allowing the study of volcanic disasters to move from the analysis of a small number of examples in recent times to the study of a large number of examples through all times. In so doing, the record provides the potential of searching for predictable responses to volcanic disasters by human societies in general and by specific kinds of societies impacted by specific kinds of volcanic events in particular. That is, the analysis of prehistoric volcanic disasters may provide a better means of searching for theoretical statements concerning how people respond to those disasters; whereas the sparse historical record might suggest hypotheses concerning these responses and might allow some tests of those responses to be conducted, only the archaeological record can provide the large series of events needed to test those hypotheses across a wide variety of settings in both time and space. In an important sense, the archaeological record provides a series of replicated experiments on the effects of volcanism on people that is not available from modern times.

In short, the use of the archaeological record in the study of the impacts of volcanic disasters on people affords the only opportunity available to examine the long-term impact of these events on human societies, long-term in the sense of centuries or millenia, not years or decades. Using the archeological record also offers a means of expanding the sample of impacted societies available for study from the relative handful provided by the recent past to the very large, but unknown, number available from all times. In so doing, archaeological approaches increase the potential of generating valid and reliable theoretical statements concerning the responses of human societies to volcanic disasters. It is, we think, clear that such statements would have great practical import.

The investigation of the effects of volcanism on a society that can be known only through the archaeological record requires a fully interdisciplinary research design of massive scope. Such a research design is necessary because the testing of hypotheses concerning the interrelationships between volcanism and prehistoric human societies demands detailed knowledge not only of the volcanic event itself but also of the nature of the flora, fauna, and soils of the region prior to and after the event, as well as precise knowledge of human subsistence and settlement systems before and after that event. Various contributors to this volume have provided individual studies of the sort needed to understand aspects of the impact of volcanism on a prehistoric landscape: for instance, Blinman, Mehringer, and Sheppard (Chapter 13), Hevly and his colleagues (Chapter 16), and Grayson (Chapter 14) have focused on the floral and faunal records; Sheets (Chapter 17) has coupled detailed studies of tephra with analyses of the distribution of archaeological sites across space. But to gain full insight into the impact of a prehistoric volcanic event on a prehistoric human population will require that these and other approaches be combined in single regional studies of volcanism and prehistoric human ecology.

To derive such information from the earth requires interdisciplinary teams composed of geologists, botanists, zoologists and, of course, archaeologists. But it must also be admitted that the kinds of data needed to determine the relationship between prehistoric volcanic events and human populations may at times be nearly impossible to extract. What was the precise nature of the volcanic event? What effect did it have on the regional vegetation and on the regional fauna? What was the nature of the human subsistence system prior to the event? How did this system, and the distribution of archaeological materials across space, change after the event? In many instances, answers to one or more of these crucial questions may not be possible. For instance, most of the archaeological record in the Great Basin of North America is to be found on the surface of the ground, and has so far defied attempts at any but the most general form of dating. It is not enough to know that an occupation was "mid-Holocene"; one must instead be able to construct regional settlement systems for the time immediately before and immediately after the volcanic event.

In some situations—most notably those in which the ages of the exposed surfaces on which the archaeological record sits are greater than the volcanic events in question—the chronological controls necessary to interrelate the volcanic and archaeological records are not currently available.

While there may be some situations in which detailed studies of the impacts of volcanism on prehistoric human societies may not be possible because of the nature of the record and of the methods available to deal with that record, the studies presented in this volume demonstrate that there are many cases in which such studies can proceed profitably. The approach that studies of prehistoric volcanism must take in these instances seems clear. The approach must be regional, since human populations occupy regions, not points on a map, and because volcanic eruptions are regional events. The approach must be fully interdisciplinary, because the questions to be answered deal, as we have discussed, with much more than the simple direct, mechanical impact of volcanic debris on people and their settlements. Finally, the approach must pay exacting attention to chronology, for without precise knowledge of the timing of events, no causal chain can be constructed.

The archaeological record, of course, does not provide the same kind of information provided by the modern record. The call by White and Haas (1975) for long-term studies on how families recover from natural disasters is not usually a call that can be answered archaeologically, nor can the archaeological record routinely be used to address detailed questions of the effects of volcanism on social organization. Archaeological approaches cannot incorporate questionnaires of the sort employed by Hodge, Sharp, and Marts (Chapter 8), nor can they observe floral, faunal, and human responses in the detail achieved by Nolan (Chapter 10) and Rees (Chapter 9) in their studies of Parícutin. Those dealing with the archaeological record must deal with human responses to disasters in very different terms. How did the overall nature of the flora change in response to the episode? What shifts in membership of a fauna occurred? How did the distribution of archaeological materials across space change in response to the event? Can patterns in the changes of such distributions be seen, and can general statements interrelating these changes with the volcanic event itself be made? Can predictive models be built through which these general parameters of responses to volcanic events can be understood, and can these models then be confirmed on as yet unknown instances of prehistoric volcanic disasters as well as on those cases available from the modern record? These questions do not require data that can only be gathered by on-site inspection immediately after an episode of volcanism occurred, yet they clearly address major issues directly: How do entire societies and groups of societies respond to volcanic events? Indeed, it may be a benefit of the archaeological record that the researcher cannot be concerned with, for instance, the psychological aspects of readaptation after such a disaster, and must instead look only at the larger-scale aspects of such readaptation. Both sets of approaches are crucially important, the fine-grained approach possible with contemporary or nearly contemporary events, and the coarser-grained approach possible with the archaeological record, since both approaches have the potential of answering questions of a different scale. But researchers using only archaeological data have no choice but to spend all of their efforts on the coarse-grained approach, building predictive models to understand and account for the general nature of human responses to volcanic impacts on human societies. It is here that the value of the archaeological record to disaster studies lies.

Even at this early stage in the development of analyses of the interrelationship between volcanic events and prehistoric human societies, the studies presented in this volume allow some general observations to be made. A number of prehistoric eruptions are thought to have had negligible long-term effects on the societies affected, even though the initial impact may have been disastrous in a small area for a short time. Sunset Crater's eruption almost a millenium ago and its effect on the Anasazi as well as Aleutian volcanism and the Aleuts are exemplary (see Pilles, Chapter 15; Dumond, Chapter 12; Workman, Chapter 11). In the Aleutian case, the apparent social, economic, and kin networks available to disaster victims were of a larger geographic scale than the area devastated, and the same situation likely pertained in the Sunset

Crater area. Migrants could maintain their cultural and adaptive traditions while living as refugees. Thus, as soils, vegetation, and fauna recovered, people could reoccupy the area with much the same kind of social and adaptive system as existed prior to the disaster, and no drastic, long-term effects are notable in the record. Pompeii and Mt. Vesuvius represent a similar case of a disastrous eruption with negligible long-term impacts, and we suspect that Parícutin as well as post-Ilopango eruptions in central El Salvador may rank in the same general category.

However, there are some cases that may have been disastrous in their immediate effects and may also have had long-term repercussions that remain visible in the record for hundreds of years. The eruption of Thera may have delivered the *coup de grace* to Minoan civilization and allowed for the aggressive expansion of Mycenean civilization from mainland Greece some 3500 years ago. The Ilopango eruption in the third century A.D. apparently terminated the vitality of southeast Maya civilization, allowing for some of the Lowland Maya of northern Guatemala to reroute and exploit the lucrative long-distance trade route in exotic and utilitarian commodities. In an egalitarian setting, Alaskan volcanism may have overwhelmed inland Yukon groups, forcing lengthy migrations. In all of these cases, the scale of the deleterious volcanic impacts exceeded the support and adjustment capacities of the societies affected, with major and long-term repercussions for the societies involved.

We feel that these studies demonstrate that the archaeological record provides an extremely valuable, if neglected, source of information on past volcanic disasters. Not only does that record promise an opportunity to gain insight into unique events, fascinating in-and-of themselves, but that record also offers the only opportunity available to study what can be regarded as replicative experiments on the responses of human societies to volcanic disasters and to study those responses comparatively across the centuries. Such studies are bound to be expensive in terms of both time and money, but offer in return the potential of learning what can be learned in no other way.

## REFERENCES

Haas, J., R. Kates, and M. Bowden (Eds.)
   1977  *Reconstruction following disaster.* Cambridge: MIT Press.
Mileti, D., T. Drabek, and J. Haas
   1975  *Human systems in extreme environments: A sociological perspective.* Boulder: University of Colorado, Institute of Behavioral Science.
Quarantelli, E. and R. Dynes
   1972  When disaster strikes. *Psychology Today* 5:67–70.
White, G. and J. Haas
   1975  *Assessment of research on natural hazards.* Cambridge: MIT Press.
White, G. (Ed.)
   1974  *Natural hazards; local, national, global.* New York: Oxford Press.

# Subject Index

DATE DUE